The Correspondence of Johannes Hevelius

DE DIVERSIS ARTIBUS

COLLECTION DE TRAVAUX
DE L'ACADÉMIE INTERNATIONALE
D'HISTOIRE DES SCIENCES

COLLECTION OF STUDIES
FROM THE INTERNATIONAL ACADEMY
OF THE HISTORY OF SCIENCE

TOME 108 (N.S. 71)

BREPOLS

The Correspondence of Johannes Hevelius

Volume IV

The Correspondence with Stanisław Lubieniecki

Maciej Jasiński

BREPOLS

Research funded by the Ministry of Science and Higher Education of the Republic of Poland within the 'National Programme for the Development of Humanities' for the years 2015–2018

NARODOWY PROGRAM
ROZWOJU HUMANISTYKI

Publié avec le soutien du legs Christophe Scriba.

D/2021/0095/39
ISBN 978-2-503-57118-8
ISSN 1766-7593
E-ISSN 2565-9448

Printed in the EU on acid-free paper

Contents

Abbreviations

Manuscripts

BO	Bibliothèque de l'Observatoire, Paris,
BnF	Bibliothèque nationale de France, Paris,
AUL	Universiteitsbibliotheek, Amsterdam,
LUL	Universiteitsbibliotheek, Leiden,
MNW	Muzeum Narodowe, Warsaw.

Early Prints

TC	Stanisław Lubieniecki, *Theatri cometici pars prior* (Amstelodami: apud Franciscum Kuyperum, 1668),
TC2	Stanisław Lubieniecki, *Theatri cometici pars posterior* (Amstelodami: apud Franciscum Kuyperum, 1666),
TC3	Stanisław Lubieniecki, *Theatri cometici exitus* (Amstelodami: apud Franciscum Kuyperum, 1668),
Olhoff	Johann Eric Olhoff, *Excerpta ex litteris illustrium et clarissimorum virorum ad nobilissimum, amplissimum et consultissimum dominum Johannem Hevelium* (Gedani: ex officina Janssonio Waesbergiana, 1683).

Modern Editions

CJH1	*Correspondance de Johannes Hevelius. Tome I: Prolégomènes critiques*, ed. by Chantal Grell (Turnhout: Brepols, 2014),
CJH2	*Correspondance de Johannes Hevelius. Tome II: Correspondance avec la cour de France et ses agents, avec un dossier sur la querelle de la comète de 1664–1665*, ed. by Chantal Grell (Turnhout: Brepols, 2017),
CJH3	*Correspondance de Johannes Hevelius. Tome III: Correspondance avec Pierre des Noyers, secrétaire de la Reine de Pologne, 1646–1686*, ed. by Chantal Grell in collaboration with Damien Mallet (Turnhout: Brepols, 2020),

CHO *The Correspondence of Henry Oldenburg. Vol. III: 1666–1667*, ed. and transl. by A. Rupert Hall and Marie Boas Hall (Madison: University of Wisconsin Press, 1966),

OCCH5 *Œuvres complètes de Christiaan Huygens. Tome cinquième, correspondance 1664–1665* (La Haye: Martinus Nijhof, 1893),

OCCH6 *Œuvres complètes de Christiaan Huygens. Tome sixième, correspondance 1666–1669* (La Haye: Martinus Nijhof, 1895).

Figures

All the figures are in the public domain and accessible on-line at the Polona, polona.pl.

Tables

Foreword

This volume continues the publication of the correspondence of Johannes Hevelius (1611–87), an astronomer from Danzig. The series is being prepared under the auspices of the International Academy of the History of Science and the International Union of Academies. Here, I present astronomical letters exchanged between Hevelius and Stanisław Lubieniecki (1623–75),[1] a Polish nobleman and religious dissident who was briefly interested in astronomy. The correspondence is one of the largest parts of the Hevelian corpus and contains 92 letters (31 by Hevelius and 61 by Lubieniecki). Their epistolary exchange began in late 1664, when Lubieniecki inquired the Danzig astronomer about the recent cometary phenomenon, and lasted almost until Lubieniecki's death.

Because Lubieniecki is not a widely recognized person, a significant part of the Introduction has to be devoted to his biography and actions. There is an extensive scholarship covering his life and work, written mainly in Polish. In this volume, I will rely on the results of these studies.[2]

The Introduction is divided into three parts. The first presents Lubieniecki against the background of his age and the main historical events which had an impact on his life. Lubieniecki was a follower of a controversial minority religious group, the Polish Brethren. This part also contains a section devoted to Lubieniecki's astronomical work, *Theatrum cometicum*. It is a massive, three-volume publication which contains the author's astronomical correspondence with over forty people of various professions and a catalogue of cometary phenomena since the times of Noah until 1665.

The second part of the Introduction discusses the Hevelius-Lubieniecki correspondence. Not only was Lubieniecki Hevelius's frequent correspondent, but also He-

1 Even though forms such as 'Lubienietzki' can be found in the critical literature, 'Lubieniecki' is the only correct spelling variant of his family name.

2 Janusz Tazbir discussed Lubieniecki's life, family, and his principal works and achievements, and he presented his general biography: Janusz Tazbir, *Stanisław Lubieniecki. Przywódca ariańskiej emigracji* (Warszawa: Państwowe Wydawnictwo Naukowe, 1961); 2nd edition: Janusz Tazbir, *Stando lubentius moriar. Biografia Stanisława Lubienieckiego* (Warszawa: Iskry, 2003). I will refer here to the 2nd edition. I have made a small contribution to Lubieniecki studies by discussing his efforts in the field of astronomy in Maciej Jasiński, *Stanisław Lubieniecki i astronomia kometarna XVII stulecia* (Warszawa, Gdańsk: Instytut Historii Nauki im. L. i A. Birkenmajerów PAN, PAN Biblioteka Gdańska, Oficyna Wydawnicza ASPRA-JR, 2017).

Figure 1. Stanisław Lubieniecki (1623–75). Illustration from his *Theatri cometici pars prior.* Biblioteka Narodowa, Warsaw, SD XVII.4.2660.

velius was Lubieniecki's important contact. This section presents the main subjects which occurred in their letters. It also contains a list of their mutual correspondents with brief biographical notes and summaries of their letters in the *Theatrum cometicum.*

The third part of the Introduction is devoted to Lubieniecki's astronomical views and opinions. It deals with Lubieniecki's cometary considerations (which are relatively few given the volume of his book), with his polemics against astrology, and with Lubieniecki's place in the contemporary Republic of Letters.

The last part discusses the editorial principles of this edition and presents the index of the letters exchanged by Hevelius and Lubieniecki.

The letters of the Hevelius-Lubieniecki correspondence are presented (with enclosures) both in the original version, mostly in Latin, transcribed from manuscript and printed sources, and in English translation with commentary. The volume is supplemented with appendices which contain Lubieniecki's excerpts of Hevelius's works on comets, and a summary of Lubieniecki's views on the meaning of comets.

Acknowledgments

I am very grateful to Prof. Chantal Grell (Université de Versailles Saint-Quentin-en-Yvelines, Paris) and Prof. Robert Halleux (University of Liège, International Academy of the History of Science), who are in charge of the series of Hevelius's correspondence, for their patience and advice, and to Prof. Jarosław Włodarczyk (L. & A. Birkenmajer Institute for the History of Science, Polish Academy of Sciences), for his guidance and help. I owe many thanks to the referees of the volume, Prof. Andreas Kleinert (Martin-Luther-Universität Halle-Wittenberg) and Dr Philip Beeley (University of Oxford), whose reviews were indispensable for this volume to reach its present form. I also thank Dr Barbara Bienias (L. & A. Birkenmajer Institute for the History of Science, Polish Academy of Sciences) for her immense help in finishing this book. For any remaining shortcomings in this volume I alone am responsible.

INTRODUCTION

CHAPTER 1

Stanisław Lubieniecki and his *Theatrum cometicum*

Hevelius's correspondent to whom the present volume is devoted, Stanisław Lubieniecki, is generally not well known. Therefore, before proceeding to discuss his epistolary commerce with the Danzig astronomer, Lubieniecki needs to be introduced. The following chapter is divided into two parts. The first one discusses Lubieniecki's life against the background of his era. The second part is devoted to the outcome of Lubieniecki's short-lived cometary fascination: his three-volume *Theatrum cometicum*, which is one of the primary sources for his correspondence with Hevelius.

1.1. The Life of Stanisław Lubieniecki

1.1.1. Early Life and Studies

Stanisław Lubieniecki was born on 23 August 1623 in Raków, a town in southern Poland. His family belonged to the Polish nobility (*szlachta*) and used the Rola coat of arms.[1] The Lubieniecki family were followers of the Polish Brethren movement and adherents to the Unitarian religion which opposed the dogma of the Holy Trinity. Lubieniecki's loyalty to his faith turned out to be one of the landmarks in his biography.[2]

1 Whereas in western Europe noble families usually had their own coats of arms, in Poland and Lithuania one coat of arms, with its own distinct name, such as 'Rola' in the case of the Lubieniecki family, could be used by many (sometimes even a few hundred) families which were only legendarily related to one another.

2 Tazbir, *Stando lubentius moriar*, pp. 17–20. Stanisław Lubieniecki is sometimes called 'the younger' to distinguish him from his uncle and namesake (d. 1653) who was followed by his nephew in confessional polemics and the defence of his faith. As Stanisław Lubieniecki the elder plays no part in his nephew's astronomical activities and the correspondence with Hevelius, whenever I refer to Stanisław Lubieniecki, I mean Hevelius's correspondent.

Figure 2. Rola coat of arms. Illustration from Bartosz Paprocki's *Gniazdo cnoty* (Kraków: z drukarniey Andrzeia Piotrkowczyka, 1578). Biblioteka Narodowa, Warsaw, SD XVI.F.197.

The Polish Brethren were a small religious group in the sixteenth and seventeenth-century Poland and Lithuania. They originated in the mid-sixteenth century as an offspring of the Calvinist Church in Poland[3] and enjoyed the freedom of religion in Poland and Lithuania and reached the peak of their activity between the end of the sixteenth century and the 1640s. Because of their controversial approach to the Christian doctrine, they were opposed by both Catholics and Protestants. It finally led to their expulsion from Poland and Lithuania and to the fall of their religion.[4] The centre of their intellectual life was the town of Raków in southern Poland, often labelled the 'capital of the Polish Brethren.' Since 1602, it hosted the Raków Academy which is often considered to have had very high standards of teaching, both in the

3 The Polish Brethren are also called 'Antitrinitarians' or 'Unitarians' (after the main tenet of their faith), 'Arians' (after Arius, a 4th-century monk of Alexandria who started the most prominent antrinitarian heresy in the late antiquity), 'Socinians' (after Lelio Sozzini (1525–62) and Fausto Sozzini (1539–1604) who had an impact on the movement), or the 'Minor Reformed Church' (to differentiate them from the Calvinists, i.e. the 'Major Reformed Church').

4 Earl Morse Wilbur, *A History of Unitarianism. Socinianism and its Antecedents* (Cambridge, Massachusetts: Harvard University Press, 1946), pp. 282–522; Paul Fox, 'The Reformation in Poland', in *The Cambridge History of Poland from the Origins to Sobieski (to 1696)*, ed. by W. F. Reddaway et al. (Cambridge: Cambridge University Press, 1950), pp. 340–41; Józef Umiński, 'The Counter-Reformation in Poland', in *The Cambridge History of Poland*, p. 412; Nicholas Hans, 'Polish Protestants and their Connections with England and Holland in the 17th and 18th Centuries', *The Slavonic and East European Review*, 37 (1958), no. 88, 196–220 (pp. 205–09); Zbigniew Ogonowski, *Socynianizm. Dzieje, poglądy, oddziaływanie* (Warszawa: Instytut Historii Nauki im. L. i A. Birkenmajerów PAN, Oficyna Wydawnicza ASPRA-JR, 2015), pp. 1–231.

humanities and in exact sciences as well.[5] The ideas of the Polish Brethren, especially those of rationalistic religion and religious tolerance, gained them some recognition in the intellectual history of Europe.[6]

Stanisław Lubieniecki's father, Krzysztof (1598–1648), was an affluent landowner and a minister of the Polish Brethren church in Raków (where he married Katarzyna, née Filipowska, in or around 1620) and later in Lublin.[7] The future author of the *Theatrum cometicum* initially went to school in his native town. In 1638, after a group of students from the Raków Academy had been accused of profaning a wayside cross, the school, as well as the local church of the Polish Brethren, were ordered to be closed.[8] Lubieniecki continued his studies in the Polish Brethren school in Kisielin (today in western Ukraine). This school did not gain the recognition comparable to that of the Raków Academy and was forced to be closed in 1644.[9] At that time, under his father's tutelage, he began to be introduced into public life. In the Polish-Lithuanian state, the nobility was the only class allowed into public offices, and political activity was considered to be an honourable obligation of a nobleman. In the case of the Lubieniecki family, it was not only limited to the affairs of the state but included various activities devoted to the promotion and protection of their faith.[10]

In the mid-1640s, Lubieniecki was in Toruń. He probably continued his studies there, but there is no trace that he was a student at the local schools. Nevertheless, he witnessed there one of the most important events in the religious history of seventeenth-century Poland. King Władysław IV (1595–1648, King since 1632) decided to convene the representatives of Christian Churches in Toruń for *colloquium charitativum*, or a 'charitable discussion.' The intention of the meeting was to try to reinstate the unity of religion, or, at least, the peace between the confessions. The *colloquium* took place in 1645.[11] The synod of the Polish Brethren delegated its own representatives for the meeting but, despite their efforts, they were not allowed into

5 Stanisław Tync, 'Wyższa szkoła Braci Polskich w Rakowie. Zarys jej dziejów (1602–1638)', in *Studia nad arianizmem*, ed. by Ludwik Chmaj (Warszawa: Państwowe Wydawnictwo Naukowe, 1959), pp. 331–89; Hans, 'Polish Protestants and their Connections with England', pp. 207–09; Stanislas Lubieniecki (the elder), 'The Laws of the School at Raków (c. 1602)', in *The Polish Brethren. Documentation of the History and Thought of Unitarianism in the Polish-Lithuanian Commonwealth and in the Diaspora, 1601–1685. Part 1*, ed. by George Huntston Williams (Missoula: Scholars Press, 1980), pp. 71–81.

6 Hans, 'Polish Protestants and their Connections with England', pp. 209–12; Jan Woleński, 'Polish-English (British) Philosophical Contacts and Comparisons from the Fifteenth through the Eighteenth Century', in *Britain and Poland-Lithuania. Contacts and Comparison from the Middle Ages to 1795*, ed. by Richard Unger (Leiden, Boston: Brill, 2008), pp. 395–97; Ogonowski, *Socynianizm*, pp. 417–510.

7 Tazbir, *Stando lubentius moriar*, pp. 20–23.

8 Wilbur, *A History of Unitarianism*, pp. 442–55; Fox, 'The Reformation in Poland', p. 341; Hans, 'Polish Protestants and their Connections with England', p. 209; Ogonowski, *Socynianizm*, pp. 145–47.

9 Fox, 'The Reformation in Poland', p. 341.

10 Tazbir, *Stando lubentius moriar*, pp. 29–31.

11 For more information on the *colloquium charitativum* see Piotr Henryk Wojciechowski, *Il "colloquium charitativum" di Torun del 1645* (Pelplin: Bernardinum, 2000); Hans-Joachim Müller, *Irenik als Kommunikationsreform. Das Colloquium Charitativum von Thorn 1645* (Göttingen: Vandenhoeck & Ruprecht, 2004); Chantal Grell, 'Introduction', CJH3 p. 40.

the discussions, and only three confessions took part in the *colloquium*: the Catholics, the Lutherans, and the Calvinists. Young Stanisław described the futile efforts of his fellow believers in the *Fidelis relatio transactorum Thorunii*.[12] Soon after, he left Toruń.[13]

The next important step in Lubieniecki's education was his *Grand Tour*. In 1646, the synod of the Polish Brethren decided to send him for studies to France and the Netherlands, together with his coreligionist Stefan Niemirycz (before 1630–84) who was a few years younger than Stanisław. Niemirycz was a member of a wealthy and important family, and Lubieniecki was his tutor. Via Amsterdam, they arrived in France where they visited the universities in the cities on the Loire: Orléans, Saumur, and Angers. Little is known about their studies there, but jurisprudence is likely to have been a focus. At the Huguenot university in Saumur, they probably attended lectures on theology. An interesting episode of their journey was the encounter with the Sobieski brothers, Marek (1628–52) and Jan (1629–96), who were on their *Grand Tour* too. The Lubieniecki family was distantly related to the Sobieski family through Katarzyna Lubieniecka, née Sobieska, Stanisław's great-grandmother. Lubieniecki proudly emphasized this affinity, especially when Jan Sobieski was elected the King as Jan III in 1674.[14]

In 1648, Niemirycz returned to Poland, and Lubieniecki stayed in France, in Paris. His main contact in the capital city was probably Gilles Ménage (1613–92), whom he met in Angers. It was presumably Ménage who introduced Lubieniecki to Jacques Auguste de Thou the younger (1609–77).[15] In his circle, Stanisław had the opportunity to encounter and make acquaintance with many people of the French cultural life of the era. With some of them, Lubieniecki conducted correspondence in the following years, for example with the historian and translator of Latin literature, abbot Michel de Marolles (1600–87), and Ismaël Boulliau (1605–94). Having left Paris in 1649, Lubieniecki stayed for some time in the Netherlands, in Amsterdam, where he met Descartes (1596–1650), and in Leiden. He studied briefly at the Leiden University, but again there is not much information about his curriculum or the lectures he attended.[16]

1.1.2. War with Sweden

In early 1650, Lubieniecki was back in Poland. For a year, he stayed in Czerniechów (today in western Ukraine) on the estate of the Niemirycz family. In January 1652, he married Zofia Brzeska. In the same year, Lubieniecki became an assisting

12 Published in Friedrich Samuel Bock, *Historia Socinianismi Prussici* (Regiomonti: typis et impensis Io. Henr. Hartungii, 1754), pp. 115–21.

13 Tazbir, *Stando lubentius moriar*, pp. 31–35.

14 Tazbir, *Stando lubentius moriar*, pp. 36–43, see also pp. 18, 25.

15 About de Thou see CJH2 p. 261, note to Letter 54, Chapelain to Hevelius, 15 September 1668.

16 Tazbir, *Stando lubentius moriar*, pp. 43–50.

minister in Siedliska, and in 1654 he was appointed a minister in Czarkowy (both towns are in southern Poland).[17]

The Swedish invasion of Poland — known in the Polish historiography as the 'deluge' — began in 1655. It was the last of the wars between Poland and Sweden, which were waged since the end of the sixteenth century. When Polish and Lithuanian armies were busy in the east, fighting with the Cossacks and Muscovy, King of Sweden, Charles X Gustav (1622–60, King since 1654), conquered almost entire Poland in a very short time. Moreover, some aristocrats and nobility, opposing the rule of the King of Poland, Jan II Kazimierz (1609–72, King in 1648–68), sided with the Swedes. Jan II Kazimierz fled to Silesia.[18] Many members of the Polish Brethren believed that the King of Sweden could guarantee the tolerance of their faith. Stanisław Lubieniecki was among them. He left Czarkowy and together with a group of his fellow believers moved to Kraków. When the southern provinces of Poland surrendered to the Swedes, and Charles X Gustav briefly resided in the Wawel Castle in Kraków, the Polish Brethren contacted the Swedish officials and tried to convince the King to establish the law in their favour. Lubieniecki became the spokesman of the delegation to the King and praised him as the defender of the faith and as the only hope for members of the oppressed confession. The King, on his part, was kind towards the petitioners, but also very reluctant to give the Polish Brethren any substantial help or support. Lubieniecki described his actions in the first year of the war, including his efforts at the Swedish court, in the *Diariusz drogi krakowskiej i legacyjej krystyjańskiej do króla JM Karola Gustawa Anno Domini 1655* ('Diary of the journey to Kraków and of the Christian delegation to HM the King Charles Gustav Anno Domini 1655').[19] In his later years, however, Lubieniecki was not very proud of his support for the Swedish faction and considered the policy of his church towards the Swedes to be a mistake. Despite his unsuccessful efforts at the court, Lubieniecki and his fellow believers stayed in Kraków.[20]

In 1656, the tide started to turn for Charles X Gustav. Swedish troops were financed mostly by means confiscated in the occupied country, which stirred up discontent among all social classes and incited them to fight the invader. Jan II Kazimierz returned from Silesia and reclaimed Warsaw in June 1656. In early 1657, Charles X Gustav left Kraków. He handed over the control of the city to his new ally, the Duke of Transylvania, George II Rákóczi (1621–60, Duke in 1648–57 and 1659–60), who joined the war in January 1657. Charles X Gustav promised George II

17 Tazbir, *Stando lubentius moriar*, pp. 54–57.

18 Norman Davies, *God's Playground. A History of Poland. Volume 1: The Origins to 1795* (Oxford: Oxford University Press, 1981), pp. 450–51; Daniel Stone, *The Polish-Lithuanian State, 1386–1795* (Seattle, London: University of Washington Press, 2001), pp. 166–68; Damien Mallet, 'Les affaires de Pologne, 1648–1699', CJH3 pp. 701–03.

19 Published in Janusz Tazbir, 'Diariusz Stanisława Lubienieckiego (młodszego)', *Odrodzenie i Reformacja w Polsce*, 5 (1960), 201–21. The word 'Christian' in the title of this text refers to the Polish Brethren: this was the term they used to identify themselves.

20 Tazbir, *Stando lubentius moriar*, pp. 62–83.

Rákóczi the Polish throne. Later in the same year, Kraków was taken by the Polish forces.[21] The adherents to the Swedish faction were ordered to ask Jan II Kazimierz for forgiveness. Lubieniecki and other leaders of his church, however, were not among them. They left the city together with the King of Sweden in hope of receiving help from him in gaining pardon and religious freedom. Lubieniecki was subsequently appointed emissary of the Polish Brethren to the Swedish court. Together with the Swedish army, he travelled to Western Pomerania, then controlled by Sweden. Lubieniecki moved to Wolgast and soon after to Szczecin where he lived until 1660. Lubieniecki tried to convince Swedish officials to support the cause of the Polish Brethren, but his efforts were mostly ineffective. In 1660, he tried in vain to convince the Swedish delegation at the peace negotiations in Oliwa to intercede for the Polish Brethren. The peace treaty, despite the fact that it declaratively reinstated the religious laws from the pre-war period, did not help the Polish Brethren, and their situation in Poland was very tense.[22]

1.1.3. In Exile

After the war, the followers of all the Protestant Churches were considered to be the supporters of the Swedish faction, regardless of their actual actions. But it was the Polish Brethren who became a perfect scapegoat to be blamed for the collaboration with the Swedes. In 1658, the Diet in Warsaw passed a bill of their expulsion. They had to decide whether they wanted to convert to another confession or to sell their possessions and leave Poland and Lithuania within three years. In 1659, another bill was passed, and the deadline was set to 1660. Moreover, Catholicism was indicated as the only confession the Polish Brethren could convert to (initially, many of them considered conversion to Calvinism).[23] It should be stressed that even in exile, the Polish Brethren kept their Polish identity for a long time.[24]

Even though in the next few years Lubieniecki did not lose all his faith in Swedish support for the Polish Brethren and continued his efforts to revoke their expulsion,

21 Davies, *God's Playground*, pp. 453–54; Stone, *The Polish-Lithuanian State*, pp. 168–69.

22 Tazbir, *Stando lubentius moriar*, pp. 82–92, 99–103.

23 Wilbur, *A History of Unitarianism*, pp. 467–82; Ogonowski, *Socynianizm*, pp. 148–51. For the commented texts of the expulsion acts see 'Decree of Banishment of Arians or New-Baptized (1658)', in *The Polish Brethren. Documentation of the History and Thought of Unitarianism in the Polish-Lithuanian Commonwealth and in the Diaspora, 1601–1685. Part 2*, ed. by George Huntston Williams (Missoula: Scholars Press, 1980), pp. 495–96 (introductory preface to the edict of 1658); 'An Epistle Setting Forth a Brief History of the Life and Death of Andrew Wiszowaty and also of the Unitarian Churches of His Time', in *The Polish Brethren. Part 1*, p. 39 (edict of 1658); Jonas Szlichtyng, Samuel Przypkowski, *Two Letters of Exiles (1661 and 1663) Preserved by Benedict Wiszowaty in His Posthumous Edition of the Uncompleted* Historia Reformatonis Polonicae *(1685) by Stanislas Lubieniecki (d. 1675)*, in *The Polish Brethren. Part 2*, pp. 645–46 (edict of 1659), p. 646 (decree of execution of the edicts).

24 Janusz Tazbir, 'The Fate of Polish Protestantism in the Seventeenth Century', in *A Republic of Nobles. Studies in Polish History to 1864*, ed. by J. K. Fedorowicz et al. (Cambridge: Cambridge University Press, 1982), p. 215.

in 1660 he left Szczecin and moved to Copenhagen. In early 1661, he was granted an audience by King Frederick III of Denmark (1609–70, King since 1648). The King let the Polish Brethren settle in the city of Altona (today a district of Hamburg, then under the Danish rule).[25] This was not an exceptional act of the royal kindness, nor an indication that the King was interested in the Unitarian faith, as followers of some other denominations had already been allowed to practice their religion in Altona.[26]

Lubieniecki, however, initially did not settle in Altona but in Hamburg, but he moved to the former from the latter whenever his conflicts with the local clergy became too fierce. His ardent discussions and polemics with both Catholic and Protestant priests in Germany and Denmark, some of which took place in the presence of Frederick III, did not win him much favour with his contemporaries.[27] For many years he worked on his book about ecclesiastical history. He did not manage to fulfil this plan, but his work was published posthumously under the title *Historia reformationis Polonicae* (Freistadii [i.e. Amsterdam]: apud Johannem Aconium, 1685). This title is exaggerated because the book does not discuss the history of Protestant churches in Poland. Instead, it focuses on the origins of the Unitarian doctrine and the history of the Polish Brethren. Lubieniecki presents the main actors involved in the history of his church and dogmatic struggles within it. *Historia reformationis Polonicae* aims to prove that the Polish Brethren were the only ones who followed the real Christian doctrine, which was corrupted by Catholics and was not fully restored either by Luther or Calvin.[28]

In his early years in foreign countries, Lubieniecki tried to find more places where the Polish Brethren could find a haven and enjoy religious freedom. These endeavours, however, were not very successful and the plans of establishing a permanent colony of the Polish Brethren in such places as Friedrichstadt in Schleswig, or Mannheim in the Electoral Palatinate failed.[29] The only place where a considerable group of the Polish Brethren settled and kept their separateness for over a century was Transylvania, where the Unitarian faith was established and permitted since the sixteenth century.[30]

Lubieniecki also did not lose his hopes for nullifying the law of expulsion and hoped that the rebellion (*rokosz*, 1665–66) started by Jerzy Lubomirski (1616–67) could help this case. Lubomirski, powerful politician and military commander, was conflicted with Jan II Kazimierz and opposed his projects to strengthen the royal authority at the nobility's expense. Lubomirski somehow succeeded because even

25 Tazbir, *Stando lubentius moriar*, pp. 103–12.

26 Ludolph Hinrich Schmid, *Versuch einer historischen Beschreibung der an der Elbe belegenen Stadt Altona* (Altona, Flensburg: in Verlag der Gebrüder Korte, 1747), p. 47.

27 Tazbir, *Stando lubentius moriar*, pp. 113–17, 214–17.

28 Tazbir, *Stando lubentius moriar*, pp. 182–210.

29 Wilbur, *A History of Unitarianism*, pp. 494–501; Tazbir, *Stando lubentius moriar*, pp. 118–25.

30 Wilbur, *A History of Unitarianism*, pp. 483–93; Gábor Barta, 'Die Anfänge des Fürstentums und erste Krisen (1526–1606)', in *Kurze Geschichte Siebenbürgens*, ed. by Béla Köpeczi (Budapest: Akadémiai Kiadó, 1990), pp. 287–94; Felicia Roşu, *Elective Monarchy in Transylvania and Poland-Lithuania, 1569–1587* (Oxford: Oxford University Press, 2017), pp. 136–40.

though his rebellion failed and he had to ask the King for forgiveness, Jan II Kazimierz renounced his plans.[31] At the time of the sedition, Lubieniecki wrote two pamphlets, *Moriens Polonia suos et exteros alloquitur* and *Morientis Poloniae servandae ratio certissima* (both published without disclosing the place of publication and the name of publisher), favourable to Lubomirski. Concealing his authorship, Lubieniecki discussed the reasons for the calamities which fell upon Poland and suggested methods of helping the country. Among the proposals concerning taxes, monetary system, diplomacy, and army, a substantial place is devoted to the main point of Lubieniecki's programme: restitution of the religious tolerance which could allow the Polish Brethren to return from the exile. In order to make this proposal more credible, Lubieniecki pretended that the authors of both texts were Catholics.[32]

Having settled in Hamburg, Lubieniecki, expelled from his homeland and deprived of his earlier livelihood, had to find new ways of earning a living for his family (he had four sons — Samuel, born in 1652 or 1653, Teodor, born in 1654, Paweł Jan, born in 1656 or 1657, and Krzysztof, born in 1659 — and two daughters — Katarzyna Salomea, born *c.* 1662, and Gryzella Konstancja, born in 1667). He started to work as an agent of Stefan Niemirycz. Niemirycz left Poland in 1664 and gained the protection of the Elector of Brandenburg, Frederick William (1620–98, Elector since 1640), who granted him a country seat in the New March (region of Brandenburg east of the river Oder, today in western Poland), in Neuendorf (today Czarnowo). It seems that Niemirycz was rather lukewarm in matters of religion, and he left the country due to some financial reasons and because his wife, Gryzelda née Wylam, was an ardent Unitarian. After his wife's death, in 1680 Niemirycz returned to Poland as a Catholic, and in 1682 he was granted a high office of the Palatine (*wojewoda*) of Kijów (Kyiv).[33] Lubieniecki helped Niemirycz to manage his financial affairs and businesses in Hamburg.[34]

Lubieniecki's other employment was working as a political informer. He acted as an intermediary of political information from Poland, Lithuania, and Germany. Among his first employers were the King of Sweden and the King of Denmark. He was paid by them since the late 1650s. The analysis of his preserved reports to the Swedish officials, however, shows that they were probably not very useful to the addressees. It seems that his contacts were not told what kind of information he expected from them and he was not able to digest what he had received into a usable form. Nevertheless, he still tried to find new buyers of his reports. He was moderately successful, but in 1672 he became a political correspondent of Louis XIV (1638–1715,

31 Davies, *God's Playground*, pp. 343–44; Stone, *The Polish-Lithuanian State*, pp. 174–76; CHJ2 p. 508, 'Annexe III: Correspondance avec Fabien Gustmeyer', note to Letter VI, Chapelain to Gustmayer, 28 January 1669; Mallet, 'Les affaires de Pologne', CJH3 p. 704.

32 Tazbir, *Stando lubentius moriar*, pp. 128–42.

33 Janusz Tazbir, 'Niemirycz (Niemierzyc) Stefan', in *Polski Słownik Biograficzny*, vol. 22, ed. by Stefan Kieniewicz et at. (Wrocław: Zakład Narodowy im. Ossolińskich, Wydawnictwo Polskiej Akademii Nauk, 1977), pp. 820–24.

34 Tazbir, *Stando lubentius moriar*, pp. 126–27, 163–66.

King since 1643) (Letter 84). The royal benevolence towards Lubieniecki could have been a result of his flattering unpublished manifesto, written in 1671, in which he claimed that Louis XIV should be crowned the Roman Emperor.[35] Unfortunately, details of Lubieniecki's contacts with the French court and his work for the Sun King are unknown. Likewise, little is known about the contacts from whom Lubieniecki got the information for his correspondence.[36]

Lubieniecki stayed in Hamburg and in Altona until his death, where he continued his confessional polemics. His scholarly occupations also included the work on the *Historia reformationis Polonicae* and on the *Compendium veritatis primaevae*. This latter book was a summary of the author's religious views and remained unpublished until the twentieth century.[37] The book has a form of a conversation between members of Christian denominations: a Catholic, an Orthodox, a Lutheran, a Calvinist, and a Unitarian. Lubieniecki repeats there that the faith of the Polish Brethren was the original Christian faith. He also claims that the freedom of religion in Poland should be reinstated, because all the calamities which fell on the country were caused by its abolition.[38]

Lubieniecki died suddenly after a short illness on 11 May 1675. His two daughters had died two days earlier. The cause of the illness and death was declared to be poison, and was probably on account of accidental food poisoning. Although it was speculated that he was poisoned by his confessional opponents, there was no proof of this.[39]

1.2. Outline of Cometary Studies in Lubieniecki's Time

1.2.1. Overview of Cometary Theories

In the seventeenth century, ancient cometary opinions by Aristotle — that the tailed stars are terrestrial exhalations which burn in the upper parts of the elemental sphere of air — were obsolete. Tycho Brahe (1546–1601), relying on his observations

35 Tazbir, *Stando lubentius moriar*, pp. 220–23; Henk J. M. Nellen, *Ismaël Boulliau (1605–1694). Astronome, épistolier, nouvelliste et intermédiaire scientifique*, (Amsterdam: APA-Holland University Press, 1994), p. 467. The manuscript of Lubieniecki's manifesto is kept together with his *Theatrum cometicum* in BnF Imprimés, Rés. V. 250–51.

36 Tazbir, *Stando lubentius moriar*, pp. 144–53. The analysis of Lubieniecki's political correspondence in the Swedish archives was presented in Janina Kowalik, 'Stanisław Lubieniecki: företrädare för socinianerna och svensk korrespondent', *Karolinska förbundets årsbok* (1981/2), 148–73; Janina Kowalik, 'Szwedzka korespondencja Stanisława Lubienieckiego', *Odrodzenie i Reformacja w Polsce*, 47 (2003), 177–90.

37 Stanislaus Lubieniecki, *Compendium veritatis primaevae*, ed. by K. E. Jordt Jørgensen, 2 vols (København: Akademisk Forlag, 1982).

38 Tazbir, *Stando lubentius moriar*, pp. 223–25.

39 Tazbir, *Stando lubentius moriar*, pp. 230–31.

of the comet of 1577 (C/1577 V1), proved that they are not meteorological phenomena but astronomical objects.[40]

This collapse of traditional authority led to numerous cometary notions, opinions, and theories. For example, Johannes Kepler (1571–1630), even though his beliefs about comets changed over time, was a proponent of an idea that they have to move rectilinearly because they are ephemeral bodies.[41] Pierre Gassendi (1592–1655) also assigned rectilinear motion to them.[42] English clergyman and astronomer Seth Ward (1617–89) believed that comets move on circular orbits and are eternal bodies.[43]

Another theory was authored by René Descartes (1596–1650), and it was part of his complex natural philosophy. Descartes believed that each star had its vortex of particles. When a star is entirely obscured by solar spots, it loses the ability to maintain its own vortex and starts to wander between vortices of other stars as a comet.[44]

Hevelius was one of those who presented their original cometary theories. In his opinion, comets moved in respect to the Sun, but their movement was not uniform, and their trajectories were either hyperbolic or parabolic. As far as the origin and matter of comets are concerned, the Danzig astronomer believed that they are exhalations of planets, especially of Saturn and Jupiter. They are not spherical, but discoid bodies, and they turn their flat sides towards the Sun. Even though Hevelius's ideas were considered wrong, his cometary observations were later used by Isaac Newton (1642–1727).[45] In his correspondence with Lubieniecki, however, Hevelius wrote nothing about this theory. What is more, other cometary theories are also mentioned there sparsely.

1.2.2. Comets of 1664 and 1665

In November 1664, a bright comet (C/1664 W1) appeared in the sky. This phenomenon lasted for a long time and was very bright and sublime, especially at the time when it was closest to Earth in late December. Regular observations started in

40 Donald K. Yeomans, *Comets. A Chronological History of Observation, Science, Myth, and Folklore* (New York: Wiley Science Editions, 1991), pp. 35–42. For more information about the comet C/1577 V1 see Alexandre Guy Pingré, *Cométographie ou Traité historique et théorique des comètes*, vol. 1 (Paris: Imprimerie royale, 1783), pp. 511–19; Gary W. Kronk, *Cometography. A Catalogue of Comets. Volume 1: Ancient – 1799* (Cambridge: Cambridge University Press, 1999), pp. 317–20. About Aristotle's cometary opinions see also 'Subsidia de cometis', CHJ2 pp. 483–84. About Tycho Brahe see also Victor E. Thoren, *The Lord of Uraniborg. A Biography of Tycho Brahe* (Cambridge: Cambridge University Press, 1990); Chantal Grell, 'Introduction', CJH2 p. 14.

41 J. A. Ruffner, 'The Curved and the Straight: Cometary Theory from Kepler to Hevelius,' *Journal for the History of Astronomy*, 2 (1971), 178–94 (pp. 180–83); Yeomans, *Comets*, pp. 52–57; Patrick J. Boner, *Kepler's Cosmological Synthesis. Astrology, Mechanism and the Soul* (Leiden, Boston: Brill, 2013), pp. 105–34.

42 Ruffner, 'The Curved and the Straight', pp. 183–86; Yeomans, *Comets*, pp. 62–63.

43 Ruffner, 'The Curved and the Straight', pp. 186–87; Yeomans, *Comets*, p. 63.

44 Yeomans, *Comets*, pp. 64–65; Tofigh Heidarzadeh, *A History of Physical Theories of Comets, from Aristotle to Whipple* (New York: Springer, 2009), pp. 67–77.

45 Ruffner, 'The Curved and the Straight', pp. 187–91; Yeomans, *Comets*, pp. 82–87.

mid-December. The comet appeared in the constellation of Corvus, passed a significant part of the sky, including constellations of Hydra, Puppis, Canis Maior, Eridanus, and Cetus, and it finally disappeared in Aries in March 1665.[46]

Soon after, another comet (C/1665 E1) appeared. It was less spectacular than the former one. It was visible from late March to late April 1665 and moved from Delphinus to Aries through Equuleus, Pegasus, Andromeda, Pisces, and Triangulum.[47]

These extraordinary objects in the sky triggered a shared fascination. Both professional astronomers and interested amateurs turned their eyes to the sky in order to observe and study these comets. This resulted in a number of books, pamphlets, and other texts of varying importance. The focus here is on Lubieniecki's *Theatrum cometicum*.

1.3. *Theatrum cometicum*

1.3.1. Lubieniecki's Astronomical Education and Correspondents

Lubieniecki's *Theatrum cometicum* is the primary example of his astronomical interests and is more the product of an amateurish fascination, brief and superficial, lacking any deeper insight into its subject. It comprises about 1500 pages *in folio* in three volumes. The first and the third contain Lubieniecki's astronomical correspondence, and the second is a catalogue of historical cometary and cometary-like phenomena.[48]

As far as Lubieniecki's attitude to professional astronomy of his age is concerned, it has to be stated that he was not an astronomer either by education or profession. At most, he was an amateur interested in astronomy. He conducted no observations on his own, and he admitted that he had no instruments for observing the heavens. Lubieniecki also confessed that he did not learn much about mathematics and astronomy at school.[49] There are a few passages in the *Theatrum cometicum* in which he says that he saw something in the sky, but these are only qualitative observations without any reliable data. Moreover, he did not consider his *Theatrum cometicum* to be strictly an astronomical book.

Earning his living by conducting political correspondence, Lubieniecki used his contacts to learn more about the new phenomena in the sky. He entered or re-entered (or at least tried to) into epistolary commerce with about forty correspondents

46 Alexandre Guy Pingré, *Cométographie ou Traité historique et théorique des comètes*, vol. 2 (Paris: Imprimerie royale, 1784), pp. 10–22; Kronk, *Cometography*, pp. 350–57; Chantal Grell, 'Hevelius en son temps', CJH1 pp. 67–71. See also 'Subsidia de cometis', CHJ2 pp. 451–89, and CJH3 p. 349, Letter 109, Hevelius to des Noyers, 18 December 1664, and subsequent letters.

47 Pingré, *Cométographie*, vol. 2, p. 22; Kronk, *Cometography*, pp. 357–60.

48 In the following description of the *Theatrum cometicum* I rely on Jasiński, *Stanisław Lubieniecki*, pp. 35–81.

49 Lubieniecki to Raue, 28 February 1665, TC p. 575.

of various fields of expertise. The letters he exchanged with them represent the most substantial part of his book.[50]

Among these correspondents were astronomers, such as Johannes Hevelius, Giovani Battista Riccioli (1598–1671), Albert Curtz (1600–71), Erasmus Bartholin (1616–80) or Villum Lange (1624–82); men working in other fields of the exact and natural sciences, for example the experimental natural philosopher Otto von Guericke (1602–86), the medical physician Thomas Bartholin (1625–98), or the mathematician Johannes von Leunenschlos (b. 1620). There were also men who concerned themselves with a broad spectrum of scholarly topics, such as the theologian Abraham de Grau (1632–83), the historian and philologist Rudolf Capell (1634–84), or the philosopher Jan Placentinus-Kołaczek (1630–83). Finally, there were polymaths, like the famous Athanasius Kircher (1601/2–80) and his student, Gaspar Schott (1608–66). There are also letters to and from Lubieniecki's fellow brethren: Joachim Stegmann the younger (1618–78) and Andrzej Wiszowaty (1608–78). In addition, Lubieniecki started astronomical correspondence with figures in court circles, such as the Danish courtier, Johann Melchior Røtlin (*fl.* seventeenth century) and Johann Raue (1610–79), the librarian of the Elector of Brandenburg. Lubieniecki also brought cometary and astronomical themes into the letters he exchanged with his earlier contacts. The most important person in this group was Ismaël Boulliau, but others included Gilles Ménage and two diplomats: Johann Ernst von Rautenstein (*c.* 1622–66) and Nicolaus Heinsius (1620–81).

1.3.2. Overview of the Volumes

The first volume of the *Theatrum cometicum*, titled *Theatri cometici pars prior communicationes de cometis 1664 et 1665 cum viris per Europam clarissimis habitas, eorum observationes easque accuratissimis tabulis aeneis expressas continens* (or *Theatri cometici pars prior* in brief), was published in Amsterdam in 1668. The volume consists of 966 pages and 22 unnumbered pages, and almost 60 images. It was dedicated to King Frederick III of Denmark. The main aim of the book was to present Lubieniecki's astronomical correspondence. His letters are grouped into 39 collections, or *communicationes*, as the author calls them. As the title suggests, the volume was planned to present the observations made by learned men, but this intention was fulfilled only to a small degree. There are some observations, but they are relatively scarce. For example, there are no observations from Boulliau, and the only concrete observations from Hevelius are the extracts from his booklets on comets. Excerpts and summaries from the books by others are also a part of materials from Lubieniecki's correspondents printed in the *Theatri cometici pars prior.*

Unfortunately, the volume is of very little use for a potential reader because of its disordered character. There is no table of contents and no index of subjects. Observations and other enclosures are mixed in with the letters, and it is often unclear what material is related to which letter. Moreover, a *communicatio* with one correspondent

50 For the full list of Lubieniecki's correspondents see section 3.5 of this Introduction.

can be scattered throughout the book because while the book was being printed, Lubieniecki continued his correspondence and sent in new letters and materials to the printer. For example, the correspondence with Hevelius is printed in two parts[51] (the supplements to the correspondences are called *addenda*). In some of the collections of letters, there are numbers which indicate the order of correspondence, but they do not help much in comprehending the contents of the book. Nonetheless, Lubieniecki seemed to be fairly proud of his book and the diversity of its subjects. In the preface to the *Theatri cometici pars prior*, he states that his intention was to write a sort of encyclopaedia which would discuss many subjects from various fields of expertise, such as mathematics, physics, chronology, or history.[52]

Lubieniecki also wrote the 'history of comets', or a list of all cometary phenomena which were seen since the times of Noah until 1665. Relying on earlier descriptions and enumerations of extraordinary celestial apparitions, works by historians, and chronicles, he compiled the list of 415 comets. The result of his work was published in Amsterdam in 1666 as the second volume of the *Theatrum cometicum*, titled *Theatri cometici pars posterior, sive historia cometarum a diluvio ad annum Christi 1665, historiae universalis synopsin quandam continens* (or *Theatri cometici pars posterior*). The book contains 464 pages and 8 unnumbered pages. The addressees of the dedication were George William (1624–1705) and John Frederick (1625–79), the dukes of Braunschweig-Lüneburg. As the title suggests, it was meant not only to be an astronomical book but a historical work as well. In the preface, Lubieniecki repeated this declaration from the title page. He also indicated the initial purpose of writing the *Theatrum cometicum*: he intended to prove his opinion that cometary phenomena do not cause nor precede catastrophes, calamities, or other misfortunes. According to Lubieniecki, the historical information and cometary data he gathered show that both fortunate and unfortunate events took place when comets were in the sky.[53] This approach was somewhat innovative because earlier catalogues of comets focused on the misfortunes they were said to portend. On the other hand, it was outdated because Hevelius's *Cometographia* (Danzig: auctoris typis, et sumptibus, imprimebat Simon Reiniger, 1668) presented a catalogue which discussed only the comets and hardly paid any attention to events on Earth.[54] In the preface to the *Theatri cometici pars posterior*, Lubieniecki also added that in his opinion there are no comets in the cosmological sphere of air and that they are celestial objects. Therefore, they cannot have any physical influence on Earth, nor can they cause anything, nor force people to carry out any actions.[55]

51 TC pp. 361–414, 947–55.

52 TC sig. *2r–[*6]v.

53 TC2 sig. *3r–v.

54 Adam Mosley, 'Past Portents Predict: Cometary *Historiae* and Catalogues in the Sixteenth and Seventeenth Centuries', in *Celestial novelties on the eve of the scientific revolution, 1540–1630*, ed. by Dario Tessicini and Patrick J. Boner (Florence: Olschki, 2013), pp. 1–32.

55 TC2 p. 1–2. An example of an outcome of Lubieniecki's chronological studies can be found in Letter 42A.

The third and last volume of the *Theatrum cometicum* was published in Amsterdam in 1668. It was dedicated to Philip William (1615–90), the Count Palatine of Neuburg, and Philip (1630–1703), the Count Palatine of Sulzbach. Its full title reads: *Theatri cometici exitus de significatione cometarum scitis quorundam amicorum obiectionibus, responsionibus auctoris et judiciis virorum clarissimorum expeditus et justa Rautensteiniana* (or *Theatri cometici exitus*). It is a short booklet, containing 78 pages and 8 unnumbered pages. There is no detailed preface apart from the dedicatory letter. As the title of the volume suggests, it consists of two parts. The first is the discussion of Lubieniecki's opinion about the meaning and influence of comets. The publisher of the *Theatrum cometicum*, Frans Kuyper (1629–92), wrote to Lubieniecki and asked him to elucidate various notions about the significance of comets, which can be found throughout his correspondence. As a result of the epistolary discussion, Lubieniecki briefly formulated and summarized his opinion about that matter.[56] He also discussed the subject with a few of his correspondents. These letters and Lubieniecki's correspondence with Kuyper are printed in the *Theatri cometici exitus*. However, they are not divided into *communicationes* and are not even adequately labelled and close reading is required to identify where each of the collections of letters begins.

The second part of the *Theatri cometici exitus* is an obituary of Johann Ernst von Rautenstein, Lubieniecki's correspondent and patron. Lubieniecki printed there a collection of poems dedicated to Rautenstein, a funeral oration for him, and some letters discussing his death.

1.3.3. After the Publication

Lubieniecki's book did not sell well. Moreover, as Janusz Tazbir estimated, it caused an enormous financial loss to the author. Its print run was large: there were 1015 copies produced. They were sold for 7,5 guldens each, whereas Lubieniecki's archive indicates that the publication cost him 15 729 guldens. Even if the whole print run had been sold, he would have still lost over 8000 guldens. In fact, the amount was much higher, as so many copies remained unsold that in 1681, when another comet appeared, they started to be sold again. For this purpose, they were supplied with a new title page purporting to be a second edition.[57]

Kuyper had contacts with a number of religious minorities besides the Polish Brethren. It was he who published the series *Bibliotheca fratrum Polonorum* which turned to be an influential publication and helped in promoting the Polish Brethren's tenets in western Europe.[58] Nevertheless, he and Lubieniecki were in conflict over the publication costs of the *Theatrum cometicum*. As Piet Visser remarks, it is

56 See section 3.3.3 of this Introduction.

57 Tazbir, *Stando lubentius moriar*, pp. 175–79.

58 Woleński, 'Polish-English (British) Philosophical Contacts', pp. 396–97; Ogonowski, *Socynianizm*, pp. 427–32.

highly likely that Kuyper set the high price. The conflict was settled without a lawsuit: probably there was no desire to attract the attention of the authorities to some controversial publications by Kuyper.[59]

1.3.4. *Monumentum Radzivilianum*

Lubieniecki emphasized numerous benefits stemming from his book, both in the prefaces to its volumes and in his correspondence as well. For example, he wrote to Huygens that a reader could find there not only cometary, astronomical, and mathematical discussions, but much more that was of interest and useful as well.[60] Sometimes, however very rarely, he admitted that the *Theatrum cometicum* might not be too enticing for astronomers and mathematicians, but he was sure that the general public would find it pleasant (Letter 11). In spite of these declarations, it seems that Lubieniecki realized that, in fact, his book might be not very useful. Therefore, he decided to add an index of subjects as a separate publication.

He did this on the occasion of the death of Bogusław Radziwiłł (1620–69), Lithuanian aristocrat, whom he decided to honour in a pamphlet. This booklet, published in Hamburg in 1670 and containing 28 pages and 8 unnumbered pages, was dedicated to Frederick William, the Elector of Brandenburg. The publication has two title pages with separate titles: *Monumentum Radzivlianum, in quo Theatri cometici usus multiplex, exhibito historico, ethico, politico indice, et sic adornato Politico enchiridio, magna ex parte demonstratur* and *Strena lugubris at salubris orbi Christiano ineunte anno Christi MDLXX exhibita, sive perenne monumentum commendandis magnis virtutibus magnanimi herois celsissimi, illustrissimi, excellentissimi principis, Boguslai Radzivilii* [...]. The majority — almost two thirds — of its text is an inflated encomium of Bogusław Radziwiłł's family. Lubieniecki presented at great length his genealogy, which dates back to times even before those of Charlemagne. And he also lengthily described the addressee's ties to the most prominent families and dynasties in Europe.[61] Nevertheless, in the preface to his pamphlet, Lubieniecki stated that it was meant to be a summary of the *Theatrum cometicum*. He repeated his claim that his cometary work could be beneficial for the reader, as it contained information from various fields of expertise. He admitted, however, that the lack of an index in his book might make it less useful. However, this index, as he wrote, could be found in the *Monumentum*.[62]

Despite these declarations, the benefits for the reader who used this pamphlet together with the *Theatrum cometicum* were still minimal. The main text of the *Mo-*

59 Piet Visser, 'Blasphemous and pernicious: the role of printers and booksellers in the spread of dissident religious and philosophical ideas in the Netherlands in the second half of the seventeenth century', *Quaerendo*, 26 (1996), no. 4, 303–23 (pp. 318–20).

60 Lubieniecki to Huygens, 30 October 1665, TC p. 931, OCCH5 p. 519, LUL ms HUG 45 f. 1r.

61 Stanisław Lubieniecki *Monumentum Radzivilianum* (Hamburgi: ex officina Georgii Rebenlini, 1670), pp. 1–18.

62 Lubieniecki *Monumentum Radzivilianum*, 'Lectori Benevolo Salutem' (unpaginated).

numentum is adorned with dozens of footnotes which direct the reader to relevant passages in Lubieniecki's cometary book. Almost all of them, however, refer to the *Theatri cometici pars posterior*, to historical facts accompanying the history of comets, which could be related to Radziwiłł's ancestors. The references to the astronomical correspondence from *Theatri cometici pars prior* are almost non-existent. Indeed, there are only two instances where Lubieniecki made use of his astronomical letters. These are references to contemporary monarchs who are sometimes mentioned in the first volume of the *Theatrum cometicum*. For example, Louis XIV of France is extensively referenced. The footnote to his name lists many passages and ends with the words *et alibi passim*.[63] The second use of the *Theatri cometici pars prior* occurred when Lubieniecki found some moral and ethical, or merely ornamental, passages in his correspondence. For example, he refers to the excerpts of his letters in which he mentioned old religious freedoms in Poland.[64]

In the *Monumentum*, stars and other celestial objects serve ornamental purposes only, as in the dedication to the Elector of Brandenburg where Lubieniecki, praising the virtues of the addressee, compares his probity to the constellations of Ursae which indicate the steady path.[65] Even the announced index, titled *Enchiridion politicum*, which is said to contain essential passages from both the *Monumentum* and the *Theatrum cometicum* lists no astronomical subjects and focuses only on historical, ethical, and theological matters.[66] Such a summary seems to be a clear indication that Lubieniecki himself was not sure if the *Theatrum cometicum* was really a book about astronomy. It also demonstrates that his ambition to present a highly practical compendium to the reader was not fulfilled. As comets are almost absent from the *Monumentum*, there is no discussion of the main point of his history of comets: Lubieniecki's belief that there is no connection between celestial objects and misfortunes on Earth.

1.3.5. Frontispiece of the *Theatrum cometicum*

Because Lubieniecki wanted his book to be an encyclopaedia of a wide range of topics and a historical book about comets and general events, it is quite surprising that its frontispiece focuses precisely on comets and astronomy.[67] The engraving is divided into two unequal parts. The upper part (about one third of the picture) pre-

63 Lubieniecki *Monumentum Radzivilianum*, p. 12.

64 Lubieniecki *Monumentum Radzivilianum*, p. 20.

65 Lubieniecki *Monumentum Radzivilianum*, 'Magno et incluto belli et pacis artibus heroi [...] Friderico Wilhelmo' (unpaginated).

66 Lubieniecki *Monumentum Radzivilianum*, sig. Hr–I2v.

67 The frontispiece of the *Theatrum Cometicum* was discussed, or partially discussed, in James Howard Robinson, *The Great Comet of 1680. A Study in the History of Rationalism* (Northfield: Press of the Northfield News, 1916), pp. 18–19; Volker Weidemann, *Theatrum Cometicum. Hamburg und Kiel im Zeichen der Kometen von 1664 und 1665* (Göttingen: Vandenhoeck & Ruprecht, 1987), pp. 5–7; Inga Elmqvist Söderlund, *Taking possession of astronomy. Frontispieces and illustrated title pages in 17th-century books on astronomy*, (Stockholm: The Center for History of Science at the Royal Swedish Academy of Sciences, 2000), pp. 181–82. Because there were some misinterpretations and mistakes in these works,

sents various aspects of the heavens. In the middle of the top edge, there is the divine light (labelled *Creatoris in Gloriam*) and two hands stretching from the sky, one with an olive branch, another with a flaming torch.[68] The accompanying text, *bona bonis, mala malis* ('good for the good people, bad for the bad people'), is a summary of Lubieniecki's views on the influence of comets.[69] In the middle, an old man is holding two books labelled *ratio* ('reason') and *revelatio* ('revelation'). The accompanying text is the famous quotation of unknown origin, but attributed to Ptolemy: *sapiens dominabitur astris* ('a sage will rule the stars').[70] The old man or the sage, according to the label, is standing on the Earth which for its part is labelled *omne solum forti, solum caelum Christiano patria est* ('the whole earth is a homeland for a strong man, for a Christian only heaven').[71] Behind him, there are the signs of the zodiac from Aquarius to Virgo (unevenly distributed) on the background of which the comets of 1664 and 1665 actually moved. Next to the zodiac, there are the Sun and the Moon and an inscription which describes the meaning of the stars: *serviunt, non saeviunt* ('they serve, they do not frown').

The lower part of the frontispiece is an architectonical composition. On top of it the celestial and terrestrial globes are depicted (labelled *maior non sine minori*, 'the greater one not without the smaller one') together with a quotation from Jeremiah (10, 2): *a signis coeli nolite metuere, quae metuunt gentes*, 'be not dismayed at the signs of heaven; for the heathen are dismayed at them.'[72] Below, there is a cartouche with a cityscape, adorned with various astronomical and mathematical devices. The central part is given to the title, sided by the personifications of Knowledge (or Urania) and Prudence (labelled *scientia non sine prudentia*, 'knowledge not without prudence'). The lower part contains Lubieniecki's name written in anagram form: *Satis in ulna Iesu lucebis* ('you will shine enough in the arms of Jesus'), which makes *Stanislaus Lubieniecius*,[73] and the words which he used as his motto: *Stando lubentius moriar* ('I will rather die standing'). There is also the information about the place (Amsterdam) and the year of the publication (1667), the publisher (Frans Kuyper), the painter (Matthias Scheitz, *c.* 1625–1700), and the engraver (Sebastian Stoopendaal, 1637–1707). There was one frontispiece for all the volumes, the only difference being the title set in the middle.

and none of them discussed the wider context of Lubieniecki's astronomical interests, I present my own brief description of the frontispiece.

68 In his letter to Rautenstein, 3 October 1665, TC p. 161, Lubieniecki compares the 'torches of comets' and the 'torches of war', emphasising that the latter burn even if the former ones are not in the sky.

69 See section 3.3.1 of this Introduction.

70 On early usage of this phrase see Theodore Otto Wedel, *The Mediaeval Attitude toward Astrology* (New Haven, London, Oxford: Yale University Press, Humphrey Milford, Oxford University Press, 1920), pp. 135–37.

71 A combination of Ovid, *Fasti*, 1, 493, and the ancient Christian tradition according to which the heaven is the only homeland of Christians.

72 All biblical translations are given after: *The Bible. Authorized King James Version*, ed. by Robert Carroll and Stephen Prickett (Oxford: Oxford University Press, 2008).

73 In some copies, there is *statis* erroneously printed instead of *satis*.

Figure 3. The frontispiece of Lubieniecki's *Theatri cometici pars prior*. Biblioteka Narodowa, Warsaw, SD XVII.4.2660.

CHAPTER 2

The Hevelius-Lubieniecki Correspondence

This chapter outlines the epistolary commerce between Johannes Hevelius and Stanisław Lubieniecki.[1] First, it presents the data regarding the letters. Then, it discusses the main topics of the correspondence and tries to answer the question of why the Danzig astronomer was so interested in contacting Lubieniecki and kept on doing so for a very long time. The chapter also places Hevelius's letters to Lubieniecki against the background of other collections of letters from the *Theatrum cometicum*.

Hevelius played a vital role in Lubieniecki's epistolary discussions. The Danzig astronomer was one of very few whose epistolary commerce with the author of the *Theatrum cometicum* was quite regular and frequent. Moreover, Hevelius was willing to take part in Lubieniecki's debates about various cometary features. Nevertheless, Hevelius's letters were usually reactions to Lubieniecki's inquiries, and it was the author of the *Theatrum cometicum* who started almost all of those cometary discussions. The Danzig astronomer wrote *sua sponte* to Lubieniecki only when he wanted to inform him about his new discoveries.

2.1. The Number of Letters

The correspondence between Hevelius and Lubieniecki comprises 92 letters.[2] Of these, 31 are by Hevelius and 61 by Lubieniecki. The first letter was sent on 19 December 1664, the last one — on 22 December 1673 (both were sent by Lubieniecki).

1 I presented a very introductory version of the discussion of the Hevelius-Lubieniecki contacts, based only on the *Theatrum cometicum*, in the paper titled: *Correspondence between Stanisław Lubieniecki and Johannes Hevelius in Lubieniecki's 'Theatrum Cometicum'* at the conference 'The Practice of Scholarly Communication: Correspondence networks between Central and Western Europe, 1550–1700', organized by the Institute of Philosophy, Academy of Sciences of the Czech Republic, Prague, 18–19 September 2014. I discussed also the main features of the Hevelius-Lubieniecki correspondence in Maciej Jasiński, 'Stanisław Lubieniecki and Johannes Hevelius: (Extra)ordinary "Men of Letters"', *Quarterly Journal of the History of Science and Technology*, 64 (2019), no. 1, 125–37.

2 After closer analysis of the correspondence, the number of letters given in Klaus-Dieter Herbst, 'Alphabetical list of the correspondents', CJH1 p. 226, that is 95, had to be decreased.

Almost all the letters from the years 1664–67 are printed in the *Theatrum cometicum*. In *Communicatio Gedano-Heveliana* there are 67 letters, of these 24 are by Hevelius and 43 by Lubieniecki. (Because of various reasons, discussed below, the number of letters printed in Lubieniecki's book does not equal the real number of letters from that period.) The correspondence is also mixed in with several enclosures: letters written by others, various observations, etc. Most of them were enclosed in Lubieniecki's letters. There is only one letter from Hevelius which contained enclosures (Letter 83). The language of the correspondence is exclusively Latin, but some of the enclosures are in French or German.

As is suggested by the number of letters from the first three years of the correspondence published in Lubieniecki's book, the distribution of letters is extremely uneven. Table 1 presents the number of letters sent each year.

Table 1. The letters of the Hevelius-Lubieniecki correspondence in yearly distribution.

Year	Letters from Lubieniecki	Letters from Hevelius	Total
1664*	3	1	4
1665	22	11	33
1666	10	7	17
1667	12	5	17
1668	1	1	2
1669	3	1	4
1670	0	0	0
1671	2	2	4
1672	3	2	5
1673	5	1	6
Total	61	31	92

* December only.

The given numbers lead to the conclusion that Lubieniecki was much more interested in contacting Hevelius than *vice versa*. They also seem to endorse the impression that Lubieniecki's astronomical interests were rather superficial and that he conducted his correspondence mainly for the sake of the publication of his *Theatrum cometicum*. This conclusion is also supported by the length of the letters. Those at the beginning of the correspondence are much longer, with the first thirty letters covering almost half of the manuscripts of the entire volume.

Nevertheless, Hevelius and Lubieniecki were significant correspondents for each other. Hevelius's correspondence with Lubieniecki is the fourth largest in the Hevelian corpus after his those with Pierre des Noyers (259 letters), Ismaël Boulliau (204

letters), and Henry Oldenburg (118 letters).[3] As far as Lubieniecki's correspondence is concerned, unfortunately, there is no single corpus, or at least a catalogue, of all his known letters. But as for his astronomical correspondence in the *Theatrum cometicum*, the Hevelius's *communicatio* is the third largest, after the correspondences with Johann Ernst von Rautenstein (172 letters) and Nicolaus Heinsius (125 letters). It should be remarked that neither of these men was a scholar: they were courtiers and diplomats. And we ought to remember that Lubieniecki's astronomical correspondence was only a part, and probably a relatively small part, of his whole correspondence which consisted mainly of letters on political topics.

2.2. Overview of the Correspondence

2.2.1. Lubieniecki Reaches Hevelius

When Lubieniecki wrote to Hevelius for the first time, he was reaching a reputed member of the Republic of Letters and, therefore, tried to persuade the Danzig astronomer to enter the epistolary commerce with him. In his first letter, Lubieniecki referred to Ismaël Boulliau as their mutual friend and correspondent. As if he wanted to prove his contacts with the French astronomer, Lubieniecki enclosed a letter from Boulliau (Letter 1A). With his first two letters to Hevelius, he also sent large amounts of opinions and observations he had received from his correspondents and asked the addressee for his views. In the subsequent correspondence, he continued this practice and informed Hevelius about what he received.

Lubieniecki did not sign his letters to Hevelius with his real name but with a pseudonym. In this correspondence, he did not explain why he concealed his identity. It may be speculated that this was somehow related to his political and confessional activities as both of these could make enemies for him and his correspondents in Poland. Lubieniecki signed his earlier letters to Hevelius as 'de Rolis,' and the later ones as 'Rolitsius' (or 'Rolitzius'), often together with the first letters of his given and family names, i.e. 'S. L. Rolitsius.' These pseudonyms were derived from the name of the Lubieniecki family's coat of arms, Rola. Lubieniecki explained its etymology in a letter to Boulliau but without discussing the reasons why he used it.[4] In two of his letters to Hevelius, Lubieniecki's signature is even more cryptic with the letters 'T. N.', which expands to *tibi notus* (Letters 28 and 29). In his first letter to Hevelius, Lubieniecki informed him that he would like to keep their correspondence secret, although this did not stop him from asking the addressee a few months later to agree upon the publication of their letters (Letter 18). Lubieniecki seemingly did not always hide his identity, as he informed Hevelius that in Hamburg he could freely use his

3 Herbst, 'Alphabetical list of the correspondents', pp. 253–78. Hevelius's correspondence with des Noyers is published in *Correspondance de Johannes Hevelius. Tome III: Correspondance avec Pierre des Noyers, secrétaire de la Reine de Pologne, 1646–1686*, ed. by Chantal Grell in collaboration with Damien Mallet (Turnhout: Brepols, 2020).

4 Lubieniecki to Boulliau, 4 April 1665, TC p. 477, BnF ms FF. 13031 f. 229v.

family name (Letter 37). Nevertheless, his real identity was known because the *Theatrum cometicum* was published under his real name. It was also known to Hevelius: in the manuscript copy of his first letter to Lubieniecki, he noted both his pseudonym and his family name (Letter 3). There were only a few letters signed by Lubieniecki both with his real name and the pseudonym. All of them are in the last part of the correspondence (Letters 86, 88, and 92).

In his letters to Hevelius, Lubieniecki often expressed his utmost esteem for the addressee.[5] Lubieniecki praised Hevelius with quotations from classical poets (Letter 15) and compared his works to those of historical and mythological heroes of astronomy: to Hipparchus (Letters 82 and 92), Atlas (Letters 24, 82, and 84), and Hercules (Letters 15, 24, 82, 84, and 92).[6] And even though these praises were very general and usually did not refer to any specific achievement of Hevelius, Lubieniecki highly appreciated the cometary opinions of the Danzig astronomer and often quoted them. It is difficult to point out any other correspondent of Lubieniecki in the *Theatrum cometicum* whom he held in such high esteem.

Because Lubieniecki often wrote in a flamboyant and convoluted style, he sometimes got lost in his lengthy and multilevel phrases. In one letter, he realized that a sentence was so complicated that, in fact, it was grammatically incorrect and impossible to understand. Therefore, he sent to Hevelius a brief letter in which he explained what should be added and changed for the phrase to make sense (Letter 16). It should be noted that the result was unsatisfactory, and the sentence is still very obscure.

2.2.2. Hevelius's Reactions to Lubieniecki's Letters

Hevelius, on his part, was less effusive in his correspondence with Lubieniecki. His letters are not only fewer but generally shorter as well. Apart from typically polite expressions, there are no lengthy praises of Lubieniecki or his work. Hevelius initially seemed reluctant to share his astronomical opinions with Lubieniecki (Letter 3). Nevertheless, Hevelius entered epistolary commerce with Lubieniecki and continued it for many years. This could have been caused by the fact that Hevelius lived in a relatively remote part of Europe and decided to use some help to promote his astronomical works. He expressed this himself in a letter to Lubieniecki: 'I know almost no one who works on mathematics and solid philosophy in Poland, Lithuania, or Livonia' (Letter 64).

Hevelius informed Lubieniecki about the comets of 1665 and 1672 and sent him his first observations of them (Letters 10 and 83). Nevertheless, the Danzig astronomer was neither willing to share everything with Lubieniecki nor to engage in every discussion started by him, for example about historical appearances of comets

5 This was amply remarked by Tadeusz Przypkowski, 'Zainteresowania matematyczno-astronomiczne Braci Polskich', in *Studia nad arianizmem*, pp. 415–16.

6 According to a common interpretation of the myth, Atlas was the founder of astronomy, and Hercules learned the celestial science from him, see Volker R. Remmert, *Picturing the Scientific Revolution*, transl. by Ben Kern (Philadelphia: Saint Joseph's University Press, 2011), pp. 130–44.

(e.g. Letter 30). The only further extensive observational data from Hevelius in his *communicatio* are Lubieniecki's summaries of Hevelius's booklets about comets of 1664 and 1665: *Prodromus cometicus* (Danzig: autoris typis, et sumptibus, imprimebat Simon Reiniger, 1665) and *Descriptio cometae* (Danzig: autoris typis, et sumptibus, imprimebat Simon Reiniger 1665) (see Appendices 1 and 2).

2.2.3. Hevelius's Publications

On his part, Lubieniecki encouraged Hevelius to publish the booklets about the two comets: *Prodromus cometicus* and *Descriptio cometae*. He even offered to help Hevelius in printing the *Prodromus cometicus* (Letters 8 and 17). When they were ready, Hevelius asked Lubieniecki to disseminate them among his learned friends and to convey a number of copies to booksellers in Hamburg (Letters 17 and 49).

Nevertheless, it seems that Hevelius quickly lost his trust in Lubieniecki's ability to help him in circulating his publications. When the *Cometographia* was published, Lubieniecki offered his services to the Danzig astronomer (Letters 74 and 76), but Hevelius did not reply. A few years later, when Hevelius finished his *Machinae coelestis pars prior* (Danzig: auctoris typis, et sumptibus, imprimebat Simon Reiniger 1673), he sent a copy to Lubieniecki but did not ask for his help in distributing it (Letters 90 and 92).

Lubieniecki was eager to help his correspondent with his books. He also tried to protect Hevelius's interests: when he learned from Isaac Gruterus that there was an anonymous, unauthorized edition of the *Prodromus cometicus*, he informed Hevelius about it and tried to learn more about the matter (Letters 35 and 36).[7] Prompted by Lubieniecki, Gruterus was unable to obtain any details about the publication from the printer,[8] probably because there was, in fact, no such edition of Hevelius's *Prodromus* (Letters 42, 42A, and 42B).

2.2.4. Hevelius's Conflict with Adrien Auzout

Lubieniecki also tried to help Hevelius in the conflict with the French astronomer Adrien Auzout (1622–91). Auzout questioned Hevelius's observations of the comet in February 1665. As a result of subsequent discussions, the judges appointed by the Royal Society: Robert Boyle (1627–91), Robert Moray (1608/9–73), William Petty (1623–87), and John Wallis (1616–1703), pronounced that it was the Danzig astronomer who was in the wrong. Realizing that his reputation and the favours of royal benefactors, including the King of France, were at stake, Hevelius strove to persuade the learned world to his point of view, but his efforts were in vain. Even though

7 Gruterus to Lubieniecki, 12 October 1665, TC p. 866; Lubieniecki to Gruterus, 2 February 1666, TC p. 871.

8 Lubieniecki to Gruterus, 16 April 1666, TC p. 875; Gruterus to Lubieniecki, 31 May 1666, TC p. 876; Lubieniecki to Gruterus, 8 June 1666, TC p. 877.

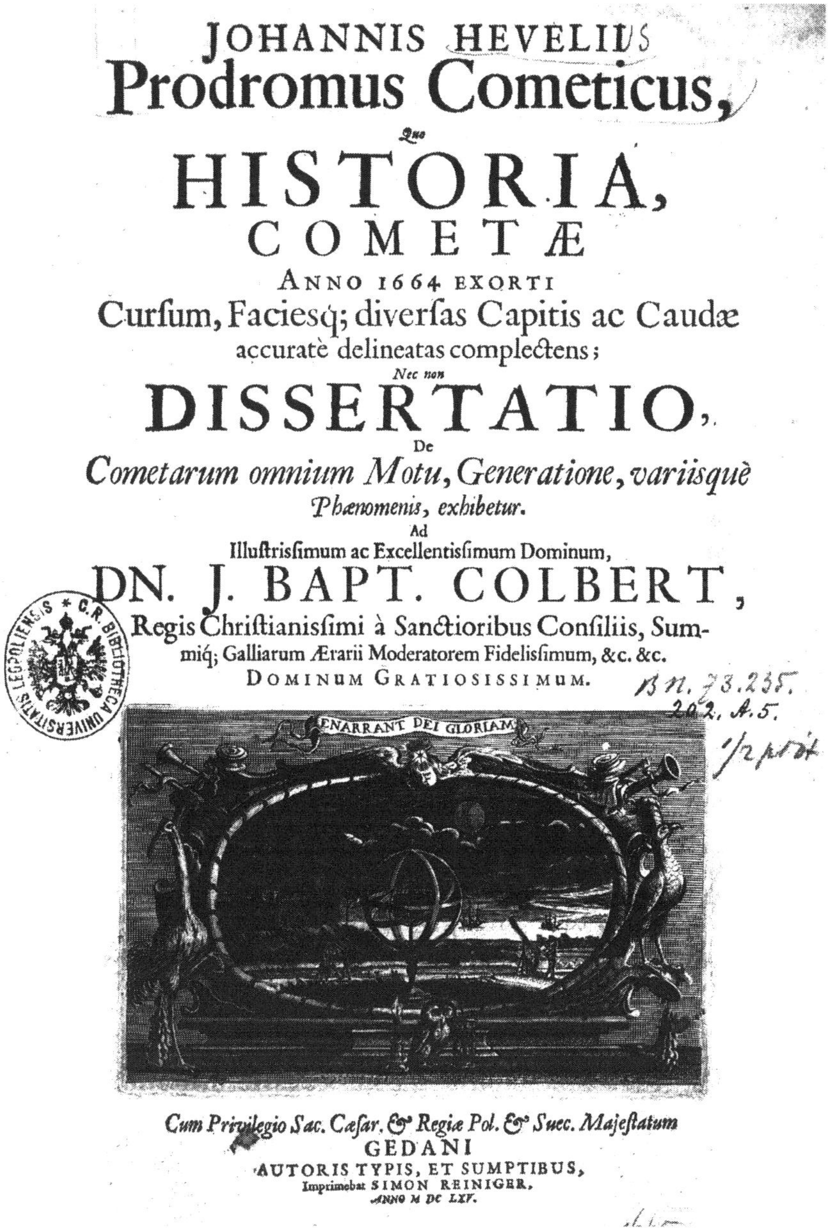

JOHANNIS HEVELII
Prodromus Cometicus,
Quo
HISTORIA,
COMETÆ
ANNO 1664 EXORTI
Curſum, Facieſq; diverſas Capitis ac Caudæ
accuratè delineatas complectens;
Nec non
DISSERTATIO,
De
Cometarum omnium Motu, Generatione, variisquè Phænomenis, exhibetur.
Ad
Illuſtrisſimum ac Excellentisſimum Dominum,
DN. J. BAPT. COLBERT,
Regis Chriſtianisſimi à Sanctioribus Conſiliis, Summiq; Galliarum Ærarii Moderatorem Fidelisſimum, &c. &c.
DOMINUM GRATIOSISSIMUM.

Cum Privilegio Sac. Cæſar. & Regiæ Pol. & Suec. Majeſtatum
GEDANI
AUTORIS TYPIS, ET SUMPTIBUS,
Imprimebat SIMON REINIGER,
ANNO M DC LXV.

Figure 4. Title page of Johannes Hevelius's *Prodromus cometicus*. Biblioteka Narodowa, Warsaw, mf. B9601.

JOHANNIS HEVELII

DESCRIPTIO

COMETÆ

ANNO ÆRÆ CHRIST. M. DC. LXV. EXORTI,

Cum genuinis Obſervationibus, tam nudis, quàm enodatis, Menſe Aprili habitis

GEDANI.

Cui addita eſt

MANTISSA

Prodromi Cometici,

Obſervationes omnes prioris Cometæ M DC LXIV, ex iisq; genuinum motum accuratè deductum, cum notis, & animadverſionibus exhibens.

Ad

SERENISSIMUM

LEOPOLDUM,

Etruriæ Principem.

Cum Privilegio Sac. Cæſareæ, & Reg. Poloniæ, ac Sueciæ Majeſtatum.

GEDANI,

AUTORIS TYPIS, ET SUMPTIBUS,

Imprimebat SIMON REINIGER, *Anno* M DC LXVI.

Figure 5. Title page of Johannes Hevelius's *Descriptio cometae anno aerae Christianae MDCLXV exorti*. Biblioteka Narodowa, Warsaw, mf. B9628.

Auzout finished his astronomical career after 1668, he managed to cast a shadow on Hevelius's good name.[9]

Lubieniecki's engagement in the conflict between Hevelius of Auzout is usually mentioned only cursorily. In fact, his role was rather insignificant. As a correspondent both of Hevelius and Auzout, Lubieniecki decided to help them to settle the dispute in a friendly and peaceful manner. He suggested that he could assist in the mediation which could be conducted by Ismaël Boulliau and Christiaan Huygens. Lubieniecki pointed out that these learned men would be suitable mediators due to their proficiency in mathematics and astronomy and because they were friends to both Hevelius and Auzout (Letter 35).[10] The reaction of the participants of the quarrel to Lubieniecki's suggestion was far from enthusiastic. Auzout did not answer at all: the letter with Lubieniecki's offer of mediation is the last one in his *communicatio*. It should be noted, however, that Lubieniecki's contacts with Auzout were much less frequent than those with the Danzig astronomer.

Hevelius's reply to Lubieniecki's suggestion was very vague. He agreed that experienced astronomers were the right people to judge this case (he did not mention, however, Boulliau and Huygens by name), but first of all, he had to prepare his answer to Auzout's objections, so that both sides could be heard out on equal terms (Letter 36). Later, he repeated that he had asked the world of learning to suspend their judgments until his answer was finished (Letter 39). When he wrote his reply to Auzout — which took the form of an appendix to his *Descriptio cometae*, titled *Mantissa Prodromi cometici*[11] — he informed Lubieniecki that his dispute with Auzout could be finally settled. He added, however, that there were some people who had already judged the case. Undoubtedly, he alluded to the astronomers of the Royal Society who had not waited with expressing their opinion for Hevelius's answer (Letter 47). Lubieniecki, in the prepared extract of the *Descriptio cometae*, added the information about the quarrel.[12] Hevelius, however, was not satisfied with the suggestion of such a summary of his book. He considered it to be too brief and too general (Letter 51).

Not only did Lubieniecki fail at persuading Hevelius and Auzout to his offer of mediation, but he also did not convince Huygens and Boulliau to be the mediators either. Huygens did not answer Lubieniecki's letter.[13] What is quite intriguing, in

9 Grell, 'Hevelius en son temps', pp. 99–107; Grell, 'Introduction', CJH2 pp. 66–90. Detailed chronology of the Hevelius-Auzout dispute and the most important sources pertaining to it are presented by Chantal Grell in 'Subsidia de cometis', CJH2 pp. 451–89. Also on this subject see Steven Shapin, *A Social History of Truth. Civility and Science in Seventeenth-Century England* (Chicago, London: The University of Chicago Press, 1994), pp. 266–91; Nausicaa Elena Milani, 'The Prodromus Cometicus in the Académie des Sciences and the Royal Society: The Hevelius-Auzout Controversy', in *Johannes Hevelius and his Gdańsk*, ed. by Marian Turek (Gdańsk: Gdańskie Towarzystwo Naukowe, 2014), pp. 195–208. See also CJH3 p. 355, note to Letter 112, Hevelius to des Noyers, 19 February 1665, and following letters.

10 Lubieniecki to Auzout, 31 October 1665, TC pp. 857–58.

11 Johannes Hevelius, *Descriptio cometae anno aerae Christianae MDCLXV exorti* (Gedani: autoris typis et sumptibus, imprimebat Simon Reiniger, 1666), pp. 57–188.

12 See Appendix 2.

13 Lubieniecki to Huygens, 30 October 1665, TC p. 932, OCCH5 p. 520, LUL ms HUG 45 f. 1v.

Lubieniecki's correspondence with Boulliau, not a single reference to Boulliau as a mediator is to be found, even though Auzout and his quarrel with Hevelius frequently occur in this *communicatio*.[14] It is possible, however, that there are some lacunae in Lubieniecki's correspondence with Boulliau. Another explanation is that Lubieniecki hoped that Boulliau would learn of his plan from a letter to Auzout, which Lubieniecki asked him to pass on.[15]

2.2.5. Olhoff's Edition of Hevelius's Letters

Hevelius profited from his correspondence with Lubieniecki in one more way. Some of Lubieniecki's letters were printed by his relative, Johann Eric Olhoff (1650–1710), in the *Excerpta ex litteris illustrium et clarissimorum virorum ad nobilissimum, amplissimum et consultissimum dominum Johannem Hevelium* (Gedani: ex officina Janssonio Waesbergiana, 1683). This publication was devised by Hevelius to promote his works and his plan to publish his correspondence.[16]

In the booklet, there are six Lubieniecki's letters to Hevelius (Letters 2, 24, 28, 48, 84, and 92). Lubieniecki praises the Danzig astronomer there and extolls his planned *Cometographia*:

> Believe me, everyone's minds, ears, and eyes are turned towards it as towards a wonder of astronomical talent, and everyone avidly anticipates to see it as soon as possible. And do not think I flatter you. Because I adduce not only the opinion of the whole astronomical senate (where you are an unparalleled leader) but also that of the Republic of Letters. (Letter 28)

In Lubieniecki's letters printed by Olhoff, there is also a remark about Hevelius's discussion with Auzout: having received the *Descriptio cometae*, Lubieniecki commends the Danzig astronomer on his answer to Auzout (Letter 48). This selection of letters is evidence of genuine esteem in which Lubieniecki held Hevelius.

Given the fact that Olhoff had at his disposal two versions of the majority of Hevelius's correspondence with Lubieniecki — the manuscript letters and their printed edition in the *Theatrum cometicum* — some conclusions can be drawn. He often preferred to use the printed versions: letters in his book usually present the same textual variations as their counterparts from the *Theatrum cometicum* (Letters 2, 24, and 28). Nevertheless, it seems that in one case he used the manuscript as the basis for his edition, even though there was a printed version he could use (Letter 48). Of course,

14 For example: Lubieniecki to Boulliau, 18 July 1665, TC p. 529, BnF ms FF. 13031 f. 310r; Lubieniecki to Boulliau, 12 September 1665, TC p. 534, BnF ms FF. 13031 f. 224r; Lubieniecki to Boulliau, 19 September 1665, TC p. 535, BnF ms FF. 13031 f. 325r.

15 Lubieniecki to Boulliau, 31 October 1665, TC p. 537, BnF ms FF. 13031 f. 331r.

16 Grell, 'Hevelius en son temps', pp. 131–32; CJH3 p. 642, note to Letter 246, des Noyers to Hevelius, 19 May 1684.

he had to use manuscripts when publishing letters which were not printed in Lubieniecki's book (Letters 84 and 92).

Apart from the letters from Lubieniecki, Olhoff reprinted from the *Theatrum cometicum* a selection of other letters from or to Lubieniecki in which Hevelius is mentioned.[17]

2.3. Hevelius's Input to Lubieniecki's Cometary Discussions

Hevelius took a major part in two of Lubieniecki's epistolary discussions: about the number of comets and about the new phenomenon seen in Andromeda. In both these instances, Lubieniecki agreed with Hevelius's point of view and shared it with his correspondents.

2.3.1. Number of Comets in December 1664

The discussion about the number of comets seen in December 1664 arose because the comet was initially seen in the early morning. Then it disappeared and only later reappeared in the evening. Most of Lubieniecki's correspondents had no doubts and were sure that both phenomena were the appearances of the same comet: he received such an opinion, among others, from Hevelius (Letter 6), Kircher,[18] and Friedrich Büthner (1622–1701).[19] Proponents of the opposite view were few. It was the opinion of Johann Ernst von Rautenstein[20] and Adam Olearius (*c.* 1599–1671).[21] Lubieniecki sent them Hevelius's opinion about only one comet.[22] His persuasion, however, was not very successful: there was no response from Rautenstein about this matter, and Olearius did not change his view.[23]

Even though Lubieniecki received many opinions about the number of comets, it was Hevelius's view which he clearly valued the most. Whenever Lubieniecki discussed the matter in his letters, he usually quoted only Hevelius's view, as we can see, for example, in the letters to Rautenstein and Olearius.[24] He sometimes mentioned only the opinion about one comet and Hevelius as its advocate.[25]

17 Olhoff pp. 92–110.

18 Kircher to Lubieniecki, 8 May 1665, TC p. 751.

19 Büthner to Lubieniecki, 21 March 1665, TC p. 802.

20 Rautenstein to Lubieniecki, 12 January 1665, TC p. 42.

21 Olearius to Lubieniecki, 15 January 1665, TC p. 636.

22 Lubieniecki to Rautenstein, 21 January 1665, TC p. 43; Lubieniecki to Olearius, 23 January 1665, TC pp. 636–37.

23 Olearius to Lubieniecki, 3 February 1665, TC p. 637.

24 Lubieniecki did a similar thing, among others, in the letters to Pál Jászberényi, 11 February 1665, TC p. 690, to Tobiasz Morsztyn, 17 February 1665, TC p. 506, and to Johann Raue, 28 February 1665, TC p. 576.

25 Lubieniecki to Sivers, 3 March 1665, TC p. 921; Lubieniecki to Büthner, 16 April 1665, TC p. 807.

2.3.2. Phenomenon in Andromeda

The number of comets in 1664 and 1665, in fact, was not a very serious problem. But there was another frequent subject concerning their observations, which triggered more discussions. Because two comets were seen in a short time, there was a common expectation that there would be more celestial phenomena. Such an object was found in the constellation of Andromeda. This part of the sky was near the routes of both comets and, therefore, was frequently observed. In Lubieniecki's correspondence, the information that something unusual occurred in the sky appears three times. Today, based on the descriptions sent by his correspondents, we know that the object they found was the Andromeda Galaxy (M31). In fact, they were not the first to observe it. Already in the tenth century, it was depicted by 'Abd ar-Rahman as-Sufi (903–86) in his star atlas. At the beginning of the seventeenth century, Simon Marius (Mayr, 1573–1624) observed it through a telescope and published his observations in *Mundus Jovialis anno MDCIX derectus ope perspicilli Belgici* ([Nürnberg]: sumptibus et typis Johannis Lauri, 1614).[26] It was also mentioned by Giovanni Battista Hodierna (1597–1660)[27] and, soon after the cometary phenomena of 1664 and 1665, by Ismaël Boulliau.[28] Boulliau received from Lubieniecki some information about the observations of this phenomenon conducted by his correspondents, but he did not refer to them in his book.

The first correspondent who sent Lubieniecki the information about the phenomenon in Andromeda was Abraham de Grau. He recounted that he saw a dark, faint dot which seemed to be growing slowly and seemed not to change its position much.[29] In fact, both of these things were impossible and were probably caused by considering the phenomenon to be a new comet. In his later letter, de Grau wrote that the object did not change its position.[30] Lubieniecki answered de Grau that Athanasius Kircher had sent him some observations from Rome which also describe a new phenomenon near Andromeda.[31] Actually, Lubieniecki made a mistake because the observation in question was sent to him by Albert Curtz (Letter 16B). Kircher was one of many correspondents whom Lubieniecki asked about these observations and the new pheno-

26 Jerzy Dobrzycki, Jarosław Włodarczyk, *Historia naturalna gwiazdozbiorów* (Warszawa: Prószyński i S-ka, 2002), pp. 101–03; Daniel Schultz, *The Andromeda Galaxy and the Rise of Modern Astronomy* (New York: Springer, 2012), pp. 39–43.

27 Giovanni Battista Hodierna, *De admirandis coeli caracteribus*, in Giovanni Battista Hodierna, *De systemate orbis cometici deque admirandis coeli characteribus* (Panormi: typis Nicolai Bua, 1654), p. 48.

28 Ismaël Boulliau, *Ad astronomos monita duo. Primum de stella nova, quae in collo Ceti ante annos aliqua visa est. Alterum de nebulosa in Andromedae cinguli parte borea ante biennium iterum orta* (Parisiis: apud Sebastianum Mabre-Cramoisy, 1667), pp. 14–17. An often repeated opinion that Boulliau observed the M31 in 1661 (e.g. see Richard Hinckley Allen, *Star-Names and their Meanings* (New York: G. E. Stechert, 1899), p. 39) is probably an error which was caused by mistyping the year 1667 as 1661.

29 De Grau to Lubieniecki, 21 March 1665, TC p. 490, LUL ms PAP 15 f. 1v.

30 De Grau to Lubieniecki, 12 June 1665, TC p. 500.

31 Lubieniecki to de Grau, 9 June 1665, TC p. 499.

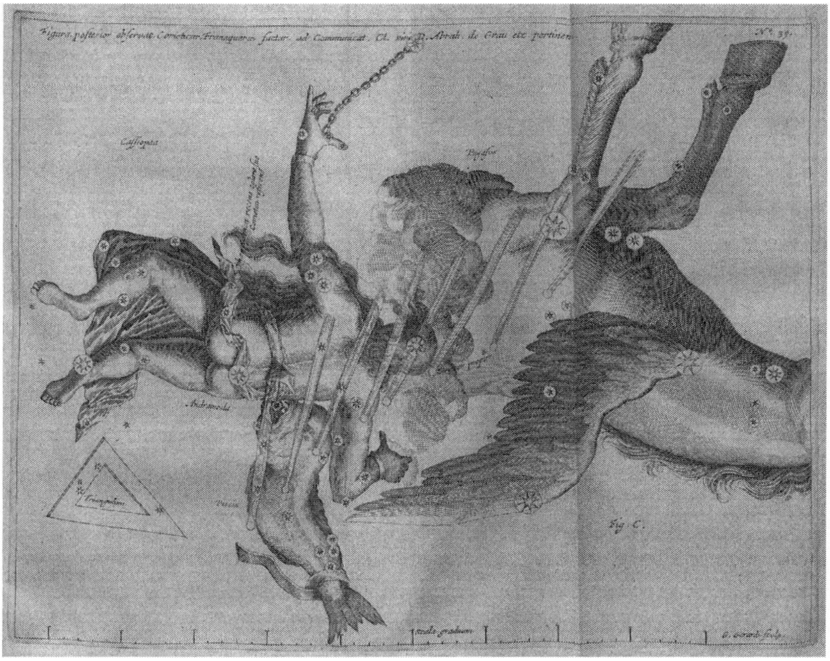

Figure 6. The constellation of Andromeda with the positions of the comet and the phenomenon in Andromeda (the Andromeda Galaxy) in de Grau's *communicatio* in Lubieniecki's *Theatri cometici pars prior*. Biblioteka Narodowa, Warsaw, SD XVII.4.2660.

menon,[32] among them Giovanni Battista Riccioli[33] and Johannes Hevelius (Letters 16 and 23). And it was Hevelius who sent the only detailed reply about that matter. Lubieniecki was able to read that the phenomenon in question was not a comet, but probably an undescribed star; Hevelius refrained from further considerations (Letters 19 and 26). Sometime later, the Danzig astronomer added that it was an object described by Simon Marius (Letter 30). Having received Hevelius's opinion, Lubieniecki sent it to his correspondents, including de Grau[34] and Curtz,[35] but received no answer.

A few months later, Lubieniecki received new information about the phenomenon in Andromeda, this time from Nicolaus Heinsius. He conveyed to him the observations and opinions of Olof Rudbeck (1630–1702) who considered the phenome-

32 Lubieniecki to Kircher, 24 June 1665, TC p. 753; Lubieniecki to Kircher, 19 August 1665, TC p. 756.

33 Lubieniecki to Riccioli, 17 June 1665, TC p. 702.

34 Lubieniecki to de Grau, 18 September 1665, TC p. 523.

35 Lubieniecki to Curtz, 2 July 1665, TC p. 222.

non to be a comet.[36] Lubieniecki again sent this information to his correspondents, and this time he managed to receive more answers. Hevelius defined the object in question not as a comet, but as a 'nebulous star' (Letter 36), and Lubieniecki agreed with his opinion (Letter 37). In the *Theatrum cometicum*, there are also answers from Christian Busmann (1610–81), who was sceptical about the new phenomenon,[37] and Ismaël Boulliau, who stated that the object in question was not a comet but a star which had been described long before.[38] Lubieniecki sent their opinions to Heinsius a few times and asked for Rudbeck's answer to them.[39] He also declared that he agreed with Hevelius's statements.[40] Heinsius finally answered that Rudbeck admitted his mistake.[41]

The last of Lubieniecki's correspondents who informed him about the phenomenon in Andromeda was Samuel Reyher (1635–1714). He remarked that the object he saw looked like a cloud or a comet, but he stated nothing certain about its nature (Letter 52B). In his reply, Lubieniecki informed Reyher about the conclusions of his earlier discussions concerning the phenomenon. He wrote that Hevelius considered it to be a star and that this object had been described by Simon Marius.[42] Lubieniecki also sent this information from Reyher to Hevelius, stating that, in his opinion, the phenomenon in question was the same as the one previously discussed (Letter 52). Hevelius agreed with this view (Letter 54), but he seemingly misunderstood Reyher's statements and wrote that it was a mistake to consider the phenomenon to be something new (Letter 56). When Lubieniecki informed Reyher about it,[43] he answered that he had written nothing of this kind and that he had suspended his opinion about the nature of the phenomenon (Letter 62A).[44]

2.3.3. Hevelius's Motives

Hevelius's involvement in these two astronomical discussions allows for determining what interested him the most in Lubieniecki's letters. The Danzig astronomer was eager to participate in deliberations which considered actual objects in the sky: recent comets and other phenomena which could be observed. In such matters, he had his own views and tried to convince Lubieniecki to them. The Danzig astronomer succeeded, and Lubieniecki became the promoter of his astronomical views. It

36 Heinsius to Lubieniecki, 6 November 1665, TC p. 321.

37 Busmann to Lubieniecki, 20 November 1665, TC p. 667.

38 Boulliau to Lubieniecki, 11 December 1665, TC p. 539, BnF ms FF. 13026 f. 302r.

39 Lubieniecki to Heinsius, 24 November 1665, TC p. 323; Lubieniecki to Heinsius, 8 December 1665, TC pp. 323–24; Lubieniecki to Heinsius, 15 December 1665, TC p. 324; Lubieniecki to Heinsius, 22 December 1665, TC p. 325; Lubieniecki to Heinsius, 19 February 1666, TC p. 337.

40 Lubeiniecki to Heinsius, 23 February 1666, TC p. 337.

41 Heinsius to Lubieniecki, 17 February 1666, TC p. 338; Heinsius to Lubieniecki, 10 March 1666, TC p. 342, LUL ms PAP 15 f. 1r.

42 Lubieniecki to Reyher, 4 November 1666, TC p. 886.

43 Lubieniecki to Reyher, 26 April 1667, TC sig. [Vvvvv2 bis r].

44 Reyher to Lubieniecki, 30 April 1667, TC sig. [Vvvvv2 bis r–v].

is highly probable that this was one of the reasons why Hevelius entered into this correspondence in the first place: he could use Lubieniecki's help in spreading his views. This conclusion is also corroborated by the fact that the Danzig astronomer was distributing his books with Lubieniecki's help, and that he asked for Lubieniecki's assistance with an alleged publication of an unauthorized edition of the *Prodromus cometicus*.

Nevertheless, when Lubieniecki touched upon subjects which were not based on solid, observational data, such as historical comets (Letters 27, 43, and 66) or cometary predictions (Letters 5 and 9), Hevelius was less willing to engage in the discussions. In such cases, he either answered briefly or not at all.

2.4. Overlap of Hevelius's and Lubieniecki's Networks

Among Lubieniecki's correspondents in the *Theatrum cometicum*, there are sixteen people who were Hevelius's correspondents as well.[45] Most of them were practitioners of astronomy or mathematics, and their letters exchanged with Lubieniecki constitute the most interesting part of his book. What is more, they were often mentioned in his letters exchanged with Hevelius. The following section presents the key features of their correspondence with Lubieniecki. The descriptions do not reflect on how important these people were for the science of their times. They rather focus on these scholars' input to Lubieniecki's astronomical and cometary discussions and to his *Theatrum cometicum*. The list is arranged alphabetically.

2.4.1. Erasmus Bartholin

Erasmus Bartholin (born 13 August 1625 in Roskilde, died 4 November 1698 in Copenhagen) was a Danish mathematician and physician, the brother of Thomas Bartholin, a physician. In 1656, he became the professor of mathematics at the University of Copenhagen; in the following year he took the chair in medicine. He was also the royal physician. In mathematics, he continued the Cartesian tradition and worked on the theory of equations. He also worked on optics and conducted cometary observations. By order of King Frederick III, he worked — unsuccessfully — on an edition of Tycho Brahe's writings.[46]

Lubieniecki started correspondence with Erasmus Bartholin after receiving no answer to the letters sent to his brother.[47] In his letters, he repeatedly asked Erasmus Bartholin for his observations and opinions and sent him information from other

45 For the list of Hevelius's correspondents see Herbst, 'Alphabetical list of the correspondents', pp. 253–78. For the list of Lubieniecki's correspondents in the *Theatrum cometicum* see section 3.5 of this Introduction.

46 C. D. O'Malley, 'Bartholin, Erasmus', in *Dictionary of Scientific Biography*, vol. 1, ed. by. Charles Coulston Gillispie (New York: Charles Scribner's Sons, 1970), pp. 481–82.

47 Lubieniecki to Erasmus Bartholin, 12 January 1665, TC p. 430. For the information about Lubieniecki's correspondence with Thomas Bartholin see section 3.3.4 of this Introduction.

correspondents.[48] In return, Bartholin sent Lubieniecki some of his observations of the comet of 1664,[49] informed him about a few historical observations of comets found in Icelandic annals[50] (but he did not enter into a detailed discussion about the history of comets), and about the early history of the university in Copenhagen.[51] Having received relatively few materials from his correspondent, Lubieniecki printed in the *Theatrum cometicum* some passages from his book *De cometis anni MDCLXIV et MDLXV opusculum* (Hafniae: apud Danielem Paulli, 1665).[52]

2.4.2. Ismaël Boulliau

Ismaël Boulliau (born 28 September 1605 in Loudun, died 25 November 1694 in Paris) was a French astronomer and mathematician. He came from a Huguenot family, but as a young man he converted to Catholicism and was ordained a priest. In his astronomical works, he supported the Copernican system of the universe. Apart from publishing his own works, he was an editor of the works of ancient mathematicians and astronomers, including those on astrology. He conducted a vast scholarly correspondence.[53]

Boulliau's contacts with Lubieniecki started long before Lubieniecki became interested in comets, and only astronomical excerpts from their correspondence were published by Lubieniecki.[54] Nevertheless, their letters in the *Theatrum cometicum* contain relatively few astronomical data. Boulliau expressed it himself when Lubieniecki asked him if he could publish their correspondence in the *Theatrum cometicum*.[55] Boulliau did not conduct observations of the comet of 1664[56] and sent to

48 Lubieniecki to Erasmus Bartholin, 12 January 1665, TC pp. 430–31; Lubieniecki to Erasmus Bartholin, 17 February 1665, TC p. 432; Lubieniecki to Erasmus Bartholin, 31 March 1665, TC p. 433; Lubieniecki to Erasmus Bartholin, 17 April 1665, TC pp. 434–34; Lubieniecki to Erasmus Bartholin, 28 August 1665, TC pp. 440–41; Lubieniecki to Erasmus Bartholin, 13 October 1665, TC p. 442; Lubieniecki to Erasmus Bartholin, 7 December 1665, TC p. 448; Lubieniecki to Erasmus Bartholin, 26 January 1666, TC pp. 449–50; Lubieniecki to Erasmus Bartholin, 11 February 1666, TC p. 450; Lubieniecki to Erasmus Bartholin, 19 March 1666, TC p. 451; Lubieniecki to Erasmus Bartholin, 14 December 1666, TC p. 955; Lubieniecki to Erasmus Bartholin, 7 July 1667, TC p. 959; Lubieniecki to Erasmus Bartholin, 2 August 1667, TC pp. 959–60.

49 Erasmus Bartholin to Lubieniecki, 27 May 1665, TC pp. 436–38.

50 Erasmus Bartholin to Lubieniecki, 26 September 1665, TC p. 441.

51 Erasmus Bartholin to Lubieniecki, 25 December 1666, TC p. 955.

52 TC pp. 956–59; cf. Erasmus Bartholin, *De cometis anni MDCLXIV et MDLXV opusculum* (Hafniae: apud Danielem Paulli, 1665), pp. 16–45.

53 Carl B. Boyer, 'Boulliau, Ismael', in *Dictionary of Scientific Biography*, vol. 2, ed. by. Charles Coulston Gillispie (New York: Charles Scribner's Sons, 1970), pp. 348–49; Nellen, *Ismaël Boulliau*; for a discussion of contacts between Boulliau and Hevelius see Nellen, *Ismaël Boulliau*, pp. 469–96; Grell, 'Introduction', CJH2 pp. 15–20 *et passim*; Grell, 'Introduction', CJH3 pp. 20–23 *et passim*.

54 For a detailed discussion of the Boulliau-Lubieniecki correspondence see Nellen, *Ismaël Boulliau*, pp. 435–67.

55 Boulliau to Lubieniecki, 22 May 1665, TC p. 483, BnF ms FF. 13026 f. 291v.

56 Boulliau to Lubieniecki, 23 January 1665, TC pp. 469–70, BnF ms FF. 13026 f. 283r.

Lubieniecki only some positional data.[57] Boulliau believed (and repeated in his letters) that comets are of ethereal matter and in nature are similar to the Milky Way.[58] When Lubieniecki asked him to elaborate on various cometary opinions,[59] Boulliau replied that most of them are based on conjectures.[60] Lubieniecki also inquired about historical observations of comets.[61] Boulliau answered that it is difficult to write a history of comets. The appearance of tailed stars, contrary to eclipses, cannot be calculated from astronomical tables, and their description must rely on the authors who claimed that they had seen them. He added that it had never happened that a comet returned after being seen once.[62] Both correspondents agreed that astrological predictions are pointless.[63]

2.4.3. Friedrich Büthner

Friedrich Büthner (born 11 July 1622 in Opočno in Bohemia, died 13 February 1701 in Danzig) was a mathematician. Since 1663 he held a chair in mathematics and astronomy at the *gymnasium* in Danzig. He specialized in various fields of mathematics, such as trigonometry and arithmetic.[64]

Büthner's answers to Lubieniecki's letters, when compared to other *communicationes*, were rather long and detailed. In his opinion, nature and matter of comets are rather topics suitable for a natural philosopher, not for an astronomer. Nevertheless, he supposed that tailed stars are fiery objects.[65] As far as astrology is concerned, Büthner stated that it should have its place among sciences but only when it is based on natural causes, without anything added. Büthner believed that comets are both natural phenomena and messages sent from God. They call people to piety, but they also foretell political changes, births of rulers, and various calamities.[66] Because comets

57 Boulliau to Lubieniecki, 2 January 1665, TC pp. 468–69; Boulliau to Lubieniecki, 23 January 1665, TC p. 470, BnF ms FF. 13026 f. 283r.

58 Boulliau to Lubieniecki, 13 February 1665, TC p. 471, BnF ms FF. 13026 f. 284r; Boulliau to Lubieniecki, 6 March 1665, TC p. 472, BnF ms FF. 13026 f. 285v; Boulliau to Lubieniecki, 1 May 1665, TC p. 480, BnF ms FF. 13026 f. 289r.

59 Lubieniecki to Boulliau, 13 June 1665, TC p. 485, BnF ms FF. 13031 ff. 277v–278v.

60 Boulliau to Lubieniecki, 26 June 1665, TC p. 527, BnF ms FF. 13026 f. 294r.

61 Lubieniecki to Boulliau, 22 August 1665, TC p. 533; Lubieniecki to Boulliau, 18 April 1666, TC pp. 543–44; Lubieniecki to Boulliau, 8 May 1666, TC p. 545.

62 Boulliau to Lubieniecki, 21 May 1666, TC pp. 545–46.

63 Boulliau to Lubieniecki, 6 March 1665, TC p. 472, BnF ms FF. 13026 f. 286r; Lubieniecki to Boulliau, 21 March 1665, p. 475.

64 Christian Gottlieb Jöcher, *Allgemeines Gelehrten-Lexicon*, vol. 1 (Leipzig: in Johann Friedrich Gleditschens Buchhandlung, 1750), coll. 1466–67; Marcin Grulkowski, 'Büthner Friedrich', in *Encyklopedia Gdańska*, ed. by Błażej Śliwiński (Gdańsk: Fundacja Gdańska, 2012), p. 153; CJH3 p. 361, note to Letter 116, Hevelius to des Noyers, 7 April 1665.

65 Büthner to Lubieniecki, 20 January 1665, TC pp. 799–800.

66 Büthner to Lubieniecki, 21 March 1665, TC pp. 803–06; Büthner to Lubieniecki, 27 May 1665, TC pp. 815–17; Büthner to Lubieniecki, 13 February 1666, TC p. 834.

are subject to divine influence, it is impossible to predict them.[67] Lubieniecki and Büthner also discussed chronological matters and historical cometary phenomena.[68]

2.4.4. Andreas Concius

Andreas Concius (born 25 November 1628 in Narzim/Narzin, near Działdowo (Soldau), died 16 May 1682 in Königsberg) was a German mathematician. Since 1654 he was the professor of mathematics in Königsberg. He also worked on geography and calendar studies.[69]

Lubieniecki sent one letter to Concius, but he received no answer. In the letter, Lubieniecki asked the addressee about his cometary observations and opinions, described the data he obtained from his other correspondents and criticized astrological predictions.[70]

2.4.5. Albert Curtz

Albert Curtz (born 1600 in Munich, died 19 December 1671 in Munich) was a German Jesuit, mathematician, and philosopher. He was a teacher of mathematics and ethics in Dillingen, and later a preacher at St Stephen's Cathedral in Vienna. In 1649, he became the provost of the Jesuit college in Neuburg at Danube. Later, he taught in Eichstätt and Luzern. He finally settled in Munich. He conducted studies of architecture. He supported Brahe's geoheliocentric system of the universe.[71]

Lubieniecki's correspondence with Curtz is very asymmetrical. Lubieniecki asked him many detailed questions about the number of observed comets, their nature, and their astrological significance.[72] Even though Curtz shared with Lubieniecki an observation from Rome,[73] his answers were brief, or he did not reply at all.[74] This incited Lubieniecki to make a cutting remark on their writing styles when he compared

67 Büthner to Lubieniecki, 17 June 1665, TC p. 819–20; Büthner to Lubieniecki, 16 December 1665, TC p. 829.

68 Büthner to Lubieniecki, 17 June 1665, TC pp. 821–22; Lubieniecki to Büthner, 1 January 1666, TC p. 831; Büthner to Lubieniecki, 13 February 1666, TC p. 833; Lubieniecki to Büthner, 26 February 1666, TC pp. 834–35; Büthner to Lubieniecki, 26 March 1666, TC pp. 836–37; Lubieniecki to Büthner, 6 April 1666, TC pp. 837–39; Lubieniecki to Büthner, 15 January 1667, TC p. 839; Büthner to Lubieniecki, 19 February 1667, TC pp. 840–41; Lubieniecki to Büthner, 18 March 1667, TC p. 842; Büthner to Lubieniecki, 9 April 1667, TC p. 963; Lubieniecki to Büthner, 9 September 1667, TC pp. 964–65.

69 Jöcher, *Allgemeines Gelehrten-Lexicon*, vol. 1, coll. 2044–45; Karl Christian Bruhns, 'Concius: Andreas', *Allgemeine Deutsche Biographie*, vol. 4 (Leipzig: Verlag von Duncker & Humblot, 1876), p. 437.

70 Lubieniecki to Concius, 13 March 1665, TC pp. 585–88.

71 Jöcher, *Allgemeines Gelehrten-Lexicon*, vol. 1, col. 2264; Georg Westermayer, 'Curtz: Albert', in *Allgemeine Deutsche Biographie*, vol. 4, pp. 654–55.

72 Lubieniecki to Curtz, 11 April 1665, TC pp. 212–14; Lubieniecki to Curtz, 27 May 1665, TC pp. 215–19; Lubieniecki to Curtz, 2 July 1665, TC pp. 219–23.

73 TC pp. 214–15.

74 Curtz to Lubieniecki, 7 May 1665, TC p. 214; Curtz to Lubieniecki, 21 August 1665, TC p. 223.

them to Athens and Sparta.[75] Lubieniecki also printed astronomical excerpts from Curtz's correspondence with Johann Ernst von Rautenstein, together with some observations from various places.[76]

2.4.6. Nicolaus Heinsius

Nicolaus Heinsius (born 26 July or 29 August 1620 in Leiden, died 7 October 1681 in The Hague) was a Dutch diplomat. In 1649, he established contacts with the court of Queen Christina of Sweden (1629–89, Queen in 1632–54); in 1650, he moved to Stockholm. By the Queen's order, he continued his scholarly journeys. In 1654, he was appointed a Dutch representative at the Swedish court. In consecutive years, he fulfilled various diplomatic missions, including the embassy at the Muscovite court. He was interested in ancient literature. He studied and collected manuscripts.[77]

The correspondence between Heinsius and Lubieniecki, even though abundant, does not contain many important cometary and astronomical remarks. Their letters were mainly focused on political issues. Heinsius informed Lubieniecki about the works of scholars from Uppsala: Jonas Fornelius (1635–1679),[78] Johannes Schefferus (1621–79),[79] and Olof Rudbeck. Heinsius was also the one who persuaded Lubieniecki to send a letter to Huygens.[80]

2.4.7. Christiaan Huygens

Christiaan Huygens (born 14 April 1629 in The Hague, died 8 July 1695 in The Hague) was a Dutch mathematician, physicist, and astronomer. His works and achievements span a number of fields: he conducted the first observations of Saturn's moon, Titan, and identified Saturn's ring; he also invented the pendulum clock. He studied the nature of light and argued that it moves like a wave. Between 1664 and

75 Lubieniecki to Curtz, 23 September 1665, TC pp. 223–24; see also Letter 92.

76 TC pp. 209–12.

77 *Bibliotheca Heinsiana, sive catalogus librorum quos magno studio et sumtu, dum viveret, collegit Nicolaus Heinsius* (Lugduni Batavorum: apud Johannem de Vivie, 1682); Christian Gottlieb Jöcher, *Allgemeines Gelehrten-Lexicon*, vol. 2 (Leipzig: in Johann Friedrich Gleditschens Buchhandlung, 1750), coll. 1456–57; [anonymous], 'Heinsius: Nicolaus', in *Allgemeine Deutsche Biographie*, vol. 11 (Leipzig: Verlag von Duncker & Humblot, 1880), pp. 656–60; John A. Sibbald, 'The *Heinsiana* — Almost a Seventeenth-Century Universal Short Title Catalogue', in *Documenting the Early Modern Book World. Inventories and Catalogues in Manuscript and Print*, ed. by Malcolm Walsby and Natasha Constantinidou (Leiden, Boston: Brill, 2013), pp. 141–59; for a discussion of contacts between Heinsius and Boulliau see Nellen, *Ismaël Boulliau*, pp. 375–92. See also Grell, 'Introduction', CJH2 p. 27.

78 Fornelius to Heinsius, 24 February 1664, TC pp. 261–62; Fornelius to Heinsius, 24 April 1665, TC pp. 271–72.

79 Schefferus to Heinsius, 17 September 1665, TC p. 314.

80 See section 3.4.2 of this Introduction.

1681, he resided in Paris, where he was a member of the Academy and where he received a stipend from Louis XIV.[81]

Huygens received two letters from Lubieniecki[82] but did not answer either of them. In his letters, Lubieniecki asked the addressee for his cometary opinions which could be later included in the *Theatrum cometicum*. He also mentioned Hevelius's dispute with Auzout and congratulated Huygens on Louis XIV's benevolence towards him.[83]

2.4.8. Athanasius Kircher

Athanasius Kircher (born 2 May 1601 or 1602 in Geisa, died 28 November 1680 in Rome) was a German Jesuit polymath. He was a teacher of mathematics and oriental languages in Würzburg. After the Thirty Years' War began, he moved to Avignon, and next, in 1633, to Rome. In 1638 he became the professor of mathematics at the *Collegio Romano*. When he resigned from the post, he started independent studies. He worked on almost all fields of contemporary *scientiae*, including astronomy, magnetism, and optics. His legacy — however often unoriginal — is still being studied.[84] Hevelius visited Kircher in Avignon during his *grand tour* in 1631.[85]

Kircher's replies to Lubieniecki's letters and inquiries were rather brief. Kircher sent to Lubieniecki a booklet about comets — *Fisiologia nuova della natura delle comete* (Roma: nella stamperia del Varese, 1665) — which was written by his student, Gioseffo Petrucci, but, as Kircher emphasized, he presented Kircher's own cometary views. He also added that Petrucci followed what Kircher had written in his *Itinerarium exstaticum* (Romae: typis Vitalis Mascardi, 1656).[86] Lubieniecki printed excerpts from Petrucci's book translated into Latin in the *Theatrum cometicum*.[87] Kircher also provided Lubieniecki with an observation of Saturn, which is said to be the first one to present the Cassini Division in the planet's rings.[88]

81 H. J. M. Bos, 'Huygens, Christiaan', in *Dictionary of Scientific Biography*, vol. 6, ed. by. Charles Coulston Gillispie (New York: Charles Scribner's Sons, 1972), pp. 597–613; C. D. Andriesse, *Huygens. The Man Behind the Principle* (Cambridge: Cambridge University Press, 2005).

82 Lubieniecki to Huygens, 30 October 1665, TC pp. 931–33, OCCH5 pp. 518–24, LUL ms HUG 45; Lubieniecki to Huygens, 16 February 1667, TC p. 933, OCCH6 pp. 109–12, LUL ms HUG 45.

83 For more information about Lubieniecki's correspondence with Huygens see section 3.4.2 of this Introduction.

84 Hans Kangro, 'Kircher, Athanasius', in *Dictionary of Scientific Biography*, vol. 7, ed. by. Charles Coulston Gillispie (New York: Charles Scribner's Sons, 1978), pp. 374–78; *Athanasius Kircher. The Last Man Who Knew Everything*, ed. by Paula Findlen (New York, London: Routledge, 2004).

85 Grell, 'Hevelius en son temps', pp. 58–59.

86 Kircher to Lubieniecki, 8 May 1665, TC p. 750; Kircher to Lubieniecki, 25 July 1665, TC p. 754.

87 TC pp. 757–59.

88 Kircher to Lubieniecki, 25 July 1665, TC p. 755, fig. 18; see Tadeusz Przypkowski, 'Pierwsze poprawne widoki Jowisza i Saturna z r. 1664 ogłoszone przez Stanisława Lubienieckiego', *Sprawozdania z Czynności i Posiedzeń Polskiej Akademii Umiejętności*, 52 (1951), no. 6, 536–37.

2.4.9. Michael Kirsten

Michael Kirsten (born 25 January 1620 in Moravia, died 3 March 1678 in Hamburg) was a German physician. He studied in Rostock, in Szczecin, and in Frankfurt an der Oder. In 1648, he settled in Hamburg. In 1655, he became the professor of mathematics in Hamburg; in 1660, he became the professor of physics and poetics. He published works on medicine and anatomy and wrote Latin poems.[89]

There is only a brief exchange on the meaning and influence of comets in the *Theatri cometici exitus*. Kirsten did not present his own views about this matter but remarked that Lubieniecki's opinion concerns rather theology than natural philosophy.[90]

2.4.10. Villum Lange

Villum Lange (born 15 January 1624 in Helsingør, died 22 May 1682 in Copenhagen) was a Danish mathematician and astronomer. In 1643, he graduated from the faculty of physics and medicine in Herlufsholm. In 1650–51, he was the professor of mathematics and astronomy at the University of Copenhagen from which he graduated in 1653, after a few years of studying abroad. Between 1656 and 1658, he was a tutor of Prince Christian (1646–99, King since 1670). In the following years, he was a courtier and diplomat. His studies concerned chronology, calendars, and comets.[91]

Lubieniecki's correspondence with Lange is one of the most recent in the *Theatrum cometicum*, as all the letters come from 1667. Lange informed Lubieniecki that comets of 1664 and 1665 were two different bodies, discussed optical issues concerning the comet's tail, and expressed his support for astrology.[92]

2.4.11. Johann Müller

Johann Müller (born in 1611 in Hamburg, died 24 May 1671 in Hamburg) was a German lawyer and mathematician. In 1635 he graduated in jurisprudence from Basel. Since 1660, he was a professor of mathematics in Hamburg.[93]

Müller believed that comets move rectilinearly and their movement is independent of the movement of the Earth.[94] He provided Lubieniecki with the observations he received from Rome and Vienna.[95] In his letters printed in the *Theatrum*

89 Jöcher, *Allgemeines Gelehrten-Lexicon*, vol. 2, coll. 2105–06.

90 Kirsten to Lubieniecki, 29 March 1667, TC3 p. 47.

91 C. F. Pechüle, Poul Heegaard, 'Lange, Villum', in *Dansk biografisk leksikon*, vol. 13, ed. by C. F. Bricka et al. (København: J. H. Schlutz Forlag, 1938), pp. 596–99.

92 Lange to Lubieniecki, 13 September 1667, TC pp. 942–45.

93 Hans Schröder, *Lexikon der hamburgischen Schriftsteller bis zur Gegenwart*, 8 vols (Hamburg: Perthes-Besser u. Mauke 1851–83), access online: schroeder.sub.uni-hamburg.de [accessed 19 June 2018].

94 Müller to Lubieniecki, 22 December 1664, TC pp. 845–46.

95 TC pp. 849–54.

cometicum, Müller supported the opinion about one comet seen in December 1664, and he only briefly mentions that he changed his earlier view.[96] Nevertheless, Lubieniecki sent previous versions of his letters to Hevelius (Letters 5B, 5C, 5D, and 5E), which significantly differ from their printed counterparts: initially Müller was adamant about two comets and presented many arguments to support this opinion.

2.4.12. Henry Oldenburg

Henry Oldenburg (born *c.* 1618 in Bremen, died 5 September 1677 in London) was the secretary of the Royal Society of London. He studied in Bremen and Utrecht, and later tutored the youth from wealthy families. In 1653, he was appointed by the authorities of Bremen as an emissary to Cromwell. He stayed in England, initially as a diplomat. Later, he joined the group which formed the Royal Society. As the secretary of this institution, he conducted their correspondence and edited the *Philosophical Transactions*. He did not make his own experiments or observations, but he promoted new discoveries throughout Europe.[97]

Oldenburg hoped Lubieniecki could somehow help in establishing contacts between scholars from Germany and Poland and the Royal Society (Letters 57A and 61A).[98] He also asked Lubieniecki to buy and send him a few books (Letter 57A). In his letters, Oldenburg briefly mentioned some studies and experiments conducted by the Royal Society (Letters 57A, 59A, and 61A). Lubieniecki wrote to Oldenburg about an unusual phenomenon he had seen in Hamburg (Letter 57B).[99]

2.4.13. Adam Olearius

Adam Olearius (born *c.* 1599 in Aschersleben, died 22 February 1671 in Schleswig) was a German orientalist and natural philosopher. In 1633–39, by order of the Duke of Schleswig-Holstein, Frederick III (1597–1659, Duke since 1616), he was a member of the embassy to Muscovy and Persia. There, he started to study oriental culture, literature, and languages. After his return, he was at the ducal court, where he served as an astronomer, astrologer, and librarian.[100]

96 Müller to Lubieniecki, 5 January 1665, TC p. 874.

97 A. Rupert Hall, 'Oldenburg, Henry', in *Dictionary of Scientific Biography*, vol. 10, ed. by Charles Coulston Gillispie (New York: Charles Scribner's Sons, 1974), pp. 200–03; Marie Boas Hall, *Henry Oldenburg. Shaping the Royal Society* (Oxford: Oxford University Press, 2002); CJH3 p. 377, note to Letter 127, Hevelius to des Noyers, 29 May 1666.

98 Oldenburg to Lubieniecki, 23 July 1666, TC p. 908, CHO pp. 179–80.

99 In his correspondence with Hevelius, Lubieniecki initially made a mistake in Oldenburg's family name and called him 'Oldendorpius'.

100 Christian Gottlieb Jöcher, *Allgemeines Gelehrten-Lexicon*, vol. 3 (Leipzig: in Johann Friedrich Gleditschens Buchhandlung, 1751), coll. 1048–49; Julius Opel, 'Olearius: Adam', in *Allgemeine Deutsche Biographie*, vol. 24 (Leipzig: Verlag von Duncker & Humblot, 1887), pp. 269–77; Grell, 'Introduction', CJH2 pp. 105–06.

Olearius sent Lubieniecki a few brief observations of the comet of 1664.[101] As far as his cometary opinions are concerned, Olearius stated that he published them anonymously in a booklet *Tertius interveniens, gesetzt zwischen denen, so den Cometsternen allzu große Kraft und Wirckung zuschreiben und daher die Leute allzu furchtsam machen. Und denen, die den Cometen gar keine Krafft zueigen, ihre Bedeutung ganz verwerffen und die Leute gar zu sicher machen wollen* ([s.l.]: [s.e.], 1665).[102] Lubieniecki answered that he knew that publication.[103]

2.4.14. Jan Placentinus-Kołaczek

Jan Placentinus-Kołaczek (born 5 June 1630 in Leszno, died 11 March 1683 in Frankfurt an der Oder) was a Polish philosopher and mathematician. His family were the members of the Bohemian Brethren from Silesia. Since 1653, he was the professor of mathematics in Frankfurt an der Oder; in 1658 and 1665, he was the provost. He was an ardent Cartesian. He supported the view about the influence of celestial bodies on human life, and in his opinion, this idea was grounded in Descartes's philosophy. Due to mental illness, in 1667, he resigned from the academic post. He recovered by 1671, as we learn from Lubieniecki's letter to Hevelius (Letter 80).[104]

Placentinus-Kołaczek informed Lubieniecki that he published a booklet about comets — *Astronomica et astrologica observatio cometae terribilis, mense Februario huius currentis anni adhuc spectati* ([s.l.]: auff Kosten des Autoris, Drucks Erasmus Rösner, 1665) — and asked him for help in its distribution, namely to send copies to the kings of France, Denmark, and Sweden.[105] On his part, Lubieniecki praised Placentinus-Kołaczek's book.[106] In his letters, Placentinus-Kołaczek discussed the movement of the comet and cometary influence on the Earth and human lives,[107] and he tried to defend astrology by referring to the authority of the Bible.[108] In his opinion, the Holy Scripture does not condemn true astrology of Egypt and Chaldea but rather their religion and worship of celestial objects instead of God. He made a

101 Olearius to Lubieniecki, 15 January 1665, TC p. 636; Olearius to Lubieniecki, 3 February 1665, TC pp. 637–38.

102 Olearius to Lubieniecki, 3 February 1665, TC p. 637; Olearius to Lubieniecki, 30 October 1666, TC p. 961.

103 Lubieniecki to Olearius, 12 November 1666, TC p. 961.

104 Ludwik Chmaj, 'Jan Placentinus-Kołaczek. Nieznany kartezjanin XVII wieku', *Archiwum Historii Filozofii i Myśli Społecznej*, 1 (1957), 71–81; Pietro D. Omodeo, 'Central European Polemics over Descartes: Johannes Placentinus and His Academic Opponents at Frankfurt on Oder (1653–1656)', in *History of Universities*, vol. XXIX/1, ed. by Mordechai Feingold (Oxford: Oxfrord University Press, 2016), pp. 29–64; CJH3 p. 273, note to Letter 69, Hevelius to des Noyers, March 1655.

105 Placentinus-Kołaczek to Lubieniecki, 1 April 1665, TC p. 549; Placentinus-Kołaczek to Lubieniecki, 19 April 1665, TC p. 555.

106 Lubieniecki to Placentinus-Kołaczek, 17 April 1665, TC pp. 550–53.

107 Placentinus-Kołaczek to Lubieniecki, 19 April 1665, TC pp. 555–61.

108 Placentinus-Kołaczek to Lubieniecki, 30 April 1666, TC pp. 569–73.

distinction between the real astrology — reading the stars according to the laws of nature, because God reveals his will through celestial objects — and the superstitious astrology, which is susceptible to pagan cults and contradicts divine and natural laws. Among the practitioners of true astrology, Placentinus-Kołaczek placed prominent biblical figures, including Moses. In his answer, Lubieniecki agreed that the Bible allows some methods of interpreting celestial phenomena and denies others. Nevertheless, all of the biblical interpreters of the heavens were prophets, and nowadays, Lubieniecki says, there is no one endowed with the gift of prophesying. Therefore, no one can be a real astrologer now, and mankind should not try to read too much from the heavens.[109]

2.4.15. Giovanni Battista Riccioli

Giovanni Battista Riccioli (born 1598 in Ferrara, died 25 January 1671 in Bologna) was an Italian Jesuit astronomer and geographer. He lectured in Parma and Bologna. In the field of exact sciences, he was an autodidact. He conducted experiments on falling bodies and pendulum. He took observations of the Moon (he is the author of the naming system of lunar objects which is still in use), solar spots, and planets.[110]

Lubieniecki sent Riccioli many letters with the information from his correspondents, but he received only a few short answers in which Riccioli excused his silence with his poor health and overwhelming responsibilities.[111] In his letters, Riccioli criticized opinions about the Earth's movement and the existence of the vacuum.[112] Lubieniecki provided Riccioli with an extract of Hevelius' *Prodromus cometicus*[113] and with lengthy discussions about chronology and history of comets.[114] Riccioli briefly replied to Lubieniecki's chronological considerations.[115]

109 Lubieniecki to Placentinus-Kołaczek, 18 June 1666, TC pp. 573–74.

110 Luigi Campedelli, 'Riccioli, Giambattista', in *Dictionary of Scientific Biography*, vol. 11, ed. by. Charles Coulston Gillispie (New York: Charles Scribner's Sons, 1975), pp. 411–12; Christopher M. Graney, *Setting Aside All Authority. Giovanni Battista Riccioli and the Science against Copernicus in the Age of Galileo* (Notre Dame, Indiana: University of Notre Dame Press, 2015).

111 Riccioli to Lubieniecki, 15 July 1665, TC p. 729; Riccioli to Lubieniecki, 8 March 1666, TC p. 733; Riccioli to Lubieniecki 1 May 1666, TC p. 742.

112 Riccioli to Lubieniecki, 28 April 1665, TC p. 700. About the vacuum see also Grell, 'Introduction', CJH3 pp. 35–48.

113 TC pp. 705–28.

114 Lubieniecki to Riccioli, 24 June 1665, TC p. 703; Lubieniecki to Riccioli, 29 July 1665, TC pp. 728–29; Lubieniecki to Riccioli, 10 February 1666, TC pp. 730–32; Lubieniecki to Riccioli, 17 February 1666, TC p. 733; Lubieniecki to Riccioli, 7 April 1666, TC pp. 734–38; Lubieniecki to Riccioli, 14 April 1666, TC pp. 738–42; Lubieniecki to Riccioli, 26 May 1666, TC pp. 742–45.

115 Riccioli to Lubieniecki, 8 March 1666, TC p. 734; Riccioli to Lubieniecki, 1 May 1666, TC p. 742.

2.4.16. Olof Rudbeck the Elder

Olof Rudbeck the Elder (born 1630 in Västerås, died 17 September 1702 in Uppsala) was a Swedish physician and natural philosopher. At the age of 20, he presented one of the first descriptions of the lymphatic system. He lectured medicine in Uppsala. He also worked on botany and history.[116]

Lubieniecki sent a joined letter to Rudbeck and Jonas Fornelius, asking them about their cometary views, but he received no answer from either of them.[117] Lubieniecki received several Rudbeck's opinions from Heinsius, including a long letter addressed to Heinsius, in which Rudbeck pointed out the difficulties in discussing the nature of comets, criticized astrological predictions, discussed shapes of comets and their tails, and presented observations of the comet of 1664.[118]

116 Sten Lindroth, 'Rudbeck, Olof', in *Dictionary of Scientific Biography*, vol. 11, pp. 586–88.
117 Lubieniecki to Rudbeck and Fornelius, 21 July 1665, TC pp. 347–48.
118 Rudbeck to Heinsius, July 1665, TC pp. 348–59.

CHAPTER 3

Lubieniecki and the Heavens

The present chapter discusses the main aspects of Lubieniecki's opinions about comets and other celestial objects, as well as his place in the contemporary world of learning, excluding the features already discussed in the context of his correspondence with Hevelius. It is divided into three parts. The first presents his astronomical views. The second is devoted to his opinions on the meaning and influence of comets and stars. The third discusses Lubieniecki's place in the Republic of Letters. Because the *Theatrum cometicum*, or more specifically, the author's correspondence it contains, is almost the only trace of Lubieniecki's astronomical activities, this chapter is chiefly based on those letters.[1]

3.1. Lubieniecki and the *Theatrum cometicum* in the Critical Literature

The critical literature about Lubieniecki's astronomical activities and opinions is rather scarce, which is quite understandable given the fact that his contribution to contemporary science was negligible. Moreover, it is challenging to analyse Lubieniecki's views on comets and astronomy. Because of the convoluted structure and substantial volume of the *Theatrum cometicum*, it is often difficult to decide which of the presented views were genuinely his. In the following section, I present some of what can be found about Lubieniecki and his *Theatrum cometicum* in the critical literature. It is not a discussion of *status quaestionis*, as majority of the following contain rather brief mentions or, at most, introductory remarks.[2]

Lubieniecki's *Theatrum cometicum* is mentioned by Pierre Bayle in his *Pensées diverses écrites à un docteur de Sorbonne à l'occasion de la Comète qui parut au mois de*

1 In the description of Lubieniecki's activities in the field of astronomy, I rely on various passages from: Jasiński, *Stanisław Lubieniecki*. My studies on Lubieniecki's astronomical activities were funded by the National Science Centre (Narodowe Centrum Nauki), Poland, within the PRELUDIUM 8 programme (decision number DEC-2014/15/N/HS3/01305).

2 I mostly repeat here what I wrote in Jasiński, *Stanisław Lubieniecki*, pp. 81–98.

décembre 1680. Bayle notices astrological predictions and discussions of events that accompanied comets in Lubieniecki's book.[3]

Jean-Étienne Montucla lists the *Theatrum cometicum* among books about historical cometary phenomena but does not think highly of it. He concludes that Lubieniecki's book is devoted to general history rather than to the history of comets. He also mocks Lubieniecki's opinion that comets bring good things to good people and bad things to the bad.[4]

A similar summary of the *Theatrum cometicum* is presented by Jean-Baptiste Joseph Delambre who also states that Lubieniecki's book presents a large amount of historical data.[5]

In his discussion of seventeenth-century cometary opinions, James Howard Robinson limits his examination of Lubieniecki's book to a discussion on the *Theatrum cometicum's* frontispiece and on the collection of observations from the beginning of the *Theatri cometici pars prior*,[6] and on the foreword of this volume. He remarks that Lubieniecki believed that comets could portend both bad and good things and that Lubieniecki was inclined to present information about various misfortunes and unusual events.[7]

In his description of Lubieniecki's confessional and theological activities, K. E. Jordt Jørgensen briefly discusses the *Theatrum cometicum*. He presents some of Lubieniecki's most distinguished correspondents and his ideas about cometary influence.[8]

Henk J. M. Nellen, who observes Lubieniecki as one of the most important correspondents of Ismaël Boulliau, also presents the main features of the *Theatrum cometicum*. He calls it Lubieniecki's key publication and mentions its origins and contents. Nellen observes that Boulliau, unlike many other correspondents of Lubieniecki, was willing to answer some of his questions about comets, but he did not want his letters to be published without alterations. Nellen also noted Lubieniecki's opinion that comets are portents sent by God.[9]

Susanna Åkerman devotes a passage to Lubieniecki and his *Theatrum cometicum* in her study *Queen Christina of Sweden and her Circle*. She observes a plethora of topics touched upon in the book and a large number of Lubieniecki's correspondents.[10]

3 Pierre Bayle, *Pensées diverses écrites à un docteur de Sorbonne à l'occasion de la Comète qui parut au mois de décembre 1680* (Rotterdam: chez Reinier Leers, 1683), pp. 102, 115.

4 Jean-Étienne Montucla Montucla, *Histoire des mathématiques*, vol. 2 (Paris: chez Ch. Ant. Jombert, 1758), p. 585.

5 Jean-Baptiste Joseph Delambre, *Histoire de l'astronomie moderne*, vol. 2 (Paris: Mme Ve Courcier, 1821), pp. 538–39.

6 TC pp. 1–18.

7 Robinson, pp. 18–20.

8 K. E. Jordt Jørgensen, *Stanisław Lubieniecki. Zum Weg des Unitarismus von Ost nach West im 17. Jahrhundert* (Göttingen: Vandenhoeck & Ruprecht, 1968), pp. 92–95.

9 Nellen, pp. 455–59, 464.

10 Susanna Åkerman, *Queen Christina of Sweden and her Circle. The Transformation of a Seventeenth-Century Philosophical Libertine* (Leiden, New York: Brill, 1991), pp. 237–38.

Nevertheless, Åkerman's assertion that it was Queen Christina who financed the publication of the *Theatrum cometicum* is absent from other critical literature, and I was unable to trace it in any sources.[11]

In her brief analysis of the *Theatrum cometicum*, Sara Schechner Genuth referrs only to the collection of various observations in its first volume and to Lubieniecki's correspondence with Oldenburg. Based on this, she claims that Lubieniecki was an enthusiast of omens and predictions, the fact which is one of the arguments supporting the conclusion that astrological interpretations of comets were discarded in England and France earlier than in Germany, Italy, and Poland.[12]

Tabitta van Nouhuys lists the *Theatrum cometicum* among early modern catalogues of comets. She remarks that Lubieniecki deliberated whether comets can cause or foretell any events and observes his almost absolute disbelief in cometary astrology. She also notes the difference between the *Theatrum cometicum* and Hevelius's *Cometographia*, as the latter was devised to facilitate astronomical calculations.[13]

Andrea Gualandi gives a longer description of the *Theatrum cometicum*. In his opinion, Lubieniecki's book was a typical product of his age which preferred erudition to accuracy. Gualandi remarks that comets and their chronology are not the only subjects of the *Theatrum cometicum*. He briefly discusses some of its subjects, for example Otto von Guericke's cometary theory, but does not evaluate them in the context of the whole book.[14]

Inga Elmqvist Söderlund adds a few words about the first volume of the *Theatrum cometicum* to her discussion of its frontispiece. She notes some of Lubieniecki's correspondence and a variety of topics pervading his book. She also states that Lubieniecki probably worked in haste.[15]

Discussing Hevelius's epistolary contacts, Chantal Grell mentions Lubieniecki's belief in the divine origin of comets and their both beneficial and harmful influence. She also notes that Lubieniecki's correspondence was vast and that the thematic scope of the *Theatrum cometicum* was not limited to tailed stars.[16] Chantal Grell devoted also a passage to Lubieniecki's book in the discussion of the dedications of Hevelius's books.[17]

Whereas Lubieniecki's astronomical achievements are of little importance for the general history of astronomy, they are sometimes considered noteworthy for the history of astronomy in Poland. He is often mentioned next to Hevelius as one of the

11 Åkerman, pp. 172–73.

12 Sara Schechner Genuth, *Comets, Popular Culture, and the Birth of Modern Cosmology* (Princeton: Princeton University Press, 1997), pp. 119, 120, 129.

13 Tabitta van Nouhuys, *The Age of Two-Faced Janus. The Comets of 1577 and 1618 and the Decline of the Aristotelian World View in the Netherlands* (Leiden: Brill, 1998), p. 19.

14 Andrea Gualandi, *Teorie delle comete. Da Galileo a Newton* (Milano: FrancoAngeli, 2009), pp. 203–11.

15 Elmqvist Söderlund, p. 181.

16 Grell, 'Hevelius en son temps', pp. 79–80.

17 Grell, 'Introduction', CJH2 pp. 85–87.

most important Polish astronomers of the seventeenth century.[18] Until recently, in Polish critical literature, Lubieniecki's *Theatrum cometicum* has not been thoroughly studied.[19] The most influential and frequently quoted discussion was a passage in Tadeusz Przypkowski's analysis of astronomical and mathematical activities of the Polish Brethren.[20] It was meant as an invitation to more in-depth studies on Lubieniecki's astronomy — a task which remained unfulfilled for a long time. Przypkowski's assertions about Lubieniecki sometimes seem exaggerated. He is right in concluding that the *Theatrum cometicum* was not very important for the seventeenth-century astronomy and that Lubieniecki was not inclined to astronomical speculations. On the other hand, his opinion that the artistic value of the *Theatrum cometicum* exceeded that of Hevelius's publications is not supported by any research and difficult to defend. Przypkowski also tends to mock those claims by Lubieniecki's correspondents which he found irrational or against the mainstream of the development in astronomy — for example, the opinion that angels steer comets. Most of the subsequent Polish publications devoted to Lubieniecki's astronomical activities rely on Przypkowski.[21]

3.2. Lubieniecki and Astronomy

3.2.1. Learned Authorities

In his correspondence, Lubieniecki often refrained from stating anything of what he believed. It seems very probable that he did not have any own opinions about these subjects. If he had had them, he would have certainly shared them with his correspondents, as he often did in the case of astrology. It is almost certain that Lubieniecki had no prior knowledge about comets or astronomy when he started his astronomical correspondence. As he emphasized, he wanted to learn from the knowledgeable, stating that it is not inappropriate or shameful to ask when one does not know something.[22]

18 Michał Kamieński, *Rola astronomów polskich w nauce światowej. The Role of Polish Astronomers in World Science* (Warszawa: [s.e.], 1933), pp. 2, 8; Jan Gadomski, *Zarys historii astronomii polskiej* (Kraków: Polska Akademia Umiejętności, 1948), p. 12.

19 I discussed Lubieniecki's astronomical endeavours in Jasiński, *Stanisław Lubieniecki*.

20 Przypkowski, 'Zainteresowania matematyczno-przyrodnicze', pp. 410–16. Przypkowski repeated his statements in Tadeusz Przypkowski, 'Astronomowie współcześni Heweliuszowi', in *Historia astronomii w Polsce*, vol. 1, ed. by Eugeniusz Rybka (Wrocław: Zakład Narodowy Imienia Ossolińskich, Wydawnictwo Polskiej Akademii Nauk, 1975), pp. 292–97.

21 For example, Janusz Tazbir in his discussion of the *Theatrum cometicum* focuses on the origin of the book, on some features of its distribution, and on Lubieniecki's correspondents. As far as Lubieniecki's views on comets are concerned, Tazbir refers to Przypkowski's conclusions; see Tazbir, *Stando lubentius moriar*, pp. 167–81.

22 Lubieniecki to Heinsius, 3 April 1665, TC p. 265; Lubieniecki to Leunenschlos, 4 July 1665, TC p. 629.

The most crucial feature of Lubieniecki's approach to astronomy was his respect for the authority of the learned. This was not limited to his contemporaries. More often, Lubieniecki seemed to prefer more ancient authors and relied on their writings. Sometimes, this reached absurd lengths, as in Lubieniecki's letter to Otto von Guericke where, on just a few pages, Lubieniecki referred to dozens of philosophers, astronomers, and theologians.[23] To some of his correspondents, this approach to natural sciences — reading old texts rather than observing natural phenomena — seemed inappropriate. In his letter to Lubieniecki, Johann von Leunenschlos regretted that even sagacious and bright men limit their activities to learning *ex solis librorum indicibus*.[24] The word *index* could be understood twofold: in its modern sense, it could mean an index of a book,[25] or a list of books devoted to a subject.[26] Either way, it seems that Leunenschlos wanted to emphasize that someone who studies nature should not rely on earlier authorities. This idea was becoming widespread at the time, especially when the Royal Society of London assumed *nullius in verba* as their motto.

3.2.2. Lubieniecki's Readings

The extent to which Lubieniecki depended on the opinions of earlier astronomers and philosophers becomes apparent when we take into consideration his main astronomical reading. The only book about astronomy which he regularly quotes is Giovanni Battista Riccioli's *Almagestum novum* (Bononiae: ex Typographia Heredis Victorii Benatii, 1651), an epitome of the geocentric worldview. The fact that Lubieniecki knew this book, read it, and held it in very high esteem is evident not only from the numerous quotations and references in his letters,[27] but also from his letter to Riccioli.[28]

In his book, Riccioli amassed a large amount of information about various astronomical opinions presented throughout history and about the scholars who supported them.[29] This was the part of the *Almagestum novum* which undoubtedly interested Lubieniecki the most. It is highly probable that whenever Lubieniecki quotes or refers to an earlier view about an astronomical topic, he referred to Riccioli's book.

23 Lubieniecki to Guericke-father, 15 August 1665, TC pp. 459–62.

24 Leunenschlos to Lubieniecki, 12 April 1665, TC p. 626.

25 For example, Lubieniecki used this meaning of the word *index* in the preface to his list of subjects touched upon in the *Theatrum cometicum* and in the *Monumentum Radzivilianum*: Lubieniecki, *Monumentum*, sig. Hr.

26 Krzysztof Pomian, 'De la lettre au périodique: la circulation des informations dans les milieux des historiens au XVIIIe siècle', *Organon*, 10 (1974), 25–43 (p. 34).

27 Lubieniecki to Heinsius, 3 April 1665, TC pp. 265–67; Lubieniecki to Morsztyn, 6 June 1665, TC p. 508; Lubieniecki to Boulliau, 11 July 1665, TC p. 528, BnF ms FF. 13031 ff. 280r–v; Lubieniecki to Guericke-father, 15 August 1665, TC p. 460; Lubieniecki to Reyher, 4 November 1666, TC p. 886.

28 Lubieniecki to Riccioli, 1 April 1665, TC p. 697.

29 For example, various cometary opinions are discussed in Giovanni Battista Riccioli, *Almagestum novum*, pars posterior tomi primi (Bononiae: ex Typographia Haeredis Victorii Benatii, 1651), pp. 35–40.

An example of Lubieniecki's dependence on Riccioli can be found in the discussion between Lubieniecki and Boulliau about Galileo's cometary views. Boulliau mentioned to Lubieniecki that Galileo considered the opinion that the matter of comets is the same as that of sunspots.[30] Lubieniecki answered that he had never heard about it and, according to Riccioli, Galileo considered comets to be terrestrial exhalations.[31] Boulliau replied that the opinion in question was one of a few pondered upon by Galileo and that he wrote about it in his *Il saggiatore* (Roma: appresso Giacomo Mascardi, 1623).[32]

Another publication which Lubieniecki quoted, but to a much lesser extent,[33] was the second part of the *Singularia* by Andreas Libavius (Francofurti: impressa typis Ioannis Saurii, impensis Petri Kopfii, 1599) which is likewise a collection of notions and remarks from all times.

When speaking about Lubieniecki's astronomical readings, one notable omission of a book needs to be mentioned. In a letter to Hevelius (Letter 67), Lubieniecki wrote that he had finally received the *Astronomia Philolaica* by Boulliau (Parisiis: sumptibus Simeonis Piget, 1645) which had been sent to him by the author two years earlier. In fact, this book had been published twenty years before then, soon before Lubieniecki went to France where he made an acquaintance with Boulliau. If Lubieniecki had been genuinely interested in astronomy, he would almost certainly have known the book written by his friend many years earlier.

3.2.3. Observations

When compared to the number of letters in the *Theatri cometici pars prior*, the observations which Lubieniecki received are relatively few. Nevertheless, even if their number cannot be considered substantial, there were enough of them to provide sufficient material to investigate. Lubieniecki tried to synthesize these data and to present them in a form which could be useful for an astronomer. He prepared two tables in which he gathered the information from those observations. The first and the longer one is a lengthy list of dates and places of observations, and the names of the observers, which he extracted from the reports about the first comet.[34] Still, Lubieniecki does not provide any index which would help in locating these observations in the volume. The time of the table ranges from 19 November 1664 to 20 February 1665. This table, however, is not helpful to anyone who would like to learn about the path of the comet, because it does not contain any positional data. Lubieniecki, on

30 Boulliau to Lubieniecki, 26 June 1665, TC p. 527, BnF ms FF. 13026 f. 294r.

31 Lubieniecki to Boulliau, 11 July 1665, TC p. 528, BnF ms FF. 13031 ff. 280r–v; cf. Riccioli, *Almagestum novum*, p. 37.

32 Boulliau to Lubieniecki, 31 July 1665, TC p. 531; BnF ms FF. 13026 f. 297r; cf. Galileo Galilei, *Il saggiatore* (Roma: appresso Giacomo Mascardi, 1623), pp. 112–13.

33 Lubieniecki to Boulliau, 11 April 1665, TC p. 479; Lubieniecki to Morsztyn, 6 June 1665, TC p. 508; Lubieniecki to Guericke-father, 15 August 1665, TC p. 459.

34 TC pp. 937–39.

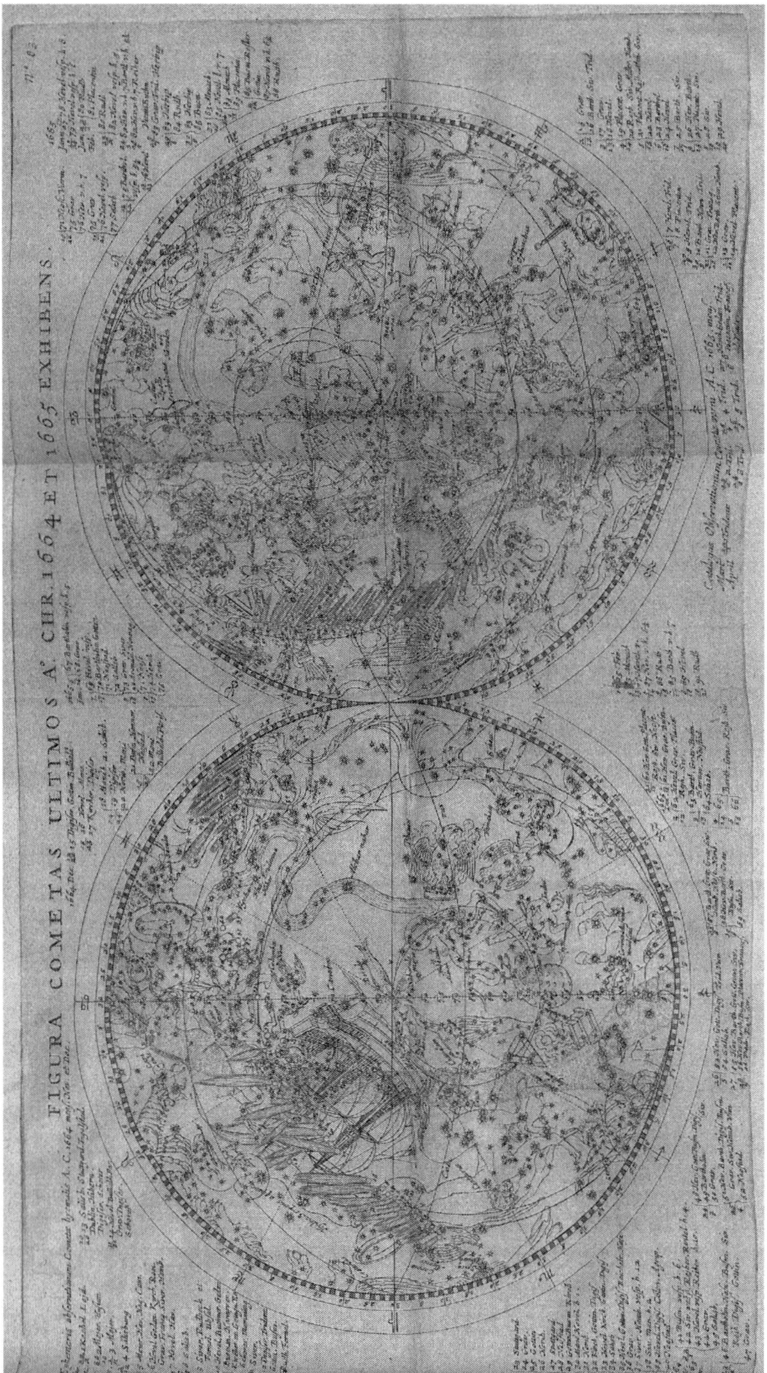

Figure 7. The routes of the comets of 1664 and 1665 from Lubieniecki's *Theatri cometici pars posterior*. Biblioteka Narodowa, Warsaw, SD XVII.4.2660.

the other hand, probably found this brief information to be sufficient for his own purposes. In the *Theatri cometici pars posterior*, in the description of the comets of 1664 and 1665, he placed a large and very detailed illustration of the routes of the two comets against the background of the constellations. Again, there is no numerical data, but such an illustration could somehow be used by those who studied these phenomena. This map of the Heavens lacks exact numbers, but positions of the comet are described with the names of the observers and the places of observations. It should be noted that there are some discrepancies between the table in the *Theatri cometici pars prior* and the information in the picture in the *Theatri cometici pars posterior*: sometimes data present in one place are absent from another.

The second table in the first volume of the *Theatrum cometicum* would be much more useful. It contains the exact positional data which Lubieniecki received from his correspondents.[35] He lists various notions about the position of the comet on given dates. Unfortunately, the timespan of this table is very short: from 21 to 30 December 1664. Therefore, the possible benefit from its use would have been somewhat limited. Lubieniecki probably chose these days to be the subject of a more detailed analysis because there was a discussion about the number of comets in December 1664.[36]

The results of Lubieniecki's efforts to synthesize the data he collected were rather mediocre. Moreover, he finished this task very late, as he placed the tables in the last *communicatio* in the *Theatri cometici pars prior*, among the letters he exchanged in mid-1667 with Villum Lange. In the *Theatri cometici pars posterior*, in the description of the comets of 1664 and 1665, in addition to the map, he informed the reader about the tables which were at the very end of the first volume of his book.[37]

Even though Lubieniecki conducted no continuous astronomical observations, because he lacked the abilities and the precise instruments, there are a few comments in his letters published in *Theatri cometici pars prior* that he saw something notable in the sky. It was extraordinary phenomena that attracted his attention the most. There are two instances which need to be mentioned. The first observation was conducted by Lubieniecki on 16 July 1665. He saw three blazing objects in the sky which lasted for some time before they merged into one and disappeared soon after. Lubieniecki presumed that it was a kind of burning exhalation. He described the phenomenon in his letter to Johann Ernst von Rautenstein[38] and depicted it on the engraving printed in the *Theatrum cometicum*. He gave no exact data of the observation apart from the hour it began (after 10 o'clock). The position of the phenomenon has to be inferred from its relation to the cityscape in the engraving.

Another observation of an unexpected object in the sky is presented by Lubieniecki in his letter to Oldenburg, dated 5 February 1667 (Letter 57B). On 1 January,

35 TC pp. 939–41.
36 See section 2.3.1 of this Introduction.
37 TC2 p. 440.
38 Lubieniecki to Rautenstein, 18 July 1665, TC p. 129; see also Letter 27C.

at 1 o'clock in the afternoon, he saw a rainbow-coloured orb with outstretching rays. This time, Lubieniecki tried to be more accurate in his description, hence the position and the direction of the rays were given in relation to the cardinal directions and the districts of Hamburg. He also discussed the mechanism of the generation of rainbows and concluded that this was not the case of the phenomenon in question. Apart from these two observations, throughout the *Theatrum cometicum*, there are just a few shorter remarks in which Lubieniecki claimed that he had seen a comet (or, less often, another celestial object) in the sky. These are, however, only brief passing mentions, and they cannot prove that he conducted any continuous, methodical astronomical studies.

3.2.4. Heliocentric Universe

It has been often repeated, especially in Polish criticism, that Lubieniecki was an ardent and convinced proponent and promoter of the heliocentric worldview.[39] This statement has no factual support and is indeed entirely wrong. The analysis of the arguments which are presented to sustain this view shows that they do not prove what they were intended for. Among them, there is an oft-quoted reference to Lubieniecki's letter to Hevelius in which the sender concluded that the addressee was a supporter of the heliocentric theory. Lubieniecki also recalled that he had seen the tomb of Copernicus in Toruń:

> So when you speak about the heavens, even with a hasty quill, you state — together with the Pythagoreans and Copernicus that best Hipparchus of your Prussia (whose epitaph I saw in the cathedral in his native city of Toruń, as I remember)[40] — that it does not move and that the Earth moves rather than the heavens because you are sure of your matchless observations like Atlas who supports the firmament. (Letter 24).

In fact, in this letter, Lubieniecki remained neutral and expressed no opinion of his own.

Another argument was the presence of the depictions of the heliocentric system in the *Theatrum cometicum*. There are such engravings in the *communicationes* of Samuel Reyher and Friedrich Büthner, but they only prove that these two astronomers supported the Copernican theory. Lubieniecki printed these illustrations together with their letters, without any comments.

There are passages in the *Theatrum cometicum* which prove that Lubieniecki suspended his opinion about the Copernicus's theory and decided to side neither with the Copernicans nor with the anti-Copernicans. A clear statement of this posi-

39 Przypkowski, 'Zainteresowania matematyczno-przyrodnicze', pp. 414–15; Przypkowski, 'Astronomowie współcześni Heweliuszowi', pp. 293, 295–196; Barbara Bieńkowska, *Kopernik i heliocentryzm w polskiej kulturze umysłowej do końca XVIII wieku* (Wrocław: Zakład Narodowy im. Ossolińskich, Wydawnictwo Polskiej Akademii Nauk, 1971) (*Studia Copernicana*, vol. 3), pp. 162–64.

40 In fact, Copernicus was buried in Frombork (Frauenburg), not in Toruń.

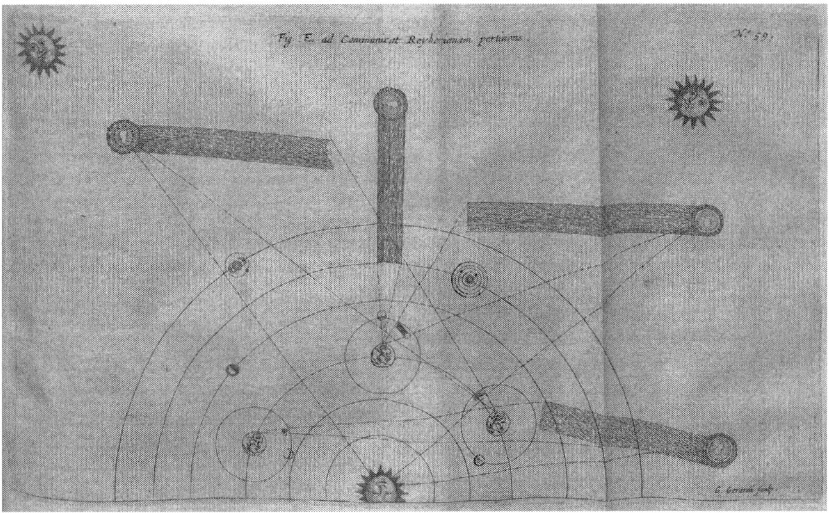

Figure 8. The comet in the heliocentric system. Illustration from Reyher's *communicatio* in Lubieniecki's *Theatri cometici pars prior*. Biblioteka Narodowa, Warsaw, SD XVII.4.2660.

tion can be found in his letter to Leunenschlos, in which, after having discussed the varying opinions on the movement of the Earth, Lubieniecki concluded: *quam ego controversiam nolo facere meam*.[41] A similar situation occurred in his letter to Riccioli. Lubieniecki referred various attitudes to the Copernican theory and eventually remarked that the steadfast hope of salvation is more important than the movement or immobility of celestial bodies.[42] One could presume that Lubieniecki did not want to antagonize his correspondents by trying to confirm the opinions they disagreed with, but such a presumption is invalid, as Riccioli was an opponent of the heliocentric system, whereas Leunenschlos supported it.[43]

3.2.5. The Natural Place of Comets

In the *Theatrum cometicum*, there is only one cometary statement which Lubieniecki declared as his and which represented an important part of his worldview. It was the idea that comets are not sublunary phenomena but that their proper place is above the Moon. He emphasized that there are no (or at least only extremely few) comets in the cosmological sphere of air. He made a passing mention of this view in

41 Lubieniecki to Leunenschlos, 4 July 1665, TC p. 629.
42 Lubieniecki to Riccioli, 17 June 1665, TC p. 701.
43 Leunenschlos to Lubieniecki, 12 April 1665, TC pp. 623, 625.

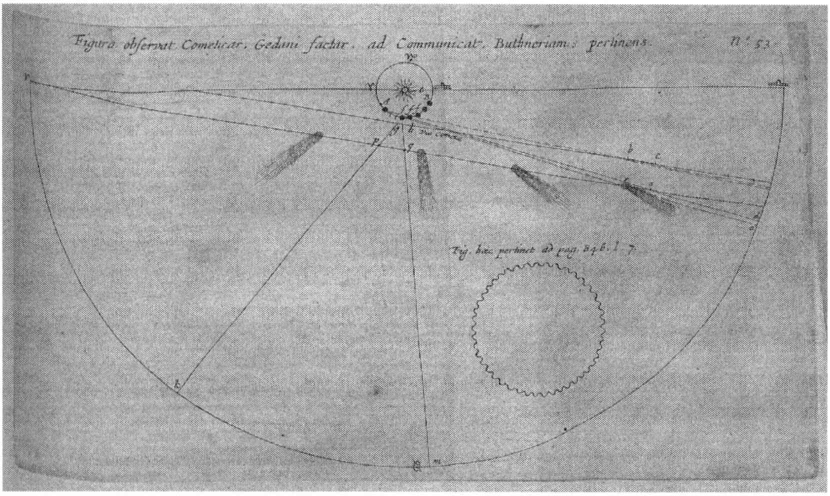

Figure 9. The comet in the heliocentric system. Illustration from Büthner's *communicatio* in Lubieniecki's *Theatri cometici pars prior*. Biblioteka Narodowa, Warsaw, SD XVII.4.2660.

the preface to the *Theatri cometici pars posterior*,[44] but he discussed it in detail in the *Theatri cometici exitus*.[45]

Lubieniecki adopted this idea as a part of his worldview due to astrological, not astronomical reasons. He believed that comets cannot have any, even physical, influence on Earth and, therefore, they must be removed from our planet. He discussed this view in detail in the summary of his opinions on cometary influence in the *Theatri cometici exitus*.[46]

It should be noted, however, that Lubieniecki made no specific claim that comets are celestial objects. His statements in this matter are sometimes vague, as in his letter to Hevelius where he suggests that it is not obvious to 'compare bodies [i.e. comets] to celestial bodies' (Letter 27). In a letter to Isaac Vossius (1618–89), Lubieniecki states that the knowledge of comets (*cometologia*) is similar to meteorology.[47] Such opinions, however, are sparse.

44 TC2 sig. *3r–v.

45 Lubieniecki to Kuyper, 29 September 1666, TC3 p. 9.

46 See section 3.3.3 of this Introduction.

47 Lubieniecki to Vossius, 1 February 1667, TC p. 934. About Vossius see *Isaac Vossius (1618–1689) betwen Science and Scholarship*, ed. by Eric Jorink and Dirk van Miert (Leiden, Boston: Brill 2012) (*Brill's Studies in Intellectual History*, vol. 214); Grell, 'Introduction', CJH2 p. 28.

3.3. Lubieniecki and Astrology

3.3.1. Effects of Comets

Lubieniecki had a clear and adamant opinion about the meaning, significance, and influence of comets. He opposed any deliberations that comets can physically affect the Earth in any way. He also mocked astrological predictions and interpretations drawn from the movement of comets against the background of the constellations and the signs of the zodiac. Nevertheless, it would be wrong to conclude that Lubieniecki's view on celestial objects was entirely founded on reason. He was inclined to ethical and historiographical interpretations of comets. He was also sure that they are heaven-sent messengers of divine origin and that their purpose is to announce God's will and admonitions to mankind.

The most crucial aspect of Lubieniecki's views on the meaning of comets is the phrase *bona bonis, mala malis*. These words summarize his belief that comets bring 'good things to the good, bad things to the bad.' Not only did Lubieniecki put this motto at the beginning of his book, but he also made it an essential assertion of his worldview in his correspondence.[48] Nevertheless, as all neat catchphrases, *bona bonis mala malis* is meaningless without a contextual analysis within the framework of its author's statements. And these, unfortunately, are often vague and contradict one another. Sometimes, while describing his views on cometary influence, Lubieniecki emphasizes that he does not believe in any way in connecting events on Earth with phenomena in the sky. He adds that there are many misfortunes around the world but, unfortunately, they happen all the time, regardless of whether there is a comet in the sky or not.[49]

More often, Lubieniecki's statements on this topic seem less definite. He repeats the opinion that after a comet appears, both good and bad things may happen on Earth. And in order to prove this view, he compiles his catalogue of comets and the historical events which accompanied them. In his letters, he stresses that good and evil on Earth are inseparable, and they always succeed each other.[50] Whenever his correspondents seem to defend the notion of the comets as the portents of misfortunes, Lubieniecki ardently opposes their views and tries to convince them that such an opinion is wrong.[51] He adds that the words *bona bonis, mala malis* describe only

48 Lubieniecki to Boulliau, 27 February 1666, TC p. 543, BnF ms FF. 13031 f. 383r; Lubieniecki to Ménage, 27 February 1666, TC p. 861. Lubieniecki very rarely used a different wording to express this idea, as he did in the letter to Hermannus Langius, 7 December 1665, TC p. 583, 'Has enim non tantum minis malos et improbos horrore ferire, sed et bonis spes, probis gaudia ferre ex divinae aequitatis ratione omnino est mihi persuasum.'

49 Lubieniecki to de Grau, 9 June 1665, TC p. 499; Lubieniecki to Rautenstein, 3 October 1665, TC p. 161; Lubieniecki to Auzout, 24 October 1665, TC p. 855.

50 Lubieniecki to Huygens, 30 October 1665, TC pp. 932–33, OCCH5 pp. 522–23, LUL ms HUG 45 ff. 1v–2r; Lubieniecki to Gruterus, 3 November 1665, TC pp. 866–67.

51 Lubieniecki to Büthner, 6 April 1665, TC p. 808; Lubieniecki to Busmann, 26 April 1665, TC pp. 657–60.

the ethical meaning of comets and not the physical one. This statement, even though important for Lubieniecki's worldview, seems not to be emphasized enough in the *Theatri cometici pars prior* but is often recalled in the *Theatri cometici exitus*.[52]

In the first volume of the *Theatrum cometicum*, Lubieniecki, although rarely, seems to present a more tangible interpretation of *bona bonis, mala malis*. In these few passages, he does not understand the notion of entanglement of good and evil only metaphorically but suggests that good events in one place are equalled by bad events somewhere else. In order to exemplify this, he makes allusions to political circumstances.[53] Sometimes these references are very precise, as in a letter to Christian Busmann who supported the view of comets as signs of misfortunes. Lubieniecki sent him a lengthy description of what happened in Poland and Lithuania after the comet of 1618 appeared, writing both about the victories and failures.[54] In my opinion, such passages in Lubieniecki's correspondence served as arguments in discussion with those who presented a very pessimistic view on comets and their meanings. After more careful consideration, he probably realized that such opinions, somehow indicating a direct link between comets and Earth, contradict his statements of no possible cometary influence and, therefore, he ceased to repeat them.

3.3.2. Ridiculed Prognostication

Even if Lubieniecki sometimes seemed to connect celestial phenomena with the events on Earth, he always opposed astrological predictions, and he believed that it is impossible to foresee any future occurrences on their basis. An example of this approach to astrology can be found in his mocking of a prediction he received from Bremen and published in the *Theatri cometici pars prior*.[55] According to the prophecy, because the comet of 1664 moved through Hydra, Orion, Eridanus, Cetus, Aries, and Taurus — which are constellations related to water and aggression — there would undoubtedly be a naval war. And, the prediction concludes, the comet portended that the war would be between England and the Netherlands.

Lubieniecki inventively expanded this brief prognostication by adding references to other constellations through or next to which the comet passed. He suggested that there should also be a mention of Leo and of the Canis Maior, as these animals, predatory and fierce, can be easily associated with warfare. Moreover, rapacious Corvus can also be considered a belligerent constellation. On the other hand, as Lubieniecki remarked, there were constellations of Cancer, an animal which moves backwards, and Lepus, which is not especially brave. Therefore it is probable that the soldiers

52 Lubieniecki to Kuyper, 1 June 1666, TC3 pp. 2–3; Lubieniecki to Kuyper, 16 July 1666, TC3 p. 6; Lubieniecki to anonymous (Niemirycz), 1 January 1667, TC3 p. 31; Lubieniecki to anonymous (Noricus), 26 January 1667, TC3 p. 36.

53 Lubieniecki to anonymous (Niemirycz), 3 April 1665, TC pp. 20–21; Lubieniecki to Rautenstein, 7 October 1665, TC p. 163.

54 Lubieniecki to Busmann, 26 April 1665, TC p. 660.

55 TC p. 11. I could not identify the source of this prognostication.

during the war will not be very courageous; this being also the possible effect of young Gemini. Furthermore, the constellation of Crater induced Lubieniecki to a conclusion that abusing wine might make the armies less bellicose. But it was the constellation of Virgo in which Lubieniecki's mockery reached its peak, because in his opinion it was the least warlike of the constellations, even if we consider the example of the Amazons. It seems that Lubieniecki thoroughly enjoyed his derision of astrological predictions. He sent it with ironical comments to a few of his correspondents, each time in a slightly different form, referring to a selection of non-warlike constellations.[56]

For some of Lubieniecki's correspondents, the irony was not obvious. This was the case with Hevelius, who received the prognostication twice in one letter, in the main text and in the enclosure (Letters 7 and 7A). He must have misunderstood Lubieniecki's attitude towards astrology and decided to admonish him:

> Seeing that you are somewhat interested in those astrological vanities, I, together with other friends, very benevolent towards you (which I have nearly forgotten), would like to suggest that you abstain entirely from predictions and prophecies. Because one can easily trip when he deals with such things. I ask you not to take it as an offence, as it has been said with a sincere heart. (Letter 26)

In his answer, Lubieniecki thanked Hevelius for this admonition and promised that he would be very cautious whenever he encountered astrological prognostications (Letter 27). In fact, he would probably have done so even without Hevelius's warning because this actually was his attitude towards astrology, but he limited himself to a well-mannered answer and did not explain in detail his opinion about the meaning of comets to the addressee. In the correspondence in the *Theatri cometici pars prior*, however, Lubieniecki omitted both these passages, Hevelius's warning and his answer to it. It might be speculated that Lubieniecki did it on purpose as he did not want his readers to get a false impression that he supported astrology of any kind.

3.3.3. Declaration of Lubieniecki's Opinions

As said before, Lubieniecki's only astronomical view of his own concerned the natural place of comets.[57] He believed them to be situated above the Moon, in the celestial region. Therefore, being far above the cosmological sphere of air, they could

56 Lubieniecki to Jászberényi, 11 February 1665, TC p. 691; Lubieniecki to de Grau, 17 February 1665, see Letter 7A; Lubieniecki to Büthner, 24 February 1665, TC pp. 801–02; Lubieniecki to Raue, 28 February 1665, TC p. 577; Lubieniecki to Leunenschlos, 28 February 1665, TC pp. 622–23; Lubieniecki to Rautenstein, 4 March 1665, TC p. 49; Lubieniecki to Concius, 13 March 1665, TC p. 586; Lubieniecki to Heinsius, 20 June 1665, TC pp. 283–84.

57 See section 3.2.5 of this Introduction.

not have any influence on the Earth.[58] Lubieniecki did not investigate this subject in relation to astronomy but to astrology. Because this opinion is almost entirely absent from his correspondence in the *Theatri cometici pars prior*, it seems that it had not been part of Lubieniecki's worldview before the comet of 1664 appeared. Lubieniecki discussed this idea chiefly in his correspondence in the *Theatri cometici exitus*.[59] Therefore, because of Frans Kuyper's request, Lubieniecki expressed his opinion about the vanity of astrology much clearer: he stated that comets cannot affect the Earth and emphasized that their only possible significance belongs to the sphere of ethics.

The correspondence with Kuyper led Lubieniecki to write the summary of his astrological views which also appeared in the *Theatri cometici exitus*.[60] This is the only place in the *Theatrum cometicum* in which he presents his own views and beliefs in a systematic and organized way. The summary consists of 30 points which Lubieniecki calls *theses*, and is titled *Fundamenta sententiae contra determinatam astrorum efficentiam et pro indifferenti ac aequa bonis malisque cometarum significatione*. He probably prepared this *résumé* in the first half of 1667.

Lubieniecki begins this summary with theological considerations. He states that astrology of any kind is contradictory to the Bible and the Christian religion. Moreover, the idea that stars can force men to some actions cannot be reconciled with God's omnipotence and human free will. Then, he proceeds to the main point of his view on astrology: he emphasizes that the only purpose of the stars is to emit light and that they cannot affect the Earth in any way. Even if there were a mechanism for such an influence, they are so far away from us that they would not be able to have any effect. The only celestial bodies which can influence the Earth are the Sun, thanks to its natural light, and the Moon because of the reflected sunlight. The planets, even though they also reflect the sunlight, do not affect the Earth. Lubieniecki stressed that the nature of the potential influence of stars would be impossible to define.

According to Lubieniecki, the meaning of comets has to be considered supernatural because they are signs from God, and the only correct way of their interpretation is ethical. He remarks that even if a comet had any natural influence, it would be good because nature is good. Nevertheless, even if there were comets in the cosmological sphere of air, they could affect that sphere only. In other words, they could affect the weather and they could not have any direct influence on men. Moreover, comets are bright, beautiful, and rare phenomena, therefore it seems obvious that one should consider their potential effects to be good.

58 There are a few passages in the *Theatrum cometicum* in which Lubieniecki was not entirely consistent in this view. In his letter to Curtz, 23 September 1665, TC pp. 224–27, he admitted that comets may somehow influence the Earth but they cannot affect people. Such remarks, very sparse, do not change the coherent opinions he presented in his conclusions and summaries and may be rhetorical devices used to induce the addressee to discuss the subject.

59 Lubieniecki to Kuyper, 29 September 1666, TC3 p. 9; Lubieniecki to anonymous (Niemirycz), 30 September 1666, TC3 p. 27; Lubieniecki to anonymous (Niemirycz), 1 January 1667, TC3 p. 31.

60 See Appendix 3.

Further on, Lubieniecki points out why comets cannot be a subject of astrological predictions. First, because it is better not to know future misfortunes. Second, even the wise men, philosophers, and astrologers of all epochs did not attribute only harmful effects to comets. In addition, there is no reason why a comet in a specific part of the sky would have any special meaning or consequence, if we consider that old descriptions of comets may vary in presenting their positions. There is no concord among astrologers in their interpretations of cometary phenomena. According to their opinions, an alleged influence of a comet lasts for many years after the phenomenon was seen. In fact, as Lubieniecki claims, good and bad things also occur when there are no comets in the sky.

As far as the physical nature of comets is concerned, Lubieniecki repeats that they are above the cosmological sphere of air and that therefore they cannot affect the Earth. The question whether they be eternal or ephemeral objects had been a subject of many scholarly disputations. Nevertheless, the difficulty in analysing their nature should not lead us to ignore these celestial bodies. They should not be feared, but instead, they have to be considered the signs of God and of divine justice which gives good things to the good and bad things to the bad.

3.3.4. Thomas Bartholin on Comets

As it is clear from Lubieniecki's theses against cometary astrology, he did not limit his criticism to astrological predictions, but he denied the possibility of any cometary or celestial influence on the Earth as well. A case study of this approach is his reaction to the book by Thomas Bartholin in which the author tried to convince the readers that comets are the most noxious possible phenomena and there is nothing in them to be admired.[61] Bartholin described comets using medical terms and from a physician's perspective, but his opinion about these phenomena seemed to rely on a well-grounded superstition which considered them to be portents of misfortunes. According to Bartholin, comets are ulcers in the sky. Given all the changes which happen there, it has to be concluded that there are heavenly diseases. He added that the medical wisdom of all ages admits that there can be some connection between the movement of celestial bodies and human health. As one of the examples, Bartholin stresses the effects of lunar eclipses and of the full Moon. Considering the nature of comets, he relied on two authorities: Aristotle, who considered the comets to be terrestrial exhalations, and Tycho Brahe, who proved that they are far beyond the Moon. Bartholin combined these two views: the comets are between the planets, but they are made of foul exhalations from the Earth and other celestial bodies. They are inhaled by the sky and cause its sickness. The comets are symptoms of these diseases, ulcers or tumours which swell and heal very slowly. As it is clearly suggested by the number of recent cometary phenomena, Bartholin says, the amount of this morbid

61 Thomas Bartholin, *De cometa consilium medicum cum monstrorum nuper in Dania natorum historia* (Hafniae: apud Matthiam Godicchenium, 1665).

secretion is certainly very large. Moreover, there is a circulation of these particles between the Heavens and the Earth. Nevertheless, he did not describe this mechanism in detail. He only listed some comets and the misfortunes they brought. He also gave an account of some abnormal medical cases he recently encountered, especially focusing on the instances of hydrocephaly. He concluded that their number is a proof of a very severe condition of the Earth and the sky. He added that he would like to be able to consider it a crisis after which the patient's condition would improve, but he sees no grounds for such an assumption. Knowing all these things, no one should admire the beauty of comets, and their sight should rather be dreaded. As a possible cure, Bartholin recommended moderation, avoiding excesses and strong emotions, and trusting in God. Apart from his book, Bartholin described the opinion about comets in a letter to Lubieniecki (this letter is printed in Bartholin's book as a sort of an abstract).[62]

In his answer to this letter, the author of the *Theatrum cometicum* stated that, in his opinion, the comets are not the portents of misfortunes, and good and evil things have always happened and will happen, even if there are no extraordinary apparitions in the sky. He remarked that there was no way to disprove Bartholin's theory, but it is not the only possible explanation of cometary phenomena. One may argue, Lubieniecki says, that a fiery nature of comets burns all the foul exhalations. He concludes that, in his opinion, comets should not be feared, but they also should not be disregarded because they are signs from God.[63]

3.3.5. Comets Sent by God

The interpretation of comets as divine messengers was one of the most important features of Lubieniecki's view on the meaning of comets. Emphasizing their origin from God, he stated that it is improper for a Christian to be afraid of them.[64] Lubieniecki had no doubts that comets are sent to mankind by God. Nevertheless, his opinions are not entirely consistent. Sometimes he limited his discussion to a mere consideration that comets are divine signs.[65] In some of his letters, however, his description of the origin of these signs was more detailed. He presented an opinion that God directs the comets himself[66] or that the angels manage them by his order.[67] Set-

62 Thomas Bartholin to Lubieniecki, 11 May 1665, TC pp. 435–36; Thomas Bartholin, *De cometa*, pp. 150–54.

63 Lubieniecki to Thomas Bartholin, 21 October 1665, TC pp. 442–47.

64 Lubieniecki to Rautenstein, 2 May 1665, TC p. 86; as Lubieniecki remarked in this letter, apart from a Christian, also a wise man should not be afraid of comets, according to the ancient saying *sapiens dominabitur astris* (see section 1.3.5 of this Introduction).

65 Lubieniecki to Røtlin, 30 December 1665, TC p. 419; Lubieniecki to Broussel, 27 May 1665, TC p. 229.

66 Lubieniecki to Raue, 28 February 1665, TC p. 577; Lubieniecki to Gruterus, 8 December 1665, TC p. 869.

67 Lubieniecki to Kuyper, 1 June 1666, TC3 p. 3; Lubieniecki to Wiszowaty, 25 October 1666, TC pp. 619–20.

ting aside the divine mechanism of comets, Lubieniecki was sure about the message they carry. He stressed that their only possible significance is moral and ethical. Their aim is to encourage people to become more pious, to improve their deeds, and to repent their sins.[68] Sometimes he added that the punishment from God is imminent, and the true penitence is the only way to avoid it (Letter 7). More often, however, his statements were milder, and he wrote that comets are the signs of both God's anger and his mercy.[69]

God endowed men with free will. This statement is elemental for many Christian denominations. For the Polish Brethren, free will was the foundation of man's morality. Without the ability to choose between good and evil, man would not be responsible for his actions.[70] When discussing the inability of comets to influence people, Lubieniecki emphasized that they cannot affect men in any direct way nor irresistibly force them to do anything.[71] He hoped that free will would induce people to prayer and penance,[72] yet he realized that free will is often abused and people choose evil.[73]

3.3.6. Lubieniecki's Authorities against Astrology

The Bible was not the only authority on which Lubieniecki relied in his discussions about the meaning of comets. He also referred to more contemporary authors. In a letter to Tobiasz Morsztyn, Lubieniecki stated that his opinions agreed with those of Libavius and Riccioli, and he added extracts from their books: *Singularia* and *Almagestum novum*, respectively.[74] Libavius considered extraordinary celestial phenomena to be natural and concluded that astrological opinions are arbitrary and groundless. Celestial objects may be signs from God, but they are not causes of events on Earth. Similarly, Riccioli stated that if comets are celestial objects, they have to be regarded as Godly signs, and if they are in the atmosphere, below the Moon, they are natural phenomena.

Another text about the meaning of comets to which Lubieniecki referred was Marcello Squarcialupi's *De cometa in universum atque de illo, qui anno 1577 visus*

68 Lubieniecki to Rautenstein, 7 January 1665, TC p. 40; Lubieniecki to Büthner, 24 February 1665, TC p. 801; Lubieniecki to Busmann, 10 March 1665, TC p. 644.

69 Lubieniecki to de Grau, 9 June 1665, TC p. 499.

70 Ogonowski, *Socynianizm*, p. 253.

71 Lubieniecki to Huygens, 30 October 1665, TC p. 932, OCCH5 p. 521, LUL ms HUG 45 f. 1v. Despite this consideration, Lubieniecki sometimes pondered on a view that comets may somehow induce people to some actions, but he immediately added that celestial bodies may cause some inclinations, but they cannot compel anyone to do anything. It is our free will and the knowledge of the laws of God which prevent this influence. A man who is especially protected from such an influence is a wise man because *sapiens dominabitur astris* (see section 1.3.5 of this Introduction). See Lubieniecki to Busmann, 19 April 1665, TC p. 950; Lubieniecki to Busmann, 26 April 1665, TC p. 657–58.

72 Lubieniecki to Rautenstein, 7 January 1665, TC p. 40.

73 Lubieniecki to Curtz, 23 September 1665, TC p. 224.

74 Lubieniecki to Morsztyn, 6 June 1665, TC pp. 508–17.

est, opinio Marcelli Squarcialupi Plumbinensis ad amplissimum et sapientem virum Andream Dudithium, caesaris consiliarium. It was published in a compilation titled *De cometis dissertationes novae* which contained the texts written on the occasion of the comet of 1577 by Thomas Erastus (1524–1583), Simon Grynaeus (1493–1541), Andreas Dudithius (1533–89) and Squarcialupi (d. 1599).[75] They all agreed, even though their opinions about the nature of comets were firmly grounded in earlier natural philosophy, that comets cannot be used for exact prognostications and interpretations, and they should not be linked to the misfortunes on Earth. Squarcialupi's statements were the most radical ones. He disagreed with the Aristotelian view and considered comets to be astronomical objects, not meteorological phenomena. Therefore, he denied the assumption that they could affect the Earth — even the weather. He also stressed that scholars differ in their views on comets and remarked that it is easy to ascribe bad events to comets.[76] In his letters, Lubieniecki referred to this book and stated that he agreed with Squarcialupi's views.[77]

3.4. Lubieniecki's Astronomical Correspondence

3.4.1. Lubieniecki as a Scholarly Informant

In his letters, Lubieniecki asked learned men for their opinions and observations. But the outcome was insufficient to satisfy his desire for knowledge and to provide an adequate amount of material for his planned book. Therefore, he sent the data he had received from one correspondent to another. He hoped that they would answer him and present their opinions about the information he had sent them. In his description of the *Theatrum cometicum*, Tadeusz Przypkowski rightly remarked that Lubieniecki sometimes tried to sharpen the conversation: he conveyed the statements he had received to the correspondents about whom he knew that they would disagree with the information he passed to them. For example, in a letter to Riccioli, he listed a number of opinions in favour of the movement of the Earth.[78]

Lubieniecki decided to serve the Republic of Letters as an intermediary of scholarly, mainly astronomical, news and information. In the second half of the seventeenth century, such intermediaries facilitated contacts between scholars and sometimes tried to arrange and publish the outcome of their correspondence.[79] Living

75 Thomas Erastus et al., *De cometis disserationes novae* ([Basel]: ex officina Leonardi Ostenii, 1580). About this book: C. Doris Hellman, *The Comet of 1577: its Place in the History of Astronomy* (New York: AMS Press, 1971), pp. 355–58.

76 Tadeusz Przypkowski, 'Astronomia poza Krakowem w drugiej połowie XVI wieku', in *Historia astronomii w Polsce*, vol. 1, pp. 203–04; Mosley, 'Past Portents Predict', pp. 18–30.

77 Lubieniecki to Stegmann, 20 September 1665, TC p. 608; Lubieniecki to Kuyper, 29 September 1666, TC3 p. 9.

78 Przypkowski, 'Zainteresowania matematyczno-przyrodnicze', pp. 414–15; Lubieniecki to Ricciolі, 17 May 1665, TC p. 701.

79 Peter Burke, *A Social History of Knowledge. From Gutenberg to Diderot* (Cambridge, Malden: Polity, 2000), p. 25.

in a major port city and having a European network of correspondents, the author of the *Theatrum cometicum* could become one of the focal points on the map of the scholarly correspondences of his era. However, as we shall see below, Lubieniecki did not become a reputed and recognized scholarly informant.

Lubieniecki's service as an intermediary of scholarly information was only moderately successful. His correspondents were usually not very inclined to discussing in detail all the views and remarks he had sent them. In fact, Lubieniecki's correspondence with Hevelius is somewhat untypical, because Hevelius often asked for Lubieniecki's services in disseminating his discoveries and books, especially in their early letters.

Lubieniecki was not affiliated with any respectable scientific institution, and he was unable to build up a reputation in the Republic of Letters comparable to that of, for example, Henry Oldenburg. Moreover, his interests were not strictly mathematical and astronomical, but rather historical and theological. These topics were beyond the scope of interests of early modern scientific institutions. Oldenburg states it clearly in his letter to Lubieniecki:

> I would like you to inquire for me if there are any learned men in Cologne, Frankfurt, Leipzig, and Augsburg who would like to conduct correspondence with us. Be assured that we will send them in return something that they would like to ask us about in the subject of mechanics, anatomy, optics, astronomy, physics, chemistry, etc. We leave theology, jurisprudence, politics, metaphysics, and similar topics for schools and churches. (Letter 57A)

Moreover, Lubieniecki allegedly valued more the polemics in defence of his faith than his activities in the field of astronomy. He did not mention the *Theatrum cometicum* in his testament in which he recounted the most important events of his life.[80] In his biography, published in the *Historia reformationis Polonicae*, there is only short information about the dedication of the *Theatrum cometicum* to the King of Denmark and Lubieniecki's troubles with his publisher.[81]

3.4.2. Unanswered Letters

Lubieniecki was a correspondent of some of the most prominent scholars of his time, and it can be presumed that he was somewhat proud of his correspondence and its publication. Among his *communicationes*, there are such which are not 'exchanges' *sensu stricto*: five of them contain only the letters sent by Lubieniecki. In the eyes of his contemporaries, not receiving an answer from a very distinguished person was

80 AUL ms Ar 50^{1}.

81 Stanisław Lubieniecki, *Historia reformationis Polonicae* (Freistadii [i.e. Amsterdam]): apud Johannem Aconium, 1685), sig. [*6]r; see also the translation: 'The Life of Stanislas Lubieniecki, Polish Knight', in *The Polish Brethren. Part 2*, p. 526.

not considered shameful. On the contrary, failed attempts at contacting someone prominent were presumably a thing to be proud of. Pierre Gassendi published his letters sent to Galileo, but they probably remained unanswered.[82]

Among those from whom Lubieniecki never received a reply were Isaac Vossius, Christiaan Huygens, Andreas Concius, Samuel Carolus Kechelius von Hollesteyn, Olof Rudbeck, and Jonas Fornelius (views and observations of the last two, however, were conveyed to Lubieniecki by Nicolaus Heinsius). There is also no response from Peter Lambeck (1628–80), but Lubieniecki sent him a letter as late as in May 1667, so even if Lambeck's answer had arrived, it would have probably been too late to print it in the *Theatrum cometicum*.

The reasons why these learned men ignored Lubieniecki's messages and decided not to answer his letters remain unknown. We only have limited access to how Huygens responded to Lubieniecki's inquiries. Lubieniecki wrote his letters to Huygens, having been persuaded to do so by Nicolaus Heinsius.[83] Initially, Lubieniecki was very reluctant. He answered Heinsius that he would not dare to write to such a renowned man and that he would appreciate it very much if Heinsius began this conversation on his behalf.[84] As the reason for his unwillingness, Lubieniecki gave the fact that he had had no prior contacts with Huygens and that he knew him only by name. The Dutch scholar was undoubtedly the most prominent of all to whom Lubieniecki sent astronomical letters. Nevertheless, there is no information that he was so awe-struck by any other of his correspondents.

Heinsius fulfilled this request, and Huygens learned about Lubieniecki's plan to collect and publish cometary observations and opinions.[85] Huygens was not too eager to start the cometary discussions and answered that he had not written anything about comets that could be considered important.[86] Heinsius informed Lubieniecki that he had written to Huygens. He also gave him the details of their correspondence.[87] Lubieniecki finally surrendered to Heinsius's persuasion and decided to send a letter to Huygens himself.[88] When he wrote it, he sent a copy to Heinsius for approval.[89]

82 Grell, 'Hevelius en son temps', pp. 128–29.

83 Heinsius to Lubieniecki, 7 July 1665, TC p. 290.

84 Lubieniecki to Heinsius, 21 July 1665, TC p. 291.

85 Heinsius to Huygens, 19 August 1665, OCCH5 p. 437.

86 Huygens to Heinsius, 15 September 1665, OCCH5 pp. 479–80. Huygens was unwilling to announce his cometary speculations and they were not published until the 20th century, see Yeomans, *Comets*, pp. 87–93.

87 Heinsius to Lubieniecki, 5 August 1665, TC p. 300; Heinsius to Lubieniecki, 19 August 1665, TC p. 306; Lubieniecki to Heinsius, 1 September 1665, TC p. 307; Lubieniecki to Heinsius, 8 September 1665, TC p. 308; Heinsius to Lubieniecki, 14 October 1665, TC p. 317.

88 Lubieniecki to Heinsius, 27 October 1665, TC p. 317.

89 Lubieniecki to Heinsius, 3 November 1665, TC p. 319; Heinsius to Lubieniecki, 11 November 1665, TC p. 322, *Clarorum virorum epistolae centum ineditae* (Amstaeledami: prostant apud Sebastianum Petsoldum, 1702), p. 233.

Lubieniecki sent two letters to Huygens, on 30 October 1665 and on 16 February 1667.[90] He never received an answer.[91] In his later correspondence with Huygens, Heinsius once again mentioned Lubieniecki and his planned book,[92] but there was no reaction from the Dutch scholar. Heinsius wrote to Lubieniecki about his letter to Huygens.[93] In his later correspondence with Heinsius, Lubieniecki asked him for help in contacting Huygens.[94] The Dutch scholar, however, mentioned Lubieniecki's publishing plans in a letter to his brother, Constantijn, and there he declared his intention to fulfil Heinsius's and Lubieniecki's request.[95]

3.4.3. Otto von Guericke on the Generation of Comets[96]

Because Hevelius was one of the main participants of Lubieniecki's epistolary discussions about comets, the most important of them — concerning the number of comets and the phenomenon in Andromeda — have been presented in the context of Lubieniecki's correspondence with the Danzig astronomer.[97] There is one recurring topic in Lubieniecki's correspondence, especially in its earlier part, a discussion in which Hevelius was not particularly involved: Otto von Guericke's cometary theory. Lubieniecki put a lot of effort in disseminating it among his correspondents. Inspired by his experiments with vacuum and by recent cometary phenomena, Guericke presented his own theory of the nature and origin of comets: he considered the tailed stars to be parts of the air which were lifted by strong storms to the vacuum around the Earth; there, they were illuminated by solar rays (Letter 2D).

Lubieniecki learned about this idea from Guericke's son and namesake. From him, Lubieniecki got to know more about his father's works and experiments. The *communicatio* also contains his father's observations, passages from his yet unpublished book *Experimenta nova (ut vocantur) Magdeburgica de vacuo spatio*,[98] and a des-

90 TC pp. 931–33, OCCH5 pp. 518–24, OCCH6 pp. 109–12, LUL ms. HUG 45.

91 One could say that if Huygens had answered Lubieniecki's second letter, it would have been too late to publish it in the *Theatrum cometicum*, and, therefore, the case of Huygens's hypothetical answer cannot be concluded. In my opinion, such a suggestion is wrong. If Lubieniecki had received any answer from the Dutch scholar, it would have been highly probable that he mentioned it in his correspondence with Hevelius, but there is no information about it.

92 Heinsius to Huygens, 13 January 1666, OCCH6 pp. 3–4.

93 Heinsius to Lubieniecki, 13 January 1666, TC p. 330.

94 Lubieniecki to Heinsius, 16 January 1666, TC p. 332; Lubieniecki to Heinsius, 12 April 1667, TC3 p. 75; Lubieniecki to Heinsius, 31 May 1667, TC3 p. 76.

95 Huygens to Constantijn Huygens, 9 September 1667, OCCH6 pp. 148–49.

96 I discuss Guericke's cometary theory and Lubieniecki's attitude towards it in Maciej Jasiński, 'Otto von Guericke's cometary theory in Stanisław Lubieniecki's correspondence', *Journal for the History of Astronomy*, 51 (2020), no. 2, 131–51.

97 See section 2.3 of this Introduction.

98 TC pp. 245–48; see Otto von Guericke, *Experimenta nova (ut vocantur) Magdeburgica de vacuo spatio* (Amstelodami: apud Joannem Janssonium a Waesberge, 1672), sig. **3v, pp. 53, 57–58, 62–63.

Figure 10. Otto von Guericke's instrument for predicting winds. Illustration from Lubieniecki's *Theatri cometici pars prior*. Biblioteka Narodowa, Warsaw, SD XVII.4.2660.

cription and depiction of an instrument for predicting winds (in modern terms: the barometer) invented by him.[99]

Lubieniecki seemed to have been very interested in Guericke's cometary theory, even though he never stated that he agreed with it,[100] and it was incompatible with his opinion about the lack of connection between the comets and the Earth. He decided to contact its author in order to ask him for more details. Lubieniecki also found some parallels, mainly ancient (but not only), to Guericke's idea and wanted to share them with him.[101] In the answers to his letters, Guericke described his theory in detail, related it to the results of his experiments with the vacuum, and supplemented it with some additional information.[102] He was not very interested, however, in the parallel opinions found by Lubieniecki and answered that his theory was new and genuine because ancient ideas considered comets to be terrestrial exhalations and not parts of the air. He also reiterated that comets were not the main topic of his investigations. Despite all this, his *Experimenta nova* profited from the correspondence with Lubieniecki. A passage from the book devoted to comets contains only a reprint (with minor changes) of his *communicatio* from the *Theatrum cometicum*.[103]

Lubieniecki wrote about Guericke's theory to many of his correspondents but with only very limited success. Either they were not interested in its details, or they were openly sceptical about it. Hevelius's reply to Lubieniecki's information, namely that the theory could be easily refuted but the letter is not a proper place for it (Letter 9), is very typical. Some of Lubieniecki's correspondents simply ignored Lubieniecki's questions about Guericke's theory and wrote nothing about it in their responses.[104] Others rejected it for various reasons. Johann Müller carefully considered the possible routes of comets in Guericke's theory and answered that it is contradictory to the real movement of comets.[105] Giovanni Battista Riccioli, on the other hand, disagreed with Guericke's idea because it was necessary to assume the movement of the Earth around the Sun in order to accept it.[106] Therefore, because Guericke and other correspondents were not too interested in this cometary theory, Lubieniecki ceased to write about it in his letters.

99 Guericke-son to Lubieniecki, TC pp. 250–51.

100 The only strong assertion about Guericke's theory can be found in Lubieniecki's letter to Adam Olearius (23 January 1665, TC p. 637) where it is described as probable and worth further consideration.

101 Lubieniecki to Guericke-father, 14 March 1665, TC pp. 453–54; Lubieniecki to Guericke-father, 15 August 1665, TC pp. 459–62; see also Letter 7A.

102 Guericke-father to Lubieniecki, 1 August 1665, TC pp. 456–59; Guericke-father to Lubieniecki, 12 April 1666, TC pp. 463–65.

103 Guericke, *Experimenta nova*, pp. 184–97; see also the commentary in the modern edition: Fritz Krafft, *Fünftes Buch*, in Otto von Guericke, *Neue (sogennante) Madgeburger Versuche über den leeren Raum*, ed. and transl. by Hans Schimank et al. (Düsseldorf: VDI-Verlag, 1968), 'Erläuterungen und Register', pp. 296–302.

104 Lubieniecki to Raue, 28 February 1665, TC p. 576; Lubieniecki to Kircher, 18 March 1665, TC p. 749.

105 Müller to Lubieniecki, 22 December 1664, TC pp. 844–45.

106 Riccioli to Lubieniecki, 28 April 1665, TC p. 700.

3.5. Letters in the *Theatrum cometicum*

The following table presents the information about the number of letters in Lubieniecki's *Theatrum cometicum*. After the names of the correspondents ('Name'), in other columns are presented: the timespan of their correspondence in the *Theatrum cometicum* ('Timespan'), letters sent by Lubieniecki to his correspondents ('From L.'), letters sent by the correspondents to Lubieniecki ('To L.'), and the page range in the *Theatrum cometicum* where the relevant letters can be found ('Pages'). Enclosures and other important features of the collections of letters are mentioned in the footnotes. Whenever the identification of a correspondent is hypothetical, I notice this fact and grounds for such identification. I was unable to identify two of Lubieniecki's correspondents whose letters are at the beginning of the *Theatrum cometicum*.

Table 2. The list of Lubieniecki's correspondents in the *Theatrum cometicum*.

Name	Timespan	From L.	To L.	Pages
Auzout Adrien	28.IX–31.X.1665	2	1	TC 855–58
Bartholin Erasmus	12.I.1665–16.VIII.1667	19	7	TC 430–35, 436–42, 447–51, 955–61[107]
Bartholin Thomas	19.XII.1664–21.X.1665	3	1	TC 429–30, 435–36, 442–47
Blum Joannes	16.XII.1665–10.III.1667	6	6	TC 895–906, TC3 44–47
Boulliau Ismaël	13.XII.1664–31. VII.1666	38	20	TC 467–86, TC 527–48
Broussel Pierre	25.VI–5.XII.1665	4	3	TC 229–36
Busmann Christian	28.II.1665–20.X.1666	11	10	TC 641–85[108]
Büthner Friedrich	19.XII.1664–9.IX.1667	13	12	TC 797–842, TC 963–65
Capell Rudolf	6.XII.1665–10.II.1667	2	3	TC 889–94
Concius Andreas	13.III.1665	1	0	TC 585–88
Curtz Albert	2.I–23.IX.1665	4	2	TC 209–27[109]

107 Enclosures: TC 956–59 (Bartholin's cometary considerations). There is one *communicatio* with both Erasmus and Thomas Bartholin.

108 Enclosures: TC 668–72 (observations).

109 Enclosures: TC 209–12 (Curtz's correspondence with Rautenstein and various observations), 214–15 (observation from Rome).

Name	Timespan	From L.	To L.	Pages
Grau Abraham de	19.XII.1664–6.IV.1667	10	7	TC 487–503, 519–26, TC3 77[110]
Fornelius Jonas	25.VII.1665	1	0	TC 347–59[111]
Gruterus Isaac	4.VIII.1665–18. VIII.1668	14	14	TC1 863–80, TC3 69–73, 83–84
Guericke Otto von (father)	14.III.1665–29.IV.1666	5	3	TC1 453–66[112]
Guericke Otto von (son)	23.XII.1664–1.VIII.1665	4	6	TC1 236–51[113]
Heinsius Nicolaus	22.XI.1664–25.VI.1667	68	57	TC 253–45, TC3 53–57, 75–77[114]
Hevelius Johannes	19.XII.1664–16. VIII.1667	43	24	TC1 361–414, 947–55[115]
Huygens Christiaan	30.X.1665–16.II.1667	2	0	TC 931–33
Jászberényi Pál	27.XII.1664–13.X.1666	6	3	TC 687–96[116]
Kechelius van Hollesteyn Samuel	19.XII.1664–10.II.1665	3	0	TC 589–96[117]
Kircher Athanasius	18.III–19.VIII.1665	5	2	TC 747–59[118]
Kirsten Michael	20.III–5.V.1667	4	4	TC3 47–52
Kuyper Frans	24.V.1666–30.VI.1667	4	4	TC3 1–20
Lambeck Peter	25.V.1667	1	0	TC3 77–78
Lange Villum	15.VI–4.X.1667	4	1	TC 936–45[119]

110 Enclosures: TC 495–99, 521–23 (observations).

111 Lubieniecki's letter was addressed both to Fornelius and Rudbeck.

112 Enclosures: TC 454–55 (Guericke's letter to his son).

113 Enclosures: TC 238–39 (Guerick-father's observation and outline of his cometary theory), 239–41 (Guericke-father's cometary considerations), 243–49 (extract from Guericke-father's *Experimenta nova* and his letter), 249 (Guericke-father's letter).

114 Enclosures: TC 261–62 (Fornelius's letter), 271–72 (Fornelius's letter), 314 (Schefferus's letter); in TC3, the letters are mixed in with poems in Rautenstein's honour.

115 Enclosures: TC 366–69 (Müller's letter), 373–83 (observations), 388–93 (extract from the *Prodromus cometicus*), 949–50 (extract from the *Descriptio cometae*).

116 Enclosures: TC 688–90, 692–93 (observations).

117 Enclosures: TC 592–96 (observations).

118 Enclosures: TC 757–59 (Kircher's cometary considerations).

119 Enclosures: TC 937–41 (summaries of cometary observations).

Name	Timespan	From L.	To L.	Pages
Langius Hermannus	9.XI.1665–7.XII.1665	2	1	TC 581–83
Leunenschlos Johann von	28.II.1665–24.VIII.1666	5	2	TC 621–33
Ménage Gilles	19.IX.1665–27.II.1666	2	1	TC 859–61
Morsztyn Tobiasz	19.XII.1664–13.VII.1665	5	4	TC 505–19[120]
Müller Johann	19.XII.1664–25.II.1665	0	8	TC 843–54[121]
NiemiryczStefan[122]	28.III.1665–1.I.1667	6	6	TC 18–29, TC3 23–32
Noricus Johann Ernst[123]	29.XII.1666–9.IV.1667	3	3	TC3 32–40
Oldenburg Henry	20.VII.1666–13.IV.1667	4	3	TC 907–13
Olearius Adam	XII.1664–18.XI.1666	5	5	TC 635–40, 961–62[124]
Placentinus-Kołaczek Jan	19.XII.1664–18.V.1666	6	6	TC 549–74[125]
Rautenstein Johann Ernst von	17.XII.1664–2.I.1666	87	85	TC 39–208[126]
Raue Johann	28.II.1665–29.I.1666	3	1	TC 575–81
Reyher Samuel	15.XI.1665–12.VII.1667	8	8	TC 881–sig. [Vvvvv2 bis v], 965–66[127]
Riccioli Giovanni Battista	1.IV.1665–26.V.1666	15	4	TC 697–745[128]

120 Enclosures: TC 511–17 (extracts from Libavius's and Riccioli's books).

121 Enclosures: TC 848 (Philippus Miller's letter), 849–54 (observations).

122 In the TC labelled 'anonymous *ex Marchia*'. Identification is based on the fact that he was Lubieniecki's main contact in this part of Brandenburg.

123 Identification is based on the supposition that the anonymous *ex Norico* and *amicus Noricus*, identified in CHO p. 408, are the same person.

124 Enclosures: TC 638 (observations).

125 Enclosures: TC 555–61, 569–73 (Placentinus's cometary considerations).

126 Enclosures: TC 88 (Curtz's letter), 92 (a poem), 107–08 (observations).

127 Enclosures: TC sig. [Vvvvv2 bis v] (extract from Marius's *Mundus Iovialis*), 965–66 (extract from a lecture in Leipzig). Lubieniecki's first letter was sent not to Reyher but to the professor of mathematics in Kiel. Reyher answered it on 29 October 1666 when he took the chair of his late predecessor.

128 Enclosures: TC 705–28 (extract from Hevelius's *Prodromus cometicus*), 731–32, 736–37 and later mixed in with letters (chronological considerations).

Name	Timespan	From L.	To L.	Pages
Røtlin Johannes Melchior	19.XII.1664–13.I.1665	6	5	TC 415–28
Rudbeck Olof	25.VII.1665	1	0	TC 347–59[129]
Schott Gaspar	24.V.1665–9.VI.1666	8	6	TC 761–96[130]
Stegmann Joachim	24.XII.1664–20.IX.1665	4	2	TC 597–609[131]
Sivers Heinrich	18.XII.1664–28.II.1667	19	22	TC 913–30[132]
Vossius Isaacus	1.II–26.V.1667	2	0	TC 934–35
Wiszowaty Andrzej	25.IX.1665–25.X.1666	2	2	TC 610–20[133]
Ysing Conrad	undated	0	1	TC3 21–23
unidentified correspondents:				
anonymous from Paris	27.III–23.V.1665	1	1	TC 29–34
anonymous from Hamburg	24–25.V.1665	0	2	TC 34–38

129 Enclosures: TC 348–59 (Rudbeck's letter to Hensius). Lubieniecki's letter was addressed both to Fornelius and Rudbeck.

130 Enclosures: TC 762–90 (observations).

131 Enclosures: TC 598–601 (observations). There is one *communicatio* with both Stagmann and Wiszowaty; letters to Stegmann seem to be addressed both to him and to Wiszowaty.

132 Enclosures: TC 918–21, 923–95 (observations), 929–30 (about Mira Ceti).

133 Wiszowaty's letters were published in the Polish translation in 'Dwa listy Wiszowatego do S. Lubienieckiego', introduction by Zbigniew Ogonowski, transl. by Irmina Lichońska, *Studia i Materiały z Dziejów Nauki Polskiej*, 4 (1956), 347–66.

CHAPTER 4

Editorial Principles

The corpus of Hevelius's correspondence in the Bibliothèque de l'Observatoire in Paris covers almost the full scope of the Hevelius-Lubieniecki correspondence. It contains original letters received by the Danzig astronomer and drafts or copies of the letters he sent.[1] Letters he exchanged with Lubieniecki are in volumes C1, v. 6–11.

The printed edition of Lubieniecki's correspondence with Hevelius in the *Theatrum cometicum* covers about 2/3 of the whole volume. This edition exhibits some changes when compared to the letters from the Bibliothèque de l'Observatoire. Most of them are rather minor, such as the addition of names of Lubieniecki's contacts in Hamburg in Letter 1. Sometimes more important differences occur, as some of the letters were printed in two parts (Letters 12, 24, and 27), and two separate letters, or parts of letters, were printed as one letter (Letters 23 and 24).

The Bibliothèque nationale in Paris has the original manuscript of one of the letters of the correspondence (NAL 1641) (Letter 73). It belongs to the collection of reclaimed letters stolen by Guglielmo Libri (1802–69).[2] The Bibliothèque nationale also keeps the so-called *grande copie* of the whole volume of the correspondence. Hevelius's correspondence with Lubieniecki is in the volume Latin 10348. By a curious coincidence, it covers mostly the letters absent from the *Theatrum cometicum*. It is probable that these letters were in the missing part of the copy.[3] Being a copy of the manuscripts from the Bibliothèque de l'Observatoire, it usually follows its text.

1 For the history and the present state of the corpus of Hevelius's correspondence see Harald Siebert, 'De Dantzig à Paris', CJH1 pp. 148–78; Susan Keyes, 'Description of the manuscripts', CJH1 pp. 221–46 (discussion of the collection in the Bibliothèque de l'Observatoire pp. 221–32). The corpus of Hevelius's correspondence is accessible on-line at the Alidade (Accès en Ligne aux Instruments, Documents et Archives De l'astronomiE), alidade.obspm.fr [accessed 10 August 2019].

2 Keyes, 'Description of the manuscripts', p. 244. For the information about Libri and his actions see Susan Keyes, 'The thefts of Guglielmo Libri (1802–1869) and their consequences', CJH1 pp. 179–219.

3 For the information about Hevelius's letters in the Bibliothèque nationale in Paris see Keyes, 'Description of the manuscripts', pp. 232–38, 241–46. Hevelius's letters from the Bibliothèque nationale in Paris are accessible on-line at the Gallica, gallica.bnf.fr [accessed 10 August 2019].

Manuscript of one of the letters presented in this volume — Hevelius's letter to Boulliau about the comet of 1672 (Letter 83A) — is in the Muzeum Narodowe in Warsaw (ms 518/3).

As discussed above,[4] six of Lubieniecki's letters were printed by Johann Erich Olhoff in his collection of selected Hevelius's letters. They follow either the text printed in the *Theatrum cometicum* or the text of the manuscripts from the Bibliothèque de l'Observatoire.

A few of the enclosures to the correspondence in this volume, namely Lubieniecki's correspondence with Oldenburg, have appeared in a modern scholarly edition.[5] Some letters have also been published in the editions which are not of scholarly nature (see Table 3). These, however, have not been considered in this volume.

In most cases, the transmission of a letter contains at least two sources: either a manuscript from the Bibliothèque de l'Observatoire and printed letters from the *Theatrum cometicum*, or two manuscripts: from the Bibliothèque de l'Observatoire and the Bibliothèque nationale in Paris.

The letters in the present volume are arranged chronologically. It seems that all the letters of the Hevelius-Lubieniecki correspondence have been preserved, and there are no traces of missing letters.

The letters in this volume are dated in the Gregorian calendar ('new style') or with both Julian ('old style') and Gregorian calendars in the case of letters which were dated as such in the manuscript. Whenever only Julian date ('old style') was given, Gregorian date is added in square brackets. Hevelius used only the Gregorian calendar in his correspondence. Lubieniecki was not entirely consistent. Most often, he either used the Gregorian calendar or dated his letters with both Julian and Gregorian dates. Nevertheless, four times he used only the Julian calendar (Letters 13, 37, 78, and 82). Eight letters are dated without indicating the calendar (Letters 55, 59, 80, 87, 88, 89, 91, and 92). Given the frequency of Lubieniecki's use of calendars, it might be presumed that the dates of these letters are Gregorian. One of the letters is undated (Letter 68).

All preserved and identified enclosures have been included in this edition. This does not apply to a few cases in which Lubieniecki was only conveying to Hevelius letters written by others, for example by Martin Fogel (1634–75) (Letters 84 and 91). The exception was made here for Müller's letter to Hevelius, as it was referred to in Lubieniecki's letter to Hevelius and printed in the *Theatrum cometicum* (Letter 8A).

Whenever possible, the text of this edition was based on the letters from the Hevelian corpus in the Bibliothèque de l'Observatoire. Only two letters in the present volume are absent from the Bibliothèque de l'Observatoire: Letters 4 and 73 — the former is known from the *Theatrum cometicum* only, and the manuscript of the latter is in the Bibliothèque nationale in Paris, discussed above. The manifestation which

4 See section 2.2.5 of this Introduction.

5 *The Correspondence of Henry Oldenburg. Vol. III: 1666–1667*, ed. and transl. by A. Rupert Hall and Marie Boas Hall (Madison: University of Wisconsin Press, 1966).

has been the base for the transcription is given in bold in the transmission in the heading. Each letter which contained enclosures is followed by them.

Original spelling, capitalization, and punctuation have been preserved with the following exceptions: diacritics in Latin have been omitted; orthography of u/v has been standardized according to modern usage; the letter 'j' has been used at the beginning of the word before a vowel (except for 'i') and the letter 'i' has been used in all other places; titles of books have been written in italics and quotations have been placed in quotation marks. Astronomical symbols and other symbols have been translated into words. Abbreviations in honorific forms at the beginning and at the end of the letters, as well as the most often ligatures, contractions, and abbreviations, have been silently expanded. Other abbreviations have been expanded in square brackets. Page breaks in the manuscript manifestation which is the base for the transcription have been marked by '//'. Whenever a word in the manuscript was illegible, it has been marked as '[illegible]'.

The translation aimed for a text which is intelligible for a modern reader, but which remains close to the Latin original. Sometimes this has involved dividing the original, convoluted text into shorter passages or changing the syntactic order of a phrase. Proper names have been translated into their modern or standardized forms. Whenever a person has not been identified, a Latin form of their name has been left in the translation.

Footnotes to the original text contain the critical apparatus. Footnotes to the translation are editorial notes concerning the contents of the letters.

The critical apparatus omits insignificant discrepancies between manifestations, such as differences in orthography and punctuation. In a very few cases, variants absent from the manuscripts from the Bibliothèque de l'Observatoire have had to be introduced to the transcription. Headings and signatures were usually omitted in the *Theatrum cometicum*. With a few exceptions, this has not been noted in the apparatus.

Abbreviations in the apparatus:

in marg.	text in the margin
supr.	text above the main line
om.	text omitted in one of the manifestations
add.	text added in a different hand
del.	text deleted in manuscript
corr. ed.	text corrected by the editor

Table 3. Index of the letters of the Hevelius-Lubieniecki correspondence.[6]

No	Date	Sender	BO	TC	BnF	other sources	modern
1.	64 12 19	Lubieniecki	C1, v. 6, unnumbered (second after C1, v. 6, 907)	361			
1A.			C1, v. 6, unnumbered (first after C1, v. 6, 907)		FF. 13026, ff. 277r–v		
1B.			C1, v. 6, 899	843			
1C–1H.			C1, v. 6, 898	1 (1G.)			
2.	64 12 26	Lubieniecki	C1, v. 6, 910	362		Olhoff: 100	
2A–2D.			C1, v. 6, 911	1 (2A.), 2 (2C.), 238–39 (2D.)			
3.	64 12 28	Hevelius	C1, v. 6, 900	362–63			
4.	64 12 30	Lubieniecki		362			
4A.			C1, v. 6, 914	415–17			
5.	65 01 13	Lubieniecki	C1, v. 6, 913	363			
5A–5E.			C1, v. 6, 912	239 (5A.), 846–48 (5B–5E.)			
5F.			C1, v. 6, 911				
6.	65 01 27	Hevelius	C1, v. 6, 915	364			
7.	65 10 20	Lubieniecki	C1, v. 7, 949	364–65			
7A.			C1, v. 7, 950	488–90			
8.	65 02 27	Lubieniecki	C1, v. 7, 951	369–70			

6 Only Gregorian dates are given in the table. For enclosures, both the date and the sender are omitted. The table is based on: Susan Keyes, Klaus-Dieter Herbst, 'Chronological list of letters', CJH1 pp. 279–528, with some changes introduced after the analysis of the correspondence.

No	Date	Sender	BO	TC	BnF	other sources	modern
8A.			C1, v. 6, 1460 (part)	366–69			
9.	65 04 03	Hevelius	C1, v. 7, 952/A.80	370–71			
10.	65 04 07	Hevelius	C1, v. 7, 960	371			
11.	65 04 14	Lubieniecki	C1, v. 7, 972	371–74			
12.	65 04 24	Lubieniecki	C1, v. 7, 973	374–75, 375			
13.	65 05 06	Lubieniecki	C1, v. 7, 974				
13A.			C1, v. 7, 974	434			
14.	65 05 13	Hevelius	C1, v. 7, 975	375–77			
15.	65 05 19	Lubieniecki	C1, v. 7, 989	383–85			
16.	65 05 22	Lubieniecki	C1, v. 7, 998				
16A.			C1, v. 7, unnumbered (first before C1, v. 7 991)	214			
16B.			C1, v. 7, 991	214–15			
17.	65 05 26	Lubieniecki	C1, v. 7, 999	385–86			
17A–17D.			C1, v. 7, 1000	275–76 (17C.)			
18.	65 05 29	Lubieniecki	C1, v. 7, 1001	386			
19.	65 06 02	Hevelius	C1, v. 7, 991	393–94			
20.	65 06 02	Lubieniecki	C1, v. 7, 1002	386–87			
21.	65 06 05	Lubieniecki	C1, v. 7, 1008	387–88			
22.	65 06 10	Hevelius	C1, v. 7, 1003	396			
23.	65 06 12	Lubieniecki	C1, v. 7, 1009	394–95			
23A.			C1, v. 7, 1009	113			
24.	65 06 16	Lubieniecki	C1, v. 7, 1014	395–96, 396–97		Olhoff: 101	
25.	65 06 17	Hevelius	C1, v. 7, 1004	397–98			

No	Date	Sender	BO	TC	BnF	other sources	modern
26.	65 07 08	Hevelius	C1, v. 7, 1015	400			
27.	65 07 14	Lubieniecki	C1, v. 7, 1030, 1029	398–400, 400–01			Archenhold, p. 156 (part in German translation)
27A.–27C.			C1, v. 7, 1031	127 (27A.), 128 (27B.)			
27D.			C1, v. 7, unnumbered (between 1033 and 1034)				
28.	65 08 25	Lubieniecki	C1, v. 7, 1032	401–02		Olhoff: 102–03	
29.	65 08 28	Lubieniecki	C1, v. 7, 1033	402–03			
29A.			C1, v. 7, 1033	665–66			
30.	65 09 05	Hevelius	C1, v. 7, 1034	403			
31.	65 09 11	Lubieniecki	C1, v. 7, 1037	405–06			
32.	65 09 15	Lubieniecki	C1, v. 7, 1038	406			
32A.			C1, v. 7, 1038				
33.	65 09 29	Hevelius	C1, v. 7, 1039	407			
34.	65 10 06	Lubieniecki	C1, v. 7, 1070	407			
35.	65 11 17	Lubieniecki	C1, v. 7, 1071	407–08			
36.	65 11 28	Hevelius	C1, v. 7, 1073	408–09			
37.	65 12 08	Lubieniecki	C1, v. 7, 1072	409–10			
38.	66 01 29	Lubieniecki	C1, v. 7, 1093	410–11			
39.	66 02 13	Hevelius	C1, v. 7, 1094	411			
40.	66 02 26	Lubieniecki	C1, v. 7, 1097	412			

No	Date	Sender	BO	TC	BnF	other sources	modern
41.	66 04 02	Lubieniecki	C1, v. 7, 1098	412–13			
41A.			C1, v. 7, 1098	929–30			
42.	66 04 16	Lubieniecki	C1, v. 7, 1109	413			
42A.–42B.			C1, v. 7, 1109	544, 873			
43.	66 04 17	Hevelius	C1, v. 7, 1099	413–14			
44.	66 04 30	Lubieniecki	C1, v. 7, 1110	414			
45.	66 05 22	Hevelius	C1, v. 7, 1111	947			
46.	66 06 01	Lubieniecki	C1, v. 8, 1123	947			
47.	66 06 23	Hevelius	C1, v. 8, 1124	947–48			
48.	66 06 29	Lubieniecki	C1, v. 8, 1149	948		Olhoff: 113–14 (part)	
49.	66 07 24	Hevelius	C1, v. 8, 1140	948			
50.	66 09 10	Lubieniecki	C1, v. 8, 1152	948–49			
51.	66 10 20	Hevelius	C1, v. 8, 1153	951			
52.	66 11 02	Lubieniecki	C1, v. 8, 1160	951			
52A			C1, v. 8, 1160	picture no. 54 (after p. 884)			
52B.			C1, v. 8, 1160	885			
53.	66 11 09	Lubieniecki	C1, v. 8, 1161	951			
54.	66 12 18	Hevelius	C1, v. 8, 1162	951			
55.	67 01 25	Lubieniecki	C1, v. 8, 1185	952			
56.	67 02 08	Hevelius	C1, v. 8, 1186	952			
57.	67 02 18	Lubieniecki	C1, v. 8, 1201	952			

No	Date	Sender	BO	TC	BnF	other sources	modern
57A.–57B.			C1, v. 8, 1202	909–10 (57A.), 910–12 (57B.)		CHO: 303–04 (57A.), 323–26 (57B.)	
58.	67 03 19	Hevelius	C1, v. 8, 1203	953			
59.	67 03 22	Lubieniecki	C1, v. 8, 1205				
59A.			C1, v. 8, 1205	912		CHO: 345	
60.	67 04 12	Lubieniecki	C1, v. 8, 1206	953			
61.	67 01 15	Lubieniecki	C1, v. 8, 1207				
61A.			C1, v. 8, 1207	912		CHO: 345	
62.	67 05 06	Lubieniecki	C1, v. 8, 1208	953			
62A.			C1, v. 8, 1208	[Vvvvv2 bis r]			
63.	67 05 31	Lubieniecki	C1, v. 8, 1214	953–54			
63A–63B.			C1, v. 8, 1214	[Vvvvv2 bis r–v] (63B.)		TC2: 438 (63A.)	
64.	67 06 07	Hevelius	C1, v. 8, 1209	954			
65.	67 07 26	Hevelius	C1, v. 8, 1215	954			
66.	67 07 26	Lubieniecki	C1, v. 8, 1217	954			
66A.			C1, v. 8, 1217			TC2: 132, 447	
67.	67 08 02	Lubieniecki	C1, v. 8, 1218	954–55			
68.	undated	Lubieniecki	C1, v. 9, 1276,		Lat 10348, v. 9, 41–42		
68A.			C1, v. 9, 1276	945	Lat 10348, v. 9, 41		
69.	67 08 10	Hevelius	C1, v. 8, 1219	955			

No	Date	Sender	BO	TC	BnF	other sources	modern
70.	67 08 16	Lubieniecki	C1, v. 9, 1275	955	Lat 10348, v. 9, 39–40		
70A.			C1, v. 9, 1276	960	Lat 10348, v. 9, 40		
71.	67 08 19	Lubieniecki	C1, v. 9, 1277		Lat 10348, v. 9, 42		
72.	68 06 06	Hevelius	C1, v. 8, 1233, C1, v. 9, 1278		Lat 10348, v. 9, 42–43		
73.	68 06 19	Lubieniecki			NAL 1641, 78r, Lat 10348, v. 9, 170		
74.	69 03 12	Lubieniecki	C1, v. 9, 1360		Lat 10348, v. 9, 168–69		
75.	69 04 20	Hevelius	C1, v. 9, 1362		Lat 10348, v. 9, 170–71		
76.	69 06 25	Lubieniecki	C1, v. 9, 1358		Lat 10348, v. 9, 221–22		
77.	69 07 12	Lubieniecki	C1, v. 9, 1359		Lat 10348, v. 9, 222–23		
78.	71 05 02	Lubieniecki	C1, v. 10, 1447/81		Lat 10348, v. 10, 145–47		
79.	71 08 04	Hevelius	C1, v. 10, 1448/84		Lat 10348, v. 10, 149–50		

No	Date	Sender	BO	TC	BnF	other sources	modern
80.	71 10 12	Lubieniecki	C1, v. 11, 1481/6		Lat 10348, v. 11, 13–14		
81.	71 12 05	Hevelius	C1, v. 11, 1482/7		Lat 10348, v. 11, 14–15		
82.	72 01 19	Lubieniecki	C1, v. 11, 1443/19		Lat 10348, v. 11, 29–31		
83.	72 03 11	Hevelius	C1, v. 11, 1494/20		Lat 10348, v. 11, 31–32		
83A.			C1, v. 11, 1490/17		Lat 10348, v. 11, 27–28	MNW: ms 518/3	Bersohn, pp. 20–21
83B.			C1, v. 11, 1491		Lat 10348, v. 11, 28–29		
84.	74 04 05	Lubieniecki	C1, v. 11, 1518/46		Lat 10348, v. 11, 59–61	Olhoff: 131–32	
84A.			C1, v. 11, 1518/46, C1, v. 11, 1519/47		Lat 10348, v. 11, 62–63		
85.	72 06 11	Hevelius	C1, v. 11, 1520/48		Lat 10348, v. 11, 63		
86.	72 07 22	Lubieniecki	C1, v. 11, 1537/67		Lat 10348, v. 11, 93–94		
87.	73 01 24	Lubieniecki	C1, v. 11, 1572/86		Lat 10348, v. 11, 125–26		
88.	73 04 24	Lubieniecki	C1, v. 11, 1582/91		Lat 10348, v. 11, 128–29		

No	Date	Sender	BO	TC	BnF	other sources	modern
89.	73 05 02	Lubieniecki	C1, v. 11, 92		Lat 10348, v. 11, 129		
90.	73 08 29	Hevelius	C1, v. 11, 1596/111		Lat 10348, v. 11, 158–59		
91.	73 09 08	Lubieniecki	C1, v. 11, 1610/128		Lat 10348, v. 11, 182		
92.	73 12 22	Lubieniecki	C1, v. 11, 1623/141		Lat 10348, v. 11, 198–200	Olhoff: 137–38	Archenhold, p. 156 (part in German translation)

Modern sources:

Archenhold Friedrich Simon Archenhold, 'Johannes Hevelius. Ein Gedankblatt zum 300. Geburtstag', *Das Weltall*, 11 (1911), no. 10, 137–60.

Bersohn Mathias Bersohn, *Kilka słów o Janie Heweliuszu, astronomie gdańskim z wieku XVII-go oraz o jego korespondencyi* (Warszawa: druk P. Laskauera & W. Babickiego, 1898).

CORRESPONDENCE

1.

9/19 December 1664, Lubieniecki to Hevelius

BO: C1, v. 6, unnumbered (second after C1, v. 6, 907)
TC: 361

Enclosures:
1A. Letter from Boulliau to Lubieniecki, 5 December 1664
1B. Letter from Müller to Lubieniecki, 19 December 1664
1C–1G. Observations of the comet in early December 1664
1H. Excerpt of a message from Hamburg

Salutem et Officia

Nobilissime, Amplissime, Humanissime Vir. Te mihi non tantum de fama nominis clarissimi per orbem,[1] sed et de facie ac scriptis Tuis, quibus Astronomiam, illam maxime sublimem Philosophiae partem, egregiis scriptis illustras, notum esse, vehementer gratulor. Tibi pariter ut notus fiam, studium literarum humaniorum sed et commercium literarum, quod cum doctis hic illic exerceo, facit. Alliciunt et celebratissimae virtutes Tuae, praesertim singularis humanitas cum insigni et excellenti rei Astronomicae scientia, qua itidem delector, coniuncta. Nolim tamen meum literarum Tecum commercium, quod ambio et a Tua Humanitate mihi polliceor, aliis innotescere ex causis quas scio esse justas.[2] Nosti quae vincula Viros literarum bonarum amantes coniungant. Patere itaque et Tu, Vir Clarissime, Tibi haec aptari, ut quantum per seria Tua negotia et sublimia studia licebit, mecum fido literarum privatarum vinculo jungaris, et sic etiam Clariss[imo] Ismaeli Bullialdo, Amico meo, et ut puto, Tuo quoque, Integerrimo.[3] Cum quo cum assiduum literarum colam commercium, quod et nuperae epistolae exscriptum exemplum hic adiunctum docet, si et Tu idem desideres, vel, quod potius credo, possideas et exerceas, offero Tibi meam qualemcunque operam in literis vestris mutuo curandis.[4] Accedit his quae me ad scribendum impulerunt rationibus, singulare incentivum Cometae, quem ante aliquot dies observavimus,[5] de quo quid nonnulli Amici mei, Viri in hac urbe eruditione praestantes ante biduum judicaverint, adiuncta edocebit pagina.[6] Velim mihi

1 de...orbem : de fama per Orbem clarissima **TC**.

2 aliis...justas : Tibi aliis, iisque gravioribus, studiis et negotiis occupato, negotium et molestiam facessere **TC**.

3 et ut...Integerrimo : juxtaque Tuo singulari et integerrimo **TC**.

4 operam...curandis : in literis vestris mutuo curandis operam, fidam certe et industriam. **TC**.

5 observavimus : observavimus nudis tamen oculis **TC**.

6 de quo...pagina : De quo quid a nonnullis Amicis meis hac in Urbe degentibus, Viris eruditione praestantibus (Joan[ne] Alb[erto] Huswedelio, Joanne Garmers, Philosophiae et Medicinae Doctoribus,

de his tuam sententiam aperias, quam si velis cum Viris Doctis in hac urbe degentibus,[7] quibus cum // assiduo ore et scripto loquor, et cum Doctiss[imo] Bullialdo communicabo; sin autem, integram silentii fidem, pro jure amicitiae, servabo. Literas Tuas velim mihi inscriptas, speciali involucro, quod nomen Domini Martini von der Eyghen praeferat, tegas. Vale Vir Excellentissime et age feliciter diuque bono publico et me favore Tuo peosequi ne dedignare, qui sum

Clarissimo Nomini Tuo
omnibus studiis et officiis addictissimus

De Rolis
manu propria

Post Scriptum. Imminente novo Anno, precor, Vir Eximie, ut exeuntem annum feliciter finias et novum, qui appropinquat, prospere inchoes et peragas, et in plures amplius annos bene valeas. Tum ut omnes homines, qui praesertim Christiano nomine censeri gaudent, hoc igne caelesti admoniti cum novo anno novam vitam agere velint, judicia Dei horrenda, quae tales ignes praenunciare solent, effugiant!

Hamb[urgi] 9.19. Dec[embris] 1664

Iterum Post Scriptum. Mitto et quae Brema de Cometa jam jam accepi: ut et literas Clariss[imi] Mülleri, quas mihi remitti suo tempore peto.[8]

Greetings

Most noble, greatest, and most gentle Sir! I greatly rejoice because I know you not only by your name (which is famous throughout the world), but also by your countenance and excellent writings with which you illuminate astronomy — the most sublime part of philosophy. Studies in the noble arts and my correspondence — which I conduct with learned men from various places — may make me equally known to you. Your renowned virtues are likewise appealing, especially your extraordinary kindness entwined with exceptional and outstanding knowledge of astronomy, which also delights me. I would like our correspondence (which I seek and for which I hope, knowing your kindness) to remain secret due to some reasons which I know

et Inclytae huius Reipublicae Physicis Ordinariis: tum Joanne Müllero Matheseos Professore) coram et scripto accepi, accipe et Tu paucis. **TC**.

7 quam si velis...degentibus : quam libenter, cum viris illis doctrina et humanitate praestantibus **TC**.

8 sin autem...suo tempore peto : Vale, Vir Excellentissime, et age feliciter diuque, bono publico, et me favore Tuo prosequi de dedignare. Ad Dominum Hevelium Reip[ublicae] Gedan[ensis] Senatorem et Mathematicum Excellentissimum **TC**.

to be justified. You know what bonds men who like good letters. So you as well, most learned Sir, if only your serious obligations and sublime studies allow you, can begin a friendly exchange of letters with me and with the most renowned Ismaël Boulliau,[9] the surest friend of mine and, as I suppose, of yours too. Because I conduct continuous correspondence with him, which is also testified by the last letter enclosed to this one, if you wish the same, or — what I find more probable — you have already started the same, I offer you my small help in passing your letters. On top of these reasons inciting me to write, there is one more extraordinary incentive, viz. the comet which we observed a few days ago.[10] What some of my friends — who excel in erudition in this city[11] — have said about it, is given on the enclosed page. I would like you to give me your opinion, which (if you wish) I shall pass to the learned men who live in this city, and with whom I have both personal and epistolary contacts, and to the most learned Boulliau as well. But, if so is your wish, I will remain silent, as is demanded by the principles of friendship. I would like you to hide your letters written to me in particular disguise bearing the name of Master Martin von der Eyghen.[12] Farewell, most excellent Sir, and act successfully and long for the public good, and do not refuse to grant your favour to me, to

renowned Sir,
Yours most humbly devoted,

De Rolis
by my own hand

Postscript. Since the new year is near, I wish you, excellent Sir, a joyful ending of this year and an auspicious beginning and the rest of the new one which is approaching; and may you be in good health for many years to come. May all people, especially those who are happy to be called by the name of Christians,[13] afraid of the heavenly fire, begin a new life in the new year, and that the horrible judgements of God (usually announced by such fires) vanish.

In Hamburg, on 9/19 December 1664

9 See section 2.4.2 of the Introduction.

10 See section 1.2.2 of the Introduction.

11 Although only one of the enclosed observations from Hamburg bears the name of the observer (Letter 1B), Lubieniecki refers here to the people he listed in the printed copy of this letter in the TC: Johann Albert Huswedel (1618–74, German physician; in 1672, after an invitation from Charles XI, he moved to Stockholm), Johannes Garmers (1628–1700, German physician), and Johann Müller (see section 2.4.11 of the Introduction).

12 A citizen of Hamburg in whose house Lubieniecki and his family lived since 1664, see Janusz Tazbir, *Stando lubentius moriar. Biografia Stanisława Lubienieckiego* (Warszawa: Iskry, 2003), p. 125.

13 The Polish Brethren called themselves Christians and emphasised that all other names for them were coined by others.

Second postscript. I am also sending what I have just received from Bremen about the comet, together with a letter from the most famous Müller,[14] which I would like to be sent back to me.

1A.

Letter from Boulliau to Lubieniecki, 5 December 1664

BO: C1, v. 6, unnumbered (first after C1, v. 6, 907)
BnF: FF. 13026, 277r–v

Lutetiae Parisiorum, die 5. Dec[embris] 1664[15]

Non multum hic occurrunt nunc scribenda. Febri continua, paroxysmis tertianae duplicis asperiori facta, hactenus Regina laboravit: quanto moerore omnium vix dici potest, cum clementiam ac summas illius virtutes cuncti venerentur, ac immaturo fato ne abripiatur nobis, metuant. Vinum emeticum ipsi exhibere Medici decreverant: verum ipsa haurire noluit. Assiduis precibus, ut illam Deus nobis servet, omnes poscimus, speramusque morbum remissurum, ipsamque incolumem evasuram.

Dominus Fischkarding Angliae Magnatum unus donis pretiosissimis a Rege donatus, in Angliam renavigavit. Ad Regem Angliae Rex noster Dominus de Ruvigni vicissim misit. De renovandis foederibus nunc agitur. Novas illis conditiones appositas Angli vellent. Nos pristinis ipsos acquiescere cupimus. Ubi disceptatum erit, quid decernetur, ab utrisque rescibimus. Illi Domino Magnati Anglo,[16] praeter pactiones de non largiendis utrinque Legatis vel Ablegatis donis factas, munera Rex noster largitus est: an Anglus id vicissim sit imitaturus, incertum.

Comes Dietrichstenius Caesaris ad Regem gratiarum agendarum causa ablegatus, salutatis Rege, Regina Matre et Delfino cistulam adamantibus ornatam intusque Regis depicta imagine pretiosiorem factam, decem flor[enorum] millibus aestimatam dono accepit. Fulcquetius tandem[17] super omnibus accusationum capitibus interrogatus a

14 See section 2.4.11 of the Introduction.

15 Lutetiae...1664 : Generosimo Viro Domino Stanislao Lubieniesio de Lubienietz Ismael Boulliau Salutem Plurimam Dicit. Hac Septimana literas a te accepi Novemb[ris] 15 scriptas (sciptas *del.*) datas notulamque eiusdem mensis die 22. scriptam, adiunctae erant chartae plures quibus multa (multa *del.*) rerum praesentium varii casus et eventus describuntur; de omnibus tibi gratias ago; ad te porro cum mihi per alias occupationes licet, scribere non praetermitto nec frequentius hoc praestare possum. **BnF**.

16 Domino...Anglo : Domino Fischkarding **BnF**.

17 Fulcquetius tandem : Tandem Fulcquetius **BnF**.

Judicibus fuit, hodieque[18] ultimus Judicum consessus est hanc ob causam habitus. De sententiis ferendis et damnando aut absolvendo reo agitur. Factio valida reo favet, et quid decreturi sint Judices, extra coniiciendi dextre terminos positum.

De pace aut ad longos annos inter Hispanum et Lusitanum Reges induciis pactis nuncii crebri ad nos allati, ut compositas esse credamus, faciunt. //

Cum Summo Pontifice concordes nunc vivimus, et Avenionensium adversus Vice Legatum seditio pacem Pisis initam non corrumpet. Rex tamen, ut Summus Pontifex cum illis agat clementer ac benigne,[19] laborat.

Amicum Nostrum Illustriss[imum] N[otum][20] ac R[everendum] P[atrem] Curtium, quorum nomine mihi pl[urimam] salutem adscribis[21], officiosissime resaluto[22], quibus de eorum erga me benevolentia valde sum obstrictus. Ab utroque responsum expecto[23].

In Paris, on 5 December 1664[24]

This time, there is not much to write about. The Queen[25] is still ill, since lasting fever[26] has become more severe because of paroxysms of accompanying tertian fever.[27] It is a thing not easy to talk about because of everyone's sorrow, since all adore her kindness and supreme virtues, and are afraid that she could be taken from us too soon. Doctors had decided to give her emetic wine, but she did not want to drink it. We all ask God in constant prayers that He would save her for us. We hope the sickness will remise and that she will fully recover.

18 hodieque : et hodie **BnF**.

19 agat...benigne : clementer ac benigne agat **BnF**.

20 N[otum] : de Rautenstein **BnF**.

21 quorum...adscribis *om.* **BnF**.

22 resaluto : saluto **BnF**.

23 Ab utroque...expecto : Tibi etiam omnia fausta precor. Vale et cetera. 5 Decemb[ris] 1664 Paris[iis]. **BnF**. Cum Domino Broussel omnes chartas communicavi. *in marg.* **BnF**.

24 This letter is an answer to Lubieniecki's letters and news accounts sent to Boulliau on 15 November 1664 (BnF FF. 13031, 34r–40v) and 22 November 1664 (BnF FF. 13031, 41r–42v).

25 Anne of Austria (1601–66), the wife of Louis XIII, Queen consort of France in the years 1615–43, Queen Mother since 1643, and Queen Regent of France in the years 1643–51 during the minority of her son, Louis XIV. See also CJH2 p. 179, note to Letter 22A, Chapelain to Hevelius, 2 June 1665.

26 Boulliau refers here to *febris continua*, a type of fever whose symptoms do not change significantly before it is entirely cured. See discussion in Bartolomeo Castelli, *Lexicon medicum Graeco-Latinum* (Basileae: impensis Joh. Jacobi Genathi, 1628), pp. 85–86.

27 'Tertian fever' (*febris tertiana*) has symptoms which intensify every three days. See Castelli, *Lexicon medicum Graeco-Latinum*, p. 156.

Master Fitzharding,[28] one of the English aristocrats who had been given precious gifts by the King, sailed back to England. In return, our King has sent Master de Ruvigny[29] to the King of England.[30] The prolongation of treaties is now being discussed. The English would like to add some new conditions. We want them to be content with the previous ones. When the negotiations are over, we will ask both sides about the decision. Against prior agreements regarding not giving gifts to emissaries and envoys, our King gave presents to that English aristocrat. It is uncertain whether the English King will do the same.

Count Dietrichstein[31] (who was sent by the Emperor[32] to express his gratitude to the King), having greeted the King, the Queen Mother, and the Dauphin,[33] was given a casket worth ten thousand florins, decorated with diamonds and — to enhance its value — with the image of the King inside. Fouquet[34], at last, was interrogated by the judges about all the points of the accusation, and today is the last sitting of the court. The sentence is about to be pronounced, and he will be pleaded either guilty or innocent. The leading faction supports the accused, and it is beyond doubt that the judges' decision will be rightful.

As far as the peace, or long-term ceasefire, between the Kings of Spain[35] and Portugal[36] is concerned, many emissaries have made us believe it was signed.[37]

28 Charles Berkeley (1630–65), 1st Viscount Fitzhardinge of Berehaven (created 1663), 1st Earl of Falmouth (created 1664), cavalry officer. In November 1664, King Charles II sent him on a confidential mission to France.

29 Henri de Massue, 1st Marquis de Ruvigny (*c.* 1605–89), French diplomat and the ambassador to the English court (1674–77). After retiring in 1685, he stayed in England. His mission to England concerned alliances before the Anglo-Dutch war.

30 Charles II (1630–85), King of Scotland 1649–51, King of England, Scotland, and Ireland since 1660.

31 Most probably Ferdinand Joseph von Dietrichstein (1628–98), courtier and Privy Councillor of Emperor Leopold I.

32 Leopold I Habsburg (1640–1705), Holy Roman Emperor since 1658.

33 Louis (1661–1711), the oldest son of Louis XIV, known as the 'Great Dauphin.'

34 Nicolas Fouquet (1615–80), French Superintendent of Finances in the years 1653–61. In 1661 he was accused of improper administration of the state budget and after a three-year trial he was condemned to banishment. However, Louis XIV changed the sentence to life imprisonment. See also CJH2 p. 151, note to Letter 9, Chapelain to Hevelius, 28 June 1663; CJH3 p. 359, note to Letter 115, des Noyers to Hevelius, 13 March 1665.

35 Philip IV (1605–65), King of Spain since 1621, King of Portugal between 1621–40. See also Chantal Grell, 'Introduction', CHJ2 p. 95; CHJ2 p. 223, note to Letter 41, Chapelain to Hevelius, 15 May 1667.

36 Alfonso VI (1643–83), King of Portugal since 1656.

37 The Spanish-Portuguese war (known as the 'Restoration War'), which began in 1640, ended in 1668 with the treaty of Lisbon. In the treaty, the Habsburg King of Spain recognized the independence of Portugal under the rule of the House of Braganza.

We now live in peace with the Pope[38], and the Avignon sedition against the vice-legate[39] will not break the peace made in Pisa.[40] The King, however, tries to persuade the Pope to deal with them clemently and gently.

I kindly return the greetings to our friend, the most famous one whom you know, and to the Reverend Father Curtz[41] whose greetings you sent to me. I am obliged to them for their benevolence towards me. I expect answers from both of them.

1B.

Letter from Müller to Lubieniecki, 9/19 December 1664

BO: C1, v. 6, 899
TC: 843

Domino De Rolis. Hamburgi.

Vir Excellentissime et Studiosissime Salve

De Cometa nuper exorto observat[iones] Mathematico dignas proferre nequeo, quia Instrumenta his speculationibus necessaria mihi desunt. Interim tamen die 4/14 Decembris mane hora tertia prima vice eum vidi exorientem, et in gradu 6 Librae existere taliter qualiter observavi, cauda existente admodum prolixa; ipse vero Cometes ei stellae, quae est in rostro Corvi quasi adhaerebat, paulo tamen illa borealior.

Secunda vice vidi eundem Cometam ascendentem supra horizontem die 7/17 decembris (intermediis enim noctibus caelum erat nubilum) mane ante horam tertiam, ubi eius amplitudo ortiva erat 40 plus minus graduum et quidem Australis. Idem tum Australior factus longius a rostro Corvi recesserat, ut recta jungens rostrum Corvi cum Cometa videretur, dupla rectae connectentis stellas in collo et rostro dicti Corvi.

Tertia vice Cometa mihi fuit conspectus hesterno mane[42] ubi jam adhuc longius abierat a rostro Corvi versus Meridiem, ita ut recte ante dictae habere viderentur ad invicem proportionem duplam sesquialteram, hoc est ut 5 ad 2. Cauda apparebat longior.

38 Alexander VII (Fabio Chigi, 1599–1667), Pope since 1655.
39 In 1664, there was an unrest in Avignon against the papal vice-legate, Alessandro Colonna.
40 The treaty of Pisa, signed in 1664, ended the hostility between the King of France and the Pope. The papacy yielded to French demands, including the royal privilege of appointing bishops.
41 See section 2.4.5 of the Introduction.
42 mane : mane die 8. 18. Decembris quae erat Jovis **TC**.

Hoc mane caelum non fuit serenum.

Haec sunt, Vir Excellentissime[43] Domine et Fautor pl[urimum] honorande, quae cum Dominis Hevelio et Bullialdo, Astronomis summis et Amicis meis nunquam non venerandis, Tuo nomine communicare poteris. Eosdem etiam viros summos salvere per literas rogo, et simul nunciare observationes // circa Cometam a me habitas examine tantorum Virorum non esse dignas, quia Machinis necessariis nostrum adhuc caret Gymnasium. Vale, Divinae Tutelae etiam atque etiam commendatus.

Dabam hic Hamburgi, die 9/19 Decemb[ris] Anno 1664

Tuo Clarissimo Nomini
addictissimus

Joh[annes] Mullerus

To Master De Rolis. In Hamburg.

Greetings, most excellent and learned Sir!

I cannot provide you with the observations of the latest comet which could be suitable for a mathematician since I do not have the necessary instruments for such inquiries. Nevertheless, I believe I saw it for the first time rising on 4/14 December, at 3 o'clock in the morning, and it was in 6° of Libra, and its tail was very wide. The comet itself seemed to stick to the star which is in the Beak of Corvus, but slightly more towards the north.

The second time I saw it, the comet was ascending over the horizon on 7/17 December (on the days between, the sky was clouded at night) in the morning, before 3 o'clock, when its height was about 40° south at the time of appearing. It became more southern and moved away from the Beak of Corvus, so the line between its beak and the comet seemed to be twice longer than the line which connects the stars in the Neck and the Beak of the aforementioned Corvus.

I saw the comet for the third time yesterday morning when it had already moved further away from the Beak of Corvus towards the south, in such a way that the aforementioned lines seemed to have a 'dupla sesquialtera' proportion to each other, which is 5 to 2. The tail appeared longer.

That morning the sky was not clear.

43 Excellentissime : Praecellentissime **TC**.

That is all what you, most excellent Sir, Master, and most honourable benefactor, can convey in your letters to Masters Hevelius and Boulliau, the best astronomers and my friends, for whom I always have great respect. I also ask you to send my greetings in the letters to these great men and to inform them that my observations of the comet are not worthy of their opinion, since our school still lacks necessary instruments. Farewell, you who are always entrusted to divine protection!

Given at Hamburg, on 9/19 December 1664.

Yours devoted,

Johann Müller

1C.

Observations of the comet in early December 1664

BO: C1, v. 6, 898[44]

INFORMATIONES ET OBSERVATIONES
virorum Doctorum De cometa insigni
qui primis diebus m[ensis] Dec[embris] S[tylo] V[etere]
Anno 1664 apparere
caepit

[illegible] Hamburg, d[ie] 7 Dec[embris] S[tylo] V[etere] 1664

Den 3te dieses verzähleten etzliche Soldaten wie Sie beÿ gehaltener Wacht morgens zwischen 4. und 5. Uhr einen ungewöhnlichen Stern, alß einen Cometen gesehen haben. Dieses verursachte mich und viel andere Sontags früe umb 5. Uhren hinn umb Sued-Osten einen solchen ungewöhnlichen Stern observiren. Sein Schwantz vergleich sich einen Pfauenschwantz und streikte Er solchen nach den Westen Nach 5. Uhr, da mit dem annoch übrigen Mondenlicht auch ein nebel entstunden[45] kann Er uns auß den augen den 5. Uhr und 6. kundte Er wegen dicker schneelufft nicht gesehen werden. Man verlanget zu vernehmen was von andern orten hiervon einkommet.

44 Each part of the enclosure is printed separately here.
45 [illegible] *del. post* entstunden **BO**.

Information and observations of the notable comet which started to appear in the first days of December 1664, old style, by the learned men

Hamburg, on 7 December 1664, old style

On the third of this month, some soldiers recounted that they had seen an extraordinary star (resembling a comet), when they were standing their guard, between 4 and 5 o'clock in the morning. It encouraged me, and many others, to observe that extraordinary star in the south-east, on early Sunday morning, at about 5 o'clock. Its tail was similar to that of a peacock, and it was stretching westwards until 5 o'clock; then, when the moonlight faded and the mist appeared, we could not see the comet with our eyes at 5 and 6 o'clock because of a dense snowfall. One would wish to know what news about it will arrive from other places.

1D.

Observations of the comet in early December 1664

BO: C1, v. 6, 898

2 Hamburgi d[ie] 7. Dec[embris] 1664

Heri a Te rogatus tribus verbis ad Te perscripsi ea quae ab aliis de Cometa a me nondum viso acceperam: conspici eundem in signo Cancri, neque in meridiano ultra 13. gradum elevari. Sed cum nocte praeterita caelum esset sudum, adscendi ego una cum pluribus aliis editam aedis Cathedralis turrim ferens mecum globum et alia quaedam huic instituto peragendo necessaria. Deprehendi itaque Cometam illum in signo Librae medio inter Corvum et Hydram loco, atque primum quidem cum sole ad occasum ferri, sed et videri, quasi peculiarem habeat motum, quem tamen qualis sit deprehendere nondum potui. Caeterum putatis rarionibus, eum ultra 18. gradum in meridiano esse deprehendi, ut et caudatum et quidem cauda visus ratione aliquot ulnas longa, non crinitum nec barbatum esse. Caetera mox investigabimus, favente caelo, quorum te participem reddam. His interim contentus vale.

2. In Hamburg, on 7 December 1664

You asked me yesterday about the comet; I will briefly describe what I learned about it from others (though I have not seen it yet): it is visible in the sign of Cancer, and it is not exceeding 13 degrees in the meridian. But last night, when the sky was clear,

I (along with many others) climbed the high tower of the cathedral; I had a globe and other instruments necessary for this purpose. And then I saw that comet in the sign of Libra, placed in the middle between Corvus and Hydra; initially, it was going westwards with the Sun, but it seemed to have had its own movement as well which I have not been able to determine so far. Moreover, having made some calculations, I noted that it was rising over 18 degrees in the meridian, and it had neither hair nor beard, but a tail which seemed a few ells long. We will soon make further observations, if the heavens allow it, and we will inform you about them. Meanwhile, be satisfied with these, and farewell!

1E.

Observations of the comet in early December 1664

BO: C1, v. 6, 898

3. Hamburgi d[ie] eodem

Cometam de quo certior fieri cupis, scito me vidis-//se nocte, quae fuit inter diem Saturni et solis, hora tertia et sequentibus usque ad Ortum Solis in Sydere quidem Caelesti, quod vocatur Corvus in Dodecatemorio Librae. Caudam habet satis prolixam et longam decem pedes oculorum judicio adaequantem. Hoc mane eundem vidi, sed cauda breviori quod tamen tempestati, non decremento aut immunitioni ipsius caudae tribuendum puto, imo credo. In plaga coeli Orientali horis matutinis, quoties tempestas favet, optime conspicitur et adhuc per decendium duraturum eundem satis tuto crediderim, nisi forte adhuc diutius.

3. In Hamburg, on the same day

Know that I saw that comet (about which you would like to learn) on the night from Saturday through Sunday, at 3 o'clock and in the consecutive hours until the sunrise, in the celestial constellation of Corvus in the sign of Libra. It has quite a broad tail which seems to be 10 feet long by the judgement of my eyes. This morning, I saw the same comet, but its tail was shorter; I think (and even believe) that it was caused by the weather and that the tail was not diminished or shortened. The comet can be seen in the eastern part of the sky, in the morning, whenever the weather allows it, and I believe that it will last for at least ten days, if not longer.

1F.

Observations of the comet in early December 1664

BO: C1, v. 6, 898

Hamburgi d[ie] 8 Dec[embris] 1664

Observatum est Cometam hunc proprio motu praeter illum cum universo communem ad Occasum cuncta vi primi mobilis rapientem gaudere. Apparuit enim primum in rostro Corvi, unde progressum Hydram versus fecit notabilem. Caudam habet bene longam estque ipse praegrandis, ac illum insignem qui Anno huius seculi XVIII arsit nec certe impune visus fuit magnitudine aequat nisi et superet. Verum tamen non est ita lucidus et clarus sicut ille fuit.

In Hamburg, on 8 December 1664

It was observed that this comet had its own movement apart from that common one — the one together with the universe — which carries everything westwards with the force of the primum mobile. At first, the comet appeared in the Beak of Corvus whence it passed a significant path towards Hydra. It has a very long tail, and the comet itself is enormous; its size equals (if not surpasses) the size of that famous comet which was shining in the eighteenth year of this century[46] (and whose appearance was not without consequences). But this one is not as bright and clear as the other one was.

46 There were three cometary phenomena in 1618: C/1618 Q1 (seen in August and September), C/1618 V1 (seen in November and December), and C/1618 (seen from November 1618 through January 1619), see Alexandre Guy Pingré, *Cométographie ou Traité historique et théorique des comètes*, vol. 2 (Paris: Imprimerie royale, 1784), pp. 4–9; Gary W. Kronk, *Cometography. A Catalogue of Comets. Volume 1: Ancient – 1799* (Cambridge: Cambridge University Press, 1999), pp. 333–41. See also 'Subsidia de cometis', CJH2 p. 464, and Chantal Grell, 'Introduction', CJH3 p. 21.

1G.

Observations of the comet in early December 1664[47]

BO: **C1, v. 6, 898**
TC: 1

Bremae die 17/7 Decembris 1664

Vere nobis novus cometa lucet, cauda satis lata et longa, colorem non adeo accurate observare licuit ob nebulosam circa horizontem caelum. Hydram occupavit sub et ante rostrum corvi hydrae insidentis. Ortus est ante horam 3. matut[inam] plaga inter Meridiem et Orientem media. Meridiem tenuit hora 6. matut[ina], sic subsequentibus diebus hora 4. aut 5. aut 6. videri coelo sereno poterit. Vicinitas horizontis, vix nimirum ultra 10. gradus ascendit, efficit plerumque ut nonnisi e nebulis eluceat. Locis quo meridionalibus, eo magis super verticem elevatur.

In Bremen, on 17/7 December 1664

The new comet is indeed shining with quite a long and wide tail; its colour could not be precisely determined, because the sky near the horizon was clouded. It was in Hydra, under and rising before the Beak of Corvus, which is near Hydra. It rose at about 3 o'clock in the morning between the south and the east. It was in the south at 6 o'clock in the morning, so in the following days it will be possible to see it at 4, 5, or 6 o'clock, if the sky is clear. Its proximity to the horizon (it ascends only slightly more than 10°) makes it shine usually through the mist. The more southern its place is, the more it rises towards the top.

47 The same text appears in 2A.

1H.

Excerpt of a message from Hamburg[48]

BO: C1, v. 6, 898

Extrait d'une Gazette de Hambourg le 7 decembre 1664.

Quelques Soldats comterent le 3 de ce mois qu'ils avoient vue pendant qu'ils etoient de garde le matin entre 1 et 2 heures une etoile extraordinaire, comme une Comete. Ce qui fait que moi et plusieurs autres avons observée ici le Dimanche[49] matin à 5 heures vers Sud-Est la dite etoile extraordinaire. Sa queue etoit comme une queue de Pan, et elle[50] etoit placée vers l'Oest après 5 heures, outre la clarté de la Lune il y avoit encore de nues qui nous empechoient de la[51] voir, on ne pouvoit plus la voir[52] entre 5. et 6. heures, à cause que l'aire etoit plein de neges. On est inpassiant de savoir ce qu'on saura[53] de cela d'ailleurs.[54]

Excerpt of a message from Hamburg, 7 December 1664.

On the third of this month, some soldiers recounted that they saw an extraordinary star, like a comet, when they were standing their guard, between 1 and 2[55] in the morning. It made me and many others observe that extraordinary star on early Sunday, at about 5 o'clock, in the south-east. Its tail reminded that of a peacock, and it was placed westwards until 5 o'clock; then, when the clouds arose together with the moonlight, we could not see it with our eyes at 5 and 6 o'clock, because the air was full of snow. We are eager to know what we will learn about it from others.

48 This text is a French translation of 1C.
49 a *del. post* Dimanche **BO**.
50 il *del.* **BO**.
51 le *del.* **BO**.
52 a *del. post* voir **BO**.
53 ce *del. post* saura **BO**.
54 d'allieu *del.* **BO**.
55 The discrepancy in the hours given seems to be a mistake of the translator or copyist.

2.

16/26 December 1664, Lubieniecki to Hevelius

BO: C1, v. 6, 910
TC: 362
Olhoff: 100

Enclosures:
2A–2D. Continuation of the observations and opinions about the comet

Salutem et Officia

Nobilissime, Clarissime, Humanissime Vir. Invitavit me nuper ad Te literis compellandum singularis Tua, quae per orbem celebratur, humanitas et[1] rerum Astronomicarum peritia: incitavit autem novus Cometes, de quo variorum virorum doctorum judicia ad Te misi, quae Tibi non ingrata fore auguror, nisi quod in uno eorum elevatio Cometae ultra 18. gradum in meridiano producta, ab ipso postea auctore correcta et at 14. gradum reducta fuerit.[2] Adiungo nunc Continuationem talium[3] observationum, quamvis forte Bremensem nuper Tibi miserim,[4] antequam Parisiis, Hafnia et aliunde plura a Viris doctis accepero, quae libens meritoque Tecum communicabo, a quo quantopere mecum illi mutuam humanitatis et industriae vicem expectent, ex hac Huswedeliana vides.[5] Nec dubito eandem esse et aliis[6] mentem. Tuo enim merito omnes Te velut Astronomorum nostri seculi Principem celebrant et colunt. Expectamus autem abs Te et accuratam Cometae istius delineationem Tua diligentia et laude, cuius praestantissima instrumenta, sicut toto orbe, ita et Gottorpii, ubi ea in Ducali Bibliotheca vidi, monstrante Adamo Oleario, Viro Humanissimo, ad quem literas hoc nomine dedi,[7] magni fiunt, dignam. Unum addo antequam finio. Voluissem libenter Tibi adiuncta praeterito die Martis mittere, sed Huswedeliana[8] tam cito

1 et : (virtus homini maxime conveniens, imo eum constituens) tum vero et excellentissima totoque pariter orbe clarissima **TC Olhoff**.

2 nisi...fuerit *om.* **TC Olhoff**.

3 talium : eiusmodi **TC Olhoff**.

4 quamvis...miserim *om.* **TC Olhoff**.

5 a quo quantopere...vides : A quo quantopere mecum illorum nonnulli, quos scilicet hac in urbe degentes alloqui mihi licuit, mutuam humanitatis et industriae vixem exspectent, ex hac Husvedeliana epistola, quem adiungo, vides **TC Olhoff**.

6 esse et aliis : aliis esse **TC Olhoff**.

7 ubi ea...dedi : (ubi ea in instructissima et tot admirandis refera Bibliotheca Ducali vidi) **TC Olhoff**.

8 Huswedeliana *om.* **TC Olhoff**.

describi non potuere. Vale, Vir Eximie, in annos quam plurimos et quam felicissime. Hoc Tibi strenae loco votum sincerum mittit[9]

Clarissimi Nominis Tui
observantissimus

De Rolis

Hamburgi 16. 26. Decembris 1664

Greetings,

Most noble, renowned, and gentle Sir! Your kindness well-known in the world along with your proficiency in astronomical matters, have recently encouraged me to write a letter to you. The new comet has incited me as well. I sent you the opinions of various learned men about it, which I presume will not be unpleasant to you, despite the fact that in one of these opinions the elevation of the comet in the meridian extended to over 18° which later was corrected and reduced to 14° by the author himself. Now I enclose the continuation of those observations, even though I probably had sent you earlier the one from Bremen before I received more of those from the learned men from Paris, Copenhagen and other places. I will eagerly and dutifully share all of these things with you. The authors expect reciprocal kindness and engagement, as you may gather from Huswedel's letter.[10] Doubtlessly, this is also the will of others. Due to your merits, all admire and revere you as the prince of astronomers of our age. Hence we are waiting for your description of the comet, which will match your diligence and glory, the most notable proofs[11] of which are in high esteem, as in the whole world, so in Gottorp, where I have seen them in the Ducal Library, shown by Adam Olearius,[12] the most gracious man, to whom I also sent a letter about this matter. I will add one thing before I leave. I wanted to send you the enclosures last Tuesday, but the copy from Huswedel could not be made so quickly. Exceptional Sir,

9 mittit : mitto **TC Olhoff**.

10 See note to Letter 1. The letter in question is 2B.

11 Lubieniecki's phrase here is quite ambiguous, and it seems that the word *instrumenta* may mean both 'instruments' (or 'devices') and 'proofs', as he begins with praises of Hevelius's mastery in astronomy and then proceeds to mentioning astronomical instruments in the Ducal Library in Gottorp ('them'). The globe from Gottorp was devised by Duke Frederick III. The external side of that great instrument (over three meters in diameter) was a terrestrial globle, while its interior rotated around its axis and presented the sky as seen from Gottorp. See Chantal Grell, 'Introduction', CJH2 p. 105.

12 See section 2.4.13 of the Introduction.

stay well for many more successful years. This sincere wish is being sent to you as a New Year gift by

renowned Sir,

Yours obliged,

De Rolis

In Hamburg, on 16 /26 December 1664

2A.

Continuation of the observations and opinions about the comet

BO: C1, v. 6, 911[13]
TC: 1

Continuatio Observationum et Judiciorum a Viris doctis profectorum de Cometa insigni Anni 1664

Bremae die 17/7 Decembris 1664

Vere nobis novus cometa lucet cauda satis lata et longa: colorem non adeo accurate observare licuit ob nebulosam circa horizontem caelum. Hydram occupavit sub et ante rostrum corvi hydrae insidentis. Ortus est ante horam 3^{tiam} matut[inam] plaga inter Meridiem et Orientem media. Meridiem tenuit hora 6 matut[ina], sic subsequentibus diebus hora 4 aut 5 aut 6 videri coelo sereno poterit. Vicinitas horizontis (vix nimirum ultra X gradus ascendit) efficit plerumque ut nonnisi e nebulis eluceat. Locis quo meridionalibus, eo magis super verticem elevatur.

The continuation of the observations and opinions about the notable comet of the year 1664 made by some learned men

In Bremen, on 17/7 December 1664

13 Each part of the enclosure is printed separately here. The text of 2A appears also in 1G.

The new comet indeed shines with quite a long and wide tail; its colour could not be precisely determined, because the sky near the horizon was clouded. It was in Hydra, under and rising before the Beak of Corvus, which is near Hydra. It rises at about 3 o'clock in the morning between the south and east. It was in the south at 6 o'clock in the morning, so in the following days it will be possible to see it at 4, 5, or 6 o'clock, if the sky is clear. Its proximity to the horizon (it ascends only slightly more than 10°) makes it shine usually through the mist. The more southern is its place, the more it rises towards the top.

2B.

Continuation of the observations and opinions about the comet

BO: C1, v. 6, 911

Hamburgi die 13 23 Dec[embris] 1664

Generose Nobilissimeque Vir, Domine and Amice plur[imum] honorande.

Quanquam mei fuisset officii, simulac[14] quod ante aliquot dies promiseram, efficere, nosti tamen quae occupationum mearum series sit, ut interdum neque Amicis qui tamen nostri partem, si Platoni credimus, non minimam sibi vindicant neque mihi ipsi, quantumvis maxime cupiam, quaquam tamen vacare possim. Nunc vero (licet serius et quod Graeci dicunt fortasse κατόπιν[15] τῆς ἑορτῆς) cum mei arbitrii horulam nactus sum, aere isto me liberare constitui, et si recte memini, tertius erat Dec[embris] cum de Cometa quodam fama in urbem venit. Ad illum observandum[16] cum nonnulli essent intenti neque de loco atque apparitionis hora quidquam certi constaret, factum est ut aliqui frigidae noctis taedio gravati, tum temporis Cometa inviso quieti se reddiderint arbitrantes, famam esse hominum non satis recte judicantium. Verum cum ex his, licet pauciores, propositi tamen magis tenaces, ad matutinum usque durarent tempus, hora tertia phaenomenon apparuit quod omnibus nobis veri Cometae speciem de se praebuit qui sane meo judicio sive magnitudinem corporis et prolixitatem caudae, sive alia non exigui ponderis eo pertinentia spectes, haud parvi esse momenti videbitur, de quibus tamen iis judicium relinquo, qui penitius istis sacris imbuti Helicona propter pallidamque habitant Pyrenem. Ego qui ἀγεωμέτρητον[17] // ultro me confiteor, ea tantum quae oculis usurpavi, brevibus exponam. Tempus

14 simulaca (a *del.*) **BO**.
15 κατόπιν *corr. ed.* : καθώπιν **BO**.
16 observandum *del. post* observandum **BO**.
17 ἀγεωμέτρητον *corr. ed.* : ἀγεωμέτρετον **BO**.

erat quo prima quies mortalibus excussa paulatimque somno profundiore liberati[18] futurae diei praeludia speramus, cum ecce ad id constitutus Vigil nunciat coelum esse sine nube serenum pulcherrimaque conspici astra in quibus si quid dignum, adspectu jam rectissime judicari posse. Turrim itaque templi cathedralis aliqua parte adscensu supero petii, quo cum perveni ecce ante oculos clarissime se videndum praebuit ille, de quo hactenus dubitaveram Cometa: de cuius situ si quid judicare possum, ita habe: Erat quarta fere matut[ina] et jam per horae spatium supra horizontem fuisse videbatur, tum vero, nisi fallor, inter Meseurum et Notapelioten apparebat: australis admodum, ita ut postea observavi, ultra 14 gradum in meridiano non elevaretur. Erat sub signo Virginis si stellas videres in librae tamen Dodecatemorio et in Cauda Hydrae, Corvum inter et Craterem, ita tamen, ut corvi rostro esset propinquior. De Motu ipsius necdum certi quid statuere possum. Videtur tamen oblique Austrum versus proprio motu pergere ita ut jam vel in ipso Capricorni, vel ultra etiam, conspiciatur Tropico. E genere Caudatorum esse nemo dubitat et si cum aliis comparetur quemadmodum fando accepi, hic prolixitate caudae Anno 18 huius seculi Cometam longe superat quemadmodum ille fulgore et luce hunc multo antecessit. Qui vero Anno 1652 apparuit alterius erat generis Comatus scilicet neque cum hoc facile comparandus, quem admodum nec ille, qui Anno 1661 aquilam inter et Delphinum est visus. Hic enim obscurus erat, et magnitudinis exiguae ita etiam ut a nonnullis inter Cometas numerari vix dignus judicatus fuerit. Quid autem proprie Cometa sit, haud facile est dicere, et meo judicio facilius quid non sit. Neque enim mihi placet Pythagoreorum Sententia, qui Planetis Cometas annumerunt, illosque sub radiis Solaribus delitescere aliquando ex iis egredi instar Mercurii, aut alterius // Planetae atque se videndum exhibere autumant. Cui sententiae etiam Seneca aliquo modo affinis videtur, licet paulo aliter hac de re L[ibro] 7 *Quaest*[*ionum*] *natural*[*ium*] et ingeniose satis disserat. Multominus Aristotelis ipsiusque Sectatorum, qui exhalationem credunt ne quid de Democriti veterumque Romanorum suspicionibus dicam qui Cometas Animas mentesque Illustrium virorum arbitrati sunt: Cuius modi apotheosin C[aii] Julii Caesaris Augustus aliique crediderunt. Unde nummus aureus cum imaginae Caesaris ab altera parte Cometa insignitus adhuc visitur. Neque Angelos aut Daemones esse credimus, quae nonnullorum absurda est opinio. Eadem est ratio si de causis Cometarum loqui velimus et mea quidem sententia illi universaliter recte statuere videntur, qui Deum auctorem istius phaenomeni faciunt. Verum cum Deus causa universalissima sit, atque ad eum merito omnia referantur. Si penitius in causas secundas, quod Philosophum decet, inquiras, nihil omnino boni isti viri praestiterunt. Neque plus illi dixisse videntur qui elementarem materiam aut coelestem statuunt, aut qui viae lacteae particulam credunt aut solis maculam aliquam frustulumve pro materia Cometae abscedere sibi persuadent. Haec enim omnia cum firmo haud nitantur fundamento, veris, et quod proprium est Mathematicarum disciplinarum ἀποδείξεσιν[19] annumerare iniurium plane mihi videtur. Interea tamen verissimum

18 libertati (*primum* t *del.*) **BO**.
19 ἀποδείξεσιν *corr. ed.* : αποδείξεσι **BO**.

est insignis magnitudinis aliquod conflagrare corpus licet etiam neque de materia neque de forma istius corporis aliisve accidentibus certi quid determinare necdum liceat libere tamen, si philosophandum esset, fortasse probabiliter quaedam in medium produci possent, quae ulteriori inquisitioni ansam non obscure praeberent, ut tandem aliquando in his etiam veritas diu frustra quaesita, tanquam ex Democriti puteo in lucem protraheretur oculisque mortalium pulcherrima exponeretur. Veniet fortasse tempus ut Senecae // verbis concludam: 'quo ista quae nunc latent, in lucem dies et Longioris aevi diligentia extrahet. Veniet tempus quo posteri nostri tam aperta nos[20] ne scivisse mirabuntur.' Verum hoc tempore cum tanta sit ignorantiae apud nonnullos malignitas ut etiam quod ipsi nesciunt si alios ea de re sollicitos videant, anathemate dignos judicent a talibus abstinendum satis esse censeo. Tu iterum Vir Excellen[tissime] bene vale et ignosce si falcem in alienam immittens messem graviter alicubi cespitaverim. Si quid tum ad Te perveniet a Viris toto orbe celeberrimis Hevelio scilicet aut Bulialdo aliisque ad hoc pertinens negotium quin me quae Tua est humanitas[21] eorum omnium participem facturus sis nullus dubito. Iterum vale et fave

Tui Studiosissimo

Johanni Alberto Huswedel
Ph[ilosophiae] et Med[icinae] D[octori] Reip[ublicae] Hamb[urgensis] Phys[ico] Ordinario

In Hamburg, on 13/23 December 1664

Generous and most noble Sir, very honourable Master and friend!

It was my duty to fulfil what I had promised a few days ago. You know that my obligations were such that I did not have any spare time (although I would wish it very much) neither for friends, who demand a significant share of me, if we believe Plato,[22]

20 nos *del. post* nos **BO**.

21 humanitastis (tis *del.*) **BO**.

22 Allusion to Pseudo-Plato's *Letter to Archytas*, 'ἀλλὰ κἀκεῖνο δεῖ σε ἐνθυμεῖσθαι, ὅτι ἕκαστος ἡμῶν οὐχ αὑτῷ μόνον γέγονεν, ἀλλὰ τῆς γενέσεως ἡμῶν τὸ μέν τι ἡ πατρὶς μερίζεται, τὸ δέ τι οἱ γεννήσαντες, τὸ δὲ οἱ λοιποὶ φίλοι, πολλὰ δὲ καὶ τοῖς καιροῖς δίδοται τοῖς τὸν βίον ἡμῶν καταλαμβάνουσι' (Platonis *Omnium quae extant operum tomus tertius Graece et Latine* ([s.l.]: [s.e.], 1578), p. 358, 'but you should consider we are not born for ourselves only, but a part of our birth is demanded by the homeland, part by the parents, remaining part by friends, and a significant part is given to the times which capture our lifetime'). Its best-known quotation is in Cicero's *De officiis*, 1, 22, 'quoniam, ut praeclare scriptum est a Platone, non nobis solum nati sumus ortusque nostri partem patria vindicat, partem amici' ('as Plato has admirably expressed it, we are not born for ourselves alone, but our country claims a share of our being, and our friends a share'; M. Tullius Cicero, *De officiis*, transl. by Walter Miller (Cambridge, Mass., London: Harvard University Press, 1913)).

nor for myself. As now (although too late and after the feast, as the Greeks say[23]) I have gained a brief hour of spare time, I decided to free myself from this debt and, if I recall correctly, it was 3 December when the news about a comet came to the town. Since many people were willing to observe it, and nothing certain was known about its place, nor the time of its appearance, it happened that some of them, tired of a cold night, having seen no comet went to sleep, thinking it was a gossip invented by some people of wrong judgement. And when those who were fewer but more persistent in their undertakings waited until morning, at 3 o'clock a phenomenon appeared that showed to all of us a true face of the comet, which, in my opinion (if you look at the magnitude of its body, the length of its tail, or other significant aspects of it) will be of great importance; but I leave the judgement of these things to those who (being well acquainted with such holy things) live next to Helicon and pale Pyrene. I confess I do not know enough about geometry,[24] so I will briefly present what I saw with my own eyes. It was the time when the first sleep is shaken away from men and when we, gradually freed from deep sleep, make plans for the beginning of the next day, when the watchman, who was chosen for this purpose, announced that the sky is clear without a single cloud and the stars are perfectly visible. And if there is anything noteworthy among them, one can judge it precisely with his own eyes. Hence, I had climbed up onto a side of the cathedral tower, and when I came there, this comet, whose existence I doubted until then, showed itself before my eyes. If I can judge anything about its place, it was such: it was 4 o'clock in the morning, and it was visible for an hour over the horizon. If I am not mistaken, it appeared between Meseurus and Notapeliotes[25]. It was so close to the south that, as I later observed, it was not elevated higher than 14° in the meridian. It was in the sign of Virgo — if you looked towards the stars in the sign of Libra and the Tail of Hydra, between Corvus and Crater — however, it was closer to the Beak of Corvus. I cannot ascertain anything about its movement. It seemed to be inclined towards the south and to move with the proper motion, as if it were to become visible either in the Tropic of Capricorn or beyond it. Nobody doubts it was a tailed comet and, in comparison to other comets of this type (although I know them only from hearsay), it surpassed the comet of 1618[26] with the length of its tail. Nonetheless, the latter comet exceeded this one with brightness and light. The one which appeared in 1652[27] was a hair-type comet — and it cannot be easily

23 Plato, *Gorgias*, 447a.

24 An allusion to a well-known phrase which was said to be written on the entrance to Plato's Academy: 'ἀγεωμέτρητος μηδεὶς εἰσίτω' ('let no one unexperienced in geometry enter'). About its origin see Henri-Dominique Saffrey, *Ἀγεωμέτρητος μηδεὶς εἰσίτω. Une inscription légendaire*, 'Revue des Études Grecques', 81 (1968), no. 384–85, 67–87.

25 In the south-west.

26 See note to Letter 1F.

27 C/1652 Y1, see Alexandre Guy Pingré, *Cométographie ou Traité historique et théorique des comètes*, vol. 2 (Paris: Imprimerie royale, 1784), pp. 9–10; Gary W. Kronk, *Cometography. A Catalogue of Comets. Volume 1: Ancient – 1799* (Cambridge: Cambridge University Press, 1999), pp. 346–47. See also CJH3 p. 218, Letter 38, Hevelius to des Noyers, 24 December 1652, and subsequent letters.

compared to this one, and neither to the one which was seen in 1661[28] between Aquila and Delphinus. That one was dark and small, so some people considered it barely worth the name of a comet. It is not easy to state what exactly a comet is, and I think it is easier to say what it is not. Neither do I like the opinion of Pythagoreans who count comets as planets and assert that they hide behind the solar rays, sometimes emerge from them, like Mercury or other planets, and show themselves. Seneca also seems to lean to this opinion, although he discusses this subject slightly differently but ingeniously enough in Book 7 of the *Natural Questions*.[29] I will say less about the suspicions of Aristotle and his followers who believed in exhalations, or those of Democritus[30] and ancient Romans who thought comets to be souls and minds of famous men. Augustus and others believed in such an apotheosis of Gaius Julius Caesar;[31] hence, we can still see a coin with a comet minted on one side. Neither do we believe comets to be angels or demons, which is a senseless opinion of some people. The same is applicable if we talk about the causes of comets, and in my opinion those people who proclaim God to be the author of this phenomenon seem to be closest to truth. Because God is the universal cause, and everything truly tends to Him. If you deeply inquire into other causes — which is a philosopher's task — you will see that those men have not done anything good. Neither those who state that the comet is of elemental or celestial matter seem to be closer to truth, nor those who believe it to be a small part of the Milky Way, or those who are sure that a spot or a bit of the Sun separates and forms a comet. Since it all does not stand on a solid basis, it seems to me harmful to place it among true things and mathematical arguments which are typical of mathematical sciences. Meanwhile, it is true that a body of considerable size is burning up, although nothing can be determined either about the matter or the form of its body, or about its other qualities. But, if one could philosophise, they could probably proclaim something which gives a clear basis for further investigation, so that finally the truth about these things — sought in vain for so long — could be drawn out to light like from the well of Democritus[32] and show its beauty to the eyes of men.

28 C/1661 C1, see Pingré, *Cométographie*, p. 10; Kronk, *Cometography*, pp. 348–50. See also CJH3 p. 321, Letter 95, Hevelius to des Noyers, 8 February 1661, and subsequent letters.

29 In Book 7 of his *Natural Questions* (*Quaestiones naturales*), Seneca discussed various views on the nature of comets. He considered them as planetary, everlasting objects. His opinion was different from Aristotle's who believed them to be meteorological, ephemeral phenomena. Aristotle discussed this view and the opinions of his predecessors in *Meteorologica* (*Meteorology*), 1, 6, 342b25–343a8. These works by Seneca and Aristotle are one of the main sources of ancient cometary hypotheses. The Pythagoreans' opinion about the nature of comets is quoted after Aristotle.

30 The opinion that Democritus believed comets to be souls of famous men is not supported by ancient sources. It was mentioned by Jean Bodin, *Universae naturae theatrum* (Francofurti: apud heredes Andreae Wecheli, 1597), p. 221.

31 On the comet seen after Caesar's death see A. A. Barrett, 'Observations of Comets in Greek and Roman Sources before A.D. 410', *Journal of the Royal Astronomical Society of Canada*, 72 (1978), no. 2, 81–106 (pp. 95–96).

32 'Democritus's well' is an allusion to a saying attributed to Democritus, cited by Diogenes Laertius in *Vitae philosophorum*, 9, 72 (Hermann Diels, Walther Kranz, *Die Fragmente der Vorsokratiker*,

Let me conclude by quoting Seneca: 'when diligent research over very long periods will bring to light things which now lie hidden. There will come a time when our descendants will be amazed that we did not know things that are so plain to them.'[33] But now, since there is such great malice of ignorance of some people that if they see others who are skilled in anything they do not know about, they find them to be worthy of condemnation, in my opinion, it is sufficient to keep ourselves away from them. Farewell, most excellent Sir, and forgive me, if I stumbled too much anywhere while letting a sickle into other's harvest. If anything about this subject comes to you from the most famous men in the whole world, namely Hevelius or Boulliau[34] and others, I do not doubt (knowing your kindness) that you will include me in these endeavours. Farewell again, and be benevolent towards

The most devoted to you,

Johann Albert Huswedel
Doctor of philosophy and medicine, public physician of the city of Hamburg

2C.

Continuation of the observations and opinions about the comet

BO: C1, v. 6, 911
TC: 2

Observatio Rostochiana facta die 10 Dec[embris] S[tylo] V[etere]

Man hat den Cometen alhier observiret in plaga coeli Australi prope stellam rostro Corvi et Hydrae communem, welche ex sententia Copernici 3^{tiac} magnit[udinis]. Deßen colorem hat man alhier notiret mehr lividum alß candidum, und fast luridum.[35]

vol. 2 (Berlin: Weidmannsche Verlagsbuchhandlung, 1959), 68 B 117): 'ἐτεῇ δὲ οὐδὲν ἴδμεν· ἐν βυθῷ γὰρ ἡ ἀλήθεια' ('in fact, we know nothing, because the truth is in the depth').

33 An approximate quotation of two phrases from Seneca's *Natural Questions*, 7, 25. The exact quotation is: 'Veniet tempus quo ista quae nunc latent in lucem dies extrahat et longioris aevi diligentia,' 'Veniet tempus quo posteri nostri tam aperta nos nescisse mirentur.' (Seneca, *Naturales quaestiones. Books IV–VII*, transl. by Thomas H. Corcoran (Cambridge, Mass., London: Harvard University Press, 2004)).

34 See section 2.4.2 of the Introduction.

35 Latin tranlsation in the TC: Observatus hic est Cometa in plaga caeli Australi prope stellam rostro Corvi et Hydrae communem, quae ex sententia Copernici tetriae magnitudinis est. Color Cometae notatus est magis lividus, quam candidus, et fere luridus.

Observation in Rostock made on 2 December, old style

One has seen the comet here, in the southern part of the sky, near the star appearing both in the Beak of Corvus and Hydra, which is of the third magnitude, according to Copernicus's opinion. Its colour was noted as more livid than white, and nearly yellowish.

2D.

Continuation of the observations and opinions about the comet

BO: C1, v. 6, 911
TC: 238–39

Observatio Magdeburg[ensis] Ampliss[imi] Domini Ottonis Gerichii Cons[ulis] Magd[eburgensis]

Den 2. Dec[embris] 1664. habe ich den Comet morgens umb 5. Uhr gegen den Mittag in Corvo juxta collum eius zum ersten mahl gesehen, er hatte einen schwantz zimlich breit, doch nicht gar sehr lang. Hernach hatte man Ihn nicht wieder sehen können, weil es dunckel gewesen, biß den 12. Dec[embris]. Da habe Ich ihn umb dieselbe zeit weit niedriger versus Austrum in Hydra sub Cratere in der höhe von[36] 10. grad[us] über der Sonnen[37] gefunden, und zwar mit einen viel größen und Längern schwantz, so daß Er fast sehr importirlich anzusehen. War also über 12. grad[us] beßer nachn abend gegangen.[38]

Eiusdem judicium de Origine Cometarum perscriptum a Filio eius Domino Resid[ente] Brand[enburgensi] Hamb[urgi] 16. 26. Decembris 1664

Causa universalissima Cometarum est Deus, causa autem secunda sunt procellae. Inprimis sciendum est, quod terra et aer faciant unum corpus, aer constitutus est circumcirca terram in certo pondere et certa altitudine. Corpus hoc terraqueum cum

36 umb *del.* **BO**.

37 der Sonnen : den Horizon **TC**.

38 Latin tranlsation in the TC: Die 2. Decembris Anno 1664. Circa 5. horam matutinam versus Austrum in Corvo juxta collum eius Cometam prima vice vidi. Habuit is caudam satis latam, non valde autem longam. Postea non poterat videri, quod caelum esset usque ad diem 12. Decembris nubibus obductum. Tunc temporis vero longe inferius eum Austrum in Hydra sub Cratere in altit[udine] 10. grad[uum] supra Horizontem inveni, et quidem cum maiori et longiori cauda, ita ut fere maximi aestimandus videretur. Ultra 12. igitur gradus ulterius versus Occasum processerat.

aere simul constitutum est vel habet locum[39] in universo spatio vel vacuo, in quo sunt omnia corpora caelestia, sicut et[40] terra est corpus caeleste, terra in hoc spatio expedit suum cursum circa Solem in anno. Quando itaque veniunt singulares procellae, illae efficiunt, quod earum vi particula aeris rumpatur ex circulo aereo in istud spatium.[41] Haec part[icu]la aerea accipit splendorem a Sole sicut aurora, nubes, planetae, et quidem tali modo, quia in ipso vacuo vel spatio haeret, ubi nullum aliud creatum est. Quando igitur terra persequitur suum motum circularem et centralem, tunc apparet ista part[icu]la, quae est Cometa, et muta[ti]o terrae ratione loci facit cursum cometae et eius mutationem.[42] Haec sunt quae raptim delineare vel descibere volui. Est primus conceptus, peto ut crastino die remittatur. Ni fallor, ita est mens et opinio Domini mei Parentis.[43]

Observation in Magdeburg by the very worthy Otto von Guericke, Councillor of Magdeburg

On 2 December 1664, I saw the comet for the first time — in the morning, at 5 o'clock before noon, in Corvus, next to its neck; its tail was quite wide, but not very long. Later one could not see it since the sky was cloudy until 12 December. Then I found it at the same time much lower towards the south, in Hydra under Crater, at an altitude of 10° over the horizon, and it had a much bigger and longer tail, so it looked very prominent. It also moved over 12° towards the west.

The opinion about the origin of comets by the same man, written by his son, a Brandenburg resident in Hamburg, on 16/26 December 1664[44]

The most general cause of comets is God; the secondary causes are storms. First of all, it should be known that the Earth and the air compose one body; the air is placed around the Earth with a specific weight and a certain height. This body of earth and water, together with the air, is placed — or occupies a place — in a universal space or a vacuum where all the celestial bodies are — for the Earth is also a celestial body; the Earth in this space completes its route around the Sun in a year's time. Therefore, whenever extraordinary storms occur, with their force they cause a small part of the air separate from the aerial ring into that space. The small aerial part takes brightness from the Sun (like dawn, clouds, or planets) and does it in such a way that it sticks in

39 habet locum : locum habet **TC**.

40 et : etiam **TC**.

41 spatium *corr. ed.* : spatium spatium **BO**.

42 et mutatio...mutationem : Et motus terrae ratione loci, sicut etiam Sol sua attractione et radiis facit cursum Cometae et eius motum **TC**.

43 Est primus...Parentis : Est primus conceptus, et ni fallor, mens et opinio Domini mei Parentis **TC**.

44 See section 3.4.3 of the Introduction.

this vacuum (or space) where there are no other created things. So when the Earth makes its circular route around the centre, then that small part (i.e. a comet) appears, and the change of the Earth's position incites the course of the comet and its changes. This is what I briefly wanted to sketch or describe. It is the first draft, which I ask you to send back to me tomorrow. If I am not mistaken, this is my father's idea and opinion.

3.

28 December 1664,[1] Hevelius to Lubieniecki

BO: C1, v. 6, 900
TC: 362–63

Illustri Viro
Domino de Rolis
alias Lubenitzkii
Hamburgi

J[ohannes] Hevelius S[alutem]

Quod commercium literarium[2] mecum inire satagas maximopere quidem mihi gratulor; cum nihil mihi unquam sit dulcius, quam Praeclarissimorum Virorum consuetudine frui, sed doleo, quod adeo obstrictus, imo pene obrutus sim occupationibus tam publicis quam privatis, inprimis gravioribus studiis Coelestibus, ut non toties quoties quidem animus est, amicis hac in parte morem gerere possim; attamen quousque tantum imposterum licebit, operam studiaque mea Illustriss[imae] T[uae] D[ominationi][3] promptissime defero. Pro observationibus Cometae gratias habeo debitas. Quem hic Gedani quoque a 14 Decemb[ris] st[ylo] n[ovo] me sedulo observasse scias; sed crebrae aeris tempestates obstiterunt, quo sepius[4] visus fuerit. Initio, cum rostro Corvi adhaereret, satis erat tardus, paullo autem post multo factus est velocior, sic ut ad hunc diem usque 28 Decemb[ris] fere 56 grad[us] emensus fuerit: appropinquat jam Cani Maiori, tendit ad Leporem, Eridanum Cetumque versus. Ex Orientali factus est occidentalis. Principio altitudo eius Meridiana imminuebatur, jam vero in dies augetur. Capite hucusque fere crevit, cauda decrescente. Accuratiores observationes meae, Sextantibus, Quadrantibus, Tubisque praelongis habitae, reservantur // in *Cometographiam* nostram, quae sub praelo fervet, sed vix intra annum lucem videbit: cum opus sit satis vastum et laboriosum.[5] Interea rogo, si quae observationes huius Cometae, inprimis solidioribus Instrumentis acquisitae ad Tuas pervenerint manus, ut earum mihi facias copiam; maximopere[6] me Tibi obstringes. Vale, et me porro amare perge. Saluta meo nomine quam officiose Eximios Viros Dominos Joh[annem] Mullerum et Michael[em] Kirstenium, amicos plurimum honorandos.

1 In the TC the letter is dated 10/20 December 1665.
2 literarium : literarum **TC**.
3 Illustriss[imae] T[uae] Dominationi : Generosae Dominationi Tuae **TC**.
4 rarius *del.* **BO** : rarius **TC**.
5 cum...laboriosum *om.* **TC**.
6 maximopere : valde **TC**.

Dabam Gedani Anno 1664, die 28 Decemb[ris] st[ylo] n[ovo]

To the illustrious Sir
Master de Rolis
alias Lubieniecki
in Hamburg

Johannes Hevelius sends greetings

I am pleased and honoured because you made an effort to begin a correspondence with me, and there is nothing more agreeable for me than contacts with most famous men. However, I regret that I am bound with (nearly buried under) many duties, both public and private, primarily relating to more serious celestial studies. As a result, I cannot fulfil my friends' requests anytime I want. However, if it is possible in the future, most famous Sir, I will eagerly present you with my work and efforts. I am most thankful for the observations of the comet. You should know that I have also been observing it diligently since 14 December (new style), but intense storms in the air make it impossible to see it more frequently. In the beginning, when it was near the Beak of Corvus, it was quite slow, but soon after it became much faster, so until today, viz. 28 December, it has passed nearly 56°. It is now approaching the Canis Maior and moves towards Lepus, Eridanus, and Cetus. Out of an eastern comet, it has become a western one. At first, its height in the meridian was decreasing, but now it is growing. Its head is still increasing, while its tail is shrinking. More precise observations — conducted with sextants, quadrants, and long telescopes[7] — are kept for our *Cometographia*, which is now being printed, but will be published in about a year, since it is quite a large and difficult task. Meanwhile, if any observations of this comet — especially those conducted with better instruments — come your way, I would be most obliged if you copied them for me. Farewell, and stay benevolent towards me! Greet kindly on my behalf Master Johann Müller[8] and Master Johann Kirsten[9] — exceptional men and most worthy friends.

Given at Danzig, on 28 December 1664, new style

7 About Hevelius's instruments see Przemysław Rybka, *Instrumentarium astronomiczne Heweliusza (geneza i rozwój konstrukcji)* (Wrocław: Zakład Narodowy im. Ossolińskich, Wydawnictwo Polskiej Akademii Nauk, 1987) (*Monografie z dziejów nauki i techniki*, vol. 139). About his 140-foot long telescope, the *maximus tubus*, and other aerial telescopes see Chantal Grell, 'Introduction', CJH3 pp. 68–73.

8 See section 2.4.11 of the Introduction.

9 See section 2.4.9 of the Introduction. Hevelius made a mistake in Michael Kirsten's name.

4.

20/30 December 1664, Lubieniecki to Hevelius

TC: 362

Enclosure:
4A. Letter from Røtlin to Lubieniecki, 13 December 1664

Hamburgo Dantiscum die 30/20 Decembris 1665.

Tertias hasce intra paucos dies do ad Te, Nobilissime et Clarissime Heveli, literas, breves quidem et currente calamo multis negotiis me circumstantibus, exaratas, Tibi tamen, ut spero, gratas futuras, tum ob affectus in Te mei integritatem, tum ob adiuncta quae ab ultimis meis ad Te datis Hafnia et aliunde accepi. Interea responsum Tuum quod nobis, quae est spes et votorum nostrorum hac in re summa, jucundum spectaculum, situm scilicet et motum Cometae exhibebit, operimur. Ecce vero novum phoenomenon, et novam de hoc argumento sententiam Amplissimi Domini Ottonis Gerichii Consulis Magdeburgensis, quam ex adiunctis cognosces, ut et praedixisse eum ante duos menses Cometae istius apparitionem, beneficio statuae tempestates, quas protrudere Cometas ille statuit, indicantis. Fac ut quam primum taedio molestae expectationis nos liberes, et bene ac feliciter. Vale.

From Hamburg to Danzig, on 30/20 December 1665.

Most noble and renowned Hevelius, I am sending the third letter to you within a few days — written briefly and in haste due to my numerous duties — but I hope it will please you, both because of my constant affection towards you and the enclosures which I have received from Copenhagen and elsewhere since my last letter to you.[1] Meanwhile, we expect your answer which (as we hope and wish for) will reveal to us this pleasant spectacle, namely the place and movement of the comet. Below I outline this new phenomenon and new opinion in the discussion given by the greatest Sir, Otto von Guericke, Councillor of Magdeburg, which you will find in the enclosure. You will also learn from it that he had foreseen the appearance of the comet two

1 There is only one extant enclosure to this letter: Johannes Melchior Røtlin's letter to Lubieniecki. Røtlin (d. 1676) was a Danish courtier and diplomat. In 1656–58, he was a Danish emissary in Danzig. Later, he was a Prince Christian's (1646–99, King of Denmark since 1670) secretary and a member of the Danish Board of Trade (*Kommercekollegiet*). He was Lubieniecki's correspondent; TC comprises 11 letters they exchanged (5 from Røtlin, 6 from Lubieniecki), pp. 415–28.

months ago using the statue which announces storms[2]; storms, which are the cause of comets, as he believes.[3] Relieve us from the pain of unpleasant anticipation and farewell! Stay in health and good fortune!

4A.

Letter from Røtlin to Lubieniecki, 13 December 1664

BO: C1, v. 6, 914
TC: 415–17

Copenhague la 13 Dec[embre] 1664

Apres vous avoir remercié de Vostre derniere Communication, si bien[4] de[5] l'ord[inaire] de[6] ce qui passe en terre, que de l'Extraord[inaire] de ce qui paroist au ciel, et qui plus est en confidence sous le cachet volant de celle pour Mons[ieur] Thoma[7] Bartholini, à qui elle a esté rendue à l'instant. Je ne peux que je ne m'engage[8] avec Vous sur le sujet de la Comete, que ces jours passéz nous aussy decouvert ici,[9] justem[ent] en temps, en lieu, en cours, en figure, en grandeur, en couleur, et en tout, comme chez vous, et ailleurs on a observé, hormis la queue qu'on a trouvé[10] ici, à ce que l'oeil en peut juger, d'avoir plus d'une pique de long. Quoy qu'il soit,[11] je ne doute pas que les Astronomiens[12] ne soyent facilement d'accord touchant l'histoire de cette Comete et sa description,[13] pourveu que les Astroloques avec fondement[14] se puissent accorder touchant ses operations et effets. Je me souviens[15] lors que[16] j'estois à Dantzic d'avoir oui[17] dire à Mons[ieur] Hevelgoen[18] Grand Astronomien,[19] mais qui se mocque de

2 For enclosed materials from Guericke see Letters 2D and 5A.
3 See section 3.4.3 of the Introduction.
4 que *del. post* bien **BO**.
5 si...de : tant de **TC**.
6 de : touchant **TC**.
7 Thoma : Thomas **TC**.
8 je...m'engage : je ne puis m'empescher de m'engager **TC**.
9 que...ici : que nous vîmes aussy ces jours passez en nos quartiers **TC**.
10 trouvé : trouvée **TC**.
11 soit : en soit **TC**.
12 Astronomiens : Astronomes **TC**.
13 l'histoire...description : la description de cette Comete **TC**.
14 avec fondement *om.* **TC**.
15 souviens **TC** : souvien **BO**.
16 lors que : que lors **TC**.
17 d'avoir oui : j'oui **TC**.
18 Hevelgoen : Hevelguen **TC**.
19 Astronomien : Astronome **TC**.

l'Astrologie, qu'il ne desiroit que de voir et contempler encore[20] une Comete depuis tant d'observations que par le moyen de ses speculations fort curieuses il avoit marquées, à fin d'achever cet ouvrage qu'il en[21] a entre ses mains, et de descouvrir toute la nature[22] de cet embellissement de la nature. Pour moy mesme,[23] je ne suis curieux que de sçavoir la vraye generation, matiere, quantité, qualité, en somme, la nature de ce corps, choses seules, qui me semblent d'estre[24] susceptibles des monstrations,[25] et laissant se reste à des conjectures et observations Astrologiques plus fallibles[26] toujours qu'infallibles. Car quant à cela tout le monde sçait[27] comm'une affaire notoire, que communement on croit, que la Comete n'est qu'un mauvais augure, et ne presage que des pestes,[28] des guerres, des famines, des sterilites, en[29] un mot des maux.[30] Mais si contraire[31] on dit, que cette Comete-cy promet une grande fertilité, furiese richesse,[32] une corne d'abondance, une pluye des ducats, une gréle[33] // des nobles à rose, une glace des diamants, un vent d'ambre gris, un siecle d'or et un Paradis terrestre. Je croy que l'un est aussy difficile à demonstrer,[34] que l'autre. Certes[35] Scaliger en ses *Exercitations* à[36] certain passage ne des advoue pas que plusieurs Cometes n'ayent en la suite de grands desastres, mais aussy il n'ose pas nier que quelques Cometes n'ayent porté de fort bonnes adventures. Pierre Gassend[37] un Mathematicien Parisien va[38] plus loin,[39] la partie du quel pourtant je n'ose pas prendre de peur d'encourir un peu de profanité, en sorte que[40] je trouve bon d'alleguer ses propres termes, quand il dit[41] dans la *Vie de Peireske*, ou comme se traicté porté en Latin,[42] in *Vita Peireskii*, Lib[ro] 3, pag[inis] 245, 246: 'Non est postremum argumentum imbecillitatis humanae ad ista expavescere, quibuscum res nostrae nihil commercii habent. Mirum cur

20 encore : encor **TC**.
21 en *om.* **TC**.
22 toute...nature : toutes les qualités **TC**.
23 mesme *om.* **TC**.
24 d'estre : estre **TC**.
25 des monstrations : de demonstrations **TC**.
26 fallibles : fallibiles **TC**.
27 sçait : le sçait **TC**.
28 pestes **TC** : pe **BO**.
29 en : et en **TC**.
30 des maux : beaucoup de maus **TC**.
31 contraite : au contraire **TC**.
32 furieuse richesse : des richesses immenses **TC**.
33 gréle **TC** : gr **BO**.
34 demonstrer **TC** : demonster, Je di demonster **BO**.
35 Certes : Je sçay bien, que **TC**.
36 à : an un **TC**.
37 Gassend : Gassende **TC**.
38 va : va encor **TC**.
39 loin : loing **TC**.
40 la partie...que : Mais je n'ose pourtant pas prendre son parti de pœur de passer pour profane. C'est pourquoy **TC**.
41 quand...dit : qui sont tels **TC**.
42 ou...Latin : ou pour mieux dire **TC**.

saltem Deo non credatur dum a signis caeli nihil metuendum jubet, scil[icet] ipse signa non faciens, ex quibus non expositis quid futurum sit caveri non possit. Nisi autem Deus Cometam calamitatum signum constituat, unde nam esse in Cometa potest aut futurorum notitia, aut effectus significandi? Annon ille pergit viam suam ut naturalia caetera, parumque eius interest quid humanae gentes inter se moliantur? Ac iste quidem (il veut dire de cette Comete de l'an 1618. dont particulirement il tient son discours[43]) dictus est portendisse clades postmodum secutas. Sed nempe ante non fuerant bella, pestes ac Reges mortui? Nullae jam 20. post annis excitantur tragediae? Referunt adhuc ad Cometam. Verum erit cur pari jure exorituras referant turbas quousque novus Cometes sit qui oblivionem istius creet, tametsi ante centum annos appariturus non fuerit.' C'est ce que Gassend en raisonne, n'estant refuté que trop quant à la derniere proposition et cette prediction que depuis l'an 18. en cent[44] ans il n'en aura point. Vous m'obligerez, Monsieur, de me faire part de ce que Vous apprendrez touchant cette matiere de gens que sont entenduz.[45] [46]

43 il...discours : anni scilicet 1618 **TC**.

44 cent **TC** : cens **BO**.

45 de gens...entenduz : de ceux qui y connoissent quelque chose. **TC**.

46 Latin translation in the TC: Actis Tibi gratiis de Tua postrema communicatione, ut ordinaria eorum quae in terra geruntur, sic et de extraordinaria eorum quae apparent in caelo, eaque, quod plus est, facta in confidentia per literas ad Cl[arissimum] Thomam Bartholinum scriptas, quas mihi apertas misisti, quasque ego ei statim reddidi, non possum quin Tecum conferam de Cometa, quem nos etiam vidimus his diebus praeteritis in his terris, eodem prorsus tempore, loco, cursu, eadem figura, magnitudine, eodem colore, et per omnia eundem, sicut apud vos et alibi observatus fuit, excepta cauda, quae hic inventa est, sicut oculus de hac re potest judicare, una hasta longior. Quicquid sit, non dubio Astronomos fore facilius concordes sententiis in descriptione istius Cometae, dummodo Astrologi possint concordes fieri circa eius operationes et effectus. Memini, cum Dantisci essem, audire me dicentem Ampl[issimum] Dominum Hevelium, Magnum Astronomum, sed qui rideat Astrologiam, quod non desideraret amplius quam videre et contemplari adhuc unum Cometam, postquam tot observationes, tanquam curiosissimus orbium coelestium speculator notavit, ut possit ultimam manum imponere operi, quod in manibus habeat, et detegere omnes qualitates huius bellissimi ornamenti naturae. Quod ad me attinet, ego non sum in hac re magis curiosus, quam ut sciam veram generationem, materiam, quantitatem, qualitatem; verbo, naturam huius corporis, quae sola videntur mihi suscipere demonstrationes posse: relinquo quae restant coniecturis et observationibus Astrologicis, fallaciarum magis susceptis infallabilitatis. Nam quod ad hoc attinet, omnes sciunt, sicut rem notissimam, communiter credi quod Cometa non sit aliud quam malum augurium (*corr. ed.* : argurium **TC**), et nil praesagiat quam pestes, bella, famem, sterilitates; verbo, multa mala. Sed si e contrario dicatur Cometam istum promittere grandem fertilitatem, divitias immensas, cornucopiam (*corr. ed.* : cornucopiae **TC**), pluviam nummorum aureorum, grandinem rosarum nobilium, glaciem adamantinam, ventum ambrarum, spargentem seculum aureum et paradisum terrestrem, credo unum esse tam difficile demonstratu ac alterum. Scio bene Scaligerum in suis *Exercitationibus* alicubi non deffiteri, pluribus Cometis successime grandia infortunia, sed etiam non audere negare, aliquos cometas apportasse valde bonos successus. Petrus Gassendus Mathematicus Parisiensis, progreditur hic longius. Verum ego non audeo assumere defendendas partes eius; vereor enim ne pro profano habear. Iccirco consultum judico allegare eius propria verba, quae extant in *Vita Peireskii*, lib[ro] 3. pag[inis] 245, 246: 'Non est postremum argumentum imbecillitatis humanae ad ista expavescere, quibuscum res nostrae nihil commercii habent. Mirum cur saltem Deo non credatur dum a signis caeli nihil metuendum jubet, scil[icet] ipse signa non faciens, ex quibus non expositis quid futurum sit caveri non possit. Nisi autem Deus Cometam calamitatum signum constituat, unde nam esse in Cometa potest

Having thanked you for your last messages (both for the ordinary ones concerning things on Earth, and the extraordinary ones about what appears in the heavens) and for your trust in me showed in the unsealed letter to *Monsieur* Thomas Bartholin[47] (I delivered it to him immediately), I could not miss a chance to tell you about the comet which is visible while it is passing us by, viz. about its time, place, route, shape, size, colour and everything which was observed by you and by others; except for its tail which is here (as an eye can discern) and which seems to be a pike longer. Whatever it is, I do not doubt that astronomers will certainly agree in their opinions about the description of this comet, while astrologers will be able to decide about its influence and effects. I remember that when I was in Danzig, I listened to *Monsieur* Hevelius, a great astronomer (however he mocks astrology);[48] he had no greater desire other than to see and study another comet, so having made many observations as a sedulous observer of the sky he could finish his work and describe the nature of that natural beauty. As far as I am concerned, there is nothing I would like to learn more about than the generation, matter, quantity, quality — in short, the whole nature of this body; everything which can be demonstrated. I leave everything else to conjectures and observations of astrologers — somewhat fallacious rather than infallible. Because as far as this matter is concerned, everyone knows (as a well-established fact) that there is a common belief that a comet can be only a malicious sign and cannot foreshadow anything but plagues, wars, famine, or poor crops — in short, lots of misfortunes. But if someone said the opposite — that this comet promises high fertility, fantastic wealth, a cornucopia, rain of ducats, hail of golden coins,[49] ice of diamonds, wind of ambergris, a golden age and a paradise on Earth — I believe this opinion would be as difficult to prove as the former one. I know well that Scaliger, in his *Exercitationes*, admits that many comets were followed by great misfortunes,

aut futurorum notitia, aut effectus significandi? Annon ille pergit viam suam ut naturalia caetera, parumque eius interest quid humanae gentes inter se moliantur? Ac iste quidem (anni scilicet 1618) dictus est portendisse clades postmodum secutas. Sed nempe ante non fuerant bella, pestes ac Reges mortui? Nullae jam 20. post annis excitantur tragediae? Referunt adhuc ad Cometam. Verum erit cur pari jure exorituras referant turbas quousque novus Cometes sit qui oblivionem istius creet, tametsi ante centum annos apparituros non fuerit.' Hocce est quod Gassendus ratiocinatur, qui plusquam satis refutatus est, quantum ad postremam propositionem attinet, et istam praedictionem ab Anno huius seculo decimo octavo, intra centum annos nullum appariturum Cometam. Obligabis me, cum me facies participem eorum, quae accipies de ista materia ab illis, quibus talia cognoscere datum est.

47 Thomas Bartholin (1616–1680), Danish physician and anatomist, brother of Erasmus Bartholin (see section 2.4.1 of the Introduction). He was a promoter of dissections and one of the pioneers of the studies of the lymphatic system. He also studied blood circulation. In 1663, he retired from medical studies and focused on literature, history, and philosophy. Thomas Bartholin was Lubieniecki's correspondent; in the TC (pp. 429–30, 435–36, 442–47), there are 4 letters they exchanged (1 from Bartholin, 3 from Lubieniecki).

48 Røtlin could have contacted Hevelius during his diplomatic mission in Danzig.

49 In French: *roses nobles*, in Latin: *rosae nobiles*. This term refers to gold coins minted in England since the 15th century.

but he does not dare to deny that some comets brought many happy events as well.[50] Pierre Gassendi,[51] a mathematician from Paris, goes even further. I will not pass his opinion profanely, but I will instead quote his own words which he said in the *Life of Peiresc* (or, in Latin, in *Vita Peireskii*,[52]) in Book 3, pages 245–46: 'It is not the last instance of human stupidity to fear those things with which they have nothing in common. Amazingly, one does not trust God even though He orders not to be afraid of celestial signs, especially that He does not create signs with which He could show what one can fear. So if God does not use a comet as a sign of calamities, how can there be a notion of the future or a foreshadowing effect in a comet? Does not it go its own way, like all other natural entities, not being interested in the toils of mankind? And this one (he wants to talk about the comet of 1618,[53] because it is the subject of his discourse) is said to have portended the disasters which followed it. But were there no wars, plagues and dead kings before? Will there be no tragedies in the next twenty years? And they are said to be related to the comet. It would be correct to relate future disturbances until the new comet appears which will make this one forgotten; however, there will be no new one in a hundred years.' This is Gassendi's reasoning which has been refuted as far as his last point is concerned, i.e. the prediction that there would be no new comet in one hundred years since 1618. You will make me obliged, *Monsieur*, if you share with me what you will receive on this subject from others.

50 Julius Caesar Scaliger (1484–1558), Italian humanist, scholar and defender of Aristotelianism. The publication in question is his *Exotericarum exercitationum liber quintus decimus* (Lutetiae: ex officina typographica Michaelis Vascosani, 1557), sig. 123v.

51 Pierre Gassendi (1592–1655), French astronomer, mathematician, and philosopher. In 1631, he conducted the first observation of the transit of Mercury. He was one of the focal figures of the 17th-century scholarly correspondence. He exchanged 23 letters with Hevelius (10 sent by Gassendi, 13 by Hevelius) between 1644 and 1654, cf. Klaus-Dieter Herbst, 'Alphabetical list of the correspondents', CJH1 p. 261. See also Chantal Grell, 'Introduction', CJH2 p. 17 *et passim*.

52 Pierre Gassendi, *Nicolai Claudii Fabricii de Peiresc, senatoris Aquisextiensis, vita* (Parisiis: sumptibus Sebastiani Cramoisy, 1651). Nicolas-Claude Fabri de Peiresc (1580–1637), French astronomer and humanist. In 1610, he conducted detailed observations of the Orion Nebula. He also worked on the observations of the moons of Jupiter and on the methods of determining the geographic longitude. As an organizer of scholarly communication, he was one of the most important people of the Republic of Letters of his era.

53 See note to Letter 1F.

5.

3/13 January 1665, Lubieniecki to Hevelius

BO: C1, v. 6, 913
TC: 363

Enclosures:
5A. Continuation of the observations from Magdeburg
5B. Letter from Müller to Lubieniecki, 23 December 1664 / [2 January 1665]
5C. Letter from Müller to Lubieniecki, 24 December 1664 / [3 January 1665]
5D. Letter from Müller to Lubieniecki, 26 December 1664 / [5 January 1665]
5E. Letter from Müller to Lubieniecki, 28 December 1664 / [7 January 1665]
5F. French translation of the observations from Magdeburg

Salutem Plurimam Nobilissime, Amplissime, Clarissime Vir. Accepi ante octiduum Tuas gratissimas literas die 28. Dec[embris] datas, ad quas e vestigio tibi respondere non licuit, praesertim cum expectarem responsum aliquod a Cl[arissimo] Bullialdo, quod die Sabbathi praeterito acceptum en Tibi una cum Müllerianarum literarum apographis. Ad haec cuncta sicut et ad illa quae die 26. et 30. Dec[embris] ad Te scripsi misique, Te pro Tua humanitate, licet variis negotiis, praecipue vero gravioribus studiis caelestibus occupatum, responsurum esse confido. Inter eos qui Tuas Observationes exspectant[1] de Cometa cum delineatione, quae non potest non exacta esse et singularem artis et industriae laudem mereri, est Excellen[tissimus] Dominus Joan[nes] Ern[estus] de Rautenstein Neoburgicus Minister Status et ad Comitia Ratisb[onensia][2] Legatus, Vir et litteris egregie politus et de illis optime meritus, Tibi quoque notus, qui ea de re et ad Nobiliss[imum] Dominum Pelsium[3] Belgarum Faederatorum apud Vos Residentem per manus meas meque conscio scripsit. Quanquam non desint Hafniae et hic loci Viri Veteri usura Tecum juncti, et alioqui Tui observantissimi et studiosissimi, qui libenter id a Te mea qualicunque industria et sincera benevolentia provocato fieri vellent, et[4] sperant firmiter Te hanc oculorum et mentis oblectationem singularem illis non invisurum. Nudius certe tertius ea de re locuti sumus invicem Clarissimi Viri, Dominus Kirstenius, Dominus Mullerus, et ego, per quem illi plur[imam] salutem una cum prosperi[5] et laeti anni novi voto et constantis benevolentiae testimonio adscribunt. Mullerus vero etiam brevi se ad Te scripturum promisit. Atque is Tecum sentit, quod et ex eius epistolis ad me datis

1 exspectant *om.* **BO.**
2 ad...Ratisbonensia : ad Conventum Imperii Ratisbonensem **TC.**
3 Pelsium : Philippum Pelsium **TC.**
4 et : Iidem **TC.**
5 prosperi : felicis **TC.**

vides, Cometam hunc, qui adhuc visitur in caelo,[6] unum eundemque cum priori esse. Kirstenius in diversam potius sententiam ibat, pede tamen tremulo. Hafniae Vir quidam Astronomica scientia praeditus diversum esse hunc ab illo Cometam ex situ et motu eius probare conatur. Accepi a praedicto Domino de Raut[enstein] quendam celeberrimum in Germania Astronomum existimare unum eundemque Cometam esse. Hic Te velim sententiam Tuam brevibus confirmare et adstruere. An audiveris Hallis Saxonum aliquem Astronomum esse, qui ante biennium circiter Cometae istius apparitionem hoc tempore futurum praedixerit, quidve de talibus praedictionibus statuas (nam et Dominum Gerickium Consulem Magd[eburgensem] duobus mensibus Cometam hunc praesagiisse nuper Tibi scripsi) velim nobis indices, nec non certiores facias // de cursu Cometae, ac scilicet Maiorem vel Minorem circulum peregerit. Nam et hic diversas video esse sententias, nec cum Bullialdo Mullerum facere. Tu autem beneficio admirabilium Tuarum machinarum non minus prompte quam diligenter et certo observare id potuisti. Omnes porro gratulamur merito Tuam *Cometographiam* sub praelo sudare,[7] utinam saltem vice strenae vertente anno tam eximium opus Orbi literario[8] exhibeas! Ad quod credo me posse aliquid conferre per accuratas et exactas Doctorum Virorum Observationes, quas misi et mitto, et porro mitterem, si Tua mutua diligentia et benevolentiae vice eorum industriam porro provocare mihi liceat.[9] Vale Vir Eximie cum omnibus Tuis et age feliciter accepto sincero voto in hunc ineuntem annum et plures alios, qui omnes ut Tibi sint fausti et laeti, apprecor.[10]

Dab[am] Hamburgi 3. 13. Jan[uarii] 1665

Tuus
omni studio et observantia
deditissimus

De Rolis
manu propria

Post Scriptum. Commendo Tibi inclusas ad Cl[arissimum] Büthnerum, cum quo velim pro tua humanitate communices quae ad Te mitto. Ille, ut video ex eius ad me literis, cultu et laudibus maximis prosequitur Te, Tuo quidem merito. Hac benevolentia mea excitatus sententiam quoque suam nobis dicet. Volui hoc compendio laboris et impensae uti, de Tua aequitate persuasus. Iterum vale.[11]

6 qui...caelo : qui nuper ad Leporem conspectus adhuc visitur in caelo **TC**.
7 sub...sudare : lucem brevi visuram **TC**.
8 tam...literario : Orbi literario tam eximium opus **TC**.
9 Ad quod...liceat *om.*TC.
10 apprecor : precor **TC**.
11 Post Scriptum...vale *om.* **TC**.

Greetings, most noble, worthy, and renowned Sir! I received your most pleasant letter eight days ago, but I could not answer it immediately, mainly because I was waiting for an answer from the most renowned Boulliau.[12] The message was delivered last Saturday, and I have it here together with the copies of the letters from Müller.[13] I believe that out of your kindness, you will reply to all of those letters and also to what I wrote and sent to you on 26 and 30 December, even though you are preoccupied with many duties, especially with more serious studies of the sky. Among those who expect your observations of the comet with your description — which certainly are exact and praiseworthy because of your exquisite craftsmanship and industry — there is most excellent Master Johann Ernst von Rautenstein,[14] the minister of state of Neuburg and deputy to the Diet in Regensburg, a man both educated in arts and distinguished in other fields, whom you also know, and who wrote — with my help and to my knowledge — about this thing to the most noble Pels,[15] the ambassador of the Dutch Republic in your city. Indeed, there are many people in Copenhagen and here bound with you by old business, revering and devoted to you, who would like it to be done by you, encouraged by my effort and sincere benevolence, and who firmly hope you will not deny them that extraordinary pleasure for their eyes and minds. Three days ago we discussed this thing — most famous Master Kirsten,[16] Master Müller, and myself, with whose help they send you many greetings along with the wishes of a good and joyful New Year and the proof of our continuous benevolence for you. Müller also promised to write to you himself. He agrees with you (which can be seen from his letters to me) that this comet which is still visible in the sky is the same as the former one. Kirsten rather thinks the contrary, but without absolute certainty. In Copenhagen, a man learned in astronomy[17] is trying to prove that this comet is different from the former one judging by its position and movement. I found out from Rautenstein that some famous astronomer in Germany thinks this is the same comet.[18] Therefore, I would like you to confirm it briefly and to elaborate on

12 See section 2.4.2 of the Introduction.

13 There is no trace of Boulliau's letter to Lubieniecki sent in December 1664 or in early January 1665, as listed by Robert A. Hatch in *The Collection Boulliau (BN, FF. 13019–13059). An Inventory* (Philadelphia: American Philosophical Society, 1982). On Müller see section 2.4.11 of the Introduction.

14 Johann Ernst von Rautenstein (*c.* 1622–66), German nobleman and diplomat in the service of the Duke of Palatine-Neuburg. He was Lubieniecki's patron and correspondent; in the TC (pp. 39–208), there are 172 letters they exchanged (85 from Rautenstein, 87 from Lubieniecki).

15 Philips Pels (1623–82), Dutch diplomat. After 1643, he was an envoy to Poland and Prussia. Since 1659, he was the emissary of the Dutch Republic in Danzig. He is frequently mentioned in Lubieniecki's correspondence with Hevelius, although little is known about his contacts with either of them. He was probably one of the addressees of Lubieniecki's political reports. One of them was preserved in the Hevelian corpus (see Letter 17A). In Olhoff's *Excerpta ex litteris* [...] *ad dominum Johannem Hevelium* (Gedani: ex officina Janssonio Waesbergiana, 1683), pp. 202–03, there is a reprint of Boulliau's letter to Pels, dated 9 February 1680, concerning the fire of Hevelius's houses. See also CJH3 p. 437, note to Letter 160, Hevelius to des Noyers, 29 April 1672.

16 See section 2.4.9 of the Introduction.

17 Identified in Letter 7 as Johannes Melchior Røtlin.

18 Probably Albert Curtz, see section 2.4.5 of the Introduction.

your opinion. I would like you to tell us if you heard that there was a certain astronomer in Halle in Saxony[19] who had foreseen the appearance of the present comet about two years ago, or what you think about such predictions (for I wrote recently to you that Master von Guericke, the Councillor of Magdeburg, foresaw this comet two months earlier), and also if you could inform us about the route of the comet, namely whether it is moving on a great or on a small circle. Because here too I see different opinions, and Müller disagrees with Boulliau. You could observe it not only on the spot but also diligently and unquestionably, with your excellent instruments. We all congratulate you on the fact that your *Cometographia* is in press. May you present this great work to the world of learning with the beginning of the year as a New Year's gift. I believe I can contribute somehow to this work with accurate and perfect observations of learned men, which I have already sent and will continue to do so, as long as your diligence for me and reciprocal benevolence ask me to proceed with my efforts around them. Farewell, exceptional Sir, and it is my sincere wish that you prosper at the beginning of this year and in many more years to come, which will be, I pray, happy and joyful for you!

Given at Hamburg, on 3/13 January 1665

Yours devoted and obliged,

De Rolis
by my own hand

Postscript. I entrust to you the enclosures for the most renowned Büthner,[20] to whom I would like you to pass what I am sending to you. As I see from his letters to me, he honours you with reverence and highest praise. Exhorted by my benevolence, he will also tell us his opinion. I wanted to save some work and expense in this way, confident about your honesty. Farewell again!

19 Lubieniecki refers here to Bartholomäus Schimpffer (*fl.* 17th century) but he made a mistake in the year of his prediction. In his booklet *Kurtze Beschreibung deß dunckelen Cometen so anno 1652 den 8. Decembr. erschienen* (Hall in Sachsen: Johann Rappoldt, 1652), sig. [C iiii v], Schimpffer stated that something new would appear in the sky in December 1664 as a result of the conjunction of all the planets in Capricorn. Lubieniecki quoted the correct year of the prediction, for example, in his letter to Friedrich Büthner (see section 2.4.3 of the Introduction), sent on 16 November 1665, TC1 p. 826.

20 See section 2.4.3 of the Introduction.

5A.

Continuation of the observations from Magdeburg

BO: C1, v. 6, 912[21]
TC: 239

Observatio continuata Magdeburgica

Den 18. Dec[embris] morgens umb 3. Uhr hat er sich wiederumb sehen lassen in Argonavi, daß ist das[22] Schiff oben[23] zwischen der hintersten und mittelsten mast, etwa 3. oder 4. grad[us] zur selben Zeit über den Horizont, daß man ihn hernach umb 5. Uhr wegen der Atmosphaera nicht mehr sehen können, und war also von seinem Außgange[24] (alß den 2. Dec[embris]) biß dato 55. grad[us] circuli maximi nach Südwest gegangen, und unterdeßen[25] in allen fast 10. grad[us] niedriger versus Austrum gewiechen. Der schwanz war auch wie alle wege von der Sonnen abgewandt, doch nicht mehr so gar groß, wie am 12ten Dec[embris] Ob nun solche[26] declinatio versus Austrum daher entstehe, daß sich etwa der Comet niedriger zur erden senckte, oder ob es etwa auß seinem eigen[27] motu herrühre, kan man nicht wissen.[28]

Hamburgi d[ie] 23. Dec[embris] St[ylo] V[etere] 1664.[29]

Mitto continuationem observationum Domini mei parentis. In priori erat error, quem velim corrigi. Nam loco vocabuli Solis ponendum erat illud Horizon.[30]

21 Each part of the enclosure is printed separately here.
22 das : im TC.
23 oben : eben TC.
24 Außgange : anfange TC.
25 unterdeßen : unterdeß TC.
26 solche : sulche TC.
27 eigen : eigenem TC.
28 Latin translation in the TC: Die 18. Dec[embris] mane circa tertiam iterum se (Cometa) conspiciendum praebuit in Argo navi, superius inter anteriorem et medium malum, circiter 3. aut 4. grad[us] hoc tempore supra Horizontem, ita ut postea circa quintam horam propter Atmosphaeram amplius conspici non potuerit. Ab initio igitur suo (scilicet 2. Decembris) in hunc diem 55. gradus circuli maximi versus Notozephyrum progressus erat, et interea in universum fere 10. grad[us] ad inferiora versus Austrum decesserat. Cauda quoque erat, ut alias semper, a Sole aversa, non tamen tanta, quanta visa est die 12. Decembris. Utrum autem illa declinatio versus Austrum inde oriatur, quod forte Cometa magis deorsum ad terram se demittat, an vero illa ex proprio eius motu proveniat, sciri nequit.
29 23...1664 : 3. Januarii S[tylo] N[ovo] 1665 TC.
30 In...Horizon *om.* TC.

Es ist ietzo[31] wieder ein Newer Comet entstanden, welchen ich Gestern Abendts hora 6. usque ad 9. observiret, gehet außen Mittage nach den Abend, steiget sehr hoch auff, gehet geschwinde, und ist nicht Crinitus, hat neml[ich] gar kainen schwantz.[32]

The continuation of the observation from Magdeburg

On 18 December, at 3 o'clock in the morning, the comet reappeared in Argo Navis (which means Ship), over and between the rear and the middle mast, about 3 or 4 degrees (at that time) over the horizon, so that after about 5 o'clock it could not be seen any longer because of the atmosphere. Since its appearance (that is, 2 December) until today, it has passed 55 degrees on the great circle towards the south-east, and meanwhile, it has moved almost 10 degrees towards the south. Its tail was (as they always are) turned away from the Sun, but not like that one which was seen on 12 December. No one knows whether this declination towards the south comes from the fact that the comet descends towards the Earth, or it is caused by its own movement.

In Hamburg, on 23 December 1664, old style.

I am sending the continuation of my father's observations.[33] In the former one, instead of the word 'Sun,' there should be 'horizon.'

The new comet has risen again. I was observing it yesterday evening, from 6 to 9 o'clock. It is moving from the south towards the west, its height increases, it goes fast, and it is not hairy, viz. it has no tail.

31 ietzo : izo **TC**.

32 Latin translation in the TC: Jam iterum novus Cometa ortus est, quem ego heri versperi hora sexta usque ad nonam observavi; progreditur a Meridie versus Occasum, ascendit in altum, currit celeriter, nec est crinitus: nullam nempe habet caudam.

33 See Letter 2D.

5B.

Letter from Müller to Lubieniecki, 23 December 1664 / [2 January 1665]

BO: C1, v. 6, 912
TC: 846

Vir Praecellentissime

Cometa ille qui et nunc horis vespertinis caelo serenante conspicitur, primum se mihi obtulit die 20/30 Dec[embris] vesperi hora nona, prout hesternis meis literis significavi, et situm habuit quem iisdem expressi. Alter jam abiit, excessit, evasit, erupit, nullusque est, et in locis nostra hac Urbe meridionalibus comparet. Insignis enim ipsius moles facile evincit, eum nondum interiisse, sed abitu ad populos Australiores suo nos aspectu privasse.[34] Difficilem vero mihi moves quaestionem, an nimirum hic Cometa sit plane novus? an vero idem cum priore rescissa causa? qua de re ita sentio: Cometes qui post occasum Solis in ortu depre-//henditur, est plane novus, atque id his 2. argumentis reddo verisimile; primo quia Cometarum historiae, literarum monumentis, virorum doctorumque industria, traditae (in quibus tamen me parum versatum esse profiteor) nullibi testantur, ullum Cometam cursum suum inflexisse, aut a via, quam prius inceperat, aberrasse; Atqui huius ultimi Cometae via aberrat a ductu prioris. Ergo, conclusio facilis est. Secundo, quia hic Cometa non solum a prioris tramite multis parasangis abest, sed etiam conficit rectam quae prioris viam (si utraque concipiatur produci) fere secat normaliter, sive ad angulos rectos. Hic etenim a Notapeliote migrat ad Borrhozephyrum, quantum Ego liberis oculis judicare possum. Ille vero a Borrhapeliote ad Notozephyrum videbatur abire.[35] Quod vero hic Cometa cauda videtur carere, nihil facit ad rem, ut ideo judicetur esse alius a priore. Non minus enim hic Cometes est[36] caudatus ac prior ille Corvi,[37] quamvis cauda huius[38] non conspiciatur. Nec minus esset admiratione dignum si Cometam illum sine cauda fuisse dicerem, quam si Solem in plenilunio defecisse conarer Literatis persuadere. Causa vero cur quidam Cometarum nullam exhibent caudam, haec est; quia tanto intervallo (ratione partium circularium) a Sole, qui tum lucem tum caudam Cometae gignit, removentur, ut ei per diametrum circiter oppositi, caudas sursum

34 Alter...privasse *om.* TC.

35 qua de re...abire : Difficilem dico Te mihi proponere quaestionem, tum quia caelum a die undecima usque ad 20. Decembris St[ylo] V[etere] fuit obscurum, tum quia accurate observationes ex quibus rectius judicium formare liceret, nos destituunt: Sentio tamen hac de re ita, et dico: Cometes, qui nunc post occasum Solis ad Eridanum deprehenditur, non est novus, quamvis primo intuitu aliter senserim. TC.

36 est : nunc est TC.

37 ac...Corvi : ac antea, cum in Corvo haereret TC.

38 cauda huius : ea TC.

erigant, easque, sua mole quasi absorbeant, quo minus a Terricolis videri possint. Atque hoc tum de Cometa Anni 1652. tum de Cometa, qui adhuc in caelo apparet, tum de quibusvis similibus verum est. Quemadmodum enim illi populi, quibus Sol est verticularis, nullam umbram Vulgo dicuntur habere, quamvis aeque ac alii, suas proiiciant umbras, sed versus centrum terrae, quae eapropter ab ipsorum corporibus teguntur, ut videri nequeant, ita etiam se res habet cum Cometis, qui[39] diametrali fere distantia a Sole absunt. Comprobat hanc meam opinionem[40] experientia, quae ostendit Cometas a Sole remotissimos, vel sine cauda vel cum exigua eius portione // semper fuisse observatos. Sine dubio etiam mihi adstipulabitur is qui adhuc in Caelo lucet Cometes; Cum enim haereret in Lepore, hoc est in Dodecatemorio Geminorum, diametraliter fere opponebatur Soli in initiis Capricorni haerenti, atque ita caudam sursum porrigendo nullam ostendebat, nisi forte[41] Luna radiorum solarium reflexione simul fuerit[42] impedimento, quod quidem vero non est absimile.[43] Sed postquam Cometes longius fuerit c[ontra] s[eriem] s[ignorum] promotus (habet enim motum velocissimum, qui Telluri admodum vicinum esse arguit) versus Borrhozephyrum, itaque Soli factus vicinior, nullus dubito, modo satis diu perduraverit, quin paulatim caudam sit proditurus. Et ut tribus verbis me expediam, cauda cometae eiusque prolixitas dependet, caeteris paribus, ex aspectibus quos ille habet cum sole, sive a numero graduum, quibus haec duo corpora ab invicem removentur. Unicum hoc priore cum Cometa hic videtur habere commune,[44] quod Nucleum satis[45] Lucidum in sui medio ostendit.[46] Hisce nunc finio et Te divinae tutelae etiam atque etiam[47] commendo, Hamb[urgi] d[ie] 23 Dec[embris] Jul[iani] Anno MDCLXIV

Tuae Claritati ad quaevis officia paratissimus

J[ohannes] Mullerus

Most eminent Sir!

That comet which is now being seen in the evening (if the sky is clear), showed itself to me for the first time on 20/30 December at 9 o'clock in the evening — as I wrote in my yesterday's letter — and it was in that place which I noted there. The other

39 qui **TC** : quia **BO**.
40 opinionem : sententiam **TC**.
41 nisi forte : quin **TC**.
42 fuerit : fuit **TC**.
43 quod...absimile *om.* **TC**.
44 Unicum...commune : Cometam vero hunc eundem esse cum priore illo Corvi etiam hoc reddit verosimile **TC**.
45 satis : aeque **TC**.
46 ostendit : ostendat **TC**.
47 Te...etiam : Divinae protectioni Te iterum iterumque **TC**.

one has already left, went away, fled, broke away, disappeared, and is visible in the areas south of our city. Its considerable size makes us sure that it has not perished yet, but having moved to the southern peoples, it deprived us of its sight. Your question whether this comet is a new one or the former one with a cut tail is difficult. My opinion about this is as follows: the comet which is seen in the east after the sunset is entirely new, and I can give two arguments which make this probable; firstly, the observations of comets, conveyed in literary works and by diligence of learned men (I admit, however, that I am not very experienced in this matter), prove that a comet never inflected its route or deviated from the path which it had initially taken; and the path of the last comet deviates when compared to that of the earlier one. So the conclusion is straightforward. Secondly, because this comet is not only many leagues away from the trajectory of the first one, but also it is moving in a straight line which cuts the way of the former one (if we imagine that both stretch their routes) almost perpendicularly, viz. at the right angle. This one goes from the south-east towards the north-west, as far as I can judge with the naked eye. The former one seemed to go from the north-east towards the south-west. And the fact that this comet appears to lack a tail does not help to state whether this one is different from the former one. Because this comet has a tail like the former one in Corvus, but the tail cannot be seen. And it would not be less extraordinary if I said that this comet had no tail than if I tried to persuade the learned men that during the full Moon the Sun shines less. The reason why some comets show no tail is as follows: because they are at such a distance (speaking about fractions of a circle) from the Sun — which gives both light and a tail to a comet — that, being almost diametrically opposite, they raise their tails up and they absorb them with their magnitude, so the tails cannot be seen by the earthlings. And this is true about the comet of 1652,[48] about the comet which now can be seen in the sky, and about many similar ones. And those peoples over whom the Sun is vertical are often said to have no shadows, although they cast their own shadows like all others, but these are directed towards the centre of the Earth and, therefore, covered by their bodies, so they cannot be seen. And this is the case with the comets which are almost diametrically opposite the Sun. My opinion is confirmed by the experience which shows that the comets in their furthest distance from the Sun have always been seen without a tail or with a small tail only. Undoubtedly, my point is supported by the comet which is still shining in the sky. Because when it was in Lepus, viz. in the sign of Gemini, it was diametrically opposite the Sun, which was at the beginning of the sign of Capricorn. And so the comet showed no tail because it was stretching upwards (if the Moon was not enough of an obstacle due to the reflection of solar rays — which is not altogether absurd). But after the comet moves forward against the order of signs (because it has a swift movement which is said to happen next to Earth) towards the north-west, and then it will get closer to the Sun, I do not doubt that it will show its tail in due time, if only it lasts long enough. To state my point briefly, the tail of a comet and its size depend (among other things) on its aspect to the Sun,

48 See note to Letter 2B.

or on a number of degrees which is between these two bodies. The only thing which this comet has in common with the former one is the fact that it shows quite a bright nucleus in its centre. I will finish now, and I will constantly be commending you to divine protection. In Hamburg, on 23 December 1664 Julian.

Most renowned Sir,

Always at your service,

Johann Müller

5C.

Letter from Müller to Lubieniecki, 24 December 1664 / [3 January 1665]

BO: C1, v. 6, 912
TC: 847

Vir praecellentissime

Praecedentibus meis literis de novo Caeli phaenomeno sive Cometa meam mentem explicavi juxta Hypothesin Keplerianam, quae Cometas in recta linea ferri statuit. Si vero assumamus eos circulariter incedere, fere crediderim hunc Cometam non esse alium a priore, et confecisse hactenus arcum Circuli per Corvum, ventrem Hydrae, vela Navis Jasonis, Canem Maiorem, Leporem, partem Eridani, et branchias Ceti ducti. de quo tanto certiores essemus, nisi caelum adeo frequenter fuisset nubilum, nec soli oculorum judicio fidere necesse habuissem. Volui tamen aliquid hoc de Phaenomeno tecum communicare, ne plane Tuae Cl[aritatis] petitioni deessem. Hamburgi d[ie] 24. st[ylo] v[etere] Dec[embris] 1664.[49]

Most eminent Sir!

In my earlier letter, I explained my opinion about the new phenomenon in the sky, or the comet, according to Kepler's hypothesis saying that all comets go in straight lines. If we accepted that they move circularly, I would believe that this comet is no different from

49 Praecedentibus...1664. : Praecedentibus meis literis de novo Caeli Phoenomeno, sive Cometa meam explicavi mentem, nec quicquam causae invenio, quamobrem eam immutem, aut aliter sentiam. Interea temporis enim, quo caelum hic fuit nubilum, ille sine dubio *Navem Jasonis* et *Canem maiorem* percurrit. De quo tanto essemus certiores omnes, si caeli facies paulo frequentius fuisset conspecta. Vale. TC.

the former one and that so far it has passed an arc of a circle which leads through Corvus, the Stomach of Hydra, the Sails of Navis, the Canis Maior, Lepus, part of Eridanus and the Gills of Cetus. We would have been more sure about this if the sky had not been cloudy so often and I had not had to believe the judgement of my eyes only. I have decided, however, to share something about this phenomenon with you in order not to neglect your request, most famous Sir. In Hamburg, on 24 December 1664, old style.

5D.

Letter from Müller to Lubieniecki, 26 December 1664 / [5 January 1665]

BO: C1, v. 6, 912
TC: 847

Praecellentissime et Doctissime Domine

Hesterna die circa 9. vespertinam inter nubium hiatus aliquoties iterum videre Cometam mihi licuit, ubi eundem adhuc ab ipso cursus sui servari ductum deprehendi; haerebat ille vero prope stellas in Capite Ceti, habens declinationem borealem, et ut judicabam, pervadet rectam, quae Lucidam Mandibulae cum Media in ore jungit. Omnibus autem meis qualibuscunque observationibus probe inter se collatis, et accuratiore mentis trutina perpensis, nunc tandem[50] statuo, hunc Cometam alium non esse ab eo, quem in Corvo vidi et confecisse hactenus traiectione sua rectam lineam, quam oculus noster non animadvertens, sed ad stellas fixas expendens, credit esse arcum circuli maximi per Corvum, Hydram, Canem Maiorem, Leporem, Eridanum, Caput Ceti, Linum boreale et magnum Quadratum Pegasi[51] descripti. Caudam quoque adhuc alit, quae hesterna vespera dicta h[ora] 9. ubi Cometes transierat Meridianum, tendebat versus ortum longitudine apparente 2. pedum. Perspicio etiam ex tuis Literis[52] Te meam de Cometa mentem velle cum aliis communicare,[53] quod quidem illibenter fieri vellem. Nam, ut M[arcus] Tullius alicubi recte inquit, 'aliter scribimus quod eos solum, quibus mittimus; aliter quod multos lecturos putamus.' Vale felicissime per multos annos. Hamb[urgi] d[ie] 26. Dec[embris] Jul[iani] 1664.

Tuus quem nosti

J[ohannes] M[ullerus]

50 nunc tandem : adhuc **TC**.
51 magnum...Pegasi : Andromedam **TC**.
52 tuis Literis : literis Tuis **TC**.
53 velle...communicare : cum aliis communicaturum **TC**.

Most eminent and learned Sir!

Yesterday evening, at 9 o'clock, I could see the comet again a few times in the space between the clouds, in the place where (as I observed) it appeared thanks to keeping its course. It was near the stars in the Head of Cetus; it had a northern declination, and (as I gathered) it was moving in a straight line which connects the bright star in the Jaw of Cetus with the middle star in its mouth. Having correctly collated my all observations and having considered them with a more accurate judgement of the mind, I am finally stating that this comet is not different from the one which I saw in Corvus. And so far with its trajectory it has drawn a straight line which was not seen with our eyes but stretched to the fixed stars, and it was believed to be an arc of the great circle which goes through Corvus, Hydra, the Canis Maior, Lepus, Eridanus, the Head of Cetus, the Southern Line and the Great Square of Pegasus. It keeps its tail which pointed towards the east, and its visible length was 2 feet yesterday evening at 9 o'clock, as I said, when the comet was passing the meridian. From your letter I can see that you want to give my opinion about the comet to others; I would not like it to happen. As Marcus Tullius somewhere says it correctly: 'we write in one way what we intend to be read only by those to whom we send it, in another way — what we intend to be read by many.'[54] Farewell, and stay happy for many years to come! In Hamburg, on 26 December 1664 Julian.

The one you know,

Johann Müller

5E.

Letter from Müller to Lubieniecki, 28 December 1664 / [7 January 1665]

BO: C1, v. 6, 912
TC: 847–48

Vir Praecellentissime et Nobilissime

Gratias tibi habeo maximas, quod Literas Gedano missas mecum volueris communicare, easque nunc remitto. Ampliss[imus] Dominus Hevelius requirit observationes solidioribus machinis habitas, et quidem rectissime; quia caeterae observationes nihili aestimandae; sed ego tales, quales ille desiderat, minime polliceor. Heri vesperi hora 6. vidi ego Cometam multo borealiorem factum esse, et cum Lucida Mandibu-

54 Cicero, *Ad Familiares*, 15, 21, 4.

lae et Media in ore Ceti efficere triangulum Rectangulum isosceles, cuius angulus erat in Cometa rectus, et versus Zenith vergebant, basis sive Hypothenusa erat recta jungens dictas stellas, et horizonti erat vicinior. Post 9[nam] vespertinam existebat in una recta linea cum Media in ore Ceti et procedente trium; utraque tamen borealior conspiciebatur. Hac vespera hora 6. eum iterum contemplatus sum et cum Lucida mandibulae Mediaque in ore Ceti etiam fere conflare triangulum Rectangulum, sed non, ut heri, cuius angulus ad Cometam esset rectus, verum cuius angulus ad Mediam in ore recto aequivaleret, et Catheti denuo aequales. Vale. d[ie] 28. Dec[embris] st[ylo] v[etere] 1664.

Totus Tuus

J[ohannes] M[ullerus]

Most eminent and noble Sir!

I am very grateful that you have decided to share with me the letter which you received from Danzig; now, I am sending it back. The greatest Master Hevelius requests perfect observations made with better instruments because other observations are worthless. But I cannot promise those which he needs. Yesterday evening, at 6 o'clock, I saw that the comet became more northern and it made a right-angled isosceles triangle with the bright star in the jaw and the middle star in the Mouth of Cetus. Its right angle was at the comet, and was directed towards the zenith; its base, or hypotenuse, was a straight line between those aforementioned stars and was closer to the horizon. After 9 o'clock in the evening, the comet was in a straight line with the middle star in the Mouth of Cetus and the first one of the three stars; but both of them were more northwards. This evening, at 6 o'clock, I observed the comet again, and again it made almost a right-angled triangle with the bright star in the jaw and the middle star in the Mouth of Cetus, but not like yesterday (when the angle at the comet was right) but the angle at the central star in the mouth was right, and the catheti were equal again. Farewell! On 28 December 1664, new style.

Yours,

Johann Müller

5F.

French translation of the observations from Magdeburg[55]

BO: C1, v. 6, 911

Extrait des observation de Magdebourg

J'ai vue le comete pour la premiere fois le 2 decembre 1664 a 5 heures du matin vers midi in corvo juxta collum eius, elle avoit une queue assés large, mais pas bien longue, on ne l'a pas put voir après celà, par[56] ce qu'elle a été obscure, que le 12 decembre. Je[57] l'ai vue dans ce tems là mais plus bas versus Austrum in Hydra sub Cratere élevée[58] de 10° sur le soleil, savoir avec une queue bien plus grande et plus longue, de sorte qu'elle paroist soit bien importente. Ainsi elle etoit dejà avancée de 12 degrés vers l'occident. //

Suite des observations de Magdebourg.

On l'a encore vu le 18 decembre à heures du matin, in Argo navi, c'est à dire, en hut entre[59] le dernier et le mat de milieu, elevé environ de 3 ou 4 degrés sur l'orizon, mais on ne l'a plus vue[60] à 5 heures, à cause de l'Admospher, ainsi elle a fait un chemin depuis le commencement (le 2 decembre) jusqu'à present des 55 degrés d'un grand cercle vers le sud, et celle est descendue et tout presque de 10 degrés plus bas de versus Austrum.

An excerpt of the observations from Magdeburg

On 2 December 1664, I saw the comet for the first time — in the morning, at 5 o'clock before noon, in Corvus, next to its Neck; its tail was quite wide, but not very long. Later, one could not see it because it remained obscured until 12 December. I saw it at the same time much lower towards the south, in Hydra under Crater, at the altitude of 10° over the Sun, and it had a much bigger and longer tail, which looked very prominent. It also moved over 12° towards the west.

The continuation of the observations from Magdeburg.

55 See Letters 2D and 5A.
56 pars (s *del.*) **BO**.
57 Jaie (ai *del.*) **BO**.
58 ell *del.* **BO**.
59 les derniere et second *del. post* entre **BO**.
60 apres *del. post* vue **BO**.

On 18 December, at 3 o'clock in the morning, it was seen again in Argo Navis (which is the Ship), above and between the rear and middle mast, about 3 or 4 degrees over the horizon, but at 5 o'clock it could not be seen any longer because of the atmosphere. Since its appearance (2 December) until now, it has passed 55 degrees of the great circle towards the south, and it moved almost 10 degrees down, towards the south.

6.

27 January 1665, Hevelius to Lubieniecki

BO: C1, v. 6, 915
TC: 364

Illustri Viro
Domino De Rolis

J[ohannes] Hevelius Salutem

Pro communicatis observationibus Cometae iterum iterumque gratias Tibi debeo; inprimis arrisit multum judicium Hafnia transmissum: quaecunque enim de praedictionibus Astrologicis autor[1] disserit, in illis omnino ipsi adstipulor. Amicis honoratissimis observata mea circa hunc Cometam acquisita tam avide exspectantibus, vellem equidem lubentissimo animo morem gerere; verum cum integrum opus de Cometarum materia moliar, haud fieri potest, ut plenam descriptionem et accuratam delineationem rerum omnium in hoc Cometa deprehensarum hac vice exhibere possim. Ut autem[2] ex parte eorum desiderio satisfiat, constitui quantocyus generalem quendam huius Cometae descriptionem, cum quibusdam iconismis, typis committere; quae quamprimum edita fuerit, delineationesque manu mea aeri incisae lubens amicis omnibus transmittam. Interea haec Te scire velim, Cometam matutinum suo tempore vespertinum factum esse, sic ut unus tantummodo affulserit, ut ut alii aliter sentiant. Nam a prima apparitione continue hucusque illum observavi, absit gloria, quam accuratissime. Cursum suum instituit (uti ex prioribus // literis partim jam praecepisti) ex corvo per hydram, navim, Canem maiorem, Leporem, Eridanum ac Caput Ceti; die 23 Januarii in 26° Arietis, sub latitud[ine] 4° fere Bor[eali][3] sub stellis scilicet Arietis haerebat; In medio sui tramitis erat velocissimus, circa extremitates vero multo tardior, atque[4] nunc tardissimus; mense Februario, prout me observaturum spero, fiet stationarius, imo retrogradus, vel ut rectius dicam, directus, cum modo contra Signorum seriem incedat. Motum suum, ut ut inaequalem quoad apparentiam, revera tamen aequalem et summe ordinatum, sub circulo propemodum maximo exercuit; propemodum inquam, ut recte intelligas: siquidem omnes Cometae non adeo praecise ductum alicuius circuli maximi sequuntur; quare autem et quomodo *Cometographia* clare ostendet. Eximium Bullialdum a nobis in diversum abire, tam circa viam Cometae, quam quod diversos duos existisse arbitretur, non

1 ille *del.* **BO**.
2 Ut autem : Verum tamen, ut **TC**.
3 sub...Boreali : sub latitudine Boreali 4 fere grad[us] **TC**.
4 imo *del.* **BO**.

est quod mireris: quandoquidem[5] ex una aut altera observatione interrupta id ipsum decernere ille haud potuit; nullus dubito, cum omnia bene perpenderit, nostrasque observationes viderit, quin sententiam sit mutaturus. Atvero miror sane multum, qui fieri unquam possit, cum Astrologi glorientur, se posse Cometas adeo praecise praedicere, quod[6] famosus ille Astrologus Germanus circa pacem // Turcicam, quae eodem anno quo bellum minatur, inita est,[7] adeo[8] enormiter aberraverit? Profecto adeo multa, imo contraria verba nonnunquam funditant,[9] ut mirum esset si casu unum aut alterum non attingeret. Quam multos Cometas ex Coniunctionibus magnis praedixerunt quorum nullus sane humanis oculis ortus[10] est! De caetero, videtur Clarissim[um] Dominum Müllerum in ea esse opinione, Cometam iter suum inflexisse; sed nequaquam. Semitam enim quam initio iniit, constanter prosequutus est; quicquid vero de cauda disserit, ita omnino se se habent omnia. An nucleum istum lucidum, de quo loquitur, nudis an[11] armatis viderit oculis, libenter scirem. Ego quidem diversos quoque annotavi sed praelongo tubo.[12] Denique Num Cometa in linea recta traiicere possit, et nihilominus sub circulo fere maximo videri, suo loco fuse demonstrabitur. Saluta amicos omnes. Vale et fave.

Dabam Gedani Anno 1665, die 27 Januarii.

To the illustrious Sir,
Master De Rolis

Johannes Hevelius sends greetings

I still owe you many thanks for the observations of the comet which you shared with me. I much enjoyed the opinion which was sent from Copenhagen, although I entirely disagree with the author on what he says about astrological predictions. I would willingly like to fulfil the wishes of my most honoured friends who impatiently wait for what I observed about the comet, but since I am struggling to complete a book about cometary topics,[13] I cannot present now a full description and comprehensive discussion of all the things seen in the comet. Nonetheless, in order to satisfy their wish in part, I have decided to publish (without delay), a general description of the

5 si *del.* **BO**.
6 ut *del.* **BO**.
7 est inita est (*primum* est *del.*) **BO**.
8 ut *del.* **BO**.
9 proferunt *del.* **BO**.
10 ortus : visus **TC**.
11 vel *del.* **BO**.
12 tubo : Telescopio **TC**.
13 Hevelius refers to his *Cometographia*, eventually published in 1668.

comet with some illustrations.[14] Once it is published, and when I engrave the illustrations in brass, I will gladly send it to all of my friends. Meanwhile, I would like you to know that the morning comet has become the evening one. Thus, only one has been shining, although some people think otherwise. I have continuously been observing it with great diligence since its first appearance until now (I am not saying it to boast). It has passed the route (as you partially already know from my previous letter) from Corvus through Hydra, Navis, the Canis Maior, Lepus, Eridanus, and the Head of Cetus. On 23 January it was in 26° of Aries, at 4° north, below the stars of that constellation. In the middle of its route, it was the fastest; as it was tending towards the end, it became much slower, and now it is the slowest. In February, as I hope to observe, it will become stationary (even retrograde), or to say it more precisely — direct, when it is going to move against the order of signs. Although it seems to be moving irregularly, actually its movement is most regular and steady, nearly in a great circle. I say 'nearly,' so you can understand it correctly, for not all comets move so precisely in a great circle. My *Cometographia* will clearly show why and how it happens. You should not be surprised that exceptional Boulliau[15] holds different opinions (both about the route of the comet and about the fact that there were two comets) because he could not determine it from one or another unrelated observation. I do not doubt that he will change his opinion when he considers everything and sees our observations. And I am amazed how it is possible — since astrologers boast that they can predict comets so precisely — that this renowned German astrologer made such a great mistake about the peace with Turks which was signed the same year when there was a threat of war.[16] Indeed, he utters so many words (sometimes even contradicting each other) that it would be strange if he were not accidentally right about something. How many comets had been foretold on the basis of great conjunctions, none of which appeared to human eyes! As for other matters, it seems that the most renowned Master Müller[17] thinks that the comet is bending its course — but it is not the case. It is continuously following the path it entered, but what he says about the tail is generally correct. I would like to know whether he saw that bright core of which he writes with the naked or with the aided eye. I observed various things as well, but with a long telescope.[18] Finally, it will be more broadly discussed elsewhere, viz. whether a comet can move in a straight line and nevertheless be seen in a great circle. Greet all friends. Farewell, and stay benevolent!

Given at Danzig, on 27 January 1665.

14 Johannes Hevelius, *Prodromus cometicus* (Gedani: autoris typis, et sumptibus, 1665).

15 See section 2.4.2 of the Introduction.

16 The danger of Turkish invasion was a recurring topic of early modern astronomical predictions, see Robin B. Barnes, *Astrology and Reformation* (New York: Oxford University Press, 2016), pp. 224, 240, 246.

17 See section 2.4.11 of the Introduction.

18 See note to Letter 3.

7.

10/20 February 1665, Lubieniecki to Hevelius

BO: C1, v. 7, 949
TC: 364–65

Enclosure:
7A. Excerpt of the letter from Lubieniecki to de Grau, 20 February 1665

Salutem Plurimam Vir Nobilissime, Clarissime et Humanissime. Dum in itinere Holsatico essem, literas Tuas die 27. Jan[uarii] datas Schlosuici die 4. Febr[uarii] accepi; ad quas nunc demum reverso domum respondere licet. Hafniensis ille Amicus meus Tuusque est[1] inter Viros egregios merito collocandus, qui Te certe pl[urimi] aestimat,[2] et in illa de uno ac gemino Cometa controversia, totaque de hoc phaenomeno disputatione plus Tibi soli, quam cunctis aliis tribuit fiditque. Verba eius vel ad Te, vel ad Nobiliss[imum] Dominum Pelsium Faederatorum Belgii Ordinum apud vos Residentem perscripsi, quae haec fuere: 'J'ay appris entr'autre, que vous avez eu quelque communication touchant la Comete de la part de M[onsieur] Hevelius, Astronome dont je fais plus d'estat, que de tous Les autres de nostre temps sans reproche toûjours et sans presumption.'[3] Paria et ego cum eo facio in eo, quod curiosas Astrologorum praedictiones, credulis satis noxias, utpote his securitatem, illis metum ingenerantes, non morer. Ingeniose fateor nonnulli ex dodecatemorio Librae (cui antiqua curiositas eiusdem notae religionem et justitiam subesse voluit) bellum religionis, ex dodecatemoriis Tauri et Arietis, ut et ex signis Hydrae aquaticae, navis Jasonis, Orionis viri bellicosi, Eridani et Ceti, quae Cometes vel peragraverit[4] vel attigerit, bellum idque navale in mentes hominum praesagiis istiusmodi deducere allaborant. Sed videbimus quid de aliis dodecatemoriis et signis caelestibus, quae itidem Cometa vel permeavit, statuent et elicient iidem, quibus talia tractare et populo proponere placet. Libenter videbo, quid de Virgine et Cancro, quid item de Cratere, Cane Maiore, et Lepore dicturi sint, et[5] Suo instituto accomodaturi. Interim libenter patiar eos frui propria libertate et ingenio habenas laxare, dummodo // nec[6] ipsi curiositate abrepti et in avia delati impervestigabilia arcana tentent divinae sapientiae, nec alios a tramite recti vel desperatione vel audacia aberrare faciant. Cui magis de caelestibus quam caelesti voci, imo cui nisi huic soli tuto et citra pericu-

1 est : est Joannes Michael Röthlinius Serenissimo Principi Regio Daniae a Secretis Camerae, olim in Urbe vestra Serenissimi Regis sui negotia agens **TC** (name 'Michael' here is a mistake, as the man in question is Johannes Melchior Røtlin).

2 aestimat : facit **TC**.

3 Verba...presumption *om.* **TC**.

4 peragraverit **TC** : peragrarit **BO**.

5 et : et qua ratione haec **TC**.

6 nec **TC** : nec nec **BO**.

lum salutis credere debemus? Atqui haec nec[7] a signis caeli metuendum nobis gentium ritu, nec ea requirenda et investiganda, perversorum more esse, monet. Huic itaque obediamus; interim his signis, quae citra exspectationem votaque nostra eveniunt, ad paenitentiam veram et seriam et[8] avertendam a nobis decretam et imminentem vindictam divinam commoveri nos patiamur, non truncis lapidibusque, sed creaturis rationalibus, vi ingenii, perspicacitate judicii, libertate arbitrii, arcanorum divinae voluntatis notitia praeditis pares.[9] Quid in hanc rem ad Cl[arissimum] Abrah[amum] de Grau Matheseos in Acad[emia] Franekerana Prof[essorem] scripserim, adiuncta docet charta, quae et plura alia scitu haud indigna Tibi indicabit, inter quae videbis etiam, quanti Te Tuo merito R[everendus] P[ater] Alb[ertus] Curtius Vir magnae eruditionis,[10] faciat, quae Tecum communicari Excellent[issimus] Dominus de Rautensteinius Consiliarius et Legatus Neoburgicus[11] voluit ad elicienda plura, quae sunt ipsissima eius verba, ut et haec: 'Gratias maximas ago quod mei apud Dominum Hevelium tam honorificam mentionem G[enerosa] D[ominatio] V[estra] facere voluerit, a quo Viro tanto et tam celebri beneque de Rep[ublica] literaria merito nosci et amari non minus cupio quam a caeteris eiusem similibus': quae die 5. Febr[uarii] ad me exaravit: sed et alias honorificam et jucundam Tui mentionem faciens.[12] Equidem lubens intelliget Te velle veluti *Prodromum* quaedam *Cometographiae* Tuae in lucem edere[13], quod quidem mihi et mecum hic viventibus Amicis, tui observantissimis,[14] gratum auditu // fuit. Cl[arissimus] Mullerus, quicum Tuas literas communicavi, ad eas respondet, ut puto. A Cl[arissimo] Bullialdo nihil literarum noviter accepi. Sed eas in dies expecto, sicut et a Tua humanitate. Vale Vir Amplissime et Praestantissime, mihique porro fave.[15]

Dab[am] Hamb[urgi] 10. 20. Febr[uarii] 1665.

Tui
observantissimus et studiosissimus

De Rolis

Post Scriptum. Cl[arissimum] Büthnerum data occasione velim meo nomine salutes. Responsum hodie pro illo parabam, sed tempore exclusus perficere id non potui.[16]

7 nec *om.* **BO**.

8 et : ac **TC**.

9 non truncis...pares : non ut trunci lapidesque, sed ut a Deo creati, rationis praestantia, vi ingenii, perspicuitate judicii, libertate arbitrii, arcanorum divinae voluntatis notitia praediti **TC**.

10 Vir magnae eruditionis *om.* **TC**.

11 Excellentissimus...Neoburgicus : Illustrissimus Rautensteinius, ille politiori literatura et morum probitate Excellentissimus Consiliarius et Legatus Neoburgicus **TC**.

12 faciens : facit **TC**.

13 edere : emittere **TC**.

14 quidem...observantissimis : et nobis hic agentibus **TC**.

15 Clarissimus Mullerus...fave : Vale bono publico longum ac prospere, Vir Praestantissime **TC**.

16 Post Scriptum...potui *om.* **TC**.

Greetings, most noble, renowned, and kindest Sir! I received your letter, sent on 27 January, during my journey to Holstein, in Schleswig, on 4 February. I can finally answer it because I returned home. This our friend from Copenhagen[17] should deservedly be placed among the most excellent of men. He thinks very highly of you and, in this controversy about one comet seen twice (as in all discussions about this phenomenon), he believes you are right and would trust you rather than anyone else. I noted down his words both for you and for the most noble Master Pels,[18] the ambassador of the Dutch Republic in your city, which were as follows: 'I learned, among other things, that you received some correspondence about the comet from *Monsieur* Hevelius — the astronomer about whom I have better opinion than about all our contemporaries, always beyond reproach and blame.' I share his opinion that I should not care about curious[19] predictions of some astrologers, which are harmful to the gullible, as they bring safety to some people, but fear to others. I admit some of them work cleverly with such prophecies to imprint into people's minds a reading from the sign of Libra (to which ancient curiosity of this sign wanted to attribute religion and justice) that there will be a religious war; and a reading from the signs of Taurus and Aries, as well as from the sings of the aquatic Hydra, the Navis of Jason, Orion (the warlike man), Eridanus, and Cetus (which the comet passed or touched) that there will be a naval war. But we will see what those men, who like to discuss such things and show them to people, will state and suggest about other constellations and celestial signs, passed or touched by the same comet. I would like to know what they would say about Virgin and Cancer, and also about Crater, the Canis Maior and Lepus, and how they would apply them to their concepts. Meanwhile, I will let them work freely and use their minds at will, unless, caught by curiosity and moved into wilderness, they try to reach the unfathomable divine wisdom or try to deviate others from the righteous path out of desperation or boldness. Whom should we rather trust about celestial things, confident and sure of salvation, than solely the heavenly voice itself? And it admonishes us neither to fear the heavenly signs, as heathens do, nor to examine and investigate them, which is a custom of the perverse. Therefore, we should obey this voice. Yet, we should accept that those signs which appear against our expectations and wishes lead us to the real and true penitence and avert from us the ordained and imminent God's punishment, for we are similar not to trunks and stones, but to rational creatures, endowed with the force of the mind, the perspicacity of judgement, the freedom of will, and the knowledge about the secrets of the will of God. What I wrote about this matter to the most famous Abraham de

17 Probably Johann Melchior Røtlin, see note to Letter 4.

18 See note to Letter 5.

19 'Curiosity' was an important term in early modern scholarship and denoted eagerness to investigate the secrets of nature. On the other hand, it was sometimes condemned by more conservative circles, see Peter Burke, *A Social History of Knowledge. From Gutenberg to Diderot* (Cambridge, Malden: Polity, 2000), pp. 26, 83–84, 111. What is quite surprising, Lubieniecki uses this term in a derogatory sense, as this passage shows.

Grau,[20] professor of mathematics at the University of Franeker, the enclosed page will tell you. And it will also present you with other interesting things, among which you will see how much you are esteemed by the Reverend Father Curtz,[21] a man of great erudition; you will also learn what the most excellent Master Rautenstein,[22] the counsellor and ambassador of the Duke of Neuburg, wanted to incite you to write more about; rendered in his very own words thus: 'Thank you very much, most generous Sir, that you mentioned me so honourably to Master Hevelius, by whom (a man so famous and distinguished in the Republic of Letters), I desire to be known and loved, as by all people of his kind.' He wrote it to me on 5 February,[23] mentioning you also in other honourable and agreeable ways. He will gladly receive the news that you are publishing the *Prodromus* of your *Cometographia*, something I was also pleased to hear about, just as the friends living here who revere you very much. The most renowned Müller,[24] to whom I showed your letters, will answer them, I think. Recently, I have not received any new messages from the most renowned Boulliau.[25] But I expect them in the next few days, as I am looking forward to the ones from you, knowing your kindness. Farewell, greatest and distinguished Sir, and always be benevolent towards me!

Given at Hamburg, on 10/20 February 1665.

Yours obligingly,

De Rolis

Postscript. I would like you to pass my regards to the most famous Büthner,[26] if the opportunity occurs. Today I was preparing my answer to him, but due to lack of time, I could not finish it.

20 Abraham de Grau (1632–83), Dutch theologian and philosopher. He was Lubieniecki's correspondent; in the TC pp. 487–503, 519–26, TC3 p. 77, there are 17 letters they exchanged (7 from de Grau, 10 from Lubieniecki).

21 See section 2.4.5 of the Introduction. There is no extant enclosure with the letter from Curtz.

22 See note to Letter 5.

23 The letter is actually dated 12 February, TC p. 47.

24 See section 2.4.11 of the Introduction.

25 See section 2.4.2 of the Introduction.

26 See section 2.4.3 of the Introduction.

7A.

Excerpt of the letter from Lubieniecki to de Grau, 20 February 1665[27]

BO: C1, v. 7, 950
TC: 488–90

Excerptum ex literis ad Virum Cl[arissimum] Dominum Abrah[amum] de Grau Matheseos in Acad[emia] Franek[erana] Professorem datis

Quod[28] ad judicia Praestantissimorum Astronomorum aliorumque virorum haud vulgari eruditione instructorum attinet, ea quantum quidem ego accepi, varia esse comperio. Nec tantum circa Originem et effecta Cometarum, sed et circa eius situm, motum et ipsam eius existentiam, diversitatem[29] illam animadverto. Tot sententiis de causis et natura Cometae in diversum euntibus, quas bene nosti, ita ut eas hic recensere non sit necesse, accessit nova Viri Amplissimi Domini Ottonis Gerrichii Cons[ulis] Magdeburg[ensis] viri admirandorum inventorum laude dubio procul Tibi quoque noti. Is enim statuit praesuppositis duobus difficillimis de motu telluris et spatio vacuo principiis, Cometam ex procellis oriri et esse partem aeris vi tempestatis cum impetu avulsam et in sublime delatam motumque eius[30] nihil aliud quam motum terrae, et sic illum varium pro huius varietate in oculis nostris esse. Et caudatum quidem in suprema aeris regione, quae nihil nisi aura sit, consistere, atque caudam eius esse partem aurae per corpus Cometae radiis solaribus illustratae et Cometam ambientis (ad quam sententiam comprobandam an non illa Epigenis[31] apud Plinium l[ibro] 2. c[apite] 27 tum et Apiani apud Scaligerum excerc[itatione] 79. // posita nec non ipsius Scaliger ibidem expressa aliquid faciat, velim mihi indices Semper aliquid discere parato imo perquam cupido) crinitus vero extra auram in spatio puro huncque a terra et ambiente eam aera Separatum esse, ac proinde nullam caudam habere. Tu quid ad haec omnia dicturus sis scire aveo.[32] Effecta Cometarum investigare et praedicere Astrologorum est praesertim ubi certa et determinata effec-

27 In the TC: 7/17 February 1665.

28 Quod : Quod autem **TC**.

29 circa Originem...diversitatem : circa eius situm, motum, sed et ipsam eius existentiam, sententiarum varietatem **TC**.

30 eius : eius si crinitus sit **TC**.

31 Epigenis **TC** : Epigonis **BO**.

32 Et caudatum quidem...scire aveo : vagum quidem, si Cometa sit caudatus. Dicit autem caudatum cometam in suprema aeris regione, quae nihil nisi aura sit, consistere et vagari, atque caudam eius esse partem aurae, per corpus Cometae radiis solaribus illustratae et Cometam ambientis. Crinitum vero extra aurum in spatio puro, huncque a terra et ambiente eam aere separatum esse, ac proinde nullam caudam habere. Ad illam de Caudato dictam sententiam comprobandam annon illa Epigenis apud Libavium tract[atu] probl[ematico] de Com[etarum] nat[ura] p[agina] et Apiani apud Keckerman[um] *Syst*[*emate*] *Phys*[*ico*] cap[ite] de Cometis, cum Cardani apud Scalig[erum] exerc[itatione] 79. posita, nec

ta Cometarum huic vel illi loco pro signorum caelestium dispositione assignantur. Video nonnullos ex dodecatemorio Tauri et Arietis cornupetarum (ut et syderibus Hydrae aquaticae, navis, Orionis pugnacis, Eridani et Ceti) bellum idque navale, ex dodecatemorio vero Librae,[33] cui veteres Astrologi religionem et Justitiam subiecerint bellum sacrum praesagire. Quae mihi qui quidem caelestia haec signa non contemno, nec tamen metuo, sed ut invitamenta incitamentaque paenitentiae et justitiae divinae bona probis, mala improbis praemonstrantis documenta duco, magis ingeniosa quam solida videntur. Tanto magis cum nondum viderim, quid ex aliis dodecatemoriis signisque caelestibus, quae Cometes vel transierit, vel attigerit (ubi Virgo et Lepus in censum veniunt, ita ut eadem praesagiorum via ad effaemi-//natos Viros et galeatos Lepores, quos militia prorsus reiicit deveniendum foret) iidem illi ingeniosissime curiosi praesagiverint. Motus porro situs, et ipsa Cometae existentia diversos motus ingeniorum,[34] et situs[35] judiciorum inter eruditos effecit. Nam non sine varietate inter Boreum et Austrum motu vel rectilineo vel epicyclico[36] discurrunt et modo, hic, modo illic positum Cometam intuentur, et observatum praedicant, nisi diversae elevationi poli vel etiam errori calami talia sint tribuenda. Varietas illa in apertam erupit controversiam ita nempe, ut multi statuant et asserant unum eundemque fuisse Cometam, nobis diversis temporibus et locis primum quidem ante ortum Solis in corvo ad Austrum, dehinc autem post occasum Solis in lepore ad Aquilonem tendentem conspectum. Alii e contra[37] duos distinctos fuisse visos contendunt. Inter[38] illos, quos vidi, agmen ducit Cl[arissimus] Hevelius cuius accuratiss[imis] observationibus R[everendus] P[ater] Curtius S[ocietatis] J[esu] Neoburgi Palatinorum degens plusquam suis speculationibus rationibusque in contrarium antehac euntibus, et eorum qui se duos distinctos una eademque nocte vidisse Cometas referunt, narrationi fidit. // Sed et Cl[arissimi] viri et Astronomi eximii Ism[ael] Bullial[dus] et Erasm[ius] Bartolinus, tum vero et in hac urbe agentes Viri doctrina Insignes Joan[nes] Alb[ertus] Huswedelius M[edicinae] D[octor] ac primarius Inclytae huius Reip[ublicae][39] Physicus Ordinarius, et Joan[nes] Mullerus Mathes[eos] in Gymnasio hocce Profes-

non ipsius Scaligeri ibidem expressa aliquid faciat, velim mihi indices. Sum enim semper aliquid discere paratus, imo perquam cupidus, nempe quia aurei illius Pythagorici praecepti memor

πρῆσσε δὲ μηδὲν τῶν μὴ ἐπίστασαι, ἀλλὰ διδάσκευ
Ὅσσα χρέων καὶ τερνότατον βίον ὧδε διάξεις
Nihil illorum age quae ignoras, sed disce
Quaecunque opus est, atque ita jucundissimam vitam transiges:

[Pythagoras, *Aurea Verba*, 30–31, in *Poetae Minores Graeci* (Cantabrigiae: apud Joann. Field, 1667), pp. 420–21]

Sed et ad reliqua quid dicturus sis, scire aveo. **TC**.

33 ex…Librae : alios vero ex dodecatemorio Librae **TC**.

34 ingeniorum *del. post* ingeniorum **BO**.

35 situs : situs ut sic dicam **TC**.

36 epicyclico : epicyclico vel circulari vel vago **TC**.

37 contra : contrario **TC**.

38 Inter : Inter priores **TC**.

39 Reip[ublicae] : Civitatis **TC**.

sor. Quibus et Tu vir cl[arissime] accedere mihi videris. Velim tamen mentem tuam plenius et planius mihi expones. Diversas partes sequitur hic Clariss[imus] Michael Kirstenius M[edicinae] D[octor]. Tum[40] Adamus Olearius Seren[issimi] Ducis Holsatiae Bibliothecarius utriusque illius globi ad miraculum usque magni, quo Gottorpium Gloriatur, inventor, quicum ea de re nuper sermones sociavi. Habet autem et haec sententia[41] suas rationes, quas ego in medio relinquo; nititur et variis narrationibus eorum, qui duos distinctos Cometas, alterum mane, alterum vesperi vidisse se affirmant; de quibus et ego ex Germania et Suecia per literas accepi, tum denique judiciis eorum, qui duos fuisse Cometas affirmant, quos in Germania et Hybernia reperiri cognovi. Velim itaque mihi super his omnibus mentem tuam[42] aperias, omnia vicissim facturo quaecunque Tibi grata, mihi vero possibilia factu fuerint. Vale et me constanter benevolentiae affectu complecti perge.

Dab[am] Hamb[urgi] 20 Febr[uarii] 1665.

An excerpt of a letter sent to the most renowned man, Master Abraham de Grau, professor of mathematics at the University of Franeker

As for the opinions of distinguished astronomers and other men acquainted with sublime knowledge, I see that those which I have received differ. I see this variety not only in the opinions about the origin or the effects of comets but also about their place, movement, and their very existence. On top of this multitude of varying opinions about the causes and nature of comets (well-known to you, so there is no need to discuss them here), a new opinion was added by the most worthy Master Otto von Guericke,[43] the Councillor of Magdeburg, a wonderful inventor whom you certainly know. Basing on two suppositions about the motion of the Earth and the vacuum space, he stated that a comet originates from tempests and is part of the air, violently separated from it by the power of a storm and carried upwards, and that its movement is nothing else than the movement of the Earth. Hence, when the latter changes, the former seems to vary. Moreover, he claims that the comet of a tailed type resides in the highest part of the aerial region (which is the pure elemental air) and its tail is part of the elemental air around the comet, illuminated by the sunrays piercing through the body of the comet. (I would like you to tell me, as I am always ready and eager to learn something new, if the opinion of Epigenes[44] in Plinius's Book 7, Chap-

40 Tum : Tum et **TC**.

41 sententia **TC** : sententias **BO**.

42 super...tuam : mentem Tuam super his omnibus **TC**.

43 See section 3.4.3 of the Introduction.

44 Epigenes of Byzantium was a Greek astrologer and astronomer who lived probably in the last centuries BC (exact years are unknown). His views on comets were discussed by Seneca in Book 7 of his *Natural Questions* and all later references to Epigenes relied on this book. There is no mention about his cometary opinion in Pliny's *Historia naturalis*. Correct sources of his views are given in the letter in

ter 27, and that of Apianus[45] in Scaliger's Exercise 79, and Scaliger's own opinion in the same place, confirm in any way that view). Guericke claims that comets of a hairy type are above the elemental air in pure space, and therefore they are separated from the Earth and the air around it, hence they lack the tail. I want to know what your opinion about it all will be. To investigate and predict the effects of comets is the astrologers' task, especially when certain and defined effects are assigned to one place or another according to the disposition of the celestial signs. I see that some of the astrologers foretell a naval war basing their prophecy on the signs of the horned Taurus and Aries (and the stars of the aquatic Hydra, Navis, belligerent Orion, Eridanus and Cetus), while reading from the sign of Libra, with which the ancients associated religion and justice, they foretell a holy war. Those things seem to me rather cleverly invented than solidly rooted, as I neither contemn these celestial signs nor fear them, but I consider them to be an invitation and an incentive to repentance and the proofs of God's justice (signs which announce good things to the rightful, and bad things to the wicked). What is more, I have not seen yet what the same curious people cleverly foretold on the basis of other constellations and celestial signs through which the comet passed or which it touched (where Virgo and Lepus should be concerned, the foretellers' interpretation would lead to effeminate men and hares in helmets who were rejected from the army). The movement, the position and the very existence of the comet caused the stirring of minds and the generation of various opinions among learned men. For not without variety they disperse from north to south (running either in lines or epicycles); they behold a comet situated sometimes in one place, sometimes in another, and predict when it could be observed, unless these things can be attributed to different elevations of the pole or to miswriting. This hustle erupts into an open quarrel because many people state and assert that there was only one comet, seen by us in different times and places — at first before the sunrise in Corvus, tending towards the south, and later after the sunset in Lepus, tending towards the north. Others, on the contrary, prove that there were two different comets. Among the first group, as I gathered, the leader is the most renowned Hevelius, in whose accurate observations the Reverend Father Curtz[46] of the Jesuits, who lives in Palatinate-Neuburg, believes more than in his own speculations and opinions which earlier disagreed with that view, and in the stories of those who refer that they saw

the TC: *Singularium* [...] *pars secunda* (Francofurti: impressa typis Ioannis Saurii, impensis Petri Kopfii, 1599; the *tractatus problematicus* quoted by Lubieniecki is a part of this book) by Andreas Libavius (Libau, d. 1616), German physician and chemist, and *Systema physicum* (Hanoviae: apud Guilielmum Antonium, 1610) by Bartholomaeus Keckermann (1572–1609), philosopher and theologian from Danzig, since 1609 professor of philosophy at the *gymnasium academicum* in Danzig. Keckermann's books (many published posthumously) discussed a variety of topics, including logic, ethics, geography, astronomy, and rhetoric.

45 Petrus Apianus (1495–1522), German astronomer, mathematician and cartographer. He was the first to observe that tails of comets point opposite to the Sun. Scaliger's *Exotericatum exercitationum liber quintus decimus* (see note to Letter 4A) does not discuss Apianus's cometary opinions and I was not able to trace the source of Lubieniecki's information.

46 See section 2.4.5 of the Introduction.

two comets during one night. But the most famous men and exceptional astronomers — Ismaël Boulliau[47] and Erasmus Bartholin[48] — agree too, and these men of extraordinary learning who work in this city as well: Johann Albert Huswedel,[49] doctor of medicine and a distinguished public physician of this famous city, and Johann Müller,[50] professor of mathematics at the local school. I suppose you agree with these men, most famous Sir. I would like you to explain your opinion more fully and more clearly. As far as this matter is concerned, the most renowned Michael Kirsten,[51] doctor of medicine, thinks the opposite. So does Adam Olearius[52] (the librarian of the most serene duke of Holstein, the author of both globes of a wonderful size of which Gottorp is proud), with whom I recently had a few conversations about that thing. This opinion also has its reasons which I lay out here. It is based on various stories of those who claim that they saw two different comets — one in the morning and one in the evening — and who are mentioned in the letters I received from Germany and Sweden. That opinion also derived from the views of those who are in Germany and Ireland and who claim that there were two comets, as I learned. Therefore, I would wish you to reveal to me your opinion about it all and I, on my part, will do everything that I can which would be agreeable to you. Farewell, and continue to support me with your benevolence!

Given at Hamburg, on 20 February 1665.

47 See section 2.4.2 of the Introduction.
48 See section 2.4.1 of the Introduction.
49 See note to Letter 1.
50 See section 2.4.11 of the Introduction.
51 See section 2.4.9 of the Introduction.
52 See section 2.4.13 of the Introduction and note to Letter 2.

8.

17/27 February 1665, Lubieniecki to Hevelius

BO: C1, v. 7, 951
TC: 369–70

Enclosure:
8A. Letter from Müller to Hevelius, 17/27 February 1665

Salutem Plurimam Nobilissime, Clarissime Humanissime Vir. Ex quo ante octiduum ad Te dedi literas,[1] accepi literas a Cl[arissimo] Bullialdo die 13. huius mensis scriptas, quibus de nupero phaenomeno haec inserit, se cum desiderio expectare Tuas observationes et a Viris in Italia doctis R[everendo] P[atre] Joanne Bapt[ista] Ricciolo et Dominico Cassino factas, quandoquidem necdum ullam accuratam observationem viderit: optime autem cum Astronomis agi, cum Te Observatorem incomparabili industria instrumentisque praestantissimis instructum habeant, ex cuius Herculeis laboribus ingentia ad rem Astronomicam commoda redundatura sint: caeterum quod ad Cometae materiam spectat, satis esse si dicatur, materiam illam aetheream esse, et galaxiae non valde absimilem, et plura divinando dicenda non esse, quae demonstrari nequeant. Quamvis autem me certiorem reddiderit, se eadem ad Te praescripsisse: ego tamen nolui hic quoque officio meo deesse, dummodo addam, observasse eum Cometam ad diem 10. huius mensis, sed exiguae magnitudinis, vix sextae magnitudinis stellam adaequantis; quo[2] et Tu pro eximia Tua rerum Astronomicarum scientia et humanitate non mihi tantum, sed et illi communi Amico, cui praecipuus in Rep[ublica] literaria locus et honos ac observantia[3] debetur, satisfacias. Qua ratione et Cl[arissimi] Mulleri implebis vota, cuius en Tibi literas, eruditionis et benevolentiae in Te plenas. His quod addam, habeo nihil, nisi ut tester, nos magno cum desiderio *Prodromum* Tuae *Cometographiae* praestolari. Deus Te, Vir Excellentissime, diu incolumem bono publico esse velit! In candido pectore vovet

Tui
observantissimus et studiosissimus

De Rolis

Hamb[urgi] 17/27 Febr[uarii] 1665.

Verte. //

1 dedi literas : scripsi **TC**.
2 quo : Id vero officii libenter Tibi praesto. Ea tamen fini fateor, ut **TC**.
3 ac observantia *om.* **TC**.

Post Scriptum. Dum inclusas Mullerianas ad Te, Vir Amplissime, datas considero, video, quantum quidem ego de his judicare possum, esse magna cum cura scriptas. Caeterum et illius et omnium aliorum, etiam Clarissimorum et Praestantissimorum Mathematicorum animadverto[4] in Te solum conversos esse oculos et directam expectationem, quae Tibi gravis adversaria opposita est, facile Tua eximia scientia et industria superanda. Nec immerito. Primas enim Tibi in re Astronomica omnes tribuunt. Itaque ut videas quanto et rei Astronomicae et Tui captus sum studio, offero Tibi operam meam in edendo Tuae *Cometographiae Prodromo*, cui seorsim adderem quae de Cometa hinc inde ab Amicis accepi. Curarem autem librum nitide imprimi in Hollandia cum insertis tabellis cum cura delineatis et aeri incisis vel abs Te ipso (hoc enim utrumque ita praestat felix manus Tua,

> Ut hoc[5] non melius notuerit pingere[6] Apelles
> Fingere vel fuso doctius aere Myron)

vel secundum delineationes a Te factas a perito quopiam in Hollandia sculptore. Quod si dederis mihi tabellas aeri abs Te incisas, id Tibi speciatim numero exemplarium secundum aequitatem exsolvi curabo. Sin autem solum exemplar libri et delineationes adiunctas dederis, id etiam exemplaribus secundum eandem aequitatem Tibi pensabitur. En dextra fidesque![7] Velim tamen cum Nobiliss[imo] Pelsio[8] ea de re conferas. Iterum Vale et candorem meum boni et aequi consule, mentem autem Tuam super hac propositione mea mihi quam primum aperi.

Greetings, most noble, renowned, and kindest Sir! After having sent a letter to you eight days ago, I received a letter from the most renowned Boulliau,[9] written on the thirteenth of this month, in which he placed such things about the recent phenomenon: he eagerly expects your observations and those made by the learned men from Italy, viz. Reverend Father Giovanni Battista Riccioli[10] and Giovanni Domenico Cassini,[11] but he has not seen any precise observations yet. Nonetheless, it is excellent to find you (as other astronomers do), an observer of incomparable industry, equipped with superior instruments, whose Herculean labours will bring

4 animadverto **TC** *om.* **BO**.
5 Ut hoc : Hoc ut **TC**.
6 notuerit pingere : norit depingere **TC**.
7 En...fidesque *om.* **TC**.
8 Pelsio : Pelsio Amico communi **TC**.
9 See section 2.4.2 of the Introduction.
10 See section 2.4.15 of the Introduction.
11 Giovanni Domenico Cassini (1625–1712), Italian astronomer and mathematician. Since 1650 he was a professor of astronomy at the University of Bologna. In 1669, he moved to France and helped to establish the Paris Observatory. He studied rotational periods of Mars and Jupiter, discovered four satellites of Saturn and the division in its ring (known as the Cassini Division). Lubieniecki omitted here the first of Cassini's given names. See also Chantal Grell, 'Introduction', CHJ2 p. 29.

an abundance of significant advantages to astronomy. As for the comet, it will be enough to say that its matter is aethereal and not very different from the galaxy's, and nothing more should be guessed what cannot be proved. Even though he assured me that he wrote the same things to you, I did not want to neglect my duty here, and I would like to add that he was observing the comet until the tenth of this month, but its magnitude was low, equal to the stars of the sixth magnitude only. Hence, knowing your exceptional knowledge of astronomy and your kindness, you will fulfil not only my wish, but that of our mutual friend as well — the one who is worthy of a special place in the Republic of Letters, and of honour and reverence. For this reason, you will fulfil the wish of the most famous Müller,[12] whose letter I enclose (full of erudition and benevolence towards you). I have nothing more to add but to say that we are impatiently waiting for the *Prodromus* of your *Cometographia*. May God keep you, most excellent Sir, in good health for the public good! It is wished with a sincere heart by

Yours obligingly,

De Rolis

In Hamburg, on 17/27 February 1665.

Turn the page.

Postscript. Greatest Sir, if I were to consider the enclosed letter from Müller to you, I would see, if only I could have an opinion about it, that it is written with great diligence. I have noticed that not only his but everyone else's eyes (also of those most famous and distinguished mathematicians) are fixed on you. Their direct expectation is a serious problem to you, which you will overcome with your exceptional knowledge and industry. And there is a justified reason why everyone grants you the lead in astronomy. Thus, in order to show you what a great affection I have for astronomy and for you, I offer you my help in publishing the *Prodromus* of your *Cometographia*, to which I could add separately what I received about the comet from my friends from various places. I would take care of printing your book decently in the Netherlands with meticulously drawn inserted tables, engraved in brass either by yourself (for in both these things your auspicious hand is so skilled

that neither Apelles knew to paint better
nor Myron to make more skilfully in melted brass)[13]

12 See section 2.4.11 of the Introduction.

13 George Buchanan, *In Leonoram*, in George Buchanan, *Opera omnia*, vol. 2, part 1 (Lugduni Batavorum: apud Johannem Arnoldum Langerak, 1725), p. 412.

or prepared by an experienced engraver in the Netherlands after your drawings. If you send me the tables engraved in brass by you, I will make sure to pay you justly in an adequate number of copies. And if you send only a copy of the book with drawings, it will be paid equally justly in copies. This is my hand and word! I would like you, however, to discuss this with the most noble Pels.[14] Farewell again, and use my sincerity for what is good and just, and give me soon your answer about my offer!

8A.

Letter from Müller to Hevelius, 17/27 February 1665

TC: 366–69
BO: C1, v. 6, 1460 (part)

Hamburgo Dantiscum die 17. 27. Februarii 1665

Nobilissimo et Amplissimo Domino
JOHANNI HEVELIO
Senatori Gedanensi
JOHANNES MÜLLERUS
Salutem Plurimam Dicit

Quamvis jam dudum aere, quo Tibi, Vir Amplissime, obstrictus sum, me liberare in animo habuerim, varia tamen fuerunt, quae id hactenus praepedivere. Ut igitur diuturno silentio finis aliquando imponeretur, hisce literis operam dare volui. Principio itaque memini gratias Tuae Amplitudini me adhuc debere maximas, quod non solum humanissimis suis literis me invisere, sed et profundissimae eruditionis volumine, *De Mercurio in Sole viso*, aliisque rarissimis caeli phaenomenis me donare non fuerit dedignata. Maximopere equidem Tibi obstrictum me esse profiteor, pro hoc foetu mentis manuumque Tuarum, omnium doctorum admiratione dignissimo. Quis enim est, qui non indefessae mentis doctaeque manus nobile coniugium unico in homine obstupescat? Sed praestat haec et alia acutissimi Tui ingenii monumenta, per se aeterna, silentio praeterire, quam eorum magnitudinem

> Culpa minus subacti deterere ingeni.

Caeterum cum Te per communem nostrum Amicum has literas curantem, mearum talium qualium de Cometa nupero opinionum (quas ipsius obnixae petitiones mihi

14 See note to Letter 5.

ante tempus extorserunt) participem factum esse cognoscam, intermittere non potui, quin hisce literis paulo clarius easdem explicarem, praemissis prius iis, quae liberis oculis in Cometa sine ullo instrumentorum, sub manibus artificium adhuc adhaerentium, adminiculo, usurpavi.

Primo omnium Cometa novus et insignis conspectus fuit a militibus huius Urbis in vallo, et quidem nocte, quae fuit inter diem 2. et 3. Dec[embris] St[ylo] V[etere] anni nuper elapsi 1664. Quid ubi rescivissem ego, proxime sequenti statim nocte veritatem experiri non destiti. Vespertina enim hora undecima diei 3. Dec[embris] S[tylo] V[etere] quae erat Saturni, in publicum prodii, an vere novum aliquod caeli phenomenon adesset exploraturus. Postquam igitur horas quatuor sub dio transegissem, frigore existente intensissimo, tandem hora tertia matutina die 4/14 Dec[embris] tractus albicans supra horizontem nostrum, prope ortum Solis brumalem, ascendit, et subsequentis Cometae caudam exhibens, famam non fuisse falsam indicavit. Paulo quippe post oriebatur Cometes mole satis magna, et nucleo praelucido, adhaerens rostro Corvi, parumper tamen eo tum occidentalior, tum australior. Color caudae erat pallidus, ipsius vero Cometae nucleus colorem Veneri vel Jovi similem monstrabat. Sequenti biduo caelum apud nos nubibus tegebatur.

Secundo eundem Cometam videre mihi contigit die 7/17 Dec[embris] quae et Mercurii. Hora enim tertia matutina iterum oriebatur, et a rostro Corvi versus Notozephyrum longius remotus, faciebat cum fixis in capite et dicto rostro Corvi triangulum ad rostrum obtusangulum, quale formant cornua Arietis. Crura trianguli obtusi videbatur se habere ut 2. ad 1. hoc est, recta jungens Cometam Rostro, erat dupla rectae interiacentis Rostrum et Caput Corvi.

Tertio disiectis nubibus, maiore mole et prolixiore cauda apparuit Cometes die 8/18 Dec[embris] horis, ut solet, matutinis, et versus Notozephyrum a dictis Corvi stellis porro factus remotior, iterum cum iisdem triangulum paulo obtusiore (ut judicabam) quam antea angulo ad idem rostrum formabat, cuius crura proportionem duplam super tripartientem quartas, hoc est, quam 11. ad 4. habere putabam.

Diebus 9. et 10. Dec[embris] St[ylo] V[etere] caelum non favebat.

Quarto horrenda magnitudine et cauda alterno tanto adaucta conspexi Cometam die 11/21 Dec[embris] quae erat Dominica tertia Adventus Salvatoris nostri. Huius enim diei mane hora secunda (prout horologia huius urbis indicabant) ille supra nostrum horizontem se elevabat, et triquetro Hydrae proximus caudam videbatur terminare in tribus stellis eiusdem rectae, cor Hydrae proxime sequentibus.

Interiecto deinde tempore usque ad diem 20/30 Dec[embris] caelum erat nubibus tectum.

Cum itaque dicta die 20/30 Dec[embris] vesperi audita nona oculos in excelsum levarem, et innumeram fixarum multitudinem contemplarer, non sine stupore Cometam pallidi coloris, nucleo etiam satis lucido in Lepore deprehendebam. Quem quidem initio a priore diversum esse credebam, sed postea accuratiore mentis trutina omnibus perpensis eundem esse comperi. Secabat autem hic Cometa tunc rectam, quae fixas in dextro armo (Bayero α[15]) et infra oculum Leporis jungit, in duas partes, quarum Orientalis dupla erat partis Occidentalis. Eratque motus Cometae proprius versus Borrhozephyrum adeo celer, ut singulis horis alium situm cum fixis habere observaretur. Cauda apparebat nulla, tum ob viciniam Lunae, tum ob Solis in primo Capricorni Decano haerentis fere oppositionem, quae caudam sursum tendentem vel plane non, vel exigua magnitudine apparere causatur.

Die 21/31 Dec[embris] caelum non erat sudum.

Die 22. Dec[embris] Calend[arii] Juliani (quo etiam in sequentibus utor) Cometa vesperi hora nona efficiebat triangulum rectangulum isosceles fere cum praecedentibus humero et pede Orionis. Angulus rectus erat ad praecedentem sive sinistrum pedem Orionis. Cometa praecedebat. Cauda ob dictas rationes videbatur nulla.

Die 23. Dec[embris] vesperi Cometa adhaerebat Stellae Orientaliori trium (in globis Jansonii) in eadem recta prope mandibulam Ceti, quarum occidentalior Bayero non est informis, et vocatur ad mandibulam. Cauda ad Orientem tendens non adeo longa colore consueto erat praedita.

Die 24. Dec[embris] vesperi hora dimidia decima caelo existente sereno, Cometa cum media in ore et praecedente trium in capite Ceti formabat triangulum ad mediam in rectangulum, cuius cathetus Orientalis sive recta Cometa cum dicta media copulans, mihi videbatur tripla reliquae catheti. Cometae cauda obumbrabat quasi mediam praedictarum trium informium.

Die 25. Dec[embris] hora octava vespertina Cometa aequali intervallo distabat a media dictarum informium, et media in ore ceti, ita ut recta jungente has fixas esset paulo borealior.

Die 26. Dec[embris] caelum erat nubilum.

Die 27. Dec[embris] Cometa vesperi hora 7. cum lucida mandibulae et media in ore Ceti constituebat triangulum rectangulum isosceles. Angulus[16] rectus erat ad Cometam utraque fixa borealiorem. Cauda erat modica et coloris consueti, ut etiam aliis diebus. Hora 9. eiusdem vesperae Cometa unam rectam cum media in ore Ceti et

15 α *corr. ed. om.* TC.
16 Angulus *corr. ed.* : Triangulus TC.

praecedente trium occupabat, quam media in ore bifariam secabat. Utraque fixa erat Cometa australior.

Die 28. Dec[embris] caeli inclementia Cometam tegebat.

Die 29. Dec[embris] vesperi hora 9. rectam, quae fixam in fronte Ceti (Bayero μ[17]) et subsequentem in capillamento interiacet, ita dispescebat Cometa, ut segmentum Orientale duplum esset segmenti Occidentalis. Videbatur etiam cum lucida mandibulae et media in ore efficere triangulum ad mediam in ore rectangulum.

Die 30. Dec[embris] vesp[ertina] hor[a] 9. Cometa medium, inter dictam in fronte Ceti fixam, et accedentem in juba (eidem Bayero ξ),[18] tenebat locum.

Die 31. Dec[embris] vesp[ertina] 7. hor[a] eandem rectam Cometa cum eadem in fronte Ceti et sequente in vinculo boreali (Bayero ο)[19], faciebat, ita ut stellam in fronte ad Orientem linquens, duplo maiori intervallo distaret ab altera. Cauda ad Ortum vergebat.

Calendis Januarii huius anni 1665. caelum vesperi nubibus tegebatur.

Die 2. Jan[uarii] Cometes medium inter saepius dictam in fronte Ceti stellam et inferiorem trium in nexu boreo (Bayero π[20]) Lini Piscium occupabat.

Diebus 3. et 4. Jan[uarii] caelum non erat sudum.

Die 5. Jan[uarii] vesp[ertina] hor[a] 8. Cometa cum clara Lini borealis et sinistro cornu Arietis (β Bayero) quas sequebatur, efformabat triangulum aequilaterum: eandem etiam rectam cum fronte Ceti et dicta clara, eum inter utramque medium obtinere judicabam.

Diebus 6. et 7. Jan[uarii] caeli facies erat inconspicua.

Die 8. Jan[uarii] hor[a] 8. vesp[ertina] Cometa dextrum sive sequens cornu Arietis (Bayero α) et clara Lini bor[ealis] exhibebant triangulum ad Cometam meridionaliorem rectangulum, cuius cathetus minor, hoc est distantia Cometae a dicta clara Lini, aequali apparebat rectae eandem claram et sinistrum cornu Arietis connectenti.

17 μ *corr. ed. om.* TC.
18 (eidem Bayero ξ) *corr. ed.* : eidem Bayero TC.
19 (Bayero ο) *corr. ed.* : Bayero TC.
20 π *corr. ed. om.* TC.

Die 9. Jan[uarii] vesp[ertina] hor[a] 6. Cometa tardior factus vix locum mutasse videbatur.

Die 10. Jan[uarii] Cometa primam inter stellam Arietis et Nodum Lini positus, ab hoc tribus, ab illa una parte aberat.

Die 11. Jan[uarii] Locus Cometae insensibiliter videbatur mutatus.

Die 12. Jan[uarii] Cometa cum clara Lini bor[ealis] et antecedente cornu Arietis (Bayero β) fingebat triangulum isosceles rectangulum ad Cometam Orientaliorem.

Die 13. Jan[uarii] mutatio Cometae ratione loci fere erat insensibilis.[21]

Die 14. Jan[uarii] Cometa iterum trianguli rectanguli isoscelis figuram exhibebat, cum prima scilicet stella Arietis et clara Lini bor[ealis] ut angulus[22] ad Cometam orientaliorem esset rectus.

Die 15. Jan[uarii] vesp[ertina] hor[a] 9. Cometa in eadem fere erat recta, in qua sunt prima stella et praecedens cornu Arietis, utraque tamen fixa australior, ita ut eius distantia a prima stella Arietis esset quinque partium, a praecedente vero cornu 7. Dico Cometam fere in eadem recta fuisse cum dictis fixis: erat enim ille hac recta paulo orientalior.

Die 16. Jan[uarii] caelum nos aspectu Cometae excludebat.

Die 17. Jan[uarii] Quamvis praesentia Lunae impediret contemplationem Cometae, jam per plures dies imminuti, notabam tamen eum rectam lineam, qua die 15. Jan[uarii] erat orientalior, in dicta proportione occupare.

Die 18. Jan[uarii] caelum nubibus tegebatur.

Die 19. Jan[uarii] Cometa erat eadem recta, de qua antea, in dicta proportione occidentalior.[23]

Die 20 Jan[urarii] et seq[uentibus][24] hucusque coeli facies fere semper fuit obscura, ut Cometa quidem interdum comparuerit, sed non adeo distincte, ut de eius loco certi quid affirmare quispiam potuerit. Nunc vero plane eum disparuisse credo.

21 insensibilis *corr. ed.* : insensilis **TC**.
22 angulus *corr. ed.* : triangulus **TC**.
23 Hamburgo Dantiscum...occidentalior *om.* **BO**.
24 diebus *del. post* seq[uentibus] **BO**.

Atque haec sunt, Vir Amplissime, quae hactenus in miro et raro hoc Phaenomeno, quod certe magna et utilissima accessione rem Astronomicam locupletabit, nudis oculis[25] observare potui. Quod vero attinet ad ea, quae de eius motu ad Nobiliss[imum] Dominum de Rolis perscripsi, ipsemet ego nihili facio; quia me quasi oportuit ante tempus aliquod consignare, quo magno ipsius desiderio et petitioni morem gererem. Nec potui quicquam Astronomo dignum hac de re proferre, cum accuratae observationes, quibus Hypotheses sunt superstruendae, et juxta earum tenorem conformandae, nos destituant. An vero motus Cometarum fiant in circulis; an vero sint traiectiones, sive in rectis lineis conficiantur, adhuc sub judice lis est et manebit. Ego quidem fateor me[26] potius in illam abire sententiam, motus Cometarum fieri in rectis lineis, aut, etiamsi non fiant, posse tamen per has traiectiones salvari. Meliora tamen et certiora edoctus, hanc opinionem relinquere non dubitabo. Videtur quidem nuperum Phaenomenon velle evincere motus Cometarum fieri circulariter; sed licet id aperte indicet, non tamen de omnibus Cometis idem verum esse statim constabit. Posse autem Cometes motu proprio aequabiliter incedere in recta linea, et nihilominus videri in circulo fere maximo (dico: fere maximo, ob parallaxin) inaequabiliter moveri, sequenti diagraphe satis clare innotescet. Sit enim ex[empli] gr[atia] Cometa aliquis, qui aequali tempore aequalia spatia AC, CD, DE, EF, FG, GB, subtensae AB traiiciat, dico oculum in L. constitutum Cometae locum ad fixas (quas vulgo sphaericam superficiem situ suo exhibere credimus) expendendo, collocare in A, c, d, e, f, g et B, nec posse solo visu diiudicare, an Cometa in subtensa AB, an vero in arcu AeB moveatur, // quia utrobique in initio et fine, prope A et B, tardius, in medio celerius moveri videtur, uti[27] ex Opticis notum est, et praecedens diagramma ostendit. Nam rectarum aequalium AC, CD, DC, EF, FG, GB magnitudines apparentes sunt inaequales, quia earum aliae aliis obliquius oculo obversae, angulos ALC, LCD, DLE et c[etera] inaequales producunt. Licet inaequalitas illa[28] maior vel minor sit, pro diversa oculi L remotione. Eadem ratio est, si Cometa cicular iter motus suae periodi arcum perigeum BCD (vid[e] sequens schema) conficiat, et aequali tempore aequales arcus BE, EF, FC, CG, GH, HD absolvat. Proinde enim ut antea, initio et fine huius arcus, prope stationes scil[icet] in B et D, videbitur tardus, in medio vero, eo celerior, quo perigeo C factus fuerit vicinior. Atque eadem etiam phaenomena ostendet Cometa oculo M, si juxta divisionem subtensae BD (h[oc] e[st] inaequabiliter) feratur. Verum, ut modum huius Epistolae faciam, desino hisce (quae Tuae Ampl[itudini] sunt notissima) diutius immorari. Haec itaque, Vir Ampl[issime] occasione novi Cometae ad Te perscribere, Tuaeque censurae lubens volensque subiicere volui, nullus dubitans, quin pro singulari humanitate Tua, ea sis amice accepturus, scholasque Mathematicas doctioribus certioribus observationum Tuarum Astonomicarum Hypothesibus sis donaturus. Vale felicissime Divinae tutelae etiam atque

25 est *del. post* oculis **BO**.
26 post *del. post* me **BO**.
27 eti *del. post* uti **BO**.
28 aliud *del. post* illa **BO**.

etiam commendatus. Dabam Hamburgi XIII. Cal[endas] Martii Anno vul[raris] Aerae Chr[istianae] MDCLXV.

From Hamburg to Danzig, on 17/27 February 1665

To the most noble and worthy
Johannes Hevelius,
a Councillor of Danzig,
Johann Müller
sends greetings

As much as I was planning to free myself from my debt towards you, greatest Sir, many things have prevented it so far. Hence, in order to end my prolonged silence, I have decided to send you this letter. First of all, I remember that I still owe many thanks to your greatness, as not only did it find me worthy of sending me very kind letters, but also of giving me the volume of exceptional learning, the *Mercurius in Sole visus*, which also discusses other uncommon celestial phenomena. I admit that I was much obliged to you because of this offspring of your mind and hands — admirable to all learned men. And who would not become speechless seeing this union of tireless mind and learned hand in one man? But it would be better to pass this and other monuments of your great talent in silence, rather than to let their greatness

> be diminished due to frailty of a less educated mind.[29]

What is more, when I learned from our mutual friend (who is taking care of this letter) that he sent you my trivial opinions about the latest comet, I could not wait to explain them more clearly (emphasizing that I made them with the naked eye, without any help of instruments, which are still being constructed by craftsmen).

The new and significant comet was seen for the first time by the soldiers on the city walls.[30] It was the night between 2 and 3 December, old style, of the year 1664 which has recently ended. When I learned about it, on the following night I could not miss the opportunity to verify it. So at 11 o'clock in the evening, on 3 December (it was Saturday), old style, I tried to check in public if there really was some new celestial phenomenon to be studied. Hence, after I had spent four hours under the sky — and it was very cold — finally, at 3 o'clock in the morning on 4/14 December, a white path arose over our horizon near the place of the winter sunrise; and next, by showing the tail of the comet, it confirmed that the news was not false. Soon after, the comet

29 An allusion to Horace, *Carmina*, 1, 6, 11–12, 'laudes egregii Caesaris et tuas / culpa deterere ingeni.'

30 See Letters 1C and 1H.

arose; its size was quite large, and the nucleus was very bright. It was adherent to the Beak of Corvus, slightly to the south-west. The colour of the tail was pallid, and the colour of the comet's nucleus was similar to that of Venus or Jupiter. In the next two days, the sky here was covered by clouds.

I had a chance to see the comet for the second time on 7/17 December (it was Wednesday). At 3 o'clock in the morning, it arose again, and it was further removed from the Beak of Corvus towards the south-west. Together with the stars in the Head and the Beak of Corvus it made an obtuse-angled triangle — like the one made by the Horns of Aries. The ratio of the legs of the triangle seemed to be 2 to 1, viz. the line between the comet and the beak was two times longer than the one between the Beak and the Head of Corvus.

The comet, greater in size and with a wider tail, appeared for the third time when the clouds parted, on 8/18 December, in the morning (as usual). It was further towards the south-west from the stars in Corvus, and together with them it formed again a triangle with a more obtuse angle at the beak (as it seemed to me) than before. I thought that the legs had a 'dupla supertripartiens quartas' proportion, viz. 11 to 4.[31]

On 9 and 10 December, old style, the sky was not favourable.

For the fourth time, I saw the comet of enormous size and with a much larger tail on 11/21 December — it was the third Sunday of the advent of our Saviour. On that day, at 2 o'clock in the morning (as was indicated by the clocks in this city), it was rising over our horizon, and it was close to the Triangle of Hydra, seemingly ending in the three right stars which follow the Heart of Hydra.

Later, until 20/30 December, the sky was covered by clouds.

When on that day, 20/30 December, I heard it was 9 o'clock, and when I lifted my eyes and contemplated the innumerable multitude of stars, I was shocked when I saw the comet in Lepus. Its colour was pallid and it had quite a bright nucleus. Initially, I thought it was a different comet; later (after a more detailed analysis and having considered all the circumstances) I found out it was the same one. The comet was then cutting a straight line between the star in the right arm (α according to Bayer)[32] and the one below the Eye of Lepus in two parts; the eastern one was twice longer

31 See discussion of proportions in Giovanni Battista Riccioli, *Almagestum novum*, pars posterior tomi primi (Bononiae: ex Typographia Haeredis Victorii Benatii, 1651), pp. 504–07.

32 Müller describes some of the stars using the names introduced by Johann Bayer (1572–1625) in his *Uranometria* (Augustae Vindelicorum: excudit Christophorus Magnus, 1603). However, in the TC, some of those designations lack Greek letters. I supplemented them, and the omitted letters are noted in the apparatus. See also 'Subsidia de cometis', CJH2 p. 485, and CJH3 p. 352, note to Letter 110, des Noyers to Hevelius, 2 January 1665.

than the western one. The movement of the comet towards the north-east was so fast that in each hour it had different positions against the stars. It showed no tail both because of the neighbouring Moon and because of the near opposition to the Sun in the first 10 degrees of Capricorn; this caused the tail, which was directed upwards, to give no show, or to be very small.

On 21/31 December, the sky was not clear.

On 22 December, according to Julian calendar (I use it in the following text), at 9 o'clock in the evening, the comet made almost a right-angled isosceles triangle with the Foregoing Shoulder and the Foregoing Foot of Orion. The right angle was at the Foregoing (or Left) Foot of Orion. The comet was foregoing them. No tail was visible due to the reasons above.

On 23 December, in the evening, the comet was adherent to the eastern star of the three (in the Janszoon's[33] globes) in the same line near the Jaw of Cetus (the western of them is not considered unformed[34] by Bayer and is called 'the one in the jaw'). The tail (directed towards the east and not very long) had its usual colour.

On 24 December, at half past 9 in the evening, when the sky was clear, the comet made a triangle with the middle star in the mouth and the foregoing one of the three in the Head of Cetus. The right angle was at the middle one in the mouth; the eastern cathetus was the line between the comet and the aforementioned middle star. It seemed to me three times longer than the other cathetus. The tail of the comet looked as if it was overshadowing the middle star of the aforementioned unformed three.

On 25 December, at 8 o'clock in the evening, the comet was in the equal distance from the middle star of the aforementioned unformed ones and the middle one in the Mouth of Cetus; it was more towards the north from the line between these two stars.

On 26 December, the sky was cloudy.

33 Probably Willem Janszoon Blaeu (1571–1638), Dutch cartographer, publisher, and manufacturer of terrestrial and celestial globes. He was Tycho Brahe's student and later he pursued his interests in astronomy. In 1600, he discovered the variable star P Cygni. For more information about this star see Mart de Groot, Chris Sterken, 'P Cygni 2000: 400 Years of Progress', *Publications of the Astronomical Society of the Pacific*, 113 (2008), 258–59. Hevelius discusses it in *Mercurius in Sole visus* (Gedani: autoris typis et sumptibus, 1662), pp. 169–70.

34 'Unformed star' was a star which was not defined as part of a constellation, see John C. Barentine, *The Lost Constellations. A History of Obsolete, Extinct, or Forgotten Star Lore* (Heidelberg: Springer, 2016), p. 15.

On 27 December, at 7 o'clock in the evening, the comet made a right-angled isosceles triangle with the bright star in the jaw and the middle one in the Mouth of Cetus. The right angle was at the comet which was more towards the north than towards both these stars. The tail was average, and it had its usual colour, as in the days before. At 9 o'clock on the same evening, the comet was in one line with the middle star in the Mouth of Cetus and the foregoing one of the three — the middle one in the mouth was cutting the line into halves. Both stars were more towards the south than the comet was.

On 28 December, the turbulent sky covered the comet.

On 29 December, at 9 o'clock in the evening, the comet was dividing the line between the star in the Forehead of Cetus (μ according to Bayer) and the subsequent one in the hair. The eastern part of the line was twice longer than the western one. The comet also seemed to make a triangle with the bright star in the Jaw and the middle one in the Mouth. The right angle was at the middle one in the Mouth.

On 30 December, at 9 o'clock, the comet was in the middle between the star in the Forehead of Cetus and the subsequent one in the mane (ξ according to Bayer again).

On 31 December, at 7 o'clock in the evening, the comet formed the same straight line with the same star in the Forehead of Cetus and the subsequent star in the Northern Knot of Pisces (ο according to Bayer). The comet left the star in the forehead at its eastern side, being at a distance twice longer from the other star. The tail was directed towards the east.

On 1 January of this year 1665, the sky in the evening was covered by clouds.

On 2 January, the comet was in the middle between the star in the Forehead of Cetus and the lower one in the Northern Knot of the Cord of Pisces (π according to Bayer).

On 3 and 4 January, the sky was not clear.

On 5 January, at 8 o'clock in the evening, the comet formed an equilateral triangle with the bright star in the Cord of Pisces and the Left Horn of Aries (β according to Bayer). The comet was following those stars. I think that it also formed the same straight line with the Forehead of Cetus and the aforementioned bright one, being in the middle between them.

On 6 and 7 January, it was impossible to see the face of the sky.

On 8 January, at 8 o'clock in the evening, the comet formed a right-angled triangle with the right (or the subsequent) star in the Horn of Aries (α according to Bayer)

and the bright northern star in the line. The right angle was at the comet, which was more towards the south. The shorter cathetus (the distance between the comet and the aforementioned bright star in the line) seemed to be equal to the straight line between that bright star and the left one in the Horn of Aries.

On 9 January, at 6 o'clock in the evening, the comet (which became slower) did not seem to have changed its position much.

On 10 January, the comet was between the first star of Aries and the star in the Knot of the Cord in Pisces. It was three times more distant from the one in the Knot than from the former one.

On 11 January, the place of the comet did not change noticeably.

On 12 January, the comet formed a right isosceles triangle with the bright northern star in the line and the preceding Horn of Aries (β according to Bayer). The right angle was at the comet, which was more towards the east.

On 13 January, the change of the place of the comet was almost unnoticeable.

On 14 January, the comet again made a right-angled isosceles triangle with the first star of Aries and the bright northern one of the line; the angle at the comet was right.

On 15 January, at 9 o'clock in the evening, the comet was almost in a straight line with the first star of Aries and the star in the preceding horn, more towards the south. Its distance from the first star of Aries was five parts, and from the star in the preceding horn — seven parts. I am saying that the comet was almost in a line with those stars because it was a little towards the east from that line.

On 16 January, the sky denied us the sight of the comet.

On 17 January, even though the presence of the Moon impeded the observation of the comet, it grew less in the course of so many days; nevertheless, I noticed that it was in the same straight line from which it was removed towards the east on 15 January, at the same ratio.

On 18 January the sky was covered by clouds.

On 19 January, the comet was more towards the west from the said straight line, at the ratio mentioned above.

On 20 January and on the following days until today, the sky was constantly clouded. Even if the comet showed itself in that time, it was so indistinct that no one could

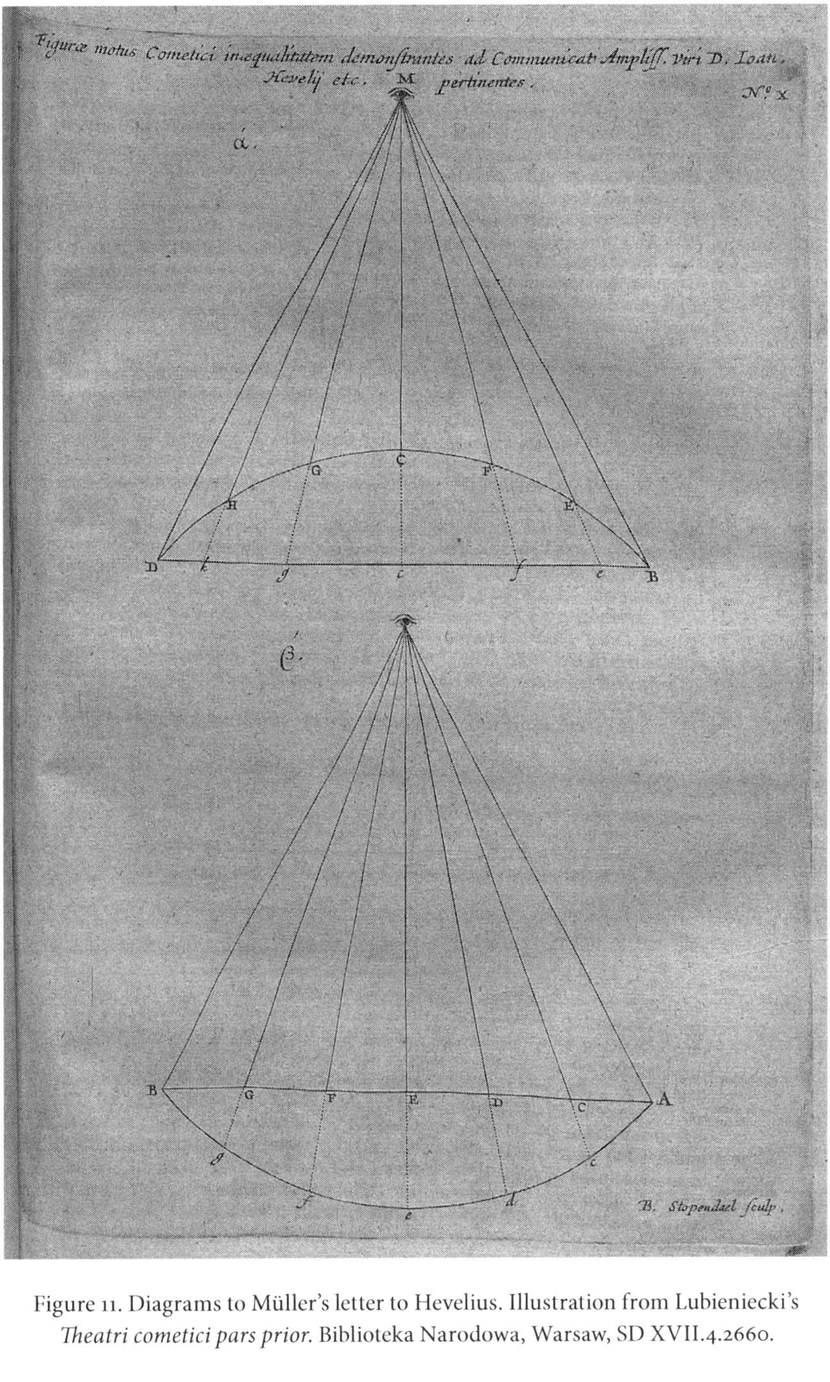

Figure 11. Diagrams to Müller's letter to Hevelius. Illustration from Lubieniecki's *Theatri cometici pars prior*. Biblioteka Narodowa, Warsaw, SD XVII.4.2660.

state anything certain about its position. I believe that it has already entirely disappeared.

And this is, greatest Sir, what I could observe so far about that miraculous and rare phenomenon whose great and useful sight will certainly enrich astronomy. As for what I wrote about its movement to the most noble Master de Rolis, I do not find it important myself. As I noted some time ago, I was only fulfilling his great desire and request. I could not present anything worthy of an astronomer about this thing because I did not have exact observations on which the hypotheses should be based and on whose results they should be formed. There has been, and will be, a quarrel whether the comets move in circles, in arcs, or in straight lines. I admit that I incline to the opinion that the movement of comets happens in straight lines or can be explained by an arc — however, it does not occur in this way. But if I learn something better and more reliable, I will undoubtedly reject this opinion. It seems that the recent phenomenon wants to persuade us that the comets move in circles. Even though it clearly seems so, it cannot be considered immediately true for all comets. The following diagram clearly shows that comets can move equally in straight lines, and nevertheless seem to move unequally almost in a great circle (I say 'almost' because of the parallax). For example, there is a comet which in equal time moves through equal distances AC, CD, DE, EF, FG, GB (parts of the chord AB), and the eye placed in L. When the place of the comet is projected on the fixed stars (which are commonly believed to represent a spherical surface with their positions), the comet is placed in A, c, d, e, f, g and B, and the observed cannot judge by their own sight whether the comet is moving in the chord AB, or in the arc AeB, because at both ends (i.e. A and B) it seems slower, and in the middle it seems faster, as is well-known from the principles of optics and as the enclosed diagram shows. The apparent lengths of equal parts AC, CD, DE, EF, FG, GB are unequal because each of them is seen by the eye at different inclination and angles ALC, CLD, DLE, and so on, seem unequal. This discrepancyis bigger or smaller because of the distance to the eye in L. And the same thing happens when the comet in a circular movement goes in arc BCD in perigee (see the attached diagram) and in equal periods moves in equal arcs BE, EF, FC, CG, GH, HD. As before, at the beginning and the end of this arc (near points B and D) it seems slow, and in the middle it seems faster, because it is closer to the perigee in C. And the comet would show the same phenomena to the eye in M, if it moved along the division of the chord BD (which is unequal). But in order to finish this letter, I leave this subject (which is well-known to you, greatest Sir). I wanted to write all of this to you, greatest Sir, on the occasion of the new comet, and to willingly submit it to your opinion. I do not doubt that (out of your exceptional kindness) you will receive it friendlily, and you will add this mathematical lecture to the learned and certain hypotheses of your observations. Farewell, be happy and always be commended to divine protection! Given at Hamburg, on the thirteenth day before the Kalends of March, in the year 1665 of the common Christian age.

9.

3 April 1665, Hevelius to Lubieniecki

BO: C1, v. 7, 952/A.80
TC: 370–71

Domino Rolis
Hamburgi

Vir Illustris.

Tibi ipsi omnino imputandum habes, quod aliquanto serius ad binas tuas multo mihi gratissimas literas respondeam. Nam, quia Tu, simul cum aliis Viris Eximiis ac Amicis honorandis a me summopere efflagitasti,[1] ut minimum aliquid antequam tota *Cometographia* absolveretur, de nupero Cometa in lucem exponerem. Ideoque sepositis aliis curis omnibus, nec non relicto ipso etiam commercio literario, nihil potius duxi, ex ipso die, quo Regiomonto Dantiscum redii, quam ut desiderio Vestro (ut ut initio plane aliter decreverim) quantum fieri hac vice, ob temporis angustiam potuerit, satisfacerem, atque sic[2] *Prodromum* quendam *Cometographiae* concinnarem. Quem, Divina annuente gratia penitus jam quoque conscripsi, atque praelo commendavi: non dubitans quin spatio unius mensis pene in publicum prodeat. In quo opusculo non solum Historiam Generalem hujus Cometae satis plenam; sed et Dissertationem de Cometarum omnium motu, generatione variisque phaenomenis habebitis. Adeo ut hisce pagellis simul omnibus Amicis, quibus per literas ita plene hactenus morem gerere nequivi, nunc ad interrogata quaestionesque omnes propositas plenius respondere possim. Breviter quidem pertractata sunt omnia, nihilominus tamen perspicue satis mentem, hypothesinque meam intelligetis. Reliqua vero quae nonnemini forsitan vel adhuc obscura, vel nimis strictim pertractata esse videbuntur, *Cometographia* nostra propediem ad umbilicum perducenda fusius clariusque elucidabit ac demonstrabit. Quod operam vero Tuam in // edendo hocce opusculo mihi adeo humaniter offerre haud es gravatus, debita gratiarum actione agnosco; atvero cum nitidam et politam Typographiam ipsemet, non sine sumptu, in aedibus meis possideam, tresque Typographos alam, facile intelligis, quod satius fuerit, ut ea omnia me praesente typis exscribantur. Nam, profecto absente Autore, nunquam Astronomica illa inprimis numeris repleta, tanta diligentia, quanta quidem debent, in publicum proferuntur. De caetero, percepi a Nobiliss[imo] Pelzio Tibi propositum esse Variorum Clarissimorum Virorum observationes de hoc Cometa simul edere; id quod, ut puto, Astrophilis haud potest haud esse ingratum, quare cum ab amico

1 efflagitasti : postulasti **TC**.
2 concinnando atque sic nimirum (concinnando *et* nimirum *del.*) **BO**.

quodam varias etiam observationes ex Hispania Italia et Germania obtinuerim, si ita visum erit, verbulo tantum significes, lubens transmittam; literas autem meas, cum nihil solidi de nupero Cometa contineant, tum ipsemet prolixius, meliorique ordine sententiam meam quantocyus *Prodromo* expositurus sum, rogo ne adiicias.[3] Clarissimos Dominum Kirstenium nec non Dominum Müllerum velim humanissime salutes, quibus autem citius non sum responsurus, quam cum a cura typographica omnino liberatus fuero. Figurae enim aeri incidendae multum mihi facessunt negotii. Domini Ottonis Gerrichii opinio de Cometis facillimo negotio refutari potest; sed non est huius loci. Fallitur sane multum, quod putet Cometam partem esse aeris vi tempestatis cum impetu divulsam; tum quod omnis motus in Cometis motu // unico Terrae annuo solum salvari possit: prout aliquanto dilucidius ex *Prodromo* nostro *Cometico* percipies. Vale et vive feliciter.

Dabam Gedani, Anno 1665, die 3 April[is] st[ylo] n[ovo].

Tuae Illustrissimae Dominationis
studiosissimus

J[ohannes] Hevelius

To Master De Rolis
in Hamburg

Illustrious Sir!

You should not blame yourself for the fact that I answer late to your two pleasant letters. Because you (together with other exceptional men and honourable friends) asked me to urgently publish something about the latest comet before the *Cometographia* is finished, thus, having put aside other tasks and even having stopped my correspondence, from the day I returned from Königsberg to Danzig (although my original plan was different), I have decided that nothing is more important than to fulfil your wish (if only time allows it) and to prepare the *Prodromus* of my *Cometographia*. With God's help, I had already completed it and sent it to the printer. Without a doubt it will be published in about a month. In that little work, you will find not only a fairly exhaustive general history of the comet but also the dissertation about the movement, generation and various phenomena pertaining to all comets. Therefore, with those few pages I can now answer more extensively all the inquiries and questions which were asked by all my friends at once, whose wishes I could not entirely satisfy with letters until now. Everything is discussed briefly, but you will understand my opinion and hypothesis clearly enough. Other things, which will pro-

3 rogo…adiicias : non judico esse adiiciendas TC.

bably seem to be either still obscure or too superficially discussed, will be elucidated and demonstrated in our *Cometographia* which will be finished soon. I am deeply grateful for your offer to help me in publishing this booklet, the proposal which was sent so kindly and eagerly; but since I have a printing press in my house (a decent one and prepared not without expenses), and as I employ three printers, you will understand more than well that all the fonts should be made in my presence, because when the author is absent, those astronomical things, full of numbers, are never published with necessary diligence. As far as other things are concerned, I found out from the most noble Pels[4] that it was suggested to you to collectively publish the observations about the comet made by various famous men. I think it will be delightful for the star lovers. I received from a friend various observations from Spain, Italy, and Germany; If you are interested, just tell me, and I will willingly pass them to you. However, I ask you not to include my letters, since they contain nothing significant about the latest comet, and very soon I will present my opinion more thoroughly and in a more organized way in the *Prodromus*. I would like you to pass my kind regards to the most renowned Master Kirsten[5] and Master Müller,[6] to whom I will not answer until I become entirely free of printing duties, since engraving the drawings in brass keeps me very busy. Master Otto von Guericke's opinion about comets can be easily refuted, but this is not the place for it.[7] He makes a big mistake in thinking that a comet is part of the air, separated violently by force from a storm, and that each and every cometary movement can be explained simply by the annual motion of the Earth, as you will learn more clearly from our *Prodromus cometicus*. Farewell, and prosper!

Given at Danzig, on 3 April 1665, new style.

Illustrious Sir,

Yours devoted,

Johannes Hevelius

4 See note to Letter 5.
5 See section 2.4.9 of the Introduction.
6 See section 2.4.11 of the Introduction.
7 See section 3.4.3 of the Introduction.

10.

7 April 1665, Hevelius to Lubieniecki

BO: C1, v. 7, 960
TC: 371

Illustri Viro
Domino De Rolis
J[ohannes] Hevelius Salutem Plurimam Dicit

Recens illud insigne Caeli phaenomenum, quod denuo clarissime summo mane affulget, in caussa est, cur hasce ad Te dederim literas. Observatus autem est novus hic Cometa hesterna die mane a hor[a] 1 30' ad hor[am][1] 4 30' a me summa diligentia, in pectore Pegasi et quidem in 14° Piscium, et Latitud[inis] Bor[ealis] 26° 30' circ[iter]. Progreditur motu directo, secundum nimirum seriem Signorum in sua orbita satis veloci. Caput eius perlucidum est albicantis coloris, quanquam aliquanto minus, ac illius nuperi: caudam eo die prae se ferebat 14° longam; tendit versus Piscem Boreum. Dominus noster Büthnerus adstruere quidem hodie conatus est, cum meis observationibus interesset, hunc Cometam illum priorem esse, qui in Ariete evanuit; sed, crede, longe fallitur. Nam plane novus est, qui sine dubio ex Serpentario vel Antinoo prodiit, motu tendens plane contrario, tum etiam sub longe diverso angulo inclinationis orbitae; de quibus autem alio tempore plura. Rogo ut quantocyus haec nova referas Domino Erasmo Bartholino cui die 4 huius mensis ad suas ad me scriptas respondi;[2] tum[3] si quid de hoc Cometa perceperis pariter nobiscum communices. Saluta amicos omnes. Vale.

Dabam raptim Gedani Anno 1665 die 7 April[is] St[ylo] n[ovo].

Johannes Hevelius sends greetings
to the illustrious Sir,
Master De Rolis

The reason why I am sending this letter to you is this peculiar new phenomenon in the sky, which again shines clearly in the early morning. The new comet was being observed by me with the greatest diligence yesterday from 1.30 until 4.30 in the morning, in the Chest of Pegasus and in the fourteenth degree of Pisces, and at the lati-

1 horam *om.* TC.
2 referas...respondi : Erasmo Bartholino, cui nuper die 4. huius mensis rescripsi, referas TC.
3 tum : et TC.

tude of about 26° 30' north. It is moving in a quite fast direct motion, along the order of the signs, in its own orbit. Its head was white-coloured and very bright, but slightly less than the former one. It bore its tail ahead, 14 degrees long. It tends towards the North Fish. Our Master Büthner,[4] despite the fact that he took part in my observations, tried to convince me that this comet is the same as the previous one which disappeared in Aries. But, believe me, he is entirely wrong, because this one is brand new and undoubtedly came from the Serpentarius or Antinous; it is moving with an altogether opposite motion and at a very different angle of inclination of its orbit. But more about these issues another time. I ask you to refer this news (as soon as possible) to Master Erasmus Bartholin,[5] whose letter I answered on the fourth of this month, and also to share with us in return whatever you may receive about this comet. Greet all friends. Farewell!

Given at Danzig, in haste, on 7 April 1665, new style.

4 See section 2.4.3 of the Introduction.
5 See section 2.4.1 of the Introduction.

11.

14 April 1665, Lubieniecki to Hevelius

BO: C1, v. 7, 972
TC: 371–74

Nobilissime, Amplissime, Clarissime Vir.

Te gravibus studiis coelestibus dies noctesque occupatum ad literas meas non protinus respondisse is miretur, qui Te non novit. Tua enim eximia humanitas Tuam officii omnibus exhibendi promptitudinem[1] extra omnem dubitationis aleam ponit. Tuae vero assiduae et sublimes occupationes facile Te responsum vel officium aliquod differentem cuivis aequo excusatum reddent. Hac in Te aequitate utuntur Cl[arissimi] et Human[issimi] Viri Kirstenius et Mullerus, quibus salutem pl[urimam] nomine Tuo retuli cum mihi commodissimo loco ad spectanda phaenomena habitanti nudius tertius adessent, quibuscum[2] cometam nuper ad nos, veluti viribus auctum (adeo cauda[3] nunc est longe quam initio maior) reversum, illis enim idem esse cum priori videtur,[4] observavi, ut et hoc mane praesente quoque Cl[arissimo] Huswedelio Tui cum omnibus ingenuis cultore, qui una cum illis pl[urimam] Tibi mittit salutem, Mullerus vero etiam hanc rudi, ut dici solet, penicillo, scilicet sine omnibus instrumentis factam et observationem et delineationem. Cometam non occidisse Italis, ut nec Tibi (loquitur autem ille de priori, nam huius nullam facit die 3 April[is] mentionem) Cl[arissimus] Bullialdus mihi nuper indicavit.[5] Credimus Te continuum eius cursum observasse. Qua de re abs Te doceri merito cupimus. Nempe ut Phocylides monet

Τὴν σοφίην σοφὸς ἰθύνει, τέχνας δ᾽ ὁμότεχνος[6]

Tanto magis quod Tu Astronomicae artis non tantum sis particeps cum tot aliis, sed et scientissimus ac consultissimus prae tot aliis. Nam Te profecto Deus nobis excellentissimum Mathematicarum

– – – rerum dedit esse Magistrum

1 promptitudinem : promptam voluntatem **TC**.
2 quibuscum : Quippe cum iis **TC**.
3 cauda **TC** : comeda **BO**.
4 illis...videtur *om*. **TC**.
5 ut nec...indicavit : Clarissimi Bullialdi nostri litterae indicant **TC**.
6 ὁμότεχνος : ὁμότεχνος Sapientiam sapiens dirigit; artes autem qui eiusdem est artis **TC**.

quem versiculum Virgilianum, ut et illum Phocylideum de Te hodie suffragantibus Viris illis Doctissimis recitavi. Atque hi jam, sed et alii gaudent plurimum Te nobis *Cometographiae* Tuae *Prodromum* promittere brevi, Deo favente,[7] praemittendum augusto illi operi, de quo jam ille orbem literatum, praecipue vero Mathematicos compellat:

Cedite Romani scriptores, cedite Graii
Nescio quid maius nascitur *Iliade*.

in suo scilicet genere suaque serie[8] signorum et phaenomenorum[9] coelestium seriem, naturam et motum describens. In quo praemittendo[10] quod mihi quoque illud expetenti partem aliquam tribuere velis pro Tua solita humanitate et in omnes honestatis et literarum amantes benevolentia facis. Crede mihi posse me gratum aliquod Tibi officium facere in commendandis (quanquam scripta Tua accuratissima nativa sua commendatione gaudere, nec mea indigere scio, nisi quod tamen // illis index et nuncius[11] opus sit) Tuis scriptis et distrahendis Tuis exemplaribus: qua de re, monitu Nobilissimi Pelzii,[12] cum praecipuis Urbis huius Bibliopolis Königio, Hörtelio et Neumano non sine successus spe locutus sum.[13] Sed et ipsum me hic illic Amicorum opera, quos mihi ubique per Dei gratiam diuturno usu et magno studio thesauri vice comparavi, non pauca exemplaria distracturum spero.

Quod ad institutum meum attinet, ita est velle me[14] varias Virorum Praeclarorum Observationes, narrationes et judicia de nupero Cometa edere in publicum, cum Amici ad id faciendum me hortentur,[15] utile quoque[16] opus hoc publice fore judicantes. Mathematicae quidem rei parum utilitatis adferet, cum tot Viri Ingeniosissimi eam scriptis doctissimis illustrent. Tu vero prae caeteris, Magne Heveli, a cuius manu inimitabili, utpote et pingenda et[17] fingenda aere scribere, et scribenda pingere et fingere docta et[18] singulari rerum Astronomicarum scientia, tum instrumentorum artificiosissimorum et assiduae diligentiae, sed et quam sub tecto Tuo politissimam et accuratissimam ac supra hominis privati fortunam assurgentem[19]

7 Deo favente : favente Deo **TC**.
8 in…serie : in sua scilicet serie **TC**.
9 phaenomenorum : phoenomenôn **TC** (the circumflex denotes the long 'o' and, therefore, the given form is the Greek genitive plural).
10 praemittendo : promittendo **TC**.
11 illis…nuncius : index et nuncius illis **TC**.
12 monitu…Pelzii *om.* **TC**.
13 non…sum : locutus sum non sine successu spe **TC**.
14 velle me : me velle **TC**.
15 cum…hortentur : Nam Amici ad id faciendum me hortantur **TC**.
16 quoque *om.* **TC**.
17 et : ac **TC**.
18 et : tum vero et **TC**.
19 assurgentem **TC** : assurgentes **BO**.

habes typographiae praestantia nonnisi singulare aliquid expectant. Sed aderunt Politica multa, Historica ac Moralia, quae Amici publice grata et utilia fore judicant. Quorum hortatibus et monitis obedio[20], ne non tantum plusquam verecunde et modeste Amicis repugnans, sed et invidus ac publici emolumenti negligens videar. Atque ita spero Te pro Tua aequitate[21] vitio mihi non daturum, quod talia tentem. Optaveram quidem Tuum *Prodromum* meo operi adiungi potuisse. Nam ille huic plur[imum] lucis attulisset. Sed excusatio Tua est justissima. Certum enim est neminem sub sole melius Tua Te ipso edere posse, praesertim cum typographiam adeo felicem domi Tuae habeas. Verum tamen puto ex consilio Amicorum integrum mihi esse aliquid tentare in istis, et illo quidem modo, qui cum commodo publico sit junctus.

Non si priores Maeonius tenet
Sedes Homerus:

ideo aliorum Poetarum, et vel illa judice *Satyrico*, ridenda poemata Ciceronis latent.

Nec si quid olim lusit Anacreon
Delevit aetas:

ut alicubi Lyricorum Antesignanus canit: cuius dives adhuc in carmine luxuriabat vena, cum Ingeniosissimi Poetarum Ovidii suavitatem Roma jam amplexa ex ea voluptatem perciperet. Nec Thaletis excellentissima Theoreticae philosophiae[22] possessio ulla ratione diminutam se esse questa est, quod simul tunc terrarum ocellus et sapientiae altrix Graecia Solonem, qui illius philosophandi rationis expers civitatem legibus et moribus condebat, et Biantem[23] in extrema pauperie Magnanimum et ditissimum, quia cunctis cum patria Urbe amissis omnia sua secum portantem, habuit. Laudi cessit et aliis // civilis ac moralis disciplinae studium, licet sola tunc Thaletis philosophia ultra usum ad contemplandum sese extulerit, ut ait Plutarchus in *Solone*. Recta quippe ratio docet in viros literarum et virtutis amantes non cadere ut bona invideant iis, qui cum illis aliquo communium studiorum vinculo juncti sunt: imo si constanti Ciceronis, Viri suo seculo omni ex parte Praestantissimi sententiae credimus, 'nemo alterius, qui suae confidit, virtuti invidet.' *Phil*[*ippica*] 10. Et tamen verum est illud eiusdem effatum: 'Plerique sunt invidi, maximeque est hoc commune vitium et pervagatum.' l[ibro] 2 *de Orat*[*ore*]. Te proinde, Amplissime Heveli, quem Res[publica] literaria, ut summum Astronomum colit, securum esse jubeo, Soli tuae gloriae tantis meritis partae nullam me Eclipsin allaturum. Et quanquam bene sciam face nulla illi opus esse, si quid tamen tenuis Musa mea

20 obedio *om.* TC.
21 Te...aequitate : pro Tua aequitate Te TC.
22 philosophiae : philosophiae contemplativae TC.
23 condebat...Biantem : condebat, coluit quod eadem. Vel quod eadem Biantem TC.

valet, dabo operam, ut me id officii quod candor ab homine ingenuo exigit in Te fecisse, et Tu fateare, et universus orbis literatus videat. Et certe, ut quod res est dicam,

Nulla meis sine Te quaeretur gloria rebus.

Nam Mathematicarum

– – – Tibi maxima rerum
Verborumque fides.

Ideoque rara est inter aliquot epistolarum a me vel ad me datarum[24] decades, quae non aequum meritis Tuis praebeat testimonium, prout[25] huius rei aliquot specimina ad Nobiliss[imum] Pelzium communem Amicum nuper misi, ut[26] ita Tuarum ad me literarum, sed a Te pro lubitu recognoscendarum exempla coram tot Viris literatis reticeri salva et Tua et mea existimatione nequeant.[27] Atque hoc nulla ratione vel invidiam vel etiam dedecus quoddam, imo vero[28]

Pulchrum eminere est inter Illustres Viros,[29]

conciliare Tibi poterit.[30] Caeterum ut et Tibi et mihi satisfiat, votaque utriusque nostrum conveniant,[31] velim mihi Tuam generalem Cometae observationem mittas, quo aeri incidi et sicut aliis, ita et illis, quas mihi ex Hispania, Italia et Germania acceptas offers (quo nomine ego Tibi gratias ago, quantas possum maximas) sub Tuo nomine edendis addi queat. Nam neque hoc Tuis luminibus, ut puto, officiet. Quod si tamen aliter Tibi videantur // ego invito Te nihil hic de Te statuam, et contentus ero solis Tuis epistolis in quibus Te ad *Prodromum Cometographiae*, demum ad hanc ipsam referes. Vale Vir Eximie, quam diutissime et prospere age bono publico, ac[32] mei in Te sinceri studii certus esto. Sum enim

Amplissime et Clarissime vir

24 datarum : adhuc exaratarum **TC**.
25 prout *om.* **TC**.
26 ut *om.* **TC**.
27 nequeant : nequeunt **TC**.
28 vero *om.* **TC**.
29 Viros *om.* **TC**.
30 conciliare...poterit *om.* **TC**.
31 conveniant : in unum conveniant **TC**.
32 ac : et **TC**.

Tui

omni cultu et officio
addictissimus

De Rolis
manu propria[33]

Hamb[urgi] 14. April[is] 1665.

Post Scriptum. Adhuc in aere Tuo fui, Nobilissime et Clarissime Heveli, nondum Tibi dato ad priores literas die 3. April[is] datas, a me vero ante triduum acceptas responso, et ecce, quae est singularis Tua humanitas tantis digna meritis, novo me Tibi officii nexu devincis. Hodie enim novas accepi literas, easque nuncias novi Cometae, quem Tu quidem alium a priore dicis esse.[34] Quod ego ex variis coniecturis, diversitate scilicet situs, motus et magnitudinis suspicabar,[35] at nunc fidem Tibi do. Nam ex veteri dicto, Artifici, tanto magis tam Excellentissimo,[36] in sua arte credendum est. Observavi ego hodierno mane ortum fuisse Cometam quadrante horae ante 2. prope ortum Solis aestivum: caudam proiiciebat, praesertim oriens, quam longissimam, existens in brachio Andromedae vix extra Tropicum Cancri ad 15. vel 16. gradum Arietis. Intuebar eum usque ad horam 4 1/2.[37] Paulo ante disparuit caput Andromedae: una cum Cometa Lucida Capella. Sed ecce paulo post hos[38]

Aureus affulget jucundo sydere terris
Juppiter, unde juvans dicitur esse Pater.

Cum Cl[arissimo] Bartholino Tuam observationem communicabo hodie, si id fieri poterit,[39] sin minus, proxime id fiet. Saluta Cl[arissimum] Büthnerum.[40] Vale.[41]

33 Sum…propria *om.* **TC**.
34 dicis esse : esse statuis **TC**.
35 suspicabar : colligebam **TC**.
36 Excellentissimo : excellenti **TC**.
37 4 1/2 : dimidiam quintam **TC**.
38 paulo post hos : post hos paulum **TC**.
39 id…poterit : si potero **TC**.
40 Saluta…Büthnerum *om.* **TC**.
41 Vale : Iterum vale **TC**.

Greetings

Most noble, worthy, and renowned Sir!

One can wonder why you have not answered my letter immediately only if they do not know you, as you are busy night and day with serious celestial studies. Due to your exceptional kindness, you strive to show your goodwill to everyone, and no one can doubt it. And every just person will undoubtedly excuse your delayed answer (or other duties), because of your continuous and superior preoccupations. Therefore, the most renowned and gentle Masters Kirsten[42] and Müller[43] are patiently waiting. I greeted them on your behalf when they visited me three days ago because I live in a place which is very convenient for astronomical observations. We all saw the comet which had returned to us — they think it is the same as the former one, grown and strengthened (its tail is bigger than it was before). I saw it again this morning with the most renowned Huswedel[44] who admires you with all his heart; Along with Kirsten and Müller, he sends you many greetings. Müller sends you also this observation and the description which he made with a crude pen, as the saying goes, viz. without any instruments.[45] The most renowned Boulliau[46] has recently informed me that the comet is still being seen by Italians, as well as by you (he says about the former one, as he made no mention about this one on 3 April).[47] We believe that you have seen all its route. We want to learn from you on this subject, of course. Because, as Phocylides[48] says,

> The wise man leads the wisdom, the one skilled in art leads the art.[49]

Moreover, not only are you a contributor to astronomy (as many other people are), but you are the most learned and most experienced one, surpassing all of them. Because you are the one whom God

> appointed as a leader[50]

42 See section 2.4.9 of the Introduction.
43 See section 2.4.11 of the Introduction.
44 See note to Letter 1.
45 There is no extant manuscript with this observation.
46 See section 2.4.2 of the Introduction.
47 See Boulliau's letter to Lubieniecki in TC pp. 477–78, BnF ms. FF. 13026 ff. 287r–288r.
48 Phocylides of Miletus (6th century BC), Greek poet and supposed author of poetical moral maxims. Lubieniecki quotes here a poem composed in the Alexandrian Jewish community in the 1st or 2nd century AD, allegedly written by Phocylides.
49 *Pseudophocylidea* (*Φωκυλίδου γνῶμαι*), 88, in *Poetae lyrici Graeci*, vol. 2, ed. by Theodorus Bergk (Lipsiae: in aedibus B. G. Teubneri, 1882), p. 93.
50 Virgil, *Aeneis*, 9, 173, transl. by H. Rushton Fairclough, in Virgil, *Aeneid VII–VII. Appendix Vergiliana* (Cambridge, London: Harvard University Press 2002). Lubieniecki changed the original plural form *magistros* into a singular *magistrum*.

for us in mathematics. Today I declaimed about you this line from Virgil, as well as the one from Phocylides, and the most learned men mentioned above agreed with it. And they (and others too) greatly rejoice because you promise us the *Prodromus* of your *Cometographia* which, with God's help, will be soon announced. It is already stirring the learned world, especially mathematicians:

> Recede, Romans, recede, Greeks,
> something bigger than *Iliad* is being born,[51]

of course in its own kind and type, describing the order, nature, and the movement of celestial signs and phenomena. As usual, you act kindly and benevolently towards all who love honesty and science because you would like to grant me — the one expecting the book — a role in announcing it. Believe me that I can fulfil a proper duty if I recommend your writings and help in circulating the copies of your books (your writings, however, have their own innate recommendation and, I think, they do not need mine, unless they require an advertiser or a messenger). I have discussed this case (as the most noble Pels[52] admonished me) with the leading booksellers of this city — Königius,[53] Härtel,[54] and Naumann[55] — and there is a chance for success. But I hope that I will distribute a significant number of copies on my own, wherever I can, with the help of my friends. With God's grace, I have made them everywhere, thanks to my long-lasting contacts and great efforts, and they are like a treasure to me.

As far as my plan is concerned, I want to publish various observations, descriptions, and opinions made by some famous men about the latest comet, because my friends encourage me to do so. They think that the work would be useful to the public. It will be of little use for mathematics since it will be adorned with elaborate writings by the most talented men. You will be placed there above others, great Hevelius, from whose inimitable hand (which is skilled in writing what will be drawn and carved in brass, and to draw and carve what will be written) and also unique knowledge of astronomy, and the advantage of your printing press (perfect and incredibly precise, which you have in your house and which surpasses what an ordinary man can wish for), they expect something extraordinary. But there[56] will also be much about politics, history, and ethics — my friends find it interesting and useful for the public. I obey their encouragement and directions. In this way, I would not only seem to oppose my friends

51 Propertius, *Elegiae*, 2, 34.

52 See note to Letter 5.

53 Probably the same Samuel Konigius mentioned in Letter 18. It is not very likely that he was related to Georg König (1657–1713) who run his printing business in Hamburg since 1693, because König's family did not originate from Hamburg but from Bautzen in Lusatia.

54 Zacharias Härtel (Hertel, d. 1678).

55 Johann Naumann (1614–68).

56 After the praises of Hevelius and his upcoming book about comets, Lubieniecki returns to discussing his planned *Theatrum cometicum*.

(more than decency and modesty allow), but I would also seem to be jealous and to disregard public benefit as well. Because you are kind, I hope you will not resent me when I ask for such a thing. I would like to include your *Prodromus* in my work because it could enlighten my book. But your refusal would be justified. It is certain that there is no one under the Sun who can publish your book better than yourself, especially because you have a great printing press in your house. I think, however, as my friends advise me, that I can try to ask you about it, aiming at common good.

Though Homer fills the foremost throne,[57]

none of the other poets, even those ridiculous poems of Cicero — as judged in the *Satyricon*[58] — remains unknown.

The songs of Teos are not mute,[59]

as the poet of the first rank sings elsewhere. His inspiration was still full of songs when Rome had already embraced the sweetness of Ovid, the most talented of poets, and was feeling its pleasure. Thales's skill in speculative philosophy did not complain that it was diminished in any way only because, then Greece (the most precious of lands and the feeder of wisdom) had Solon[60] (who, unfamiliar with the former's method of philosophying, endowed the city with laws and customs) and Bias[61] (who was the greatest and richest in extreme poverty because he left everything he had possessed for the sake of his native town and carried all his things with him). He[62] left the study of civil and ethical knowledge to others, however only Thales's philosophy exceeded the utility for the contemplation of itself, as Plutarch says in the *Life of Solon*.[63] The sound reason teaches us that the men who love sciences and virtue should not be jealous of good things which belong to those with whom they are bonded by common studies, especially if we believe in the sound opinion of Cicero (a man who was the most distinguished in everything in his time) 'no one envied the virtue of another who was confident of his own,' *Philippic* 10.[64] And indeed his opinion is true: 'Many are jealous, and this is the most common and widespread fault,' Book 2 of *On the Orator*.[65] Therefore, greatest Hevelius, revered

57 Horace, *Carmina*, 4, 9, 5–6, transl. by John Conington, in Horace, *The Odes and Carmen Saeculare* (London: George Bell and Sons, 1882).

58 Petronius, *Satyricon*, 5.

59 Horace, *Carmina*, 4, 9, 9–10, transl. by John Conington, in Horace, *The Odes and Carmen Saeculare*. The original Latin text refers directly to Anacreon of Teos.

60 Solon (*fl.* 7th/6th century BC), Athenian politician and poet, archon in 594/3. He was a reformer of the Athenian political and social life and considered the founder of Athenian democracy.

61 Bias of Priene (*fl.* 6th century BC), ancient Greek sage. Considered to be one of the Seven Sages.

62 The subject of the sentence is obscure but I presume that Lubieniecki intended it to refer to Thales.

63 Plutarch, *Life of Solon*, 3, 8.

64 Cicero, *Philippicae*, 10, 1, transl. by C. D. Yonge, in M. Tullius Cicero, *The Orations of Marcus Tullius Cicero* (London: George Bell and Sons, 1903).

65 Cicero, *De Oratore*, 2, 210.

by the Republic of Letters as the best astronomer, I want to assure you that I will not eclipse the Sun of your glory which was earned with your great merits. And even though I am aware it needs no torch, if my modest talent can serve anything, I will make sure to fulfil my duty towards you. Justice requires this from a noble man, so that you could acknowledge it and the whole learned world could see it. And, of course, to put it thus:

> No glory shall be sought for my own without you.[66]

Because in the mathematics

> you will have my greatest trust in deed and word.[67]

Therefore, there are hardly any among dozens of letters sent by me or to me, which do not give a just testimony of your merits. I have recently sent some such specimens to the most noble Pels, our mutual friend. I could not pass over in silence (in the presence of so many learned men) the examples of your letters to me, which you had allowed being presented to the public, when the opinion about you and me is safe. And it cannot cause any jealousy or any dishonour for you, but rather

> it is beautiful to surpass the famous men.[68]

Moreover, in order to fulfil our wishes and our plans, I would like you to send me your general observation of the comet. I can engrave it in brass and add to other observations, including those which you received from Spain, Italy, and Germany, and which you offer to send me (I am very grateful for that). I would publish it under your name. I hope you will not find it unpleasant. If you disagree with it, I will write there nothing about you without your consent, and I will be contented only with those of your letters in which you refer to the *Prodromus* of your *Cometographia* and to the latter book itself. Exceptional Sir, be in good health for very long, prosper for the public good, and be assured of my sincere devotion to you! Because I am,

most worthy and renowned Sir,

Your most devoted admirer,

De Rolis
by my own hand

66 Virgil, *Aeneis*, 9, 275, transl. by H. Rushton Fairclough, in Virgil, *Aeneid VII–XII. Appendix Vergiliana* (Cambridge, Mass., London: Harvard University Press 2002).

67 Virgil, *Aeneis*, 9, 279–80, transl. by H. Rushton Fairclough, in Virgil, *Aeneid VII–XII. Appendix Vergiliana*.

68 *Octavia* (authorship uncertain, attributed to Seneca the Younger), 472.

In Hamburg, on 14 April 1665.

Postscript. So far, I have been indebted to you, most noble and renowned Hevelius, because I have not answered your earlier letter yet, which was sent on 3 April, and received by me three days ago. Because your kindness equals your merits, you oblige me with a new bond of duty. Today, I have received a new letter and some information about the new comet — the one you think to be different from the former one. I suspected it on the basis of some conjectures, viz. differences in the place, movement, and magnitude, and now I believe you. According to an old saying, one should believe an artist in his art, especially the most skilled artist. I saw the comet rising today morning, at a quarter to 2, near the time of the summer sunrise. It was pushing its very long tail ahead, especially while rising, and it was in the Arm of Andromeda, a fair bit above the Tropic of Cancer in the fifteenth or sixteenth degree of Aries. I watched it until 4.30. The Head of Andromeda disappeared a moment earlier, and so did the comet with the bright Capella. And soon after them

> with the pleasant star the golden Jupiter
> shone to the Earth, that is why he is called the helping father.[69]

I will share today your observation with the most renowned Bartholin,[70] if I only can. If not, it will be done soon. Greet the most renowned Büthner.[71] Farewell!

69 I have been unable to identify the origin of this passage. It is not unlikely, however, that it was composed by Lubieniecki himself. On Lubieniecki's poetry see Stanisław Kot, '"Descriptio gentium" di poeti polacchi del secolo XVII', *Ricerche slavistiche*, 6 (1958), 151–84 (pp. 151–52, 170–77); Janusz Tazbir, 'Wiersze Stanisława Lubienieckiego', *Przegląd Humanistyczny*, 2 (1965), 113–22. See also Letter 84A. Etymology of Jupiter (*Iuppiter*) as the 'helping father' (*iuvans pater*) has been already presented by Cicero, *De natura deorum*, 2, 64.

70 See section 2.4.1 of the Introduction.

71 See section 2.4.3 of the Introduction.

12.

14/24 April 1665, Lubieniecki to Hevelius[1]

BO: C1, v. 7, 973
TC: 374–75 and 375

Salutem Plurimam Nobilissime, Amplissime, Clarissime Vir. Pauca de novitio Cometa quae habebam dicenda, et nomine Clarissimorum Virorum Kirstenii et Mulleri Tecum communicanda, ad Te ante dies 14.[2] misi. En Tibi quae rogatu Illustrissimi Rautensteinii Ministri Status et Legati ad Comitia Ratisbonensia Neoburgici[3] mittere Tibi debui a R[everendo] P[atre][4] Curtio ad Eum perscripta, quae ad Te spectant. Prior epistola haec in se habebat:

'Sum negotiosus, et inter Sacra aliaque munera vix tantum mihi superest, ut Ill[ustrissimae] D[ominationis] V[estrae] comprecer felices quas exspectamus Ferias plenasque caelestium benedictionum, quarum munere vana et falsa sint, quae in aestatem proximam de coelo et de Gallinario metuunt. Aiunt novum sidus in coelo lucere, in constellatione Andromedae, mutareque locum, magnitudinem, stationem. Dominus Hevelius sciet, qui Theoriam novae stellae in Cygni pectore identidem lucentis perspectam habet. Eam Seren[issima] Sueciae Regina putat Parenti suo natalem et mortis prodromam fuisse. Mittam illi sane *Historiam caelestem* perlibenter. Sed exemplum Augusta petendum. Hic enim promptum non habeo. Atque illud ipsum erit acephalum, numerisque non semper integris. Tamen, ut dixi, mittam.' Neoburgo die 3. April[is] 1665.

Posterior, utraque enim simul Ratisbonam erat allata et ad me missa, haec continebat:

'Temporis et tabellarii culpa est, quod priore hebdomade nullae literae adfuerunt: tanquam si Saturno et Mercurio imputarem haec vitia silentii.[5] Interim nobis Cometam denuo videre datum est mane ex Solis radiis emergentem. Stetit non longe abiunctus a Quadrilatero volantis Pegasi, poteratque hodie rectius observari si coelum favisset. Domino Hevelio mittam // exemplum *Historiae coelestis* quamprimum Augusta habuero. Sed velim ab Ill[ustrissima] D[ominatione] V[estra] excusari acephalum et abortientibus interdum litteris male habitum.' Neob[urgi] 10. Apr[ilis].

1 In the TC, this letter is printed in two separate parts. The first is dated 21 April 1665, the second — 4 May 1665. In the manuscript, the date is 14 April 1665.
2 14. : octo **TC**.
3 Ministri...Neoburgici *om.* **TC**.
4 R[everendo] P[atre] : R[everendo] D[octissimo] P[atre] **TC**.
5 silentii *corr. ed.* : silentum **BO TC**.

Hinc vides quanti Te, Magne Heveli, merito Tuo me non tantum conscio et teste, sed et indice faciant Rautensteinii, Heinsii, Bullialdi, Curtii, praeclara nomina, quae[6] se magnis meritis et egregiis ingenii facinoribus orbi jam dudum prodiderunt et feliciter pararunt prosperam sui memoriam a grata posteritate sollicite conservandam. Vides Praestantissimos quosque Astronomos ad Te velut ad vivum Oraculum, quo Urbs Vestra aeternum illustris cluet, appellare, tantum abest ut quis laudes Tuas deterere vel Tibi

– – – detrahere ausit
Haerentem capiti multa cum laude coronam.

Crede mihi magno cum desiderio (cui nihil satis festinari nosti) expectamus *Prodromum* Tuae *Cometographiae*, antequam haec ipsa conspiciendam se nobis praebeat. Ergo

– – – quid gaudia nostra moraris?

Urge opus: urget enim Te nostra expectatio, nos taedium. Siderum conditor et Rector Tibi propitius adsit, ut mirabilia eius opera oculos ad videndum habentibus exhibeas, et caelum spirantes sperantesque animas ad verum eius cultum excites! Cum autem in postremis Tuis spem feceris plura Te nobis de novo Cometa (qui a die 9/19 huius hic observatus non est, et forte heliace occiderit)[7] scripturum fuisse, haec a Te expectamus, si tamen id Tibi per Tua assidua studia (scio enim Te perdium pernoctemque Tua agere) facere licet. Cognoscere cumprimis aveo vel ex Te vel ex Cl[arissimo] Bütnero, quae nunc ei de novo isto Cometa mens sit vel demum

– – – quae nunc animo sententia surgat. //

Video enim alios quoque Mathematicos idem cum eo sentire.[8] Sed

Non nostrum inter vos tantas componere lites.[9]

Accepi[10] Cl[arissimum] Concium Regiomontanum[11] itidem asserere hunc Cometam eundem esse cum priori, quod Tu me melius scies.[12] Cl[arissimus] Abrah[amus] de Grau Prof[essor] Franequer[anus] in suis die 15. huius datis literis, quas hodie acce-

6 praeclara nomina, quae : praeclara certe in Republica litteraria nomina. Haec **TC**.

7 (qui...occiderit) *om.* **TC**.

8 In the TC, here ends the first part of the letter.

9 Sed...lites : Salutem Plurimam. En Tibi, Amplissime Heveli! quae de novo Cometa Hafnia accepi scitu digna **TC** (the beginning of the second letter).

10 Accepi : Accepi etiam **TC**.

11 Regiomontanum : Mathematicum Regiomontanum **TC**.

12 scies **TC** : scibis [*sic*] **BO**.

pi, aperte non decidit hanc quaestionem. Vocat tamen novum Cometam, qui sibi primum visus sit die 16. 26. Martii, procedens e Pegaso versus caput Andromedae. Cl[arissimus] Placentinus[13] die 3. 13. huius mensis scribit tunc Cometa in Perseo extitisse. Sed cur ego noctuas Athenas mitto? Nempe ut tester Tibi[14] integram meam sinceritatem[15] et bene de Te, Vir Amplissime,[16] merendi voluntatem. Quae etiam me permovet, ut Te certiorem reddam, me cum Cl[arissimo] Bartholino Tua communicasse. Expecto a multo tempore responsum eius ad meas priores. Illud fere exciderat Gravium et Placentinum promittere mihi certo Observationes singulares et accuratas. Expecto et abs Te promissas, tum et responsum ad meas postremas. Vale, et favere porro perge ei qui est

Nobilissimo Nomini Tuo
omni cultu et
studio addictissimus

De Rolis

Hamb[urgi] 14. 24. Apr[ilis] 1665.[17]

Greetings, most noble, worthy, and renowned Sir! Fourteen days ago[18], I sent you as much as I could possibly say about the new comet. It was passed to you on behalf of the most renowned Kirsten[19] and Müller.[20] Here is what I had to send to you as requested by the most illustrious Rautenstein,[21] the minister of the state of Neuburg and deputy to the Diet in Regensburg. It was written to him by the Reverend Father Curtz[22] and concerns you. In the first letter, it reads as follows:

'I am busy, and, among my religious duties and other tasks, there is hardly any time left to wish you, most illustrious Sir, happy holidays full of heavenly blessings (which we all hope for). By the heavenly gift, may all these things from the sky and the henhouse,[23] which people fear of next summer, turn out false and absurd. They say that

13 Cl[arissimus] Placentinus : At vero celeberrimus quidam Mathematicus **TC**.

14 tester Tibi : Tibi probem **TC**.

15 sinceritatem : fidem, sinceritatem **TC**.

16 Vir Amplissime *om*. **TC**.

17 Quae etiam me…1665 *om*. **TC**.

18 In the TC, Lubieniecki says '8 days ago.' Neither number seems to fit the order of the correspondence.

19 See section 2.4.9 of the Introduction.

20 See section 2.4.11 of the Introduction.

21 See note to Letter 5.

22 See section 2.4.5 of the Introduction.

23 I could not decipher the meaning of this passage.

there is a new star in the constellation of Andromeda[24] and that it changes its place, magnitude, and position. Master Hevelius (who has an elaborated theory about the new star in the Chest of Cygnus,[25] which is constantly shining) will know it. The most serene Queen of Sweden[26] thinks that it was a portent of both the birth and death of her father.[27] I will willingly send her the *Historia coelestis*.[28] But a copy has to be ordered from Augsburg. I do not have it here. And it will be without the title and not all the numbers will be correct. Nevertheless, as I said, I will send it.' In Neuburg, on 3 April 1665.

The second one, which had been delivered to Regensburg at the same time and sent to me, included the following:

'The time and the messenger are to blame that there was no letter last week. So I could attribute this fault of silence[29] both to Saturn[30] and Mercury.[31] Meanwhile, we could see the comet again in the morning, emerging from the solar rays. It stood not far away from the Square of the flying Pegasus, and it could have been seen more clearly had the sky allowed it. I will send a copy of the *Historia coelestis* to Master Hevelius as soon as I get one from Augsburg. I would like you, most illustrious Sir, to excuse the lack of title and some shortcomings due to miscarried letters here and there.' In Neuburg, on 10 April.

Thus you see, great Hevelius, how much such famous names as Rautenstein, Heinsius,[32] Boulliau,[33] and Curtz[34] respect you. They have already presented themselves to the world with their great merits and exceptional deeds of mind, and they have fortunately gained the perpetual remembrance which will be preserved with esteem by thankful posterity. Not only do I agree and testify this, but I point it out as well. You see for yourself that the distinguished astronomers ask you as if you were a li-

24 See section 2.3.2 of the Introduction.

25 The variable star P Cygni, see note to Letter 8A.

26 Christina (1626–89), the Queen of Sweden from 1632 until her abdication in 1654. She is considered to be one of the most learned women of her era. She was a patron of arts and sciences. She corresponded with many prominent scholars, including Pierre Gassendi and René Descartes whom she invited to Stockholm in 1649. In 1654, she converted to Catholicism and abdicated.

27 Gustav II Adolph. He was born in 1594 and died in 1632. According to Hevelius, P Cygni was seen for the first time in 1601 and faded out by 1626 or 1629.

28 *Historia coelestis* (Augustae Vindelicorum: apud Simonem Utzschneiderum 1666) was a collection of the observations from Tycho Brahe's manuscripts, edited by Albert Curtz. Its being mentioned here suggests that it was almost ready for publication in early 1665.

29 Both in the BO and TC, the passage literally reads: 'the fault of the dead' (*vitia silentum*). I presume that it is a mistake and that the intended meaning was 'the fault of silence' (*vitia silentii*).

30 As the deity of time and old age.

31 As the divine messenger.

32 See section 2.4.6 of the Introduction.

33 See section 2.4.2 of the Introduction.

34 See section 2.4.5 of the Introduction.

ving oracle. This brings eternal pride to your famous city. It is impossible that anyone could diminish your praises or

> presume to pull off the laurel crown placed upon your brow with great applause[35]

from you. Believe me, we expect and desire (which, as you know, no haste can satisfy) that the *Prodromus* of your *Cometographia* will show itself to us before the latter book is out. So

> – – – why do you delay our joy?[36]

Hasten the work, because our expectation hastens you and impatience hastens us! May the Builder and Ruler of the stars be more favourable to you so that you could present His wonderful works to those who have eyes to see, and incite the breathing and hoping souls to His real worship! Since in your previous letter you gave us hope that you would write to us more about the new comet (which has not been seen here since the ninth/nineteenth of this month and perhaps has set heliacally), we expect it from you, if only you can do it among your incessant studies (as I know you conduct them night and day). First of all, I would like to know either from you or the most renowned Büthner[37] what he thinks now about this new comet or at least

> – – – what purpose now rises in his heart.[38]

Because I see other mathematicians agree with him. But

> it is not for me to settle so close a contest between you.[39]

I have learned from the most renowned Concius[40] from Königsberg that he also thinks this comet was the same as the former one (you will know this better than I). In his letter sent on the fifteenth of this month (which I have received today), the most renowned Abraham de Grau,[41] a professor in Franeker, makes no clear statement about this question. He calls it a new comet because he saw it for the first time on 16/26 March when it was passing from Pegasus towards the Head of Androme-

35 Horace, *Satyrae*, 1, 10, 48–49, transl. by C. Smart, Theodore Alois Buckley, in Horace, *The Works of Horace* (New York: Harper & Brothers, 1863). Quoted with minor changes.

36 Martial, *Epigrammata*, 8, 21, 1.

37 See section 2.4.3 of the Introduction.

38 Virgil, *Aeneis*, 1, 582, transl. by H. Rushton Fairclough, in Virgil, *Eclogues. Georgics. Aeneid I–VI* (Cambridge, Mass., London: Harvard University Press, 2004). Quoted with minor changes.

39 Virgil, *Eclogae*, 4, 104, transl. by H. Rushton Fairclough, in Virgil, *Eclogues. Georgics. Aeneid I–VI.*

40 See section 2.4.4 of the Introduction.

41 See note to Letter 7.

da. The most renowned Placentinus[42] writes on 3/13 of this month that the comet was then in Perseus. But why do I bring owls to Athens?[43] Of course to testify my constant sincerity and my will to do a good service to you, greatest Sir. It forces me to inform you that I sent to the most renowned Bartholin[44] what I had received from you. Since I have been waiting for his answer to my earlier letter for a long time, I have nearly forgotten that de Grau[45] and Placentinus promised me some exceptional and precise observations. I am also waiting for what you promised me and for the answer to my last letter. Farewell, and still be benevolent towards the one who is

noble Sir,

Yours most obliged,

De Rolis

In Hamburg, on 14/24 April 1665.

42 See section 2.4.14 of the Introduction.
43 This proverbial phrase was first attested in Aristophanes' *Aves*, 301.
44 See section 2.4.1 of the Introduction.
45 See note to Letter 7.

13.

26 April / [6 May] 1665, Lubieniecki to Hevelius

BO: C1, v. 7, 974

Seal with Lubieniecki's arms (Rola coat of arms)

Address: 'A Mons[ieur] / Mons[ieur] Jean Hevelius Cons[eiller] / de la ville de Dantzic, et / Grand Astronomien / a Dantzic / Inclusae' ('To *Monsieur* Johannes Hevelius, Councillor of the city of Danzig and great astronomer in Danzig. Enclosures')

Enclosure:
13A. Letter from Erasmus Bartholin to Lubieniecki, 11/21 April 1665

Salutem Plurimam Nobilissime, Amplissime, Clarissime Vir.

Hospitum et Amicorum, interque eos quorundam Virorum Illustrium officiis variisque occupationibus impeditus, has literas ante octiduum mittere ad Te non potui. Expectabam etiam a Te responsum ad meas, cuius spem mihi Nobiliss[imus] Dominus Pelsius fecit. Cui quae mihi Tuo nomine significavit sufficienter respondi. Nunquam animum meum subiit haec cogitatio, Te mihi invidere aliqua ratione velle. Verum Tibi potius fidem facere volebam, me nequaquam Tibi invidere. Tenues et exigui mei conatus non subiacent invidiae. Gratulor autem mihi pl[urimum] quod et Observationes illas Hispanicas, Italicas et Germanicas ad me mittere velis, et postulato meo assenseris, ut nimirum tot aliorum Clariss[imorum] Virorum epistolis Tuae quoque insecantur, hac tantum ratione quae Tibi ipsi placuerit, ut nempe lector remittatur ad *Prodromum* et ipsam *Cometographiam*. Ita meum tenue opusculum nuncium feret per orbem, si Deus voluerit, de tuis maximis et felicissimis operibus cedro dignis. Crede mihi, omnem dabo operam, ut tibi probem fidem et industriam. Unum atque alterum exemplar *Prodromi* velim quamprimum ad me mitti cures fasciculo Lubecam ad Christianum Wernerum directo. Vale et fave constanter.

Hamb[urgi] festinanter 26. April[is] S[tylo] V[etere] 1665.

Tui
observantissimus

De Rolis
manu propria

Greetings, most noble, worthy, and renowned Sir!

Because I had been preoccupied with duties towards my guests and friends (among whom are some illustrious men) and with other responsibilities, I could not send you this letter eight days ago.[1] I also waited for your answer to my letter — the most noble Master Pels[2] made me hope for it. I have replied sufficiently to what he passed to me on your behalf.[3] It has never crossed my mind that you would envy me for any reason. I wanted to assure you that I would never envy you. My little and insignificant efforts are not worthy of any jealousy. I am glad, however, that you want to send me those observations from Spain, Italy, and Germany, and that you agree with my wish that your letters would follow those of so many other renowned men — but only in a manner you accept — so that the reader would turn to the *Prodromus* and to *Cometographia* itself. And so my little work (God willing) will spread throughout the world the news about your great and successful endeavours which are worthy of cedar oil.[4] Believe me, I will do my best to prove my sincerity and efforts to you. I would like you to send me, as soon as possible, one or two copies of the *Prodromus* in a package to Lübeck, addressed to Christianus Wernerus. Farewell, and stay benevolent!

In Hamburg, in haste, on 26 April 1665, old style.

Yours obliged,

De Rolis
by my own hand

1 Lubieniecki either refers to Letter 12 (but makes a mistake about its date), or there was another letter; or, possibly, a pack of enclosures sent on 28 April, now missing.

2 See note to Letter 5.

3 There is no correspondence with Pels in the TC.

4 Cedar oil was used to preserve books. Therefore, the phrase 'worthy of cedar oil' (*cedro dignus*) praised the texts worth keeping, see John Evelyn, *Sylva, or a Discourse of Forest Trees. A Reprint of the Fourth Edition in Two Volumes*, vol. 1 (London: published by Arthur Doubleday & Company Limited, 1908), p. 257. This proverbial expression was used, e.g. by Persius in *Satires* 1, 42.

13A.

Letter from Erasmus Bartholin to Lubieniecki, 11/21 April 1665

BO: C1, v. 7, 974 (precedes the foregoing)
TC: 434

Hafnia Hamburgum die 11/21 April[is] 1665.

Gratissimas a Te perscriptas die 21. Martii, et die 7. April[is] S[tylo] V[etere] accepi magnum humanitatis Tuae pignus, cuius solicitudine responsum diutius differre pudet. Ad praelum jam mittebatur *Historia Observationum* Cometae Anni praecedentis, cum ecce die 30. Martii, S[tylo] V[etere] nova Fax mane nobis oritur, cauda prolixiore, quam praecedentis Cometae. Consuetis organis locum eius deprehendimus medium Pegasi in 27° piscium cum latitudine Boreali 25° motum vero in consequentia signorum usque ad signi Arietis extremas partes, latitudine in dies decrescente. Contendentibus hic Viris Eruditis non diversum esse Cometam a priori me opposui, rationibus non tantum ex contrario motu (id quod in uno eodemque contingere posse arbitrabantur, sicuti in planetis Retrogradis) sed etiam ex quantitate motus, diversitate anguli intersectionis et magnitudine diametri quae adeo diminuta erat cum visum nostrum effugeret, ut si generationem Cometis concedamus, tantam molem tam brevi tempore reassumere non posset et si coaeva reliquis statuamus corpora, tam subito propinquus accedere, emenso longissimo itinere nequiverit. Amplissimi Domini Hevelii suffragium // quod significas suspicio, quem scio haec acri magnoque judicio, sicuti omnia, expendisse. Porro perquam jucundum mihi fuit audire moliri Te Sylloges operum et sententiarum de Cometa editionem, ne sine fructu posteritatis multa egregia abeant. Tantum jam indulsit Frater precibus nostris, ut *Consilium Medicum de Cometa* conscripserit quod edetur in vulgus: ad Tuas quoque quam primum respondere paratus. Literas meas Tuae voluntati permitto, *Historia* vero *Observationum* cum publici fit juris tuo quoque conceditur arbitrio. Sed cave ne egregium opus naevo hoc deformes. Salutem Tibi mittit Frater et mecum Te valere jubet.

From Copenhagen to Hamburg, on 11/21 April 1665.

I have received some pleasant letters from you sent on 21 March and 7 April, old style (a high proof of your kindness), and it is shameful to delay the answer, knowing of your efforts. *Historia observationum* of the last year's comet[5] has been sent to the printing press, while the new torch arose for us on 30 March in the morning, with a

5 Bartholin most probably refers to his work which was finally published as *De cometis anni MDCLXIV et MDLXV opusculum* (Hafniae: apud Danielem Paulli, 1665).

wider tail than the former comet. With the help of some typical instruments, we determined its place in the centre of Pegasus, in the twenty-seventh degree of Pisces, at latitude 25° north; it was moving according to the order of signs towards the furthest parts of the sign of Aries, while the latitude was diminishing. When the local learned men tried to convince me that this comet is the same as the former one, I objected — relying not only on its opposite movement (which, as they think, can happen to one body, as in retrograde movement of planets), but also on the speed of this movement, on the difference in the angle of the intersection, and on the dimension of its diameter. It was so small that when the former comet escaped our sight, the new one could not grow to such a size again in such a short time, if we assume the generation of comets. And if we state that they are as old as other celestial bodies, this comet could not appear so close and so soon, since there was too long a way for it to travel. I admire the opinion of the greatest Master Hevelius, which you have sent to me. I know that he has considered this thing, as he always does, with his sharp and great mind. As far as the rest is concerned, I was pleased to hear that you are working on the edition of the collection of works and opinions about the comet so that many superb things will remain as a gift for posterity. My brother, who is ready to answer your letter as soon as possible, agreed to fulfil our wishes and write *De cometa consilium medicum*,[6] which will be published. I leave my letters at your disposal, and when the *Historia observationum* is published it will also be left to your judgement. But be careful not to deform the splendid work with any such fault. My brother sends his greetings to you and bids a farewell together with me.

6 Thomas Bartholin (see note to Letter 4A), *De cometa consilium medicum* (Hafniae: apud Matthiam Godicchenium, 1665). See section 3.3.4 of the Introduction.

14.

13 May 1665, Hevelius to Lubieniecki

BO: C1, v. 7, 975
TC: 375–77

Domino de Rolis
Hamburgum

Generose ac Clarissime Vir

Hactenus nulla potior mihi fuit cura, quam ut Tuo aliorumque Praeclarissimorum[1] Virorum desiderio satisfacerem, promissumque solverem; conscribendo nimirum et edendo *Prodromo* nostro Cometae antecedentis; jam vero opusculo illo ad finem perducto[2] ad binas Tuas multo mihi gratissimas respondeo, *Prodromumque* etiam ipsum *Cometicum* simul hisce, et quidem per Tabellarium ordinarium, quo eo citius illius te redderem participem, Tuis subiicio oculis. Ex ipso frontispicio statim vides, quid caussae fuerit, quod hae pagellae alterius operi annecti haud potuerint;[3] attamen Tibi integrum utique esse[4] in eadem materia quidvis etiam statuere omnino[5] Tecum sentio; praesertim cum certum sit, intentionem hanc Tuam in commodum cedere publicum. Nefas igitur est, si hocce laudabile Tuum propositum quisquam vel secus interpretari vel invidere vellet: in me, profecto, tantum abest ut suspicio aliqua invidiae cadere possit (prout ipsemet nosti, et jam[6] saepius cum Nobiliss[imo] Domino Pelzio loquutus sum) ut potius suasor fuerim, quo quantocyus observationes ac sententiae quotquot extent Eruditorum[7] Virorum, eo modo, uti proposuisti a Te ederentur. Quem in finem, ipsius operis eo magis promovendi caussa etiam promissas[8] Observationes hinc inde ab amicis acquisitas, nunc simul Tibi communico, ut si quid in iis dignum notatu reperiatur Tuis adiicere possis. Perge igitur alacriter in suscepto opere, et fac ut quamprimum illud absolvas; non mihi solum, sed, crede, omnibus rerum caelestium cultoribus, cumprimis quibus Scientia naturalis promovenda curae cordique est, facies // rem gratissimam. Porro, Generalem Cometae antecedentis descriptionem nullam conscripsi; poteris tamen eam ex *Prodromo* facillimo

1 Praeclarissimorum : Clarissimorum **TC**.
2 ad…perducto *om.* **TC**.
3 annecti…potuerint : haud potuerint annecti **TC**.
4 integrum…esse : utique integrum esse **TC**.
5 omnino : penitus **TC**.
6 et jam **TC** : etiam **BO**.
7 Praeclarissimorum *del.* **BO**.
8 etiam promissas : promissas quoque **TC**.

negotio construere; simul[9] ex eo excerpere, quaecunque visum fuerit, nec moleste feram, si meas literas,[10] dummodo tanti sint pretii, Tuis adiunxeris. De reliquo, ex hocce qualicunque opusculo nunc etiam intelliges ex parte hypothesin nostram Cometarum, tum quid de variis Quaestionibus statuam, nec non qua ratione Cometarum phaenomena universa et singula recte ac debite salventur. Tibi vero aliisque Viris Eruditis, quid de his nostris contemplationibus videatur, non dubito, quin, pro eo amicitiae nexu, quo obstringimur, suo tempore lubens libereque, more Philosophico, sis expositurus. De posteriori autem Cometa quid mihi statuendum sit: an is omnino in *Cometographiam* reiiciendus; an vero pariter ex parte luci in antecessum exponendus? adhuc haereo. Quodsi tamen Tibi amicisque visum erit, posterius esse amplectendum,[11] non minus istum laborem perlibenter prima occasione suspiciam; quo accuratam aliquam illius Descriptionem et Delineationem, imo etiam ipsas nudas observationes quarum haud paucas possideo, absque omni tamen calculo, qui *Cometographiae* reservatur, in gratiam Astrophilorum praemittere possim. Cometam autem hunc a die 6 Aprilis st[ylo] n[ovo] ad 20 eiusdem Mensis exquisite maioribus nostris Instrumentis aeneis, singulis fere diebus dimensus sum; diutius quidem non licuit: quippe dies 21 et 22 April[is] nubili prorsus extiterunt; at die 23, licet caelo undique sereno tamen jam nusquam eius vestigium deprehendebatur, sic ut eo die plane se se[12] jam subduxerit. Progressus est continuo motu directo secundum scilicet // Signorum seriem, in dies tamen tardiori, a 14 grad[u] Pisc[ium], quantum ego observavi, ad 2 grad[um] Tauri, latitudinem semper diminuendo. In prima enim observatione 26° 30'; in ultima vero tantum 13 grad[us] Boream versus removebatur. Ductum itaque plane eundem infra scilicet Caput Andromedae per Piscem Boreum, quem initio die 7 Aprilis Tibi praedixeram, sequutus est. Ratione capitis, facie plane alia quam ille prior luxit: nam in huius capite, beneficio Praestantissimi Telescopii singularia quaedam deprehensa sunt, quae distincte adumbrare aerique manu mea[13] incidere,[14] Deo volente, animus est. Percepi insuper plurimos Eruditissimos[15] viros adhuc in ea esse sententia, hunc ultimum Cometam omnino cum priori esse eundem. Sed, profecto[16] miror, quod talia statuere audeant, quae omni rationi prorsus contrariantur, tum quae etiam ex solo globo artificiali plane[17] secus omnibus, quibus tantum[18] ductus tramitis tam antecedentis, quam sequentis recte cognitus est, patent. Imo adhuc magis demiror, quod Clarissimus Placentinus Cometam hunc in Pegaso extitisse

9 simulque (que *del.*) **BO**.
10 meas literas : literas meas **TC**.
11 etiam *del. post* amplectendum **BO**.
12 oculis *del. post* se **BO**.
13 aerique...mea : manuque mea aeri **TC**.
14 suo tempore *del. post* incidere **BO**.
15 Clarissimos *del.* **BO**.
16 plane *del. post* profecto **BO**.
17 omnino *del.* **BO** : longe **TC**.
18 tantum *om.* **TC**.

asserat; cum tamen nunquam ibidem extiterit,[19] nec eo pervenerit[20] motu retrogrado; sed penitus ex Serpentario, Sidere fere opposito motu directo prodierit;[21] atque sic infra Aquilam et Delphinum per pectus Antinoi[22] nares Equulei, et Pegasum ad cornua usque Arietis fere viam suam duxerit.[23] Clariss[imus] atque Doctiss[imus] Dominus Büthnerus communis noster amicus, jam folia quaedam de hocce Cometa evulgavit; ex quibus percipio,[24] sententiam suam[25] mutare atque mecum sentire, novum scilicet omnino hunc esse Cometam; // non dubito, quin pariter reliqui omnes, dummodo rem omnem recte consideraverint opinionem arreptam sint mutaturi. Nam multis evidentibus rationibus id suo loco[26] debite demonstraturus sum duos diversos apparuisse Cometas. Denique plurimum Tibi debeo, quod tam varia,[27] non solum ex literis Illustriss[imi] Domini Rautensteinii,[28] sed etiam Rev[erendi] Patr[is] Curtii, mecum communicare volueris: agnosco inde et Tuum et P[atris] Curtii prolixissimum erga me affectum; cumprimis Illustr[issimi] Domini Raustensteinii propensam erga me voluntatem, quam ut aliquando vicissim demereri possim, sedulo annitar: interea eos meo nomine quam officiosissime, et alios[29] omnes nobis bene cupientes salutes, meaque promptissima iis offeras officia, etiam atque etiam a Te contendo. Propediem per Lubecam, cum hac vice per Tabellarium (ut facile intelligis) haud fieri potuerit, illis quoque, ut et Dominis Kirstenio et Müllero, quos pariter meis verbis salutabis, *Prodromum* nostrum exhibebo; ut sic et horum Doctissimorum Virorum judicium de hocce quali quali[30] opusculo nostro, inter tot curas occupationesque festinanti calamo conscripto impetrare non nequeam. Vale et me amare perge.

Dabam Gedani Anno 1665 die 13 Maii.

Tuae Gener[ositati] and Clar[itati]
affectu propensissimus

J[ohannes] Hevelius[31]

Postscriptum.

19 Imo...extiterit : Neque vero Cometa hic, ut nonnemini, de quo scribis, visum fuit, in Perseo extitit **TC**.
20 pervenerit : pervenit **TC**.
21 prodierit : prodiit **TC**.
22 infra Delphinum perque *del. post* Antinoi **BO**.
23 duxerit : duxit **TC**.
24 video *del.* **BO**.
25 plane *del. post* suam **BO**.
26 tempore *del.* **BO**.
27 tam *del. post* varia **BO**.
28 quam *del. post* Rautensteinii **BO**.
29 quam...alios (eos *del.* **BO**) : aliosque **TC**.
30 quali quali **TC** : quale quale **BO**.
31 Dabam...Hevelius *om.* **TC**.

Has dum obsignare volebam, Tuae mihi gratissimae die 25 Apr[ilis] St[ylo] V[etere] datae exhibentur. Gratias denuo Tibi debeo pro literis Praeclarissimi Bartholini, ut puto, mecum communicatis. Libenter scirem an meas die 3 April[is] St[ylo] N[ovo] ad illum datas acceperit. Aliquot exemplaria pro Bibliopolis, praeter illa ad amicos Tibi quoque prima occasione per Lubecam transmittam; sed quodlibet exemplar dimidio minimum Imperiali sed parata pecunia ab illis[32] redimendum. Constitutum namque pretium[33] pro reliquis erit 2 flor[enis] Polon[icis] h[oc] e[est] 2/3 unius Imper[ialis]. Si igitur ex Bibliopolis aliqui adsint, qui ea conditione plura exemplaria desiderent, mittam quantocyus. Vale iterum.[34]

Oblitus sum Tibi significare nuper die 4 Maii vesp[ere] hor[a] 11 1/2 globum notabilem[35] igneum[36] ex aere hic Gedani magno cum clangore decidisse.

To Master De Rolis
to Hamburg

Generous and most renowned Sir!

So far, I have not had any more urgent concerns other than fulfilling your wish (and that of other famous men) and doing what I have promised, viz. to write and publish the *Prodromus* of the former comet. Having finished that little booklet, I can answer your two very pleasant letters; together with this one, I present *Prodromus cometicus* to your eyes. I send it via regular messenger so that you can have it sooner. On the frontispiece, you see why these pages could not be added to another book. I entirely agree, however, that you can have a different opinion in this matter, especially given the fact that undoubtedly your intention is meant for the public good. It is improper to misinterpret or envy this praiseworthy proposition of yours. It is certainly impossible that I should be suspected of jealousy (as you know yourself, and as I often said to the most noble Master Pels).[37] I would rather encourage you to publish the observations and the opinions (as soon as possible and as many as there are) of the learned men in the way you suggested. To this end, and in order to further support this work, I send you here the promised observations which I received from friends from various places, so if there is anything noteworthy in them, it could be included in your book.[38] Continue then the work you have begun, and make every effort to finish it as soon

32 ii *del.* BO.
33 commune *del. post* pretium BO.
34 Aliquot...iterum *om.* TC.
35 notabilem : haud vulgarem TC.
36 igneum : igneum, ut milites nostri certo asserunt TC.
37 See note to Letter 5.
38 It is almost certain that the observations Hevelius refers to are those printed in the TC on pages 377–83. Lubieniecki makes no references or comments about them.

as possible! It will be a pleasure not only for me but (believe me) for all the investigators of heavenly things, especially for those who take care to develop the science of nature. Moreover, I wrote no general description of the former comet, but you can easily construct it basing on the *Prodromus* and adopt from it anything you wish. And it will not be unpleasant for me if you enclose my letters (if they are worth it). As far as the rest is concerned, from that booklet you will partly learn about our theory of comets, what I think about various topics, and how each and every cometary phenomenon can be correctly and suitably explained. And I do not doubt that you and other learned men (due to the bond of friendship which connects us) will reveal to me, in due time (willingly and freely), in philosophers' manner, what you think about our deliberations. I still hesitate over what I should decide about the second comet; whether it should be entirely included in the *Cometographia*, or similarly published in parts in advance. But, if you and our friends would like me to describe the latest comet, I will willingly undertake this task at the first given opportunity so that I could present, for the sake of the lovers of astronomy, some precise descriptions and drawings of it, or even mere observations (of which I have plenty), but without any calculations, as I keep those for the *Cometographia*. I used to make careful measurements of this comet since 6 April, new style, until the twentieth of that month with our larger brass instruments almost every day. I could not do it any longer because on 21 and 22 April it was cloudy, and on the twenty-third — although the sky was entirely clear — there was no trace of the comet, so on that day it moved away. It moved with constant direct motion (viz. according to the order of signs), however every day it was becoming slower: from the fourteenth degree of Pisces, as I observed, to the second degree of Taurus, constantly diminishing its latitude. In the first observation it was 26° 30'; in the last one it moved down to only 13 degrees. It passed the route nearly as the one which I had foretold you at the beginning, on 7 April, that is below the Head of Andromeda and through the North Fish. Because of its head, it was shining entirely differently from the former comet. In its head (with the help of the best telescope), some singular objects were noticed; I plan to discuss them in detail and engrave them in brass, God willing. Moreover, I found out that many learned men still think this latest comet was the same as the former one. But I wonder why they dare to say such things which are clearly opposed to common sense and which are visible on the artificial globe for all who correctly consider the path which the first and the second comet travelled? I wonder the more as the most famous Placentinus[39] states this comet was in Pegasus, while it was never there, nor came in the retrograde movement, but it started from Serpentarius (a nearly opposite constellation) and passed below Aquila and Delphinus, through the Chest of Antinous, the Nostrils of Equuleus, and Pegasus to the Horns of Aries. The most famous and learned Master Büthner,[40] our mutual friend, has already published a text about this comet[41] from which I see

39 See section 2.4.14 of the Introduction.
40 See section 2.4.3 of the Introduction.
41 I could not idenfity this publication by Büthner.

that he had changed his mind and now agrees with me that this is an entirely new comet. I do not doubt that all others (if they only correctly reconsider this thing) will likewise change their earlier minds. In a proper place, with some evident proofs, I will definitively demonstrate that two comets appeared. Finally, I am much indebted to you for many things you sent me — not only from the letters of the most illustrious Master Rautenstein,[42] but also from the letters of the Reverend Father Curtz.[43] Thence I see the great esteem for me from you and Father Curtz. First of all, I will take every effort to return the most illustrious Master Rautenstein's immense benevolence towards me. Meanwhile, I ask you to greet them kindly on my behalf, as well as all those who wish us well, and to offer them my willing service. In the next few days, I will send our *Prodromus* through Lübeck for them, and for Masters Kirsten[44] and Müller.[45] I would like you to greet them on my behalf, too. This time, I cannot send it via messenger, as you surely understand; I would like to ask those learned men about the opinion on my little booklet which was written in haste, among many worries and duties. Farewell, and continue to be benevolent towards me!

Sent in Danzig, on 13 May 1665.

Generous and renowned Sir,

Your affectionate admirer,

Johannes Hevelius

Postscript.

When I wanted to seal it, your pleasant letter dated 25 April, old style, was brought to me. Again, I owe you thanks for enclosing (as I presume) the letter from the most renowned Bartholin.[46] I would like to know if he received my letter dated 3 April, new style. At first given opportunity, I will send you through Lübeck some copies for booksellers, along with those for friends; but each copy should be sold them for at least half an imperial in cash. The set price will be 2 Polish florins, which is 2/3 Reichstaler. So, if there are any booksellers who want more copies on these terms, I will send the books as soon as possible. Farewell again!

I have forgotten to tell you that recently, on 3 May, at 11.30 in the evening, a noteworthy fireball fell from the sky with great noise, here, in Danzig.

42 See note to Letter 5.
43 See section 2.4.5 of the Introduction.
44 See section 2.4.9 of the Introduction.
45 See section 2.4.11 of the Introduction.
46 See section 2.4.1 of the Introduction.

15.

9/19 May 1665, Lubieniecki to Hevelius

BO: C1, v. 7, 989
TC: 383–85

Salutem Plurimam Nobilissime Amplissime et Doctissime Vir. Tandem mihi hodiernus tabellarius *Prodromum* Tuum *Cometicum* attulit, opus ut a multis me indice cum taedio expectatum, ita mihi cumprimis gratissimum. Nam et multa jam ex eo, cui percurrendo jam nonnihil temporis suetis laboribus suffuratus sum, didici, et discam longe plura, plus otii nactus: et, quae Tua est singularis humanitas et in me benevolentia, aliquas mihi in illustri illo opere partes esse vis.

– Haud equidem tali me dignor honore.

Quod tamen meum consilium de praemittendo *Cometographiae Prodromo* Tibi ita placuerit, ut in illo Herculeo et Atlantico opere occupatus, hunc quoque sat molestum laborem subire volueris, est de quo mihi gratuler, rite me scilicet et feliciter officio boni Amici (quod nunquam non pro virili exequar) defunctum fuisse. Verum tam eximii operis laus Tibi in solidum[1] debetur. Multa enim profecto de Cometarum origine, situ et motu, tum et parallaxibus ac distantia a terra aliisque rebus huc spectantibus, Tuus iste praecursor dicit, ut[2] maiorem exspectationem cum taedioso desiderio, cui nil satis festinatur,[3] coniunctam, *Cometographiae* Tuae animis omnium in adventum tanti[4] Legati erectis iniecturus, si quid ego de his judicare possum, certo sit.[5]

Hic meret aera liber Sosiis: hic et mare transit
Et longum noto scriptori prorogat aevum.

Et quid ni is multos doceat multa, dum Te multa antehac non ita certo explorata in assiduis et pernocibus observationum Tuarum (quarum numerum tot centurias implentem et incomparabilem ἀκρίβειαν[6] omnis mirabitur aetas) vigiliis didicisse nunciat? Discent profecto ex hoc libro Astronomorum alioqui clarorum et praestantium multi, nisi forte quidam reperiantur contumaces, quibus volupe ideoque solemne est

1 solidum : totum **TC**.
2 ut : Ita **TC**.
3 cui...festinatur : quod omnem celeritatem accusat tarditatis **TC**.
4 tanti : tanti, ut sic dicam, **TC**.
5 sit : est **TC**.
6 ἀκρίβειαν : diligentiam **TC**.

// omni animi orisque contentione διαφυλάττειν τὴν θέσιν et contra apertam veritatem jurata placita tueri[7]

– – – et quae
Imberbes didicere senes perdenda fateri.

Eximius Bullialdus, sicut Virum illum undiquaque Doctissimum vocare soles, qui in secunda Cometae prioris apparitione primo adspectu a Te dissenserat et mox, habita die 9. Jan[uarii] Cometae observatione, operi Tuo inserta et meo quoque (quicum Vir ille Praestantissimus eandem communicarat) inserenda, Tibi consenserat, plurimum viso et lecto tuo opere gaudebit. Sed et P[ater][8] Curtius (qui mihi proxima occasione per Ill[ustrissimum] Dominum Rautensteinium non pauca operi meo inserenda, observationes scilicet Romae et Ingolstadii habitas misit) et[9] Cl[arissimus] Bartholinus aliique Viri Doctissimi et Astrophili, Tuam *Cometographiam* et hunc eius *Prodromum* solicite praestolari.[10] Quos inter erit amor ille literarum et literatorum Illustrissimus Rautensteinius, tum et per omnia Nobilissimus Heinsius, qui Tibi per me multa cum salute hanc superiori seculo de insigni illo Anni 1577. Cometa habitam Lutetia Obsevationem, si forte Tuo instituto nonnihil prodesse possit, mittit. Atque ita gratulor mihi, me Tibi pro tam egregiis observationibus operi meo inserendis (de quibus, ut et de *Prodromo Cometico*, maximas[11] Tibi ago gratias) hoc licet alienum hostimentum mittere. Tu pro solita humanitate, ut Nobilissimus Heinsius, Te id rite et grato quidem animo accepisse intelligat, facies. Commendavit enim mihi hunc libellum, ut id ad Te tuto deferatur, proximis literis die 26. Aprilis Juliani[12] datis. Quibus adiunxerat autographum Cl[arissimi] Fornelii, Prof[essoris] Math[eseos] Ubsalensis, qui Cometam in Corvo et Lepore conspectum unum eundemque fuisse concedit, difficilem[13] adsensum motu Cometae insolenter inaequali nec perfecte circulari sed per quosdam veluti anfractus perexacto,[14] excusando,[15] nec non se idoneis instrumentis destitutum esse fatendo.[16] An vero nuperrimus Cometa idem cum priori sit nec ne? haud facile judicatu esse pronunciat. Caeterum audi, Magne // Heveli, quid Clarissimus ille Heinsius de me ad me scribere sustinuerit. 'Quas tu, inquit, de Cometis collegisti Observationes, eas videre vehementer et hisce manibus contrectare desidero. Fac igitur quaeso, ne longa expectatione tanti thesauri maceremur, aut potius enecemur.' Addit porro, deprehendisse se omnino diversum

7 tueri : tueri, ne cogantur suis opinionibus nuncium remittere **TC**.
8 P[ater] : R[everendus] D[octissimus] P[ater] **TC**.
9 et : Gaudebit et **TC**.
10 praestolari : praestolantur **TC**.
11 maximas : maximas iterum iterumque **TC**.
12 die...Juliani : ante dies decem **TC**.
13 difficilem : Difficilem autem **TC**.
14 perexacto : exacto **TC**.
15 excusando : excusat **TC**.
16 fatendo : fatetur **TC**.

fuisse nuperrimum Cometam ab illo priore, cum cursum plane contrarium instituat. Cur autem ego illa, quibus Vir ille Egregius magnas spes de me concepisse se testatur, Tibi affero. Officium quidem sinceri amici, et Tibi, ut spero, gratum futurum facio.[17] Sed forte eo ipso expectationem quoque Tuam maiorem,[18] quam,[19] ut gravem adversariam nonnisi superandam[20] maximo ingenii, doctrinae et rerum Mathematicarum apparatu, quo defectus, non sine metu vobis, quibus jus amicitiae magnis meritis partum in me magnum esse agnosco, rem[21] arduam aggredior.[22] Polliceor autem mihi aequa[23] orbis literati et Christiani, de quo bene mereri studeo, benigna judicia. Caeterum gratissimum mihi fuit legere in Tua epistola, Te mihi facultatem dare non solum Tuas literas ad me de argumento Cometico scriptas operi meo inserere,[24] sed et quae velim ex Tuo libro excerpere[25] et meis adiungere.[26] Qua in re nil fiet quod non in magnam Clarissimi nominis Tui laudem cedat. Cl[arissimum] Büthnerum pl[urima] salute impertio. Et quia ex diuturno Nobilissimi Peltzii silentio eum ab urbe abesse suspicor, velim ei inclusas meas et Heinsianas cum novis quae Tibi perlegenda mitto, reddi cures.[27] Vale et de meo constanti affectu certus esto, paremque mihi retribue.

Hamb[urgi] 9. 19. Maii 1665.

Clarissimo Nomini Tuo
omni cultu et studio
addictissimus

Rolitsius

Post Scriptum. In exemplaribus distrahendis jussa Tua exequar. Hoc exemplar cum Ampliss[imo] D[omino] Consule Müllero et Clarissimis Viris Huswedelio et Mullero communicavi, missurus idem hodie in Aulam Hafniensem.[28]

Greetings, most noble, worthy, and learned Sir! Today (at last), the messenger has brought me your *Prodromus cometicus*, a work expected anxiously by many, due to my

17 facio : facio, meque non temere opus hoc aggredi, sed votis tanti Viri et aliorum obsequi testor **TC**.

18 maiorem *add.*

19 maiorem, quam : maiorem reddam. Atque hanc jam **TC**.

20 superandam *add.* **BO** *om.* **TC**.

21 rem : authoribus rem **TC**.

22 aggredior : aggredior superandam **TC**.

23 aequa *om.* **TC**.

24 inserere : inserendi **TC**.

25 excerpere : excerpendi **TC**.

26 adiungere : adiungendi **TC**.

27 Qua in re...cures *om.* **TC**.

28 Post Scriptum...Hafniensem *om.* **TC**.

recommendations, and, of course, very pleasant for me. I learned a lot from it, being robbed of some time for my usual works while browsing it, and I will learn more when I have more spare time. And you want me to have a share in this famous work, which is a sign of your exceptional kindness and benevolence towards me.

– I claim not such worship.[29]

I can congratulate myself on fulfilling a duty of a good friend decently and successfully (which I will never cease to do, if I only can), viz. making you agree with me on publishing the *Prodromus* of your *Cometographia*, even though you were so preoccupied with the latter work (worthy of Hercules or Atlas[30]) you decided to undertake another tedious task. But you surely deserve glory for such an exceptional work. This forerunner of yours discloses a lot about the origin of comets, their place, and movement, as well as of parallaxes and distances from the Earth, and other related things. Therefore, it will certainly induce (if I can have an opinion in this matter) increased expectations and tedious desire (which cannot be satisfied) in every soul waiting for the coming of such an envoy of your *Cometographia*.

This book gains money for the Sosii;[31] this crosses the sea,
and continues to its renowned author a lasting duration.[32]

And why should it not teach many people plenty of things, if it announces that you have learned plenty of things (which had not been investigated before) during continuous and all-night vigils during your observations (their number, so immense, and your incomparable diligence will be admired in all ages)? Many renowned and distinguished astronomers will learn something from this book, if they are not too stubborn people for whom it is a pleasure and duty to defend their thesis with every effort of the mind and mouth, and to protect sworn opinions against evident truth

and to confess, now they are old, that what they learned, when young, is deserving only to be destroyed.[33]

29 Virgil, *Aeneis*, 1, 335, transl. by H. Rushton Fairclough, in Virgil, *Eclogues. Georgics. Aeneid I–VI* (Cambridge, Mass., London: Harvard University Press, 2004). Quoted with minor changes.

30 See section 2.2.1 of the Introduction.

31 The Sosii brothers were booksellers in Rome in the second half of the 1st century BC.

32 Horace, *De Arte Poetica*, 345–46, transl. by C. Smart, in Horace, *The Works of Horace Translated Literally into English Prose*, vol. 2 (Philadelphia: Joseph Whetham, 1836).

33 Horace, *Epistulae*, 2, 1, 84–85, transl. by C. Smart, in Horace, *The Works of Horace* (New York: Harper & Brothers, 1863).

Exceptional Boulliau[34] (as you call this man of great learning) who initially did not agree with you about the second appearance of the former comet, having considered the observation made on 9 January (which was included in your book and will be included in mine, as this most distinguished man sent the same to me), agreed with you and will be very happy to see and read your work. But also Father Curtz[35] (who, at the first given opportunity, sent me via the most illustrious Master Rautenstein[36] some observations to be included in my work, viz. those conducted in Rome and Ingolstadt[37]) and the most famous Bartholin,[38] and other learned men and star-lovers will gladly wait for your *Cometographia* and its *Prodromus*. Among them will be the most illustrious Rautenstein (who has a great affection for the arts and artisans) and Heinsius,[39] most noble in every respect, who passes to you through me (along with many greetings) this observation of the spectacular comet of 1577,[40] from the previous century, conducted in Paris,[41] if it could only somehow help in your plan. And I congratulate myself that I could send you this as payment (however not my own) for the utstanding observations for my book (for which, as well as for the *Prodromus cometicus*, I am very grateful). Due to your usual kindness, you will decently and thankfully make the most noble Heinsius know that you have received it. He has told me to transmit this book safely to you in his last letter, sent on 26 April, old style. He enclosed the manuscript of the most renowned Fornelius,[42] a professor of mathematics in Uppsala, who agrees that the comet seen in Corvus and Lepus was one and the same, with a reservation that this is difficult to assert, because the movement of the comet was plainly irregular and not circular, and it had a few turns, and he admits he did not have proper instruments. He states that it is difficult to assert whether the last comet is the same as the former one or not. As far as the rest is concerned, I have heard, great Hevelius, what the most renowned Heinsius keeps writing to me about myself. He says: 'I desire to see and to take into my hands those observations of the

34 See section 2.4.2 of the Introduction.

35 See section 2.4.5 of the Introduction.

36 See note to Letter 5.

37 See section 2.3.2 of the Introduction.

38 See section 2.4.1 of the Introduction.

39 See section 2.4.6 of the Introduction.

40 C/1577 V1, see Alexandre Guy Pingré, *Cométographie ou Traité historique et théorique des comètes*, vol. 1 (Paris: Imprimerie royale, 1783), pp. 511–19; Gary W. Kronk, *Cometography. A Catalogue of Comets. Volume 1: Ancient – 1799* (Cambridge: Cambridge University Press, 1999), pp. 317–20.

41 In his letter sent on 5 May 1665, TC p. 272, Heinsius informed Lubieniecki that he was sending him a description of the comet of 1577 written by François Liberati. Most probably, he referred to *Discours de la comete commencee a apparoir sur Paris le XI, iour de Nouembre, mil cinq cens septante-sept, à six heures du soir* (Paris: Pour Iean de Lastre, libraire, demouran en la rue S. Iean de Latran, pres le College de Cambray, 1577).

42 Jonas Fornelius (1635–79), Swedish astronomer and mathematician. Since 1664, he was the professor of astronomy at Uppsala University, since 1669 also the professor of mathematics. He worked on the theory of planetary motions, spherical trigonometry, and geography. In the TC, there is a letter sent by Lubieniecki to him and to Olof Rudbeck (see section 2.4.16 of the Introduction; *Communicatio Ubsalensis*, TC pp. 348–59) but without Fornelius's answer.

comets which you have gathered, as soon as possible. Therefore, I ask you not to let us perish, or rather die, because of the expectation of such a treasure.'[43] He also says that he understood that the latest comet was entirely different from the former one since it moved along the opposite path. Why do I inform you that this exceptional man testifies his high hopes about me? I fulfil the duty of a sincere friend (a thing which is pleasant for you, I hope). But perhaps, at the same time, I cause your expectation to grow. As a grave enemy, it can be overcome with a great variety of talent only, with the learning and mathematics. Lacking in both, I undertake a serious task (not without anxiety) because of you, whose friendship for me, gained with magnificent merits, I know to be great.[44] I hope for just and kind opinions from the learned and Christian world whom I always want to serve. As far as the rest is concerned, it was very pleasant for me to read in your letter that you allow me not only to include in my work your letters about the cometary discussion written to me, but also to include anything I wish to borrow from your book. Nothing will happen with this that does not refer to the glory of your famous name. I pass many greetings for the most renowned Büthner.[45] And due to the long silence of the most noble Pels,[46] I presume he has left the city. I would like you to take care to send him the enclosed material from Heinsius and me together with the new enclosure which I am sending to you. Farewell, and be assured of my constant benevolence towards you, which I crave in return!

In Hamburg, on 9/19 May 1665

Renowned Sir,

Your most devoted admirer,

Rolitsius

Postscript. I will follow your directions regarding the distribution of the copies. I showed this copy to the greatest man, Councillor Moller,[47] and the most renowned men, Huswedel[48] and Müller,[49] and I will send it to the court in Copenhagen.

43 TC p. 272.

44 The Latin phrase whose translation begins with the words 'But perhaps' and ends here is extremely convoluted and it is obvious that even Lubieniecki himself got lost in its entangled structure, as the result is hardly grammatically correct. In the manuscript, there are two words added by another hand (most probably Hevelius's) which make the sentence more legible, cf. Letter 16. There are further corrections in the TC.

45 See section 2.4.3 of the Introduction.

46 See note to Letter 5.

47 Barthold Moller (1605–67), German lawyer. Since 1635, he was a Councillor of Hamburg. In 1642, he was elected the mayor of Hamburg.

48 See note to Letter 1.

49 See section 2.4.11 of the Introduction.

16.

12/22 May 1665, Lubieniecki to Hevelius

BO: C1, v. 7, 998

Enclosures:
16A. Letter from Curtz to Lubieniecki, 7 May 1665
16B. Observation of the comet of 1665 from Rome, 11 April 1665

Salutem Plurimam Nobilissime, Amplissime Clarissime Vir.

En Tibi autographum Curtianum, mox remittendum: et quam ille misit mihi observationem Romanam: tum et apographum literarum Riccioli, quas heri accepi. Nuper ad Te scripsi, sed festina manu et αὐτοσχεδίας. Ideo in apographo quod scriba meus festinatissime descripsit, animadverti insignem errorem quem corrigi peto, ita scilicet 'forte eo ipso expectationem quoque Tuam maiorem quam, ut gravem adversariam nonnisi superandam maximo ing[enii]' et cetera 'magnum esse agnosco, autoribus rem arduam aggredior, mihi video esse oppositam.' Saluta Pelsium et Büthnerum. Vale et Fave eximium seculi decus.

Tuus

Rolitsius
manu propria

Hamb[urgi] 12. 22. Maii properante manu 1665.

Greetings, most noble, worthy, and renowned Sir!

Here I am sending you a manuscript from Curtz[1] (take care to send it back soon), an observation from Rome which he sent me, and a copy of a letter from Riccioli,[2] which I received yesterday. I have recently written to you but in haste and without preparation! That is why I found a significant error (which I ask you to correct) in the copy made by my secretary, namely: 'But perhaps, at the same time, I cause your expectation to grow. As a grave enemy it can be overcome with a great variety of talent only' and so on, 'I undertake a serious task for an author, very difficult for me [...]

1 See section 2.4.5 of the Introduction.
2 There is no extant manuscript of Riccioli's letter (see section 2.4.15 of the Introduction).

I know to be great.' Pass my greetings to Pels[3] and Büthner.[4] Exceptional pride of the century, farewell, and stay benevolent!

Yours,

Rolitsius
by my own hand

in haste, in Hamburg, on 12/22 May 1665.

16A.

Letter from Curtz to Lubieniecki, 7 May 1665

BO: C1, v. 7, unnumbered (first before C1, v. 7, 991)
TC: 214

Generosissime Domine

Domine observandissime.

Invitas me Vir Generosissime ad Eclogas Coelestes de Cometis quos hic annus accendit. Ego vero nolim bis ad eundem lapidem offendere. Quadraginta octo anni sunt, quod adolescens una cum P[atre] Cisato Magistro meo grandem illum Cometam observavi, qui bellis eo tempore ortis, velut Pyrrhica Saltatione praelusit. Seria[5] tum concertatio nostris P[atribus] cum Keplero, quem sub id tempus nosse coeperam. Is litteris non unis testabatur tres eodem tempore Cometas arsisse. Riserunt[6], sed Epistolae deinde ab Oceano et Oriente ad nos missae fidem fecerunt, non vanam fuisse Kepleri assertionem. Jam denuo acriter disceptant, unusne sit Roscius in illo Theatro. Ego observationes quas hucusque[7] nactus sum, tanti non facio ut ex iis pronunciare fidentius aliquod possim.

Doctissimus Hevelius Dantisco forsitan aliquid daturus est, quo Exspectationi satisfiat. Romanas Schedas et Ingolstadiensis[8] Mathematici typum mitto. Atque una quae Roma Vir doctissimus scripsit. 11. Aprilis.

3 See note to Letter 5.
4 See section 2.4.3 of the Introduction.
5 Seria : Varia **TC**.
6 Riserunt : Arserunt **TC**.
7 hucusque **TC** : hicusque [*sic*] **BO**.
8 Ingolstadiensis **TC** : Eydstachiensis **BO**.

De fide ac pretio pronunciare, non patitur ea reverentia, quam gero erga istos Naturae interpretes: Et Brahei observata autographa identidem tractanti fastidium est aliorum conatus inspicere. Hoc agamus potius ut bona fide conferantur in publicum quae quisque inspexit; Cetera tempus aperiet et ii quibus otium et amor rerum Coelestium. Gen[erosissimae] Dominationi V[estrae] me unice commendo.

Generosissimae Dominationis Vestrae
Servus in Christo

Alb[ertus] Curt[ius]

Neab[urgi], 7. Maii 1665.
Remittendum

Most generous Sir!

Most honourable Sir!

Most generous Sir, you invite me to celestial discussions about the comets which this year has set in flames. I would not like to trip over the same stone twice. It has been forty-eight years since I (as a young man) together with my teacher, Father Cysat,[9] saw that great comet shine like in a Pyrrhic dance[10] before the wars which began then. At that time, there was a serious controversy between our Fathers and Kepler whom I got to know then. He stated in more than one writing[11] that three comets were burning at the same time. They laughed then, but later the letters sent to us from across the ocean and from the east convinced us that Kepler's statement was not absurd. And now again, they are fiercely discussing if there is one Roscius in this theatre.[12] I do not value much these observations which I have already received; so much that I cannot state anything reliable based on them.

9 Johann Baptist Cysat (Cysatus, *c.* 1587–1657), Swiss Jesuit astronomer and mathematician. He was a student and later an assistant of Christoph Scheiner whom he assisted in the observations of sunspots. He was a teacher in Jesuit schools, in Ingolstadt (where he observed the comet of 1618) and Luzern, among others. He discussed the comet of 1618 in *Mathemata astronomica de loco, motu, magnitudine et causis cometae qui sub finem anni 1618 et initium anni 1619 in coelo fulsit* (Ingolstadii: ex typographeo Ederiano, 1619).

10 Pyrrhic dance (or Pyrrhiche) was a war dance in ancient Greece, performed by armed men.

11 Curtz alludes to the title of Johannes Kepler's book *De cometis libelli tres* (Augustae Vindelicorum: typis Andreae Apergeri, sumptibus Sebastiani Mylii, 1619).

12 Quintus Roscius Gallus (*fl.* 1st century BC), Roman actor. Renowned for his talent for both comedies and tragedies, his name became a synonym of the most talented actor.

The most learned Hevelius from Danzig will probably give us something which will meet our expectations. I enclose some papers from Rome[13] and a print by a mathematician from Ingolstadt.[14] And also what a very learned man from Rome[15] wrote on 11 April.

My reverence for the interpreters of nature does not allow me to speak about truth and value. And it is unpleasant to investigate the efforts of others for someone who all the time deals with Brahe's manuscripts of observations. We should rather benevolently present to the public what anyone could see. The rest will be revealed by time and by those who have spare time and the love for astronomy. I commend myself to you, most generous Sir.

Your servant in Christ,

Albert Curtz

In Neuburg, on 7 May 1665.
To be sent back

16B.

Observation of the comet of 1665 from Rome, 11 April 1665

BO: C1, v. 7, 991
TC: 214–15

Romae, 11. Aprilis 1665.

Tertium Cometem illustriorem primo aspicimus ab 8 diebus, longam Caudam in occasum porrigit in ortum tamen brevi porrecturus; jam processit ab Equuleo ad Caput Andromedae, seu Umbilicum Pegasi in arcu Eclipticae parallelo, quantum scilicet observari potuit. Illius motus diurnus esse fere 5 gr[aduum] 30' ortum versus, eius latitudo eadem semper gr[aduum] 25. 20'. Certe in cometis natura ludere videntur; nam primus on occasum movebatur, tertius in ortum; primus in curriculo maximo, tertius in minore; primus in orbita maxime inclinata ad Eclipticam, scilicet

13 See Letter 16B.

14 In the TC, the letter is followed by a series of depictions of cometary observations. Among them, there is an anonymous observation from Ingolstadt.

15 His name is given on the depictions of the observations: Gilles-François de Gottignies (1630–89), Belgian Jesuit mathematician and astronomer, since 1661 the professor of mathematics at the *Collegio Romano*. See also 'Subsidia de cometis', CJH2 p. 457.

fere 50 gr[adus] tertius in orbe parallelo, ac proinde nullo modo inclinato; Primus motu maxime inaequali, tertius aequatissimo; Primus versus Austrum, tertius versus Boream, primus ut planetae, secundus eodem cum stellis motu, eandem semper distantiam, eundemque situm servabat, tertius, partim ut stellae, in circulo scilicet Eclipticae parallelo, partim ut planetae quorum velocitatem, excepta duntaxat Luna, superabat. Alterum est, quod his diebus in Caelo observamus, nimirum Borealem cinguli Andromedae, juxta quam secundus Cometes apparuit, ab eo loco in quo secundum omnes tabulas sita erat, deflexisse arcu fere 2. gr[adus] et alteram novam videri cum dicta extrema boreali et media cinguli triangulum fere aequilaterum facientem; equidem utraque longiore tubo spectata aliquot Stellarum congeriem exhibet, ex hoc forte aliquid novi deducemus.

In Rome, on 11 April 1665.

We have been seeing the third comet,[16] brighter than the first one, for eight days. Its long tail stretches westwards, however it will soon stretch eastwards. It has already passed from Equuleus to the Head of Andromeda, or the Navel of Pegasus, along the arc parallel to the ecliptic, as far as it could be seen. Its daily movement was nearly 5° 30' eastwards and its latitude was always 25° 20'. Nature indeed seems to be playful with the comets. Because the first one moved westwards, the third one moves eastwards. The first one moved in a great circle, the third — in a small one. The first one moved in the orbit as much inclined to the ecliptic as possible, namely nearly 50°, the third one — in a parallel orbit and, therefore, not inclined at all. The first one moved with a very irregular motion, the third one — with a very regular one. The first one moved southwards, the third one — northwards. The first one moved like the planets, the second one moved together with the stars, always keeping the same distance and the same place; the third one moved partly like the stars, viz. in the circle parallel to the ecliptic, partly like the planets whose speed it exceeded, except for the Moon's. Another thing is what we can see in the sky these days, viz. the northern star in the Belt of Andromeda (next to which the second comet has appeared).[17] The star deflected nearly 2° from the place where it was according to all astronomical tables, and another new star was seen, which (together with the aforementioned star furthest north and the central star of the belt) makes a nearly equilateral triangle. Both of them, when observed with a longer telescope, reveal a group of some stars. Probably we will deduce something new from it.

16 The comet of 1665 was called the third one by those who thought there were two comets in late 1664. See section 2.3.1 of the Introduction.

17 See section 2.3.2 of the Introduction.

17.

26 May 1665, Lubieniecki to Hevelius

BO: C1, v. 7, 999
TC: 385–86

Enclosures:
17A. Excerpts of the letters from Heinsius and Rautenstein for Pels, 13 and 18 May 1665
17B. Excerpt of the letter from Rautenstein, 18 May 1665
17C. Excerpt of the letter from Hensius, 13 May 1665
17D. A note from Lubieniecki, 26 May 1665

Salutem Plurimam Nobilissime, Amplissime, Clarissime Vir. Accepi hodie recte fasciculum Tui *Prodromi Cometici* exemplaria X. praeter ea quae Amici dono misisti, continentem. Opportune sane. Cum enim ante octiduum exemplar primum per tabellarium accepissem, copias[1] eius videndi feceram Magnifico et Amplissimo Domino Proconsuli Mullero,[2] talium avidissimo et egregie perito, ac insigni studio literas humaniores literatosque prosequenti, tum Clarissimis Viris, Huswedelio et Mullero. Sed illa ipsa die cum e vestigio exemplar illud meum ad Illustrissimum quendam Virum Maecenatem meum mittendum mihi esset, aliquot solummodo horis[3] delectationi eorum ex adspectu et cursoria lectione Tui libri perceptae datis,[4] maius desiderium in animis eorum nec non aliorum, ad quorum aures fama libri Tui pervenit, excitavi. 'Ignoti' nempe 'nulla cupido' animum incessit tangitve. Cumprimis vero Cl[arissimus] Kirstenius, postquam me *Prodromi* Tui exemplar accepisse accepisset, literulas ad me heri dedit, quibus fruendi illius copiam dari sibi ardenter postulavit. Sed mox ei per literulas meas, dehinc et voce viva heri rem, prout se habet, enarravi. Commodum itaque accidit, ut accepto hodie Tuo fasciculo, confestim ei, ut et Cl[arissimo] Mullero destinata exemplaria miserim, quod quidem utrique pergratum extitit. Muneri enim tam eximio geminam gratiam ('duplex' enim[5] 'est bonitas, simul ut accessit celeritas,' sicut vetus verbum Publii Syri, ni fallor, et experientia crebra docet) eo ipso quod tam cito illud dederis, conciliasti. Aderat autem mihi Amicus bonae notae, cum afferretur fasciculus: qui cum adesse Tui *Prodromi* exemplaria didicisset, in argumentum singularis benevolentiae duxit, potuisse se tria exemplaria redimere. Qui etiam extemplo librum inspexit, et magna parte cursorio oculo lustravit, multaque suae de Cometis sententiae, operi meo inserendae cum

1 copias : copiam **TC**.
2 Mullero : Mollero **TC**.
3 horis : horas **TC**.
4 datis : dedi. Ita **TC**.
5 enim : nempe **TC**.

bono Deo, convenientia auctoritate Tua firmari multum gavisus est. Is est Bernh[ardus] Bekmanus Vesalia-Cli-//vensis, Theologicis pariter et Philosophicis studiis excultus, qui nuper piae eruditionis specimen, *Methodum* scilicet *Conciliandi ac in unitatem religionis Christianae, concordiae et Tolerantiae fraternae, invitandi Ecclesias Protestantium omniumque verorum Christianorum, opinionibus licet discrepantium, in fundamento fidei tamen convenientium*, a biennio edidit. Is pro amicitiae, quam mecum colit, jure sententiam suam de motu et origine Cometarum, quam aliqua ratione non tantum Cartesianam, sed et novam esse deprehendo, conferet cum Tua, et quid de hac judicet, mihi indicabit.[6] Cuperet autem ille ut pariter nuperrimi Cometae *Prodromum* in lucem emitteres, ac *Cometographiae* praemitteres. Idem judicat et optat Cl[arissimus] Kirstenius, ea de re heri a me rogatus. Cl[arissimum] Mullerum eadem opera compellare et consulere rem eandem volui. Sed eum domi non offendi.[7] Caeterum et praedictus Ampliss[imus] Dominus Proconsul Mullerus desiderat Tuas praeclaras de Cometis lucubrationes quamprimum videre. Atque hoc erit, ut firmiter credo, votum et Ill[ustrissimi] nostri Rautensteinii, Nobilissimi Heinsii, et aliorum Clarissimorum Virorum, qui Te Tuaque merito magni faciunt: quorum et ego suffragiis calculum meum adiicio. *Prodromi* enim Tuae *Cometographiae* expectationem eius Tuasque laudes non deterent, taedium vero illius detergent[8]. Exemplaria Ill[ustrissimo] Rautensteinio et Reverendo Curtio destinata quamprimum fieri poterit promovebo. Quod ad opus meum attinet, et Tibi quae iterum ab Ill[ustrissimo] Rautensteinio die 18 et Nobiliss[imo] Heinsio die 13 Maii ante biduum accepi. Magis magisque perspicio, imo sentio, gravem illam adversariam, tantorum Virorum aliorumque meritis multimodis Inclytorum, Tuique ipsius, Magne Heveli, (qui mihi Unus instar omnium es[9]) expectationem mihi oppositam. Porro ex *Prodromo* isto Tuo allegare nonnulla cum Tua laude non negligam et per omnia integram benevolentiam, et Tuae mutuae in me et Tuis magnis meritis in Remp[ublicam] literariam debitam testabor. // Vide vero quid Schepferus de Rudbeckio ad Nobiliss[imum] Heinsium scribat, eum videlicet nova quaedam prorsus aliisque intentata de hoc argumento Cometico polliceri. Erigimur in novas spes, nec largorum promissorum auctorem invidia deprimemus, sed potius votis cumulatum promovebimus[10] ad scandendum laudis fastigium, in quo quemque sua ponit

Virtus repulsae nescia sordidae,

et quae, ut addit Venusinus vates:

Intaminatis fulget honoribus:
Nec sumit aut ponit secures

6 Qui etiam extemplo...indicabit *om.* **TC**.
7 offendi : inveni **TC**.
8 detergent **TC** : deterget **BO**.
9 Unus...es : hic praecipuo consendus es loco **TC**.
10 promovebimus **TC** *om.* **BO**.

Arbitrio popularis aurae.
Virtus recludens immeritis mori
Coelum negata tentat iter via:
Coetusque vulgares et udam
Spernit humum fugiente penna.[11]

Haec Tibi, Clarissime Heveli, principem in Senatu Astronomorum deputavit locum. Fac ut publico bono bene et diu valeas, me vero pari benevolentia porro complectere, qui sum

Nobilissimo Nomini Tuo
omni cultu, studio et obsequio
deditissimus

Rolitsius
manu propria

Hamb[urgi] 26. Maii 1665.

Post Scriptum.[12] Ut Tibi fidem amicitiae per omnia servem, mitto Tibi autographia 4. Tuarum literarum ad me de argumento Cometico datarum, quas[13] una cum Observationibus mihi a Te missis, meisque ad Te exaratis literis, praefixo titulo, *Communicatio Dantisco-Heveliana* (in tales enim titulos *Comm*[*unicatio*] *Lutetia-Bullialdiana*, *Hafnia Bartholiniana*, *Holmia-Heinsiana*, *Ratisbona-Rautensteiniana*, *Neoburgo-Curtiana*, *Roma-Kircheriana*, *Bononia-Riccioliana*[14] et cetera, et cetera[15] dividere constitui) operi meo inserere animus est. Tu Tua prout libet recognita ad me remitte faveque caeptis. De distrahendis tuis exemplaribus mox cum Bibliopolis agam, volente Deo. Vale.[16]

Greetings, most noble, worthy, and renowned Sir! Today, I have received the package with ten copies of your *Prodromus cometicus*, along with those you have sent as gifts for our friends. It is very convenient. Because eight days ago, when I received the first copy through a messenger, I showed it to the magnificent and greatest

11 Virtus recludens...penna *om.* **TC**.
12 Fac...Post Sricptum *om.* **TC**.
13 quas : Has enim omnes **TC**.
14 *Lutetia-Bullialdiana...Bononia-Riccioliana* : *Ratisbona-Rautensteiniana*, *Neoburgo-Curtiana*, *Holmia-Suecorum Heinsiana*, *Luteria-Parisiorum Bullialdiana* **TC**.
15 et cetera *om.* **TC**.
16 De distrahendis...vale : Fac ut publico bono bene et diu valeas, me vero pari benevolentia porro complectere **TC**.

Councillor Moller[17] who is very interested and experienced in such things and ardently supports the arts and artisans; I showed it also to the most famous men, Huswedel[18] and Müller.[19] But on that day (as I was about to send my copy to the most illustrious patron of mine[20]), only a few hours were given them to delight in browsing and hastily reading your book, and I only sparked greater desire in their souls and in the souls of those whose ears the rumours about your book have reached. 'No need of the unknown'[21] has entered the souls or touched them. First of all, the most famous Kirsten,[22] having received the information that I had received a copy of your *Prodromus*, sent me a short letter yesterday in which he eagerly asked me to let him use the book. But I immediately answered him in a brief letter (and later yesterday in person) how the things were. Thus, it was very fortunate that when I received today the package from you, I sent at once the copies destined for him and for the most famous Müller. They were both very pleased. You have gained double gratitude by such an exceptional gift and because you gave it so fast (because 'the kindness is double if accompanied by quickness,' as is taught by the ancient words of Publius Syrus,[23] if I am not mistaken, and by frequent experience). When the package was delivered, there was a very well-known friend at my place. When he found out that there are copies of your *Prodromus*, he gave proof of his exceptional good will, and he bought three copies. He also examined your book at once and quickly leafed through it, and he was delighted that many things which agree with his own opinion about comets (which will be included in my book, if God allows it[24]) are supported by your authority. That man is Bernhard Beckmann from Wesel, equally learned in theology and philosophy, who recently (two years ago) published an example of pious erudition, viz. *Methodus conciliandi ac in unitatem religionis Christianae, concordiae et tolerantiae fraternae, invitandi ecclesias protestantium, omniumque verorum Christianorum, opinionibus licet discrepantium, in fundamento fidei tamen convenientium*.[25] As he is friends with me, he will compare

17 See note to Letter 15.

18 See note to Letter 1.

19 See section 2.4.11 of the Introduction.

20 I could not identify this person with any degree of certainty. The only guess I can make is Stefan Niemirycz (see section 1.1.3 of the Introduction). In the TC, there is a correspondence with an anonymous correspondent who can be identified as Niemirycz and Lubieniecki conceals his name. On the other hand, Niemirycz lived in the New March, far away from Hamburg, and this would mean that Lubieniecki got rid of his own copy of *Prodromus cometicus*.

21 Ovid, *Ars Amatoria*, 3, 397.

22 See section 2.4.9 of the Introduction.

23 Publilius Syrus (until 19th century misspelled as 'Publius', *fl.* 1st century BC), Roman freedman and playwright. In 1st century AD, quotations from his plays were arranged into a collection of apophthegms. Lubieniecki quotes his *Sententiae*, 161.

24 There is no correspondence with Beckmann in the TC.

25 Bernhard Beckmann (*fl.* 17th century), German Protestant preacher. Lubieniecki refers to his *Methodus conciliandi ac in unitatem religionis Christianae, concordiae et tolerantiae fraternae, invitandi ecclesias protestantium, omniumque verorum Christianorum, opinionibus licet discrepantium, in fundamento fidei tamen convenientium* (Francofurti ad Oderam: typis Eichornianis, 1663).

his opinion about the movement and the origin of comets (which I understand to be somehow not only Cartesian, but a new one[26]) with yours and he will tell me what he thinks about it. He would like you to publish a *Prodromus* of the latest comet and to publish it before the *Cometographia*. The most renowned Kirsten thinks and wishes the same (I asked him yesterday). Then, I wanted to ask the most renowned Müller about his opinion. But he was not at home. The aforementioned greatest Councillor Moller also wishes to see your famous works about the comets as soon as possible. And I firmly believe that our most illustrious Rautenstein,[27] the most noble Heinsius,[28] and other prominent men who justly esteem you and your work very much will wish the same. I add my vote to theirs too. The *Prodromi* of your *Cometographia* will not diminish the expectations of the work itself, nor the praises of you, but will ease the agony of waiting. I will send the copies for the most illustrious Rautenstein and reverend Curtz[29] as soon as possible. As far as my work is concerned, I send you what I received two days ago from the most illustrious Rautenstein, sent on 18 May, and from the most noble Heinsius, sent on 13 May. I can see (even feel) that more and more is being expected of me by such men and other people (who are famous for various reasons) and of you, great Hevelius (as you stand one for all of them). Moreover, I will not neglect to quote something from this *Prodromus* of yours with your approval, and I will always testify my constant benevolence, deserved both by your mutual benevolence towards me and by your exceptional merits for the Republic of Letters. See what Schefferus[30] writes about Rudbeck[31] to the most noble Heinsius, viz. that he promises something new and what he said against the others in this cometary discussion. He hopes that we will not humiliate the author of such bountiful promises because of pure jealousy, but instead, we will proceed from the accumulated wishes to reciting the praise in the words of the poet from Venusia:

True Virtue never knows defeat:
Her robes she keeps unsullied still,
    Nor takes, nor quits, her curule seat
        To please a people's veering will.
True Virtue opens heaven to worth:

26 See section 1.2.1 of the Introduction.

27 See note to Letter 5.

28 See section 2.4.6 of the Introduction.

29 See section 2.4.5 of the Introduction.

30 Johannes Schefferus (1621–79), Swedish historian, since 1648 professor of rhetoric at the University of Uppsala. He was the pioneer of studies on the archaeology of Sweden and the history of Lapland. See also Chantal Grell, 'Introduction', CHJ2 p. 28.

31 See section 2.4.16 of the Introduction.

She makes the way she does not find:
The vulgar crowd, the humid earth,
Her soaring pinion leaves behind.[32]

The virtue has assigned you, most renowned Hevelius, the chief seat in the senate of astronomers. Be in good health for a long time for the public good and bestow equal benevolence on me, the one who is

most noble Sir,

Your most obliged admirer,

Rolitsius
by my own hand

In Hamburg, on 26 May 1665.

Postscript. In order to assure you that I always cultivate our friendship, I send you the manuscripts of your four letters to me about the cometary discussion. I intend to include them in my book together with the observations you sent to me, and with my letters written to you under the title: *Correspondence with Hevelius in Danzig* (because I decided to divide the book into such chapters: *Correspondence with Boulliau*[33] *in Paris, with Bartholin*[34] *in Copenhagen, with Heinsius in Stockholm, with Rautenstein in Regensburg, with Curtz*[35] *in Neuburg, with Kircher*[36] *in Rome, with Riccioli*[37] *in Bologna* and so on). Send them back to me, if you approve them and endorse my plans. I will soon speak with the booksellers about the distribution of your books, if God allows it. Farewell!

32 Horace, *Carmina*, 3, 2, 17–24, transl. by John Conington, in Horace, *The Odes and Carmen Saeculare* (London: George Bell and Sons, 1882).

33 See section 2.4.2 of the Introduction.

34 See section 2.4.1 of the Introduction.

35 See section 2.4.5 of the Introduction.

36 See section 2.4.8 of the Introduction.

37 See section 2.4.15 of the Introduction.

17A.

Excerpts of the letters from Heinsius and Rautenstein for Pels,[38] 13 and 18 May 1665

BO: C1, v. 7, 1000[39]

Holmiae Suecorum die 13 Maii 1665

Rempubl[icam] quod spectat, classem adornare Sueci pergunt maiore tamen, ut existimo, fama, quam conatu. Quod navem bellicam omni armamento instructam apparere in hoc portu spectaculum sit toto jam quinquennio Sexennioque tantum non insolens. Ad vicenas tamen naves bellicas fore ut excrescat classis, sunt qui jactitent, quam vere, ipsi viderint. Unam ex his destinari aiunt Ingricae profectioni, ut Oratorem Moschum inde huc transvehat, eamque jam in hoc portu paratam esse, cum quia illo discessurus sit Taubius, aulicus Regis Mareschallus, Ingricae provinciae praefecturam suscepturus. Forbesius Regni Senator graviter etiamnum aegrotat. Regni[40] Cancellarius cum aliis nonnullis procerioribus hinc discessit dies proximos[41] festos ruri exacturus, pro instituto gentis huius, quem maior Senatorum pars itidem feriatura mox subsequeretur. Felices profecto Proceres hosce, quibus Rex infans tantum otii indulget ac quietis. Suppetias quas faederati Belgae a Suecis adversus Britannos exigunt parum aut nihil promoveri videas. Quid Orator ille qui ex faederato Belgio hanc ob caussam cum potestate pleniori nobis adfuturus est propediem sit impetraturus mox videbimus. Ministrum Brandenb[urgicum] instaurato cum Suecis foedere sub aestatem hinc abiturum ad suos fert rumor. Mecklenburgicus Gustroviensis jam in procinctu est Regi crastino aut perendino valedicturus die. Invitaverat is Regem ac Reginam matrem ad suscipiendum ex sacro baptismate filium herilem. Sunt tamen qui occulte opem ab illo in hac aula imploratam contendunt adversus machinationes Ducis Christiani Mecklenburgici Evangelicis partibus haud dubie nocituras nisi tempestive huic malo occurritur. //

Ratisbonae die 18. Maii 1665

38 The enclosure to the letter is clearly divided into two parts. The first concerns various political subjects and it is almost certain that it was meant to be passed to Pels (see note to Letter 5) — at the bottom of the first page there is an annotation 'Monsieur le Resident Pels[ius]' ('Ambassador Pels'). The second part contains other excerpts of the same letters from Heinsius and Rautenstein — printed also in the *Theatrum cometicum* — and belongs to Lubieniecki's cometary discussions. The first part of the enclosure is printed here as 17A, the second — as 17B and 17C. There is also a short note — 17D — whose exact meaning is not entirely clear.

39 Each part of the enclosure is printed separately here.

40 Senator…Regni *add.*

41 hinc…proximos *add.*

Pax inter Caesares Romanum et Turcicum missis utrinque splendidis legationibus et muneribus brevi nunc rati habebitur. Utinam bono orbis Christiani! Tam Lubomirscii quam totius Poloniae res periculoso loco esse videntur, nec abs re esse metus ille, ne Turcis aliisque Barbaris florentissimum quondam Regnum dilacerandum exponatur. Vere hic si ullibi, illud Virgilianum applicari posset

– – en quo discordia cives
Produxit miseros – – –

Quis vero finis tandem tot malorum et quae revera causae fuerint Deus et dies aperient. Gallos multa adhuc ibi machinaturos rumor est, etiam exercitum missuros, licet non videam qua via, quo praetextu et quo effectu? Suecos autem impacturos in Lapidem priorem, ambigo. Lombresius ex Polonia in Galliam redit Hamburgum transiturus, et ad Duces quoque Luneburgenses se collaturus, si mediationem Regis desiderent. Quis vero Zamoiscio successit Nepoti ac filio maximorum virorum? quorum ignoro an ulla ratione attigerit dotes. Ita sunt humana! Valetudo mea per hoc tempus vernum aliquantum nutat subinde tamen iterum confirmatur, Lampadis instar cui antequam extinguatur, si quod subinde olei affundatur lumen aliquod redit, tandem tamen deficiente materia penitus aufugiturum. Talis nostra vita quae tum demum felix, si tanquam rivuli ad primum fontem ita ipsa ad primaevam originem et creatorem revertatur et sic vere vivere incipiat. Nam haec vita mors est, nisi talis sit ut ad alteram illam tendat et aeternam mortem evitet, a qua nos benignissimus Deus custodiat!

In Stockholm, on 13 May 1665

As far as the state is concerned, the Swedish tend to enlarge the fleet, but with greater publicity than effort, I think. The sight of a fully-armed warship in this port is a spectacle not seen in the last five or six years. There are some who boast that the fleet will grow to twenty warships, but they will see for themselves how much truth is in it. They say one of the ships will be intended to be sent to Ingria[42] to bring the Muscovite envoy here from there, and the ship has already been prepared in this port, because Taube,[43] marshal at the royal court, will go there to take the command of the province of Ingria. Forbesius,[44] the senator of the Kingdom, is still very ill. The Chancellor of the Kingdom[45] together with some noblemen will leave from here and go to the countryside for the next holidays (it is in the habit of this people). He will

42 Region in the north-eastern Europe between the Gulf of Finland, Lake Ladoga and Lake Peipus, in present-day Russia (main city: Saint Petersburg), in the 17th and 18th centuries disputed by Sweden and Russia.

43 Jacob Johan Taube (1624–95), Swedish officer, governor-general of Ingria since 1664.

44 Probably Arvid Forbus (1598–1665), Swedish military commander and politician.

45 Magnus Gabriel De La Gardie (1622–86), the Chancellor of Sweden in 1660–80.

soon be followed by a majority of the senators who will also have a holiday. These noblemen are really lucky because the young King[46] allows them to have such rest and ease. Reinforcements for which the Dutch Republic asks the Swedish against the British seem to be very few or none.[47] We will soon see if that plenipotentiary envoy (who will come here from the Dutch Republic in the next few days to solve this case) achieves much. The rumour has it that the Brandenburg ambassador will leave from here to his country in the summer since the treaty with the Swedish has been renewed.[48] The Mecklenburg ambassador from Güstrow is ready, and tomorrow, or the day after tomorrow, he will bid farewell to the King. He asked the King and the Queen Mother[49] to be godparents for the son of his lord.[50] There are some who claim that the Ambassador secretly asked the King in this court for help against the machinations of Christian, the Duke of Mecklenburg,[51] which will undoubtedly harm the Protestant states unless this evil is stopped in time.

In Regensburg, on 18 May 1665

The peace[52] between the Roman emperor[53] and the Turkish sultan[54] (as both sides sent splendid delegations and gifts) will be ratified soon, and we are sure about it. May it bring good to the Christian world! It seems that there is a threat for Lubomirski[55] and for the whole of Poland, and there is some reasonable fear that the formerly flourishing Kingdom will be left for the Turks and other barbarians to be torn into pieces. This line from Virgil is relevant here more than anywhere else:

> See where strife has brought our unhappy citizens![56]

God and time will reveal the end of such misfortunes and their real causes. There is a rumour that the French also have many plans there and that they will even send the army,[57] but I do not see how, under what pretext and for what purpose. I am not sure

46 Charles XI (1655–97), King of Sweden since 1660.

47 Pels refers here to the war between England and the Netherlands, see note to Letter 37.

48 Sweden and Brandenburg signed a number of treatises in the 17th century and I could not determine to which one Pels refers here.

49 Hedwig Eleonora (1636–1715), the wife of Charles X Gustav and the Queen of Sweden since 1654.

50 Gustav Adolf (1633–95) was the Duke of Mecklenburg in Güstrow since 1636. In 1664, his son, Charles (1664–88), was born.

51 Christian Louis (1623–92, ruling since 1658), the Duke of Mecklenburg in Schwerin.

52 Rautenstein probably refers here to the peace of Vasvár, made in 1664 for twenty years.

53 See note to Letter 1A.

54 Mehmed IV (1642–93), Sultan 1648–87.

55 See section 1.1.3 of the Introduction.

56 Virgil, *Eclogae*, 1, 71–72, transl. by H. Rushton Fairclough, in Virgil, *Eclogues. Georgics. Aeneid I–VI* (Cambridge, Mass., London: Harvard University Press, 2004).

57 Rautenstein probably refers here to the war between England and the Netherlands, see note to Letter 37.

if the Swedish will make the same mistake again. De Lombres[58] returns from Poland to France and will pass through Hamburg, and he will visit the Dukes of Lüneburg,[59] if they wish for a mediation of the King. Who has succeeded Zamoyski?[60] He was the son[61] and grandson[62] of the greatest men, and I do not know if their abilities could be matched in any way. Those are the fortunes of men! My health this spring fluctuates a little; sometimes it becomes stronger again, as if some oil was poured into a lamp before it fades and then the light returns; but soon it disappears entirely when the fuel runs out. Our life is such that it is happy only when it returns to its primeval origin and to the Creator (like little rivers to their original source), and then we begin to live truly. For this life is death unless it is such that it tends towards this second life and avoids the eternal death, from which the most gracious God may protect us!

17B.

Excerpt of the letter from Rautenstein, 18 May 1665[63]

BO: C1, v. 7, 1000

Excerpta literatum de Cometico Systemate
Ratisbonae 18. Maii 1665.[64]

Si editio operis de Cometa vel potius de Cometis adhuc aliquantisper retardatur, forte et plura de iis judicia aliosque effectus videbis, quorum spes nondum cessat, cum res orbis Christiani semper mobilis multisque vicissitudinibus subjecti ad magnas hic illic[65] confusiones sint praeparatae, ita ut de magnis conversionibus vix sit

58 Antoine de Lombres, French ambassador in Poland in 1655–65.

59 Augustus (1579–1666), Duke of Brunswick-Lüneburg, Prince of Brunswick-Wolfenbüttel (since 1635), and George William (1624–1705), Duke of Brunswick-Lüneburg, Prince of Calenberg (since 1648) and Prince of Lüneburg (since 1665).

60 Jan Sobiepan Zamoyski (1627–65), Polish aristocrat, Palatine (*wojewoda*) of Kijów (Kyiv) since 1658, Palatine of Sandomierz since 1659. He had no legal heirs and the inheritance of his estates and possessions had to be decided in a number of prolonged lawsuits.

61 Tomasz Zamoyski (1594–1638), Polish aristocrat and politician, Palatine of Kijów (Kyiv) since 1618, Chancellor of Poland since 1635.

62 Jan Zamoyski (1542–1605), Polish aristocrat, politician (Chancellor of Poland since 1578), and military commander (great hetman of Poland since 1581), one of the most influential people in Polish politics of his times.

63 The letter with the given date is printed in the TC, p. 103, but its text — however similar in its meaning — is entirely different from the manuscript.

64 Excerpta...1665 *add.*

65 ad magnas *del. post* illic **BO**.

dubitandum. Tot Cometae et ostenta exiguo tempore conspecta non paucas nec exiguas portendunt mutationes. Quae utinam in bonum omnes eveniant omnibus, sicut certo evenient illis qui se divinae conformare voluntati norunt. Nunc plura non licet occupato, materia et usu dextrae destituto.

The excerpts of the letters about the cometary system
In Regensburg, on 18 May 1665

If the publication of the work about a comet, or rather comets, is delayed for some more time, perhaps you will be able to see more opinions about them and some of their effects for which there is still hope, because the Christian world (which is always moving and subject to many changes) is ready for great confusions everywhere, so there should not be much doubt about great mutations. So many comets and phenomena seen in such a short time foretell many important changes. May they all have good effects for everyone, as they certainly will for those who know how to accommodate themselves to God's will. Nothing more can be done by someone who is busy and lacks the composure and strength in hand.

17C.

Excerpt of the letter from Heinsius, 13 May 1665

BO: C1, v. 7, 1000
TC: 275–76

Holmiae die 13. Maii 1665

Proxime Fornelii literas mittebam rem mathematicam tractantes quas una cum meis ad te pervenisse confido. Plura et exquisitiora ab industria Rudbeckii mihi promitto, quae tamen hactenus non comparent. Ipse Rudbeckius in hanc civitatem Ubsalia advenisse nudius tertius dicitur: quem ut necdum vidi ita hinc discessurum vix opinor, me non convento. Cum incidero in hominem, haud intermittam urgere, ut telam detexat et promissa praestet. De eo Schefferus ultimis litteris haec ad me. 'Scripsit jam ante octiduum ut opinor Fornelius, quae spectare ad propositum negotium videbantur. At Rudbeckius noster tardior est propter et labores alios aliosque, quibus obruitur, et quia collatis omnium aliorum sententiis annotaturum sese sperat, quae adhuc a nemine sunt observata in hoc toto de Cometis argumento. Dabit et picturam quae omnia reddet evidentiora. Et sperabam // fore ut nunc omnia haberet confecta, iccirco scriptionem huc usque differebam. Sed quia heri in caena mihi dixit adhuc aliqua desiderari tacendum diutius non putabam, ne me oblitum mandatorum tuo-

rum censeres. Veniet ipse Rudbeckius ad vos circa finem hebdomadis huius per dies aliquot, ubi ipse forsitan sua tibi coram tradet' et cetera. Haec ille.

Amicis Gallis cum ex me scire desiderarent,[66] si quid hic terrarum ad Cometam novum a viris harum rerum intelligentibus observatum esset, de tuis lucubrationibus salivam jam movi, eandem Italis quoque propediem moturus. Quapropter vide, ne pollicitationes meas frustreris, neve parum verax fuisse deprehendar ab iis apud quos inveni hactenus fidem. Optime autem mihi hic[67] consules, si telam inchoatam quam primum detexis. Hoc[68] si feceris, expectationem de tua eruditione singulari publicam non aequabis solum, sed antevertes quoque ac post te relinques.

In Stockholm, on 13 May 1665

I recently sent you a letter from Fornelius[69] about mathematical issues (I think it was delivered together with mine). I expect more and more exquisite things due to Rudbeck's[70] ingenuity, but he is still not ready. There is a rumour that Rudbeck came to this city from Uppsala three days ago. I have not seen him yet, and I do not think he would leave from here without visiting me. When I meet him, I will not forget to urge him to weave the fabric and finish what he has promised. Schefferus[71] wrote to me about him in his last letter: 'Fornelius wrote to me eight days ago (as I think) about what seems to concern his suggested work. And our Rudbeck is tardier both due to more and more works which bury him and because having compared the opinions of all others he hopes that he will note something that has not been stated yet by anyone in all this cometary discussion. He will also present an illustration in which everything will be clearer. And I hoped that by now he would have completed everything, therefore I have delayed my letter until now. But because yesterday at dinner he had told me that there are still some things which are needed, I decided I should not be silent any longer so that you may not presume that I forgot my obligations towards you. Rudbeck will come to you at the end of this week for a few days, so perhaps he will tell you about his work himself.' And so on — these are his words.

When the French friends wanted to learn from me if anything in this country had been stated about the new comet by the men who know a lot about these things, I made their mouth water mentioning your works, and in the next few days, I will do so to the Italians. Therefore, take care not to foil my promises or to make me seem not trustworthy enough to those whom I have by now made trustful. So you will do the

66 cum...desiderarent : quando ex me scire desiderabant **TC**.
67 mihi hic : non mihi tantum, sed Tibi quoque **TC**.
68 Hoc : Id enim **TC**.
69 See note to Letter 15.
70 See section 2.4.16 of the Introduction.
71 See note to Letter 17.

best for me, if you weave the fabric you started as soon as possible. If you do so, not only will you confirm the common assumption about your erudition, but you will also enlarge it and leave it for the future.

17D.

A note from Lubieniecki,[72] 26 May 1665

BO: C1, v. 7, 1000

Mons[ieur]

J'ay bien reçcu aujourdhui pacquet connu. Vos exemplaires seront envoyés avec la premiere occasion, Dieu voulant. Je vous recommande l'inclose pour M[onsieur] Hevelius, et de lui faire part de ce papier. Je demeure à jamais

M[onsieur]
Vostre valet

De Rolis
manu propria

26. Maii 1665:

Dear Sir,

I have received today the package you know about. Your copies will be sent at the first opportunity, if God allows it. I suggest you enclose them for Master Hevelius and inform him about this letter. I remain forever,

Sir,
Your servant,

De Rolis
by my own hand

On 26 May 1665

72 The context suggests that the addressee was Pels (see note to Letter 5) and that the whole enclosure was meant for him.

18.

19/29 May 1665, Lubieniecki to Hevelius

BO: C1, v. 7, 1001
TC: 386

Seal with Lubieniecki's arms (Rola coat of arms), address: 'A Mons[ieur] / Mons[ieur] Jean Hevelius Cons-/eiller de la ville / à Dantzic / Franco Stetin.' ('To *Monsieur* Johannes Hevelius, city Councillor, in Danzig. Franco Stetin.')

Salutem Plurimam Nobilissime Amplissime Clarissime Vir. Illa X. *Prodromi* Tui[1] *Cometici* exemplaria extemplo distraxi. Sosii nostri Samuel Konigius et Zacharias Hertelius multas mihi egere gratias[2] de aliquot exemplaribus. Clarissimi Viri Kirstenius et Mullerus depraedicare satis laudes Tuas, de tam pretioso et docto munere Tuaque insigni humanitate gloriantes, nequeunt. Cum illo satis diu heri in Bibliopolio KönIgiano locutus sum. Plenus certe est optimi in Te affectus, plenoque ore et corde laudes meritas Tibi admetitur, Teque omnes alios vias et semitas Cometarum investigandi docuisse fatetur. Verbo, tribuit Tibi quod debetur praeclaris Tuis meritis. Jam jam accipio literas a Cl[arissimo] Bullialdo die 22. Maii scriptas, quibus mihi facultatem ea quae mecum de Cometa communicavit, libro meo inserendi concedit. A P[atre][3] Kirchero singularia quaedam heri accepi. Ita et is, ut et Ricciolus et Curtius, socii celeberrimi meum illud opus augebunt ornabuntque. Tua in illo, Magne Heveli, familiam ducent.[4] Proxime Tibi missurus sum, ut spero, exemplum literarum ad P[atrem][5] Curtium // cui ad illas, nuper Tibi in originali missas, respondebo, ita ut Observationem sive Theoriam Cometae Ingolstadiensem in multis cum Tuo *Prodromo* convenientem illi exhibeam. Quod reliquum est, Humanissime Vir, illud solum ad summam votorum Tua ex parte desidero, ut primae et[6] praecipuae operis mei sub titulo *Communicationis Dantisco-Hevelianae* parti inserere mihi Tua cum bona venia liceat observationes a Te factas et doctissime fuso fictas aere, *Prodromo* additas: ea scilicet lege, ut tot exemplaria, quot requirentur, in meum usum et meis per omnia impensis imprimi cures. Hoc enim sumptus compendium facere constitui, nisi forte velis tabellam aeneam paratam mihi iis legibus, quas praescripseris, dare. Quod Te facturum non dubito, si nimirum illa non habeas amplius opus. Adiuncta Nobil[is-

1 Tui *om.* TC.

2 gratias : grates TC.

3 P[atre] : Reverendo et Celeberrimo Viro Athanasio TC.

4 ducent : ducent, etsi de ordine communicationum necdum certi quidquam statuere potuerim. Sed omnes norunt qui loco sis in orbe Astronomico habendus. Pro jure amicitiae mitto Tibi excerptum et scripto Gallico de Di-Cometa, et Tri-Cometa, ut judicium tuum de his cognoscam TC.

5 P[atrem] : R[everendum] P[atrem] TC.

6 primae et *om.* TC.

simi] Domino Pelsio remittes.[7] Cl[arissimum] Buthnerum meo nomine pl[urimum] salutabis et de responso monebis. Vale, feliciter age et constanter fave

Tuo

Rolitsio
manu propria

Hamb[urgi] 19. 29. Maii 1665. //

Post Scriptum. Accepi hodie literas a Cl[arissimo] Placentino, fatetur is errore calami Perseam pro Pegaso irrepsisse. Iterum V[ale].[8]

Greetings, most noble, worthy, and renowned Sir! I immediately distributed those ten copies of the *Prodromus cometicus*. Our Sosii,[9] Samuel Konigius[10] and Zacharias Härtel,[11] were very thankful for the copies. The most renowned men, Kirsten[12] and Müller[13] (proud of such precious and learned gift and due to your great kindness) cannot praise you enough. Yesterday I had quite a long conversation with Kirsten in Königius's bookstore. He is undoubtedly full of good will towards you, and he pours the well-deserved praises for you both from his mouth and heart, and he admits that you teach everyone the ways and paths of investigating comets. In brief, he grants you what your great merits deserve. I have just received a letter from the most renowned Boulliau,[14] written on 22 May, in which he allows me to insert into my book what he sent me about the comet. Yesterday I received something from Father Kircher.[15] Together with Riccioli[16] and Curtz[17] — his most renowned brethren — he will enlarge and decorate this work of mine. Your letters, great Hevelius, will be there at home. I will soon send you (I hope) a copy of my letter to Father Curtz. I will answer his letter which I have recently sent you in its original version. I will show him that the observation (or the cometary theory) from Ingolstadt agrees in many points with your *Prodromus*. As far as the rest is concerned, most gentle Sir, I wish only one favour from you, knowing it is a big request, but also knowing your grace: that I could insert into the first and central part of my work the observations you made, precisely

7 ea scilicet...remittes *om.* TC.
8 Post Scriptum...V[ale] *om.* TC.
9 See note to Letter 15.
10 See note to Letter 11.
11 See note to Letter 11.
12 See section 2.4.9 of the Introduction.
13 See section 2.4.11 of the Introduction.
14 See section 2.4.2 of the Introduction.
15 See section 2.4.8 of the Introduction.
16 See section 2.4.15 of the Introduction.
17 See section 2.4.5 of the Introduction.

those made in brass and enclosed to the *Prodromus*. I would publish them under the title *Correspondence with Hevelius in Danzig*. It could be done in such a way that you could print as many copies as will be needed for my use and at my expense. I decided thus to reduce the costs, unless you wanted to send me the brass printing plate, prepared in the way you described. I do not doubt that you will do so, unless you need it in the future. Send the enclosures to the most noble Pels.[18] Greet the most renowned Büthner[19] on my behalf, and ask him about the answer. Farewell, be prosperous and always benevolent towards

Yours,

Rolitsius
by my own hand

In Hamburg, on 19/29 May 1665.

Postscript. I have received today a letter from the most renowned Placentinus.[20] He admits that due to a scribal error, Perseus sneaked in instead of Pegasus. Farewell again!

18 See note to Letter 5.
19 See section 2.4.3 of the Introduction.
20 See section 2.4.14 of the Introduction.

19.

2 June 1665, Hevelius to Lubieniecki

BO: C1, v. 7, 991
TC: 393–94

Illustri Viro
Domino De Rolis
J[ohannes] Hevelius Salutem Plurimam Dicit

Exemplar *Prodromi* mei *Cometici* die 13 Maii transmissi Te rite accepisse laetor; non dubito quin nunc etiam reliqua exemplaria, tum[1] quae amicis, tum Bibliopolis[2] destinavi cura Domini Pelzii[3] acceperis. Rogo itaque ut quantocyus illis amicis Honoratissimis opusculum istud[4] transmittas, officiaque nostra porro offeras; utinam ad gustum Clarissimorum illorum Virorum sit compositum; sed dubito. Nam, fateor, festinanti calamo, ob angustiam temporis a me esse conscriptum, plurimaque leviter tantummodo esse attacta, quae fusiori dilucidationi opus alioquin habent. Verum, quae hac vice lubens praeterii, volente Deo, *Cometographia* ipsa exhibebit. Eximio D[omino] Heinsio cumprimis multum debeo pro benevolo illo erga me affectu, tum quod libellum Astrologicum de Cometa 1577 mihi communicare voluerit. Ego, profecto, hac vice amorem istum demereri aliter haud possum, quam transmittendo exempl[ari] aliquo[5] *Prodromi*, in significationem reciproci nostri affectus, quod[6] ut aequi bonique consulat, obnixe rogo. Pro communicatis literis illis, atque observatione Romana gratias habeo debitas; Tibique ea remitto, quae voluisti. Referunt illi Observatores Romani, Borealem Cinguli Andromedae juxta quam secundus Cometa (ut loquuntur) apparuit, a vero eius loco ad duos gradus recessisse, alterumque novam in vicinia videri. Verum haec observatio contrariatur reliquorum omnium observationibus: siquidem nullus omnino Cometa illorum (sive duos sive tres statuas perinde est[7]) qui a Decembri apparuerunt juxta borealem cinguli Andromedae conspectus est. Prior namque in Ariete evanuit, alter posterior sub capite, atque ita sub lucida cinguli Androm[edae][8] ad 14 circ[iter] grad[us] a boreali illa[9] iter carpsit. Deinde hanc borealem // cinguli duos grad[us] deflexisse multo citius affirmatur,

1 tum *om.* TC.
2 tum Bibliopolis *om.* TC.
3 cura...Pelzii *om.* TC.
4 istud : illud TC.
5 exempl[ari] aliquo *om.* TC.
6 quod : *Prodromo* nostro, quem TC.
7 est : erit TC.
8 sub...Androm[edae] *om.* TC.
9 a...illa *om.* TC.

quam probatur. Hoc quidem certissimum stellam illam dictam in Catalogis omnium nondum esse recte restitutam. Nam ex meis observationibus ad dimidium circ[iter] grad[um] secundum longitudinem[10] aliter sita est. Siquidem distantia Capitis Medusae ab illa stella ex calculo Tychonico provenit 25° 26' 20"; atvero ex mea observatione 25° 56' 35" observata est. Hinc facile patet longitudinem Tychonicam illius Stellae debito esse maiorem. Non propterea tamen sequitur, etiamsi locus illius stellae sit erroneus, quod stella illa debitum locum mutaverit. Plurimae enim stellae in Catalogo globisque inveniuntur, (quod tamen salvo honore omnium dico) quae correctione indigent; non quod vero loco fuerint emotae; sed quod nondum fuerint rite constitutae: ut suo loco plenius aliquando in *Machina* nostra *Coelesti* dicetur. Denique novam aliam in vicinia illius borealis apparere, dubito. Nam, poterit quidem ibidem adesse quaedam globo nondum inscripta, sed non omnino recens exorta: ut huius generis quamplurimae passim inveniuntur; ac ego etiam Globo jam meo feliciter annumeravi. Vale feliciter, et meo nomine amicos omnes saluta.

Dabam Gedani Anno 1665, die 2 Junii St[ylo] n[ovo]

Postscriptum.

Scripta jam epistola superiore, accepi tuas literas die 26 Maii scriptas, amoris et benevolentiae denuo plenas, quem[11] benevolum tuum erga me affectum, quomodo demereri possim sane, nescio. Interea gratias Tibi habeo magnas pro iis omnibus quae nobiscum communicare voluisti. At literas meas haud opus fuisset remittere: pertinet enim maxime ad veritatem promovendam, seriemque rerum ac studiorum cuiuscunque melius apertiusque intelligendam, ut ea, quae scripta sunt, scripta maneant; atque ita, meo quali quali judicio, optime facies (ut aperte cum amico candido loquar) si amicorum literas plane immutatas omnes Orbi Erudito exponas: quo cuiuscunque Autoris[12] sententia, judicium, rationesque, atque sic diversitas opinionum recte omnibus pateat. Literas itaque meas proximo tabellario remittam pleniusque respondebo. Nova illa inventa Cl[arissimi] Rudbeckii avidissime expecto. Vale.

Johannes Hevelius sends greetings
to the illustrious Sir
Master De Rolis

I am glad that you received a copy of my *Prodromus cometicus*, sent to you on 13 May. I do not doubt that you also received (thanks to Master Pels[13]) other copies which

10 secundum longitudinem *om.* **BO**.
11 quem : quem autem **TC**.
12 Autoris : Authoris **TC**.
13 See note to Letter 5.

I intended both for friends and booksellers. Therefore, I ask you to pass this little work to those most honoured friends as soon as possible and to offer them our services. I wish it were written suitably for the taste of those most renowned men, but I doubt it. Because I admit I wrote it in haste, due to lack of time, and, therefore, many things which would otherwise need more detailed explanation have been touched upon only superficially. But what I omitted willingly this time, the *Cometographia* will describe, if God allows it. In the first place, I owe much more to the exceptional Master Heinsius[14] for his benevolent affection towards me and because he wanted to send me this little astrological book about the comet of 1577.[15] I certainly cannot repay this benevolence in any other way than by sending a copy of the *Prodromus* in order to signify our likeminded attitude for the greater good. I am deservedly thankful for those letters you sent me and for the observation from Rome. I send them back as you wished. Those observers from Rome say that the northern star in the Belt of Andromeda,[16] next to which the second comet (as they call it) appeared, moved 2 degrees from its real place and that a new comet is seen in its vicinity. But that observation contradicts the observations of all others, as none of those comets (it does not matter whether you say there were two or three of them)[17] which appeared since December has been seen next to the northern star in the Belt of Andromeda. For the first one disappeared in Aries, the second one near the Head of Aries and its path was 14 degrees under the bright star in the Belt of Andromeda. Moreover, it is easier to state that this northern star in the belt has deviated 2 degrees than to prove it. It is certain that the aforementioned star has not been correctly described in any catalogue. According to my observations, there was a half-degree difference in longitude in its position. Because the distance of the head of Medusa from this star (according to the calculation by Tycho Brahe) is 25° 26’ 20”, and according to my observation it is 25° 56’ 35”. Hence it is clear that the longitude of that star, according to Tycho, is larger than it should be. But the conclusion is not that the star changed its proper place but that the position of the star was mismisgiven. For there are many stars in the catalogue and on the globes (I do not intend to insult anyone here) which need corrections, not because they have been moved from their places, but because they have not been correctly described yet, as will be said in detail in a proper place and time in our *Machina coelestis*.[18] To conclude, I doubt there appeared a new star in the vicinity of that northern one. Because there might be a star which has not been placed on the globe yet and not the one which has not appeared recently. Many similar stars are being found in many places, and I have successfully added them to my globe. Farewell, and greet all friends on my behalf!

14 See section 2.4.6 of the Introduction.

15 See note to Letter 15.

16 See section 2.3.2 of the Introduction.

17 See section 2.3.1 of the Introduction.

18 Johannes Hevelius, *Machinae coelestis pars posterior* (Gedani: in Aedibus Autoris, 1679), pp. 12–13 *et passim*.

Given at Danzig, on 2 June 1665, new style.

Postscript.

After having written the above letter, I received the one you sent on 26 May, which was full of kindness and benevolence. I do not know how I could repay you for this kindly affection for me. Meanwhile, I am very thankful for all what you decided to send me. There was no need to send my letters back because (for the sake of credibility and for understanding better and more clearly the sequence of things and all the studies) what was written should remain written. Therefore, in my opinion, you will do best (to say it frankly to a sincere friend) if you show to the world of learning all the letters from our friends without any changes,[19] so that the assumptions, judgements, and views of each author and the diversity of their opinions will be accessible to everyone. Therefore, with the next messenger, I will send you my letters back, and I will answer more comprehensively. I eagerly expect those new discoveries of the most renowned Rudbeck.[20] Farewell!

19 In the errata in the TC, p. 968, Lubieniecki explains the meaning of the word *immutatas* used by Hevelius as 'not changed' (*non mutatas*), noting in an erudite passage that it could also mean 'changed.'

20 See section 2.4.16 of the Introduction.

20.

2 June 1665, Lubieniecki to Hevelius

BO: C1, v. 7, 1002
TC: 386–87

Salutem Plurimam Nobilissime Amplissime Clarissime Vir.

Spem Tibi nuper feci missurum me esse exemplum responsi ad illas Curtianas, quas abs Te praestolor. Sed nondum illud perficere potui, ob molesta quaedam negotia, studiorumque, quorum profecto teneor desiderio, impedimenta, quae[1] mihi vel amicitiae, vel etiam officii ratio subinde literis operam dare volenti obiicit. Neque etiam inficias eo, me non tam segniorem, quam remissiorem paulo in illis vestris sublimibus mysteriis Astronomicis, per intensum aliquot dierum praeteritorum aestum redditum esse. Certius tamen nunc spero, proxime me fidem Tibi factam promissi, quam religiose, omnium honestatis ab omni aevo amantium, more custodire in parte officii pono, praestiturum, juvante Deo. Interim haec pauca accipe, quae nupera mea vota de exemplaribus Tuarum observationum *Prodromo Cometico* insertarum tot, quot libri mei exemplaria imprimentur,[2] mihi largiendis et aere meo comparandis, in memoriam revocabunt, et[3] Tibi indicabunt me jam heri, cuidam Amicorum bonam libri mei partem, quae introitum operis et vagas de Cometis hinc inde narratiunculas continet, in Hollandiam deferendam, per Dei gratiam tradidisse. Id enim Amici a me adhuc reluctante, et opus saltem ad mensem unum atque alterum dilaturo, ardentibus et validis precibus extorserunt. Jam communicationes illae ex ordine sequantur necesse est. His omnibus Tui Cl[arissimi] Nominis splendor praelucebit.[4] Fac itaque ne diu vota nostra moreris. Exemplaribus illarum Observationum non tam cito indigeo, dummodo mihi Tua voluntas quamprimum sit perspecta. // Iconismi etenim illi, etiam postquam exemplaria, de quorum numero necdum quicquam decrevi (hic nempe in ponendis illorum rationibus rerum quoque mearum rationes mihi erunt advocandae in consilium) fuerint perfecta, imprimi, saltem iis, ubi praesto fuerint, adiungi facile poterunt. Ego autem breviusculas notas, numeros iconismorum explicantes, et Tecum brevi communicandas adiiciam: quandoquidem mihi, pro singulari Tua humanitate, ea quae mihi placuerint ex opere Tuo excerpere et meo inserere integrum esse voluisti. Cl[arissimo] Mullero, ut et Huswedelio, nec non egregie docto Viro Juveni Henrico Syvers, Mathematicarum sed et aliarum Philosophicarum disciplinarum, Tuique Nominis Illustrissimi cultori eximio, his diebus adfui, cumque il-

1 quae : Haec vero **TC**.

2 imprimentur : imprimentur, (siquidem id sine Tuo incommodo fieri possit) **TC**.

3 et : Haec etiam **TC**.

4 Tui…praelucebit : Tua communicatione inter primas posita, insigne ornamentum Tui Clar[issimi] nominis splendor conciliabit **TC**.

lis de *Prodromo* Cometae nuperrimi *Cometographiae* praemittendo contuli. Consentiunt Cl[arissimo] Kirstenio et exoptant quamprimum *Prodromum* illum videre. Caeterum Mullerus iterum iterumque gratiis Tibi de munere *Prodromi* Tui actis et pl[urima] salute adscripta, petiit a me, ut Te consulerem, annon pag[ina] 7. lin[ea] 18. pro conjunctione Solis oppositio substitui debeat, cuius ibidem lin[ea] 9. fecisti mentionem? Quaesivi hac occasione paulo post ex Syversio, annon ibi putaret aliquid esse mutandum? Idem plane ad hoc respondit, de quo me Mullerus monuit. Percepi autem illos velle in nonnullis Te consulere. Hoc enim Tibi fore gratum literis Tuis indicibus affirmavi. Caeterum[5] en Tibi nuncium ab Ill[ustrissimo] Rautensteinio ad me de novo Cometa die 22. Maii viso perlatum![6] //

Cl[arissimo] Buthnero pl[urimam] salutem adscribo et impertio, ac de literis non minus a docta quam amica manu die 27 Maii Gregor[iani][7] datis, ad quas nunc respondere non possum, sed nec tam cito, tot aliis laboribus praegravatus potero, maximas ago gratias,[8] meaque prompta defero officia.[9] Vale, illustrissimum coeli Astronomici sydus, in annos quam plurimos, et radiis Tuae scientiae tenebras, quae adhuc coelum Astronomicum exasperant,[10] feliciter discute.[11]

Tuo Clarissimo Nomini
omni studio, cultu et officio
addictissimus

Rolitsius

Hamb[urgi] die 2 Junii Gregor[iani] 1665.

Post Scriptum. Pro 10. illis exemplaribus *Prodromi* 5. Imper[iales] numerabo cui et quando jusseris. Bibliopolae vendunt librum 4. flor[enis]. Mitte mihi aliquot exemplaria pro Amicis, qui mihi justum pretium libenter rependent: nec non *Selenographiam* Tuam, pro qua inde pretium mihi numerabitur. De hoc scripto Gallico velim judicium Tuum me perscribas. Inclusas commnedo ad Pelsium, cui pl[urimam] mitto salutem. Iterum V[ale].

5 Caeterum : Quod reliquum est **TC**.
6 perlatum : perlatum. De quo quae sit Tua sententia, verbo haud gravatim mone **TC**.
7 27...Greg[oriani] : 17. 27. Maii **TC**.
8 ago gratias : gratias ago **TC**.
9 officia : obsequia **TC**.
10 quae...exasperant : caelo Astronomico hic illic incubantes **TC**.
11 discute : dicute, mihique et meis studiis constanter fave **TC**.

Greetings, most noble, worthy, and renowned Sir!

I have recently given you hope that I would send you a copy of the answer to Curtz's[12] letter which I expect from you. But so far I could not do it due to some onerous duties and obstacles in my studies which I still want to complete. They are thrown in front of me when I want to work on the letters for the sake of friendship or duty. I do not deny that (due to excessive heat in the last few days) I have become not idle as such, but deferred in those sublime astronomical mysteries of yours. But now I surely hope that, with God's help, I will soon fulfil what I have promised you. I declare keeping my word to be my duty, as all who love honesty should do at all times. Meanwhile, take this short letter which will remind you of my request to make for me, at my expense, as many copies of your observations from the *Prodromus cometicus* as there will be copies of my book printed. The letter will also tell you that yesterday I gave to one of my friends a vital part of my book (with the preface and some loose descriptions of comets from here and there) to be sent to the Netherlands, if God allows it. My friends have extorted it from me with their ardent and vigorous requests, however so far I have been reluctant, and I wanted to delay the work for at least a month or two. Those collections of letters should be in the right order. The brightness of your most famous name will outshine all others. Therefore, do not postpone our wishes any longer. I do not need copies of those observations any time soon, if only you promise me your kindness. Those illustrations can be printed after the copies are finished (I have not decided about the number yet; I will have to consider the cost of their distribution); they can be enclosed to some of the copies after they are ready. I will add brief notes to explain the numbers of illustrations. I will send them to you soon because, out of your exceptional kindness, you allowed me to borrow from your work whatever I want and to include it in mine. In the last few days, I visited the most famous Müller[13] and Huswedel,[14] and also a young man very learned in mathematics and other sciences, Heinrich Sivers[15] (who admires you very much), and I discussed with them the publication of the *Prodromus* of the last comet prior to the *Cometographia*. They agree with the most famous Kirsten[16] and wish to see this *Prodromus*. As far as the rest is concerned, Müller is very thankful for the gift of your *Prodromus* and sends you many greetings, and he told me to ask you if on page 7, line 18, instead of 'conjunction with the Sun' there should be 'opposition' which you mentioned in the same place, in line 9. On that occasion, I asked Sivers if, in his opinion, anything should be changed. He gave me the same answer as Müller. I noticed they would like to ask you about some things. I answered them that it would be pleasant for you, as

12 See section 2.4.5 of the Introduction.

13 See section 2.4.11 of the Introduction.

14 See note to Letter 1.

15 Heinrich Sivers (1626–91), German mathematician, professor of mathematics, cathedral vicar, and librarian in Hamburg. He was Lubieniecki's correspondent; in the TC pp. 913–30, there are 41 letters they exchanged (22 from Sivers, 19 from Lubieniecki).

16 See section 2.4.9 of the Introduction.

you suggested in your letters. I also pass some information about the new comet seen on 22 May to you, sent to me by the most famous Rautenstein![17]

For the most renowned Büthner[18] I send and enclose many greetings, and I thank him very much for the letter sent by his learned and friendly hand on 27 May, Gregorian style, which I cannot answer now (nor any time soon) as I am overwhelmed with many other works, and I pass my high regards to him. Stay in good health for many years, the brightest star of astronomical sky, and disperse the darkness covering the astronomical sky with the rays of your science!

Most renowned Sir,

Your most devoted admirer,

Rolitsius

In Hamburg, on 2 June, Gregorian, 1665.

Postscript. For those ten copies of *Prodromus* I will count 5 Reichstalers, as you requested. The booksellers sell the book at 4 florins. Send me some copies for the friends who will willingly pay a fair price, and also your *Cometographia* for which I will take payment. I would like you to send me your opinion about that French text. I entrust you with the enclosure for Pels[19] for whom I send many greetings. Farewell again!

17 There is no extant enclosure for Rautenstein (see note to Letter 5).
18 See section 2.4.3 of the Introduction.
19 See note to Letter 5.

21.

5 June 1665, Lubieniecki to Hevelius

BO: C1, v. 7, 1008
TC: 387–88

Salutem Plurimam Nobilissime, Amplissime, Clarissime Vir.

Si quis felici suavissimarum[1] rerum peritia praestans et laude humanitatis omnibus patentis adque praestanda officia grata pronae inclytus, multorum, etiam longe dissitorum, importunam in difficilioribus quaestionibus resolvendis et solida rerum cognitione impertienda, diligentiam experiatur necesse est, Tu certe es, cui consentiens Virorum Doctissimorum vox praestantissimam rei Astronomicae tribuit scientiam. O quam heri Cl[arissimus] Kirstenius de orbita Cometae a Te monstrata et certis passibus signata, tum de motu Cometae accurate indicato aliisque, quae *Prodromus Cometographiae* Tuae attulit et explicuit, mecum una[2] gavisus est! Et quis non Tuos istos maximos et insignes labores, quos unum hominem dies noctesque exantlare potuisse merito sera mirabitur aetas, maximi faciat, gaudio cum stupore in eis concurrente? Ego omnium hic minimus, quanto studio, intra hoc breve temporis spatium ac inter continuos nec exiguos profecto labores, Tuo *Prodromo* incubuerim, testatur apographum istud epistolae nudius tertius ad P[atrem][3] Curtium exaratae, in qua comparationem Tuarum et Ingolsta-//diensium observationum instituo, et ubi quantum consentiatis dissentiatisque quadamtenus exhibeo. Quod si qua in re mentem Tuam non satis recte assecutus sum,[4] Tuum erit me monere et docere discere semper, cumprimis hic abs Te certae fidei magistro, paratum. Crede mihi non pudori[5] sed laudi id mihi ducam. Multum enim profecto interest a qua manu ictus, institutio et monitio veniat. En vero et *Prodromum* Tuum, Tuo consensu, in compendium redactum![6] Hanc epitomen iconismis Tuis addere et operi meo, Deo volente, inserere animus est. Sed vide quo candore Tecum agam, dum nihil sine Te de Te decernere volo. Sed da operam,[7] ut haec quamprimum ad me remittas, ne retardetur editio operis,[8] in cuius fronte Tua praeclare lucebunt, et agmen ducent tot communicationum de argumento Cometico aliisque huc spectantibus cum Viris doctrina Clarissimis a

1 suavissimarum : gravissimarum et gratissimarum simul **TC**.
2 una : una iterum **TC**.
3 P[atrem] : R[everendum] **TC**.
4 sum : fuerim **TC**.
5 mihi...pudori : vera dicenti; non pudori **TC**.
6 *Prodromum*...redactum : *Prodromi* Tui Tuo consensu factum quoddam compendium, quod ad observationes Tuas ei insertas attinet **TC**.
7 Sed...operam : Dabis, quae Tua est benevolentia, operam **TC**.
8 remittas...operis : redeant, ne forte retardentur labores mei **TC**.

me habitarum. Sed ut hoc compendium Tuarum observationum revisum quamprimum ad me remittas, iterum iterumque rogo.[9] A Cl[arissimo] Gravio Prof[essore] Franequ[erano] hodie multa et egregia, quae ad rhombum faciunt, accepi. En Tibi apographum epi-//stolae Kircherianae. Cl[arissimo] Büthnero sal[utem] pl[urimam] adscribo, at non rescribo, tot laboribus pressus. Excuset velim aequitas eius dilati officii necessitatem mihi invito impositam. Vale, Vir Humanissime et Praestantissime, ac porro fave

Clarissimi Nominis Tui
Cultori observantissimo

Rolitsio

Hamb[urgi] 5. Jun[ii] 1665.

Post Scriptum. Exemplaria illa *Prodromi* persolvam brevi, et ibi quidem per Amicum quendam. Plura mitte quamprimum maritimo itinere, fasciculo Lubecam directo ad Christianum Vernerum. Iterum V[ale].[10]

Greetings, most noble, worthy, and renowned Sir!

If there is a man who excels in what is most pleasant and who is renowned for praises of his kindness in fulfilling all duties for everyone (wherever they live), he has to endure answering difficult questions and providing knowledge about many things. And you certainly are such a man because the unanimous voice of all the learned men grants you the most distinguished proficiency in astronomy. The most learned Kirsten[11] and I were pleased yesterday with the orbit of the comet (demonstrated and soundly described by you), and about the precisely determined movement of the comet and other things brought to us and explained by the *Prodromus* of your *Cometographia*. And who would not revere (with joy and amazement) these great and significant works of yours? Of course the future times will wonder at how one man could work so excessively, night and day. And the copy of the letter written three days ago to Father Curtz[12] testifies how much effort I have put into other essential works.[13] In this letter I compare your observations with those from Ingolstadt and show where and to what extent you agree or disagree. If I had not followed your thought precisely enough in anything, it would be your task to admonish and teach me, as I am always ready to learn, especially from you, a teacher whom I may trust. Believe me, I will not

9 in cuius...rogo *om.* TC.
10 Post Scriptum...V[ale] *om.* TC.
11 See section 2.4.9 of the Introduction.
12 See section 2.4.5 of the Introduction.
13 There is no extant enclosure to this letter.

treat this as something to be ashamed of, but to boast. There is certainly a significant difference from whose hand a stroke, a lesson, or an admonition comes. And here is your *Prodromus*, abridged with your consent! I plan to add your illustrations to this summary and to include it in my book, God willing. But look how decently I act towards you, as I do not want to make any decision about you without you. But do take care to send the book back to me as soon as possible in order not to delay its publication. On its front page, what is yours[14] will shine splendidly, and it will lead the army of numerous collections of letters on comets and other related subjects; I have had this discussion with some very learned men. But I ask you to send me back the revised summary of your observations as soon as possible. Today, I have received a long letter on that subject from de Grau,[15] a professor from Franeker. I send you a copy of the letter from Kircher.[16] I pass many greetings for the most famous Büthner,[17] but I do not answer him because I am burdened with so much work. May this honest delayed task be excused by the obligation which was imposed on me against my will. Farewell, most gentle and distinguished Sir, and still be benevolent towards

most renowned Sir,

Your obliged admirer,

Rolitsius

In Hamburg, on 5 June 1665.

Postscript. Soon, through a friend, I will send you the money for the copies of the *Prodromus*. Send me more by ship, as quickly as possible, in a package addressed to Lübeck to Christianus Wernerus. Farewell again!

14 There is no noun after the possessive pronoun 'Tua.' The meaning of the passage is obvious, but such an omission seems unusual and is possibly Lubieniecki's mistake.

15 See note to Letter 7.

16 See section 2.4.8 of the Introduction.

17 See section 2.4.3 of the Introduction.

22.

10 June 1665, Hevelius to Lubieniecki

BO: C1, v. 7, 1003
TC: 396

Illustri Viro
Domino De Rolis

Literas illas quas voluisti Tibi remitto, et quidem immutatas. Nam, sicuti jam percepisti minime conducet *Historiae* Tuae *Cometarum*, ut ea quae semel statuta a quopiam fuere, deleantur, vel alia sententia substituatur. Haec dum scribo, binae aliae Tuae literae die 19/29 Maii[1] et 2 Junii datae mihi traduntur; ad quas quidem hac vice plene respondere haud possum; brevibus tamen haec Te scire velim: cum intelligam quod meis leviusculis conatibus tantum tribuas, ut etiam Iconismos illos *Prodromo* nostro annexos Tuo operi inserere satagas, gratias habeo de tanto eximio Tuo erga me affectu; sed doleo quod tuo desiderio, ut ut vel maxime velim satisfacere hac vice nequeam. Iconismi enim illi non sine aliquo labore, et temporis jactura, ut facile intelligis, manu mea aeri incisi, praeprimis ad *Cometographiam* pertinent, eique adhuc sunt suo tempore inserendi, sic[2] denuo[3] saepius praelo[4] subiiciendi. Idcirco cum illis adhuc maxime opus imposterum habeam, atque[5] ne consumantur, necessum est, ut operi, cui destinati sunt, integri conserventur: Imo etiamsi eos, illum in finem nunc statim praelo subiicere vellem, nullo tamen pacto id fieri licet, antequam tota *Cometographia* ad finem perducta typisque exscripta fuerit: quippe Literae illae, in Tabulis illis, A, B, C, immutandae sunt, secundum illum ordinem in *Cometographia* susceptum, qui necdum mihi constat. Adhaec, cum in // ampliori forma Figurae istae adumbratae fuerint; opus vero tuum, sine dubio in quarto prodibit, nescio, anne commodius sit delineationes illas tabulis arctioribus etiam exhibere: atque ita possunt illae, quod mihi minime erit adversum a quopiam Artifice iterum delineari aerique incidi. De reliquo recte Dominus Mullerus advertit, quod pag[ina] 7. lin[ea] 18 pro coniunctione, oppositio substitui debeat. Nam, quemadmodum pag[ina] praecedente 6. lin[ea] 19 vides, etiam ex subsequentibus, inprimis loco Cometae et Solis clare liquet nunquam mihi aliam fuisse mentem, quam quod voluerim scribere 'in oppositione'; inter festinandum vero accidit quod coniunctio scripta fuerit. Poteris igitur ea ut et marginale corrigere. Quod superest, quia ita placet, exemplaria aliquot *Prodromi* mei

1 19/29 Maii : 29 Maii S[tylo] N[ovo] **TC**.
2 atque adeo *del.* **BO** : atque sic **TC**.
3 de *del.* **BO**.
4 adhuc *del. post* praelo **BO**.
5 atque : tum **TC**.

nec non *Selenographiam* prima occasione per[6] Lubecam transmittam; cuius pretium ordinarium alias est 8 Imper[iales]. Amicis tamen ac Bibliopolis pro 6 Imperi[alibus] quodlibet exemplar concedam. De Scripto illo Gallico, de quo nonnulla mihi communicare Tibi placuit, judicium meum, usque dum totum opus videre ac perlegere continget, suspendo. Tempus docebit, an verbis facta respondeant? Vale et pristino amore et favore me prosequere.

Dabam Gedani 1665, die 10 Junii

To the illustrious Sir,
Master De Rolis

I send you back those letters you asked for without any alterations. Because, as you have noticed, deleting what was stated by anyone or adding another opinion will not benefit your *History of comets*. As I am writing it, two letters from you (sent on 19/29 May and 2 June) are being delivered to me. I cannot answer them fully this time, but I would like you to know this: when I read and find out that you esteem my small efforts so much that you would like to include the illustrations from the *Prodromus* in your work, I am very thankful for such a great admiration of me, but I am afraid I cannot fulfil your wish this time, even though I would like it very much. Because those illustrations — engraved in brass by my own hand (a painstaking and lengthy process) — as you well know, are destined mainly for my *Cometographia*, and those illustrations will be included there in due time, and the plates reused. Therefore, as I will need them in the future (and I would not like to have them frayed), they should be kept intact for the work they are meant for. And, to this end, even if I wanted to print them now, I cannot do it before the whole *Cometographia* is finished and written with fonts. Especially those letters on the plates (A, B, C) will be changed, according to the order planned for the *Cometographia*, which I have not devised yet. Moreover, since those illustrations are larger, and your work will certainly be published *in quarto*, I am not sure if it is convenient to present those images in the plates which are too narrow. But they can be (without any objection from my side) painted again by any artist and engraved in brass. As far as the rest is concerned, Master Müller[7] noted correctly that on page 7, line 18, instead of 'conjunction' there should be 'opposition'. Because, as you can gather from the preceding page 6, line 19 — and from the subsequent text as well (mainly from the position of the comet and the Sun) — it is clear that I had no other intention than to write 'in opposition', but, being in haste, I wrote 'conjunction'. So you can correct it and make a note on the margin. Moreover (since it is your wish), I will send some copies of my *Prodromus* and *Selenographia* through Lübeck as soon as possible. Its standard price is 8 Reichstalers.

6 tecum *del. post* per **BO**.
7 See section 2.4.11 of the Introduction.

I will agree for 6 Reichstalers per copy from friends and booksellers. I withhold my judgement about that French text you informed me about until I can see and read the whole book. The time will tell if the facts match the words. Farewell, and keep bestowing your benevolence and favour on me!

Given at Danzig, on 10 June 1665

23.

12 June 1665, Lubieniecki to Hevelius[1]

BO: C1, v. 7, 1009
TC: 394–95

Enclosure:
23A. Letter from von Rautenstein to Lubieniecki, 4 June 1665

Salutem Plurimam Nobilissime, Amplissime, Clarissime Vir. Parabam[2] ad Te hasce meas, ut officio responsi ad Tuas amicissimas et eruditissimas literas postrid[ie] Kal[endas] Junias Gregoriano stylo exaratas rite defungerer. Sed superveniebant[3] varia impedimenta, quae parum aberant,[4] quin me currentem sufflaminarent.[5] Sed vicit[6] ea et e medio submovit[7] Amor in Te meus unaque cultus Nomini Tuo magis meritis in Rep[ublica] Literaria clarissimo debitus. De exemplaribus *Prodromi* Tui rite curatis jam Tibi satisfeci. Observationem illam Romanam quod attinet, nudius tertius de ea monui Kircherum, ut quisquis auctor eius fuerit, Tibi nobisque hic satisfaciat. Tunc autem et ad Ricciolum literas dedi, et de hac re[8] itidem monui, ut et ille judicium suum super hac controversia ferat. Utrique *Prodromum* Tuum prodiisse indicavi ut et Italia sciat eum jam *Cometographiae* spes maximas per orbem nostrum ferre et simul strenue et feliciter in stadio laudis decurrere. Nam de eo certum Te esse jubeo, non meo, qui hisce fere nullus sum, sed omnium Doctorum Virorum, qui hac in urbe degunt, // nomine. Itaque tanto lubentius illa ex *Prodromo* excerpta et typos Cometicos a Te doctissime aeri incisos explicatura futuro operi meo, Deo volente, inseram. Quod quanto studio Ill[ustrissimus] Rautenst[einius] Amor ille literarum Literatorumque ambiat, Teque simul prosequatur, nuperrimae eius epistolae apographum testatur. En vero Tibi et Bartholinianas! Hodie eas accepi, nec non Observationes eiusdem Viri Doctissimi, ut et Dissertationem Medicam Fratris eius Thomae, Anatomicorum nostro aevo facile Principis, de cometis: cuncta operi eodem inserenda. Sed Romam revertor, ubi tamen nunquam fui. Expendas velim illam observationem, annon novum aliquod phaenomenon illic locorum fuerit visum? Certe Cl[arissimus] Gravius Franequerae lumen aliquam lucem, si quid ego hic video, adfert meae ignorantiae tenebris. Ita enim

1 In the TC, this letter and the majority of the next one are printed together and dated 12 June 1666.
2 Parabam : Dum paro **TC**.
3 defungerer…superveniebant : defungar, superveniunt **TC**.
4 aberant : abest **TC**.
5 currentem sufflaminarent : ad deferendum Tibi studium et officium meum properantem medio in cursu sufflaminent **TC**.
6 vicit : vincit **TC**.
7 submovit : submovet **TC**.
8 de…re : hanc rem **TC**.

die XI. Martii ad me scripserat: 'Visum mihi jam ante 5. imo 6. hebdomadas est aliud in coelo phaenomenon ad cingulum Andromedae Septentrionalius similitudine eadem cum ultimo Cometa, quod eo magis deprehendere potui, quod eodem tempore et illud et dictum Cometam non semel viderim. Et heri adhuc mihi, dum haec scribo, visum est. Primo apparuit sub obscuro aliquo punctulo, deinde eius species paulum crevit, mihi per tubum Opticum spectanti notabile satis apparuit. Spatium vix mutasse ad sensum aut non ita multum, dico duobus aut 2 1/2 gr[adibus] a proxima stella abesse deprehensum est.'[9] Literas Tuas nudius tertius in Bibliotheca aedis Cathedr[alis] cum Amicis[10] legi. Aderat magnus globus. Contulimus Tua cum communibus. Animadvertimus illam dimidii gradus differentiam. Scio ex Cl[arissimo] Büthnero Te Globos omnes correcturum et fixas melius fixurum.[11] Deus Tibi adsit! Vale et salve

a

Tuo Rolitsio[12]

Hamb[urgi] 12. Jun[ii] 1665.

Greetings, most noble, worthy, and renowned Sir! I was preparing this letter to you in order to duly answer your friendly and learned letter written on 2 June, new style. However, there came many obstacles which nearly blocked my plans. But they were overcome and removed by my admiration for you and my worship of your most renowned name, well-deserved in light of your great merits for the Republic of Letters. I have already informed you that I took care of the copies of your *Prodromus*. As far as the observation from Rome is concerned, three days ago I reminded Kircher[13] about it that its author, whoever he is, should fulfil our wish. I have also sent a letter to Riccioli[14] and asked him to state his opinion about this controversy. I have informed them both

9 Ita enim...deprehensum est : Quid autem hac de re die XI. Martii St[ylo] Jul[iano] ad me scripserit, adiunctum apographum docet. Cum desiderio, ne dicam taedio, Tuum ulteriorem huius rei explicationem praestolabor. **TC**.

10 cum Amicis : (cuius cura Eruditissimo viro Juveni Henrico Syvero, studiorum et meritorum Tuorum strenuo cultori, commissa est) cum Amicis magna cum voluptate **TC**.

11 Aderat...fixurum : Contulimus, dum globus insignis ad manum esset, Tua singularia cum communibus, quibus gaudemus adhuc, et mirati sumus (erant autem nostrum plures) illam haud levem, 14 fere gr[aduum] distantiam scil[icet] in designando Cometae ad boreal[em] cinguli Andromedae stellam itinere, cumque taedio illam, quam te demonstraturum promittis, viam ingredi festinamus. Globos omnes Te correcturum, et nova stellarum lumina eis illaturum, fixasque ipsas melius fixurum, jam antehac ex Cl[arissimo] Büthnero intellexi. Quas Tibi pro hoc beneficio grates nostra et postera referet aetas? Vive proinde diu feliciterque eximium rei Mathematicae mearumque, si quae sunt, vel in censum hunc referri possunt, rerum decus. **TC**.

12 Deus...Rolitsio *om.* **TC**.

13 See section 2.4.8 of the Introduction.

14 See section 2.4.15 of the Introduction.

that your *Prodromus* was published, and now Italy knows too that it raises high hopes for the *Cometographia* throughout the world and runs fast and successfully in the race for glory. For I assure you about it — not in my name, as I mean nothing in this debate, but in the name of all the learned men who live in this city. Therefore, I would willingly include in my future work those excerpts of the *Prodromus* which will explain the types of comets and which are skilfully engraved in brass by you, God willing.[15] The copy of the recent letter from the most illustrious Rautenstein[16] proves with what great effort he supports arts and artisans (and you as well). I send you a letter from Bartholin,[17] too.[18] I have received it today together with the observations by the same learned man, and also with the medical dissertation about comets[19] from his brother, Thomas, who indeed is the prince of physicians of our age; all these will be included in my book. But I go back to Rome, despite the fact I have never been there. I would like you to examine that observation in order to check if any new phenomenon was seen in those places. Surely the most famous de Grau,[20] the splendor of Franeker, brings some light to the darkness of my ignorance, if I am not mistaken. He wrote to me on 11 March: '5 or 6 weeks ago, I saw another phenomenon in the sky near the Belt of Andromeda,[21] more towards the north, which was very similar to the late comet, which I could better determine because at that time I observed both that phenomenon and the aforementioned comet more than once. And I have just seen it while I am writing this. At first, it appeared as a small point; later it grew a little. It seemed quite clear when I watched it through a telescope. Seemingly it did not change its place at all or only insignificantly, viz. it was 2 or 2 1/2 degrees from the nearest star.'[22] I read your letters with friends in the cathedral library. There was a great globe. We compared your observation with those made by others. We saw a half-degree difference. I know from the most renowned Büthner[23] that you will correct all the globes and you will place all the stars more precisely. May God be with you! Take wishes and greetings

from

your Rolitsius

In Hamburg, on 12 June 1665.

15 See Appendix 1.

16 See note to Letter 5.

17 See section 2.4.1 of the Introduction.

18 There is no extant manuscript of this enclosure.

19 Thomas Bartholin (see note to Letter 4A), *De cometa consilium medicum* (Hafniae: apud Matthiam Godicchenium, 1665). See section 3.3.4 of the Introduction.

20 See note to Letter 7.

21 See section 2.3.2 of the Introduction.

22 De Grau's letter to Lubieniecki, 21 March 1665, TC p. 490, LUL ms PAP 15.

23 See section 2.4.3 of the Introduction.

23A.

Letter from Rautenstein to Lubieniecki, 4 June 1665

BO: C1, v. 7, 1009 (follows the preceding one)
TC: 113

Ratisbonae die 4 Junii 1665.

Pro communicatis tam novis quam etiam Cometicis, maximas ago gratias, Heinsioque[24] ac Hevelio, Viris undiquaque eximiis commendari me quavis occasione peto. Gaudeo autem nos omnes in eo convenire, coincidere,[25] quod maturanda sit editio partus, quem praestantissimum ingenium vestrum jam parat. Nisi me continua negotia et taedia a suavissimo alias commercio Musarum invitum abriperent tentarem depositam chelym, si forte epigramma aliquod Operi tam excellenti adnectere possem, quasi Corollam amicitiae indicem aut comitem, Libro per tot mundi regiones inque secula futura ituro. Nullus vero dubito G[enerositatem] V[estram] alios amicos, quibus felicior vena aut maius ocium est, in partem vocaturam curaturamque ne etiam hic aliquid ornamenti venustissimo Libro decedat. Quem conspicere cum tot aliis aveo.[26] Bullialdus silentium, ut ex nuperis eius vidi minatus est. Ideoque eo minus est admirandum. //

Hinc quod scribam nihil nunc habeo, quia 'eadem semper chorda oberramus' et concludendo multa quasi ludimus, aut ludendo concludimus concordiam tentamus, reipsa nihil praestamus et in ipso concordiae[27] limine iterum discordes reddimur. Faxit tandem Deus ut tot moris et inanibus laboribus finem aliquem invenire et simul concordiam pacemque retinere possimus!

Post Scriptum. Dominus Lubomirscius, ut video censeoque tandem quoque se accomodabit Aulae. Publica privatis semper hodie praeferuntur, quicquid illa patiantur dummodo haec bene habeant.[28]

In Regensburg, on 4 June 1665.

I am very grateful for the news about the comets that you sent me, and I ask you to commend me, if you will, to those exceptional men — Heinsius[29] and Hevelius.

24 Hensioque : Bullialdo Heinsioque **TC**.
25 coincidere : et quasi coincidere **TC**.
26 aveo : aveo et constanter maneo G[enerositatis] V[estrae] Paratissimus Servitor De Rautenstein **TC**.
27 tentamus *del. post* concordiae **BO**.
28 Bullialdus...habeant *om.* **TC**.
29 See section 2.4.6 of the Introduction.

I rejoice that we all agree and consent that the publishing of the book (on which your great talent works) will be more advanced. If my constant and troublesome obligations had not dragged me away from the most pleasant contacts with the Muses, I would take again the instrument I put aside, and I would try if I could add an epigram to such an excellent work. It would be a garland of the testimony of friendship and a companion for the book which will travel through all the countries of the world in many ages to come. But I do not doubt that other friends (who are more talented and have more spare time) will be called to do so and will make sure that the most pleasant book will not lack an ornament. I desire to see it, as many others do. Boulliau[30] has posed a threat of silence, as you have seen from his last letter. Therefore, there is nothing to wonder about.[31]

I have nothing more to write, because we always 'blunder at the same string'[32] and (as if we played when we conclude or we concluded when we play) we try to reach harmony, but, in fact, we achieve nothing, and, at the very doorstep of concord, we become discordant again. May God cause that we can find an end to so many delays and pointless troubles and keep the concord together with the peace!

Postscript. Master Lubomirski,[33] as I see and presume, will finally adapt himself to the court. Public affairs today are always preferred to the private ones. It does not matter what happens to private issues if only the public ones are well.

30 See section 2.4.2 of the Introduction.

31 The context of these two sentences and the next paragraph is obscure.

32 Horace, *De arte poetica*, 356, transl. by C. Smart, in Horace, *The Works of Horace Translated Literally into English Prose*, vol. 2 (Philadelphia: Joseph Whetham, 1836). Quoted with minor changes.

33 See section 1.1.3 of the Introduction.

24.

16 June 1665, Lubieniecki to Hevelius[1]

BO: C1, v. 7, 1014
TC: 395–96 and 396–97
Olhoff: 101

Salutem Plurimam Nobilissime, Amplissime, Clarissime Vir. Ad humanissimas literas Tuas die 2. huius mensis scriptas, aliquid Tibi proxime, festina manu, ne prorsus officio deessem, respondi. Non ita tamen per omnia, ut par erat. Quod velim mihi ignoscas et hoc veluti responsi mei additamentum boni et aequi consulas. Ac primum[2] de literis Tuis, quarum autographa ad Te nuper misi, ut eas prius videas,[3] quam illae lucem. Tueri hic quoque volui officium meum, quo me cogitationes Tuas vicisse video. Neque enim hoc a me postulaveras neque exspectaveras. Quin etiamnum ut aliorum quoque Amicorum epistolas immutatas plane orbi erudito, quo cuiusque mens et diversitas opinionum tanto melius pateat omnibus, edam auctor mihi es. Sed satisfeci hic[4] officio meo erga alios quoque Amicos. Alioqui maior sententiarum discrepantia a me protracta in lucem fuisset. Qua ratione in amicitiae leges peccasse me credidissem. Fateor autem me omni studio niti, ne ullatenus a quopiam negligens alicuius, imo diligens cultor omnis officii a quolibet dicar. Puto enim id facere ad fovendam amicitiam, rem in vita civili maxime necessariam et jucundissimam, mihi praesertim hoc[5] acerbo rerum mearum[6] statu. Et propria edoctus experientia novi saepe nos vel pro jure amicitiae vel multis negotiis presso[7] calamo effundere ea quae nollemus in lucem proferri, imo volumus 'centum servata clavibus' intra claustra silentii et privatos amicitiae parietes custodiri. Nempe ut in sermone volubilis lingua, ita et in hoc literarum commercio familiari penna praepes, artificiosa ista lingua, per quam tot terrarum mariumque immensis dissiti spatiis mutuum licet mutum miscemus sermonem, haud raro mentem praecurrit. Caeterum sat bene scio nec Tibi mentem hanc fuisse, ut alicubi in amicitiae et silentii, quod multa bona habere Graeca ante Romanam docuit sapientia Vetus, peccarem: sed tantum ut cuiusque literis in medium productis sententiam eius in hoc de Cometis

1 In the TC, this letter is printed in two parts: its main text is printed together with the preceding one, dated 12 June 1665, and its postscript is printed separately, dated 16 June 1665. Olhoff reprints only the postscript from the TC.

2 Salutem...primum : Unum prope exciderat memoria **TC**.

3 videas : videres **TC**.

4 hic : jam hic **TC**.

5 hoc *om.* **TC**.

6 mearum : nostrarum **TC**.

7 presso : pressos **TC**.

argumento propalarem.[8] Verumtamen cuique de proferendis in lucem suis aurem vellere officii mei duxi. Hac ratione δευτέρας φροντίδας σοφωτέρας,[9] // ut Graecorum, posteriores curas sapientiores, ut Latinorum vetus fert verbum, habebimus. Quamvis vero nonnulli palinodiam, me solo conscio cecinerint (quod quidem ipse Augustinus, quem nonnemo Sapientissimum Sanctissimorum et Sanctissimum Sapientissimorum dicere non dubitavit, eundem supra hominem immodicis efferens laudibus. luci, mox relicturus lucem, in illis *Retractationum* libris fecit)[10] habebimus tamen sat opinionum, etiam portentificarum, dum quomodo ista Cometarum portenta fiant, moveantur, dispareant, minimum dispareant, disserentes audiemus. Quippe in tanta humani ingenii imbecillitate et rerum difficultate fieri aliter nequit, quin hallucinemur quandoque et vacillemus subinde. Eminen[tissimus] *Annalium Ecclesiasticorum* Conditor Baronius in capite fere vasti illius operis dum diversas de annis et assumptione B[eatae] Virginis castissimae ante partum, in partu et post partum (quod parergon merito hic insero) Servatoris nostri Matris in coelum sententias refert, referente Spondano in haec verba erumpere sustinuit: 'Non possumus non vehementer mirari tantae rei historiam, vel saltem tempus migrationis eius sic a scriptoribus praetermissum, ut opus fuerit, quid huic vel illi de ea re fuerit revelatum investigare. Tolerandum tamen esset, si omnes aeque in eandem sententiam conspirarent, aut si unius vel alterius anni spatio dissiderent. At cum monstrosa quaedam inter eos discrepantia reperiantur, quinam potius audiendi erunt?' Quod si ita comparatum est in rebus fidei, ubi de coelo haereditate adeundo[11] agitur, quid miremur talia in rebus Astronomicis, ubi in coelo contemplando et motu orbium eius investigando phaenomenisque in eo observandis occupamur? Unicuique tamen convenit, opus, imo necesse est certa sequi, ibi praesertim, ubi, si usquam, bis peccare non datur, quia plusquam de capite, de salute scilicet animae aeternae res est. A Te certe multi Viri docti multa in Astronomicis discent, quod nonnullos confitentes jam audivi.[12] Literas illas Tuas ante aliquot dies in Bibliotheca aedis Cathedralis (cuius cura Eruditissimo Viro Juveni Henr[ico] Syversio studiorum et meritorum Tuorum strenuo cultori commissa est) cum Amicis magna cum voluptate legi. Contulimus, dum globus insignis ad manum esset, Tua singularia cum communibus, quibus adhuc gaudemus, et mirati sumus (erant autem nostrum plures) illam haud levem 14 scilicet graduum fere distantiam in designando Cometae ad Borealem Andromedae cinguli stellam itinere, cumque taedio illam, quam Te demonstraturum promittis suam certiorem et magis rectam ingredi festinamus. Globos omnes Te correcturum et nova stellarum lumina eis illaturum, fixasque ipsas melius fixurum, jam antehac

8 propalarem : manifestarem **TC**.

9 δευτέρας...σοφωτέρας *corr. ed.* : δεύτερας φροντίδας σοφώτερας **BO TC**.

10 Quamvis...fecit) : Et cecinerunt quidem jam nonnulli palinodiam, me solo conscio. Nec mirum. Nam ipse Augustinus, quem nonnemo Sapientissimum Sanctissimorum et Sanctissimum Sapientissimorum dicere non dubitavit, luci, mox relicturus lucem, in illis duobus *Retractationum* libris, idem fecit. **TC**.

11 coelo haereditate adeundo : caelesti haereditate adeunda **TC**.

12 audivi : audivi. Sed jam, Magne Heveli! Vale et Salve. **TC**.

ex Cl[arissimo] Büthnero intellexi. Quas Tibi pro hoc beneficio grates nostra et postera referet aetas? Vive proinde diu feliciterque eximium rei Mathematicae, mearumque si quae sunt, vel in censum hunc referri possunt, rerum decus.[13] Iterum, Magne Heveli, iterumque Vale et Salve[14]

a

Tuo Rolitsio

Hamb[urgi] 16 Junii 1665.

Post Scriptum. Accepi recte cum hodierno tabellario literas Tuas die 10. currentis mensis[15] scriptas, iisque inclusa literarum Tuarum apographa antehac a me ad Te missa, in quibus nihil prorsus mutare voluisti. Adeo in iis quae de coelo, vel currente calamo commentaris, immobilis, quia[16] incomparabilium observationum Tuarum, velut coelifer Atlas certus es, sicut Terram potius quam coelum moveri cum Pythagoraeis et illo primo Prussiae Vestrae Hipparcho Copernico (cuius ego epitaphium in patria urbe Thorunensi aliquando vidisse me in aede cathedrali memini) contendis. Quod autem ad aliorum literas attinet, de iis jam Tibi superioribus meis satisfeci. Hoc autem dico, propria fultus experientia, // hic et alibi Viros Doctos sententias suas ad Tuam composuisse, nec eos minimum pudere, a Te, tanto Magistro ea quae antea ignorabant, vel in quibus haerebant, discere. Legunt enim Tua vestigia, qui Te ex observationibus nuperi Cometae orbitam Cometae penitius investigasse, et singularia quaedam ex hoc caelesti, ut sic dicam, Doctore didicisse fateris. Caeterum de quo Te nuper compellavi, etiamnum repeto, ut scilicet consideres Cl[arissimum] Gravium quoque non solos Romanos observasse Cometam alium sive aliud phaenomenon, quamvis diversis temporibus, ad cingulum Androm[edae] septentr[ionalem]. Et quod quidem Gravianam observationem attinet, affirmat illa visum esse sub initium Febr[uarii] illud phaenomenon Cometae simile, cum primus Cometa, quem alii secundum volunt esse, adhuc conspiceretur, idemque die 10. 20. Martii visum fuisse, specie quidem maius, at loco vix motum apparuisse. Quod quia verum esse dubitandum non est, si praeterea Franequerana observatio cum Romana colligari possint, vix etiam dubitare licebit alium eundem tum Franequerae, tum Romae Cometam, medium scilicet inter primum et alterum Nostrum, animadversum fuisse, ut ita secundum Te et alios ductum Tuum sequentes tres diversi, juxta alios vero quatuor, imo si, ut Ratisbona acceperam, die 22 Maii ibi visus est Cometa, quinque Cometae tam brevi temporis spatio fuerint observati. Velim quamprimum hic nobis omnem scrupulum mentibus iniectum, easque urgentem, eximas felicissima Tua

13 rerum *om.* **BO**.
14 Literas illas...Salve *om.* **TC.** (but see Letter 23 where this passage is placed at the end in the TC).
15 mensis : S[tylo] N[ovo] **TC Olhoff**.
16 quia *om.* **TC Olhoff**.

dextra.[17] Te in *Prodromo* p[agina] 7. pro Oppositione Coniunctionem Solis errore calami posuisse, quivis aequus et talium peritus, quia et antecedentia et sequentia idem docent, perspiciet. Scriptum illud Gallicum velim mihi remittas et aliquid de eo judicare velis. Aliud mihi exemplum mihi ad illud responsuro non reliqui. Typos Cometicarum Tuarum observationum libro meo, qui in folio imprimetur, libenter habuissem, si fieri id potuisset. Sed aequa et praestitu possibilia ab Amicis postulanda esse, lex antiqua et res ipsa docet. Exemplaria *Prodromi* cum desiderio expecto, et brevi Tibi 10. Imper[iales][18] ad rationes nostras[19] numerari faciam Gedani.[20] En quae Ill[ustrissimus] Rautensteinius de Te ad me scripsit! Vale.[21]

Greetings, most noble, worthy, and renowned Sir! Last time, in order not to neglect my duty, I briefly answered your very kind letter written on the second of this month. But I did not touch upon everything, as I should have. I would like you to forgive me and take this letter as supplement to my answer. Firstly, about your letters whose copies I recently sent to you so that you could see them before they are published. I wanted to fulfil my duty because I see that your opinions have convinced me. For you neither asked nor expected it from me. You convinced me to publish the letters from other friends without any changes for the sake of the world of learning, where everyone's views and various opinions will be better accessible for all. But I fulfilled my duty towards other friends as well. Otherwise, I would present too much discord of opinions. This is why I would believe that I have sinned against the principles of friendship. I admit that with all the efforts I try not to be called negligent in anything by anyone, but rather a diligent caretaker of each and every duty. I think it will strengthen the friendship, which is necessary and most pleasant in civic life, especially for me, in my current bad condition. And from my own experience, I know that often (either because of friendship or many duties) we write with a hasty quill what we do not want to be published, and we rather want it to be kept 'hundred-padlocked'[22] in silence and to be guarded within the private walls of friendship. Because, like a voluble tongue in a conversation, a flying quill (an artificial language with which we keep the conversation between us, however mute, separated with so

17 Hoc autem dico...dextra *om.* **TC**.

18 Imper[iales] : Imper[iales] thaleros **TC**.

19 ad...nostras *om.* **TC**.

20 Caeterum de quo...Gedani : et caetera **Olhoff**.

21 En quae...Vale : En quae Illustrissimus Rautensteinius de Te ad me scripsit die 8. huius mensis, quae Tibi fidem facient, quanti Te ulle faciat: qualia plura jam antehac a me habuisti. Itaque perge strenue in Tuis laboribus, quos dies noctesque in perficiendo insigni illo *Cometographiae* opere, exantlas, et ut alter Atlas tantam caeli molem sustines. Deus caeli constantem valetudinem et prosperos successus Tibi largiatur, quo praeclaris ingerii Tui monumentis, non tantum hoc nostrum seculum, sed et secutura cum magno scientiae Astronomicae et publicae utilitatis monumento decores et locupletes! Id Tibi ex animo vovet, Tuoque favori se et exiguos conatus suos commendat et caetera. **TC Olhoff**.

22 Horace, *Carmina*, 2, 14, 26, transl. by John Conington, in Horace, *The Odes and Carmen Saeculare of Horace* (London: George Bell and Sons, 1882). Quoted with minor changes.

many vast spaces of lands and seas) often outruns the thought in this friendly epistolary exchange. But I know well that you do not think that I have sinned anywhere against the friendship or the seal of silence (which is very good, as the ancient Greek wisdom, older than the Roman one, teaches us),[23] but that I only spread the opinion which someone expressed in his letter in this cometary discussion. But I have considered it my duty to pluck everyone's ears about public presentation of their opinions. This is why we will have 'second thoughts which are wiser,'[24] as the Greek proverb says. Despite the fact that some people have sung a palinode with only me as the witness (even Augustine, whom someone without any doubts called the wisest of the most saint and the most saint of the wisest, extolling him with lavish praises undeserved by any man, did it himself in his books of the *Retractationes* soon before his death),[25] we will have enough opinions, even astrological ones, while we hear various statements discussing how those cometary portents happen, move, perish or do not perish. Due to such weakness of the human mind and the complexity of the subject matter, it certainly cannot be otherwise than that sometimes we act like lunatics, and later we have second thoughts. His excellence Baronius, the author of *Annales ecclesiastici*,[26] at the very beginning of that monumental work — when he recounts various opinions about the year of the Assumption of the Holy Virgin, the Mother of our Saviour, who was chaste before, during and after the childbirth (I dutifully insert it here) — said in these words, as de Sponde[27] refers: 'We must be surprised that the relation of such an event, or at least the time of her Assumption, was overlooked by writers, and so there is a need to investigate what was said about it by one author or another. It could be tolerated, however, if everyone agreed upon the same opinion, or if there was a difference of a year or two. But since there is such a monstrous disagreement between them, who should be listened to?'[28] If such things concern the matters of faith (viz. the discussion about achieving our legacy in Heaven), why do we wonder at the same situation in astronomy where we work on contemplating the heavens and investigating the movement of their orbs, and observing celestial

23 On silence in ancient Greece see Silvia Montiglio, *Silence in the Land of Logos* (Princeton: Princeton University Press, 2000).

24 Euripides, *Hippolytus*, 436.

25 In his *Retractationes*, Augustine of Hippo (354–430) listed and discussed most of his works. He also corrected and clarified many of the passages.

26 Caesar Baronius (Caesare Baronio, 1538–1607), Italian clergyman, cardinal, and historian of the Church. He published *Annales ecclesiastici* (12 volumes, 1588–1607) presenting the history of the first twelve centuries of the Church.

27 Henri de Sponde (Spondanus, 1568–1643), French historian of the Church. Born in a Huguenot family, he converted to Catholicism in 1595. With Caesar Baronius's consent, he prepared an abridged edition of his *Annales ecclesiastici*. He also continued Baronius's work and published *Annalium ecclesiasticum eminentissimi cardinalis Caesaris Baronii continuatio ab anno MCXCVII quo is desiit ad finem MDCXL* (3 volumes, Lutetiae Parisiorum: sumptibus Dionysii de la Noue, 1641). From 1626 until his resignation in 1639, he was the bishop of Pamiers.

28 Henri de Sponde, *Annales ecclesiastici ex XII tomis Caesaris Baronii in epitomen redacti* (Lutetiae Parisiorum: sumptibus Dionysii de la Noue, 1613), p. 72.

phenomena? Everyone agrees that we should (or rather must) follow what is certain, especially if no one can sin twice because it is more than his head at stake, viz. his immortal soul. Certainly many learned men will learn much from you about astronomy — and many admit it, as I have heard. A few days ago, with friends, I read with much pleasure the letter from you in the cathedral library (the custody of which is entrusted to the most learned young man, Heinrich Sivers,[29] who much admires your studies and merits). We have compared (since there was a large globe available) the information from you with that given by others, which we have here. And we were surprised (and there were many of us) about that significant distance, viz. nearly 14 degrees, in describing the route of the comet near the northern star in the Belt of Andromeda,[30] and with difficulty we assume the value which you promised to show us as more reliable and correct. I learned earlier from the most renowned Büthner[31] that you would correct all the globes, and place on them the new lights of stars, and you would better situate the stars themselves. How can our and future age thank you for this gain? Therefore, live long and prosper, exceptional pride of mathematics and of my affairs, if there are any, or if they can be discussed here. Again and again, great Hevelius, take wishes and greetings

from

your Rolitsius

In Hamburg, on 16 June 1665.

Postscript. Today, a messenger brought me your letter, sent on the tenth of this month, and the enclosed copies of your letters which I had sent you before and in which you did not want to make any changes. So when you speak about the heavens, even with a hasty quill, you state — together with the Pythagoreans and Copernicus that best Hipparchus of your Prussia (whose epitaph I saw in the cathedral in his native city of Toruń, as I remember)[32] — that it does not move and that the Earth moves rather than the heavens because you are sure of your matchless observations like Atlas who supports the firmament. The situation with the letters from others I explained to you in my previous message. Based on my own experience, I can tell you that here and there learned men form their opinions based on yours, and they are not ashamed that they learn from you (as from a great teacher) what they did not know earlier, or where they were wrong. Those follow in your footsteps who admit that you have investigated in detail the orbit of the comet from the recent observations and that they have learned something from such a celestial teacher (let me say so). As far

29 See note to Letter 20.

30 See section 2.3.2 of the Introduction.

31 See section 2.4.3 of the Introduction.

32 In fact, Copernicus is buried in Frombork (Frauenburg).

as the rest is concerned, I repeat what I asked you last time, viz. to consider that the most famous de Grau[33] and some men in Rome have seen (at different times) another comet, or another phenomenon, near the northern star in the Belt of Andromeda. And what concerns de Grau's observation, it supports that this phenomenon, similar to a comet, was seen at the beginning of February when the first comet (which some people want to be the second one) was still visible, and again it was seen on 10/20 March, seemingly bigger, and it appeared almost in the same place. There should be no doubt about it existed, especially if the observations from Franeker and Rome can be compared; hardly anyone can doubt that another comet was seen both in Franeker and Rome between our first and second one, so according to you, and to those who follow you, there were three different comets; according to others — four; and even there were five comets in such a short period of time, if as I got to know from Regensburg a comet was seen there on 22 May. I would like you to relieve us (as soon as possible) from that headache which has entered our minds and which is painful. Everyone who is honest and experienced in such things will see that in your *Prodromus* you wrote 'conjunction' instead of 'opposition' as a result of a scribal error because the preceding and following passages say so.[34] I would like you to send me back the French text about the comet and to tell me what you think about it. I have not left a copy for myself, and I would like to reply to this text. If possible, I would greatly rejoice to have the prints of your cometary observations for my book which will be printed *in folio*. But old laws and the thing itself teach us that one should ask his friends for what is just and possible to achieve. I wait eagerly for the copies of your *Prodromus*, and I will soon send you 10 Reichstalers to Danzig to settle the account. I enclose what the most famous Rautenstein wrote to me about you![35] Farewell!

33 See note to Letter 7.

34 See Letters 20 and 21.

35 There is no extant enclosure with a Rautenstein's letter (see note to Letter 5).

25.

17 June 1665,[1] Hevelius to Lubieniecki

BO: C1, v. 7, 1004
TC: 397–98

Domino de Rolis
Vir Illustris

Nimium, sane, (ut pace Tua dicam) benevolo Tuo erga me affectui indulges, dum mihi, conatibusque meis longe maiora attribuas, quam iis omnibus, ut mihi videtur, meritus unquam fuerim. Idcirco eo magis Tibi obstringor, quo minus laudes illae, quas in me tam abunde confers, mihi jure competunt. Compendium *Prodromi* mei a Te concinnatum Tibi remitto: nihil enim inveni, quod contra mentem meam occurrat; dummodo videas, ut numeri recte sint descripti, atque correcte imprimantur, cumprimis curabis, ut Signa Zodiaci non characteribus sed literis exprimantur. Quid vero mihi de iconismis meis omnino statuendum sit, jam ex praecedentibus literis percepisti. Benigne igitur Te interpretaturum puto, quod petitioni Tuae hac vice, ob *Cometographiam*, morem gerere minime possim. Literae Tuae ad Adm[odum] Rev[erendum] P[atrem] Curtium mihi valde arriserunt. Agnosco, Te quoque haud parum[2] temporis Speculationibus Cometarum concedere Teque *Prodromum* meum oculo haud fugitivo legisse, sed omnia probe considerasse. Si quaestionem illam mihi proposuisses: cur, nimirum, P[ater] Limberer priorem Cometam barbatum, posteriorem crinitum et caudatum appellaverit? sic Tibi respondissem. Non memini quidem rationes has proferendas apud ullum aliquem me legisse Autorem; veruntamen puto[3] haud iniuria posse[4] eius generis Cometas hunc in modum distinguere. Nam, barbatus aliquis coelum suspiciens, caputque attollens, barbam an-//trorsum versus exporrigit, eo nempe, quo iter meditatur. Cauda vero est pars posterior animalis, et semper in cursu sequitur. Hinc, jure Cometa ille prior cum comam contra seriem signorum in partes scilicet anteriores, quo iter tum temporis instituebat, proiiceret, appellatur barbatus. Deinde, quia sub exitum anni elapsi, et initium currentis, Cometa a plerisque exutus omni fere jubare vel barba,[5] et solummodo radiis crinibusque in orbem cinctus visus est, merito dicitur crinitus. Ultimo, vespertinus factus, crines praelongos invertens, atque ortum versus ad partes posteriores caudam exponens, convenienter Caudatus denominatur. Quippe cum idem susceptum semel[6] iter stricte contra

1 In the TC, the letter is dated 16 June 1665.
2 parum : modicum **TC**.
3 puto : arbitror **TC**.
4 ita *del. post* posse **BO**.
5 appe *del. post* barba **BO**.
6 susceptum semel : semel susceptum **TC**.

seriem signorum perageret, comam vero post se traheret, capite cursui exposito, ut animalia caudam: ideo recte Cometa is principio barbatus, in medio sui curriculi crinitus, et sub finem caudatus appellatur. Verum quaestionem hanc Rev[erendo] P[atri] Curtio et Cl[arissimo] P[atri] Limberer dirimendam relinquo; qui Clarissimi Viri, sine dubio, rationes adhuc solidiores huius rei producent. Sed nolo diutius Te alias occupatissimum morari, qui non minus occupationibus pene obruor: cumprimis cum accedat *Desriptio et Delineatio Cometae posterioris* concinnanda aerique incidenda. Exinde si imposterum in rescribendo aliquanto tardior fuero, non nisi studiis, negotiisque id adscribas[7] velim. Vale Literarum decus atque ornamentum.

Dabam Dantisci Anno 1665, die 17 Junii

Eruditionis et Virtutum Tuarum
constans aestimator et cultor

J[ohannes] Hevelius
manu propria

To Master De Rolis
Illustrious Sir!

You indulge in your benevolence for me (let me say so) when you attribute to me and to my works much more than I have achieved in any of them, as I believe. Therefore, the less these praises (which you ascribe to me so profusely) fit me, the more I am obliged to you. I send you back the summary of my *Prodromus* which you prepared.[8] I have found nothing contrary to my intention. Only make sure that the numbers are copied without any errors and printed correctly. First of all, you should take care that the signs of the zodiac are not printed with symbols, but with letters. What is decided about my illustrations, you have already learned from the previous message. I think you will understand that I cannot answer your request this time due to my *Cometographia*. I greatly enjoyed the letter from the Reverend Father Curtz.[9] I see that you also spend much time on cometary deliberations and that you have not read my *Prodromus* cursorily but diligently considered everything. If you asked me why Father Leinberer called the first comet 'bearded' and the second 'hairy' and 'tailed',[10]

7 attribuas *del.* **BO**.

8 See Appendix 1.

9 See section 2.4.5 of the Introduction.

10 Wolfgang Leinberer (1635–93), German Jesuit mathematician, astronomer, and philosopher, teacher at the University of Ingolstadt. Hevelius refers to his *Theoria cometae mense Decembri anni MDCLXIV Januario item ac Februario anni MDCLXV Ingolstadii observati* ([Ingolstadt]: typis Joannis Ostermaieri, 1665), pp. 4–6. 'The first' and 'the second' comet refer here to various apparitions of the comet in December 1664.

I would answer you in the following manner: I do not remember that I have read these arguments in a book by any author, but I think it is justified to distinguish comets of that type in the following way. For the 'bearded' comet, when it goes to the sky and raises its head, it stretches its beard forward, viz. in the direction in which it wants to go. But a tail is a rear part of an animal and follows it in its path. Therefore, that earlier comet — because it carried its hair against the order of signs, viz. towards the previous parts in whose direction it decided to move then — is rightly called 'bearded.' Further on, because at the end of last year and at the beginning of the current one, the comet was seen by many without any glare or beard and was surrounded only by rays and hairs, it is deservedly called 'hairy.' Finally, when the comet appeared in the evening, showing its long hair and stretching its tail towards the east, towards the rear parts, it is conveniently called 'tailed.' Of course, when a comet keeps strictly the once taken route against the order of signs, and it hauls its hair behind with its head turned towards its path, as animals do with their tails, this comet at the beginning is called 'bearded', in the middle of its route 'hairy,' and at the end 'tailed'. But I leave this question to decide for the Reverend Father Curtz and the most renowned Father Leinberer because these famous men will undoubtedly present firmer arguments in this discussion. But I do not want to distract you any longer as you are preoccupied elsewhere, and I am nearly buried under my duties too, chiefly with writing and engraving in brass the *Descriptio et delineatio cometae posterioris*. Therefore, if I am late in answering in the future, I would like you to blame it on my studies and duties only. Farewell, pride and ornament of sciences!

Given at Danzig, on 17 June 1665

Admirer of your learning and virtues,

Johannes Hevelius
by my own hand

26.

8 July 1665, Hevelius to Lubieniecki

BO: C1, v. 7, 1015
TC: 400

Domino de Rolis

Vir Illustris

Ad literas tuas die 16 Junii datas, ob varia negotia brevibus tantum hac vice respondeo. Videris mihi Tu[1] cum aliis Viris Eruditis adhuc in ea haerere opinione, ad cingulum[2] Andromedae, juxta observationem Romanam et Clarissimi Gravii alium intermedium quendam illuxisse Cometam; sed ego plane aliter sentio, et persto in sententia: nimirum nonnisi duos tantum diversos a Mense Novembri anni praecedentis hucusque visos observatosque esse Cometas. Largior quidem lubens, Observatores illos Clarissimos aliquod[3] ea in caeli parte deprehendisse phaenomenum: siquidem fere eo ipso tempore, probe memini, me pariter tale quidpiam oculis meis percepisse; sed nequaquam tamen istud phaenomenon Cometam extitisse concedo. Conducet igitur ut haec quaestio magni momenti funditus examinetur ac discutiatur: quid videlicet id extiterit, quod Observatores illi Sollertissimi circa Andromedae partem borealem deprehenderint?, tum quot Cometae revera hoc anno apparuerint? Verum hae quaestiones cum literis hisce brevioribus expediri haud possint, constitui DEO favente, opusculo meo propediem edendo de posteriori Cometa eas dare demonstratum; non dubito quin Tibi aliisque satisfaciam, sententiamque meam cum reliquorum observationibus optime conciliem quin etiam commonstrem, quid istud fuerit, quod Mense Maio // tam Ratisbonae quam Warsaviae instar Cometae observatum est. Exemplaria *Prodromi* mei 16 tum 1 Exempl[ar] *Selenog*[*raphiae*]; 1 *Mercurii*; 1 *Saturni*; et 1 *Epistolarum* jam Lubecam transmisi; quae a Domino Mattis Rhodeng[4] Mercatore Hamburgensi accipies. Decem illos Imperiales hesterna die, ab Israel van Bennigen[5] hic recte accepi. Literas vero tuas, si ita visum est imposterum Tabellario tradas rogo, ut sic recta ad me perferantur, tutius erit quam si per Dominos Rhetios illas tradi cures.[6] Scriptum Gallicum cum gratiarum actione remitto; de quo tamen cum opus ipsum nondum viderim, nihil quicquam judicare possum; videntur tamen omnia valde mira, quae quomodo fieri possint, nondum satis capio.

1 Tu **TC** : Te **BO** ('Tu' in the TC is the form corrected in the errata).
2 ad cingulum **TC** : quod ad cingulum **BO** ('quod' in the TC is deleted in the errata).
3 d *del. post* aliquod **BO**.
4 a Mattis Rhodeng : a N[oto] **TC**.
5 ab...Bennigen *om.* **TC**.
6 Literas...cures *om.* **TC**.

Interea precor DEUM ut opus tuum susceptum felicissime succedat, Teque salvum et incolumem in rei literariae commodum diutissime conservet. Suasor autem Tibi sum cum aliis amicis Tibi impense faventibus, quod fere oblitus eram, cum, uti hactenus percepi non admodum Astrologicis illis vanitatibus delectaris, ut[7] a praedictionibus et Prophetiis omnino abstineas. Nam facile quisquam ea in parte in scopulum impingere potest. Quae ut ex sincero animo profluxere sic etiam ea haud male vertas rogo.[8] Vale iterum.[9]

Dabam Dantisci Anno 1665 die 8 Julii

Tui praeclari nominis
officiosus cultor

J[ohannes] Hevelius
manu propria

To Master De Rolis

Illustrious Sir!

I can now answer your letter (sent on 16 June) if only briefly, due to my numerous duties. Together with other learned men, you seem to be still convinced that another comet between those two shone next to the Belt of Andromeda, according to the observations from Rome and those conducted by the most famous de Grau.[10] My opinion is the opposite, and I persist that only two comets have been seen and observed so far since last November. I admit, however, that those famous observers had found a phenomenon in that part of the sky (as I remember that nearly at the same time I also saw something similar with my own eyes), but I entirely disagree that the phenomenon was a comet. Therefore, it will be useful to examine this question of great importance thoroughly, and to decide what that thing (seen by those skilful observers near the northern part of Andromeda) was, and how many comets appeared this year. But because these issues cannot be solved in this short letter, I decided (with God's help) to discuss them in my little work about the latest comet; it will be published soon. I do not doubt I will fulfil both your wish and the wish of others, and I will reconcile my opinion with the observations made by them, showing what that comet-like thing which was seen in May both in Regensburg and in Warsaw was. I have already sent to Lübeck sixteen copies of my *Prodromus cometicus*, one of *Selenographia*, one of *Mer-*

7 vel *del. post* ut **BO**.
8 Suasor…rogo *om.* **TC**.
9 iterum *om.* **TC**.
10 See section 2.3.2 of the Introduction and note to Letter 7.

curius in Sole visus, one of *Dissertatio de nativa Saturni facie* and one of *Epistulae*.[11] You will receive them from Master Mattis Rhodeng, a merchant from Hamburg. Yesterday I received those 10 Reichstalers from Israel van Beunigen. In the future, I would rather you give your letters to a messenger, if you could, so that they can be passed directly to me. It will be safer than handing them to Masters Rhetii[12] to be delivered to me. I am sending back the French text with many thanks, but because I have not seen the work itself yet, I cannot say anything about it. However, everything seems very strange, and I cannot grasp how it could happen. Meanwhile, I pray to God that the work you have begun proceeds successfully and that you are kept healthy and safe for the benefit of sciences. Seeing that you are somewhat interested in those astrological vanities, I, together with other friends, very benevolent towards you (which I have nearly forgotten), would like to suggest that you abstain entirely from predictions and prophecies. Because one can easily trip when he deals with such things. I ask you not to take it as an offence, as it has been said with a sincere heart. Farewell again!

Given at Danzig, on 8 July 1665

Renowned Sir,

Your affectionate admirer,

Johannes Hevelius
by my own hand

11 It is uncertain whether Hevelius is referring to *Epistulae II. Prior de motu lunae libratorio. Posterior de utriusque luminaris defectu anni 1644* (Gedani: sumptibus autoris, typis Andreae Julii Molleri, 1654) or to *Epistulae IV. I. De observatione deliquii Solis anno 1649 habita. II. De eclipsi Solis anno 1652 observata. III. De motu lunae libratorio. IV. De utriusque luminaris defectu anni 1644* (Gedani: sumptibus autoris, typis Andreae Julii Molleri, 1654).

12 This is perhaps a reference to the printing house run in Danzig by David Friedrich Rhete (*c.* 1630–94) who took over the business from his father, Georg Rhete (1600–47).

27.

14 July 1665,[1] Lubieniecki to Hevelius

BO C1, v. 7, 1030, 1029
TC: 398–400 and 400–01

Enclosures:
27A. Excerpts of Rautenstein's letter to Lubieniecki, 6 July 1665
27B. Excerpts of Rautenstein's letter to Lubieniecki, 9 July 1665
27C. Lubieniecki's description of a celestial phenomenon seen in Hamburg, 7 July 1665
27D. List of errors in the *Prodromus cometicus*

Salutem Plurimam Nobilissime, Amplissime, Clarissime Vir. Minime adhuc neglectui erat mihi debita Literis Tuis doctissimis et humanissimis die 16. Junii datis responsio. Ut enim in aliis officiis praestandis cuilibet Amicorum, praesertim Praecipuorum, qualem Te mihi et omnibus probis magna Tua merita humanitate condita fecerunt, ita et in respondendo solenne mihi est (absit jactantia verbo) agrestem et inhumanam negligentiam fugere. Sed nosti nos aliquando, quae tamen est quaedam nostra infelicitas, non esse nostri juris, nec semper officio hoc vel alio erga Amicum, etiam primi commatis, defungi posse. Atque hoc mihi infortunii a duabus hebdomadibus,[2] ex quo Tuas illas acceperam, obtigit. Nisi tamen dicam jus excusationis et veniae utrique nostrum hic competere. Nam quominus Tibi e vestigio, illis quae per communem Amicum Nobiliss[imum] Dominum Peltsium Tibi impertii cum salute officiosa minime contentus, responderem, duo effecerunt, quae ad utrumque nostrum spectant. Alterum est, quod adhuc summo studio incubuerim adornando et quam ocissime cum expedito cursore mittendo Doctissimis illis et Celeberrimis Viris Ricciolo et Kirchero *Prodromi* Tui *Cometici* prodromo, quod etiam divina gratia perfeci, adiunctis quibusdam parergis vel ex Tuis vel ex aliorum Amicorum literis, quae commode textui Tuo adiungere potui. Quo ipso et illis Amicis, et Tibi cumprimis rem gratam facere volui. Quem finem an fuerim assecutus, Vestrum erit judicare. Ego vere testari possum, me cum Tibi, quanti Tuum istum laborem faciam, tum et illis una monstrare voluisse. Alterum vero, quod exspectarim a quibusdam Amicis, quos Tuo jussu[3] ea de re monueram, ea de quibus Te consulere potius quam monere

1 In the manuscript, the letter is dated 14 July 1665 without the information about the calendar. In the TC, there are two dates: 30 June and 17 July, both without the information about the calendar. Manuscript C1, v. 7, 1030 contains the main letter, C1, v. 7, 1029 — the postscript. In the TC, the letter is printed in two parts with the letter from Hevelius (Letter 26) between them.

2 a...hebdomadibus : ab integra hebdomade **TC**.

3 quos...jussu *om.* **TC**.

vel judicium de Tuo illo scripto praeclaro (quod Tu quidem ab illis fieri voluisti) volunt: cum[4] eorum spem a multo tempore mihi fecerint.[5] Sed cum hoc praestare per alias occupationes nondum potuerint, ego diutius officii mei vicem differre nolui, ne tanti Amici negligens videar. Sunt enim et mihi praeter debi-//tam scriptionis vicem causae quae me ad scribendum impellunt. Ac primum quidem describere Tibi debeo, quae Ill[ustrissimus] Dominus Legatus Neoburgicus Rautensteinius et Nobiliss[imus] Dominus Residens Faederati Belgii[6] de Te ad me scripserunt. Quae ille die 18. Junii ad me scripsit Tecum communicanda de accepto scilicet Tuo *Prodromo* sibi gratissimo, non dubito Te ex Nobiliss[imo] Domino[7] Peltsio percepisse. Postrema, quae nudius tertius accepi, die 2 Julii data, haec fuere. 'Hevelianas literas die 16. Junii ad G[enerositatem] V[estram] datas[8] P[atri] Curtio communicabo, qui quod et mei antehac in suis ad G[enerositatem] V[estram] mentionem tam honorificam fecerit, eidem debeo, scripturus ipse ad Virum illum Praeclarissimum, quamprimum P[atris] Curtii librum illum obtinuero.'[9] Quae ab isto Holmia die 24 Jun[ii] S[tylo] N[ovo] scripta ante aliquot dies accepi haec sunt. 'Hevelianum *Prodromum* Dantisco accepi proximis diebus a Pelsio: quod munus Hevelium ne ipsum an Pelsium auctorem prae se ferat, necdum satis habeo compertum.[10] Opus ipsum avide et attente non percurri solummodo, sed et Faederatorum Belgarum Oratori Isbrandio perlegendum dedi, cum Ill[ustrissimus] Regni Suecici Cancellarius eius ante dies hosce 14. fecisset mihi copiam. De ipso opere quid sentiam quaeris? Consummata quaevis de ingenio industriaque scriptoris mihi promisissem, omnia tamen spe atque exspectatione illic maiora apparuerunt. In Faederato Belgio, quantum agnosco, alii ex aliis quotidie scriptores prodeunt,[11] qui idem argumentum sibi illustrandum proponant, et post hunc Roscium ausint saltare, utinam tam feliciter quam confidenter.'[12] Haec illi.[13] Hunc[14] ego monui nuper[15] me ex Tuis literis intellexisse, Te illi exemplum *Prodromi* Tui mittere constituisse.[16] Sed accedo ad alia, quae in *Prodromo* illo Tuo legi et notavi, et de quibus Te consulere debeo, optime[17] de Tua humanitate et aequanimitate persuasus, atque ita certo mihi promittens,[18] Te, quibus legibus literati vivant, memorem, non

4 cum : Nam **TC**.

5 fecerint : fecerant **TC**.

6 Residens...Belgii : Hensius **TC**.

7 Domino *om.* **TC**.

8 die...datas *om.* **TC** (p. 124).

9 quamprimum...obtinuero *om.* **TC** (p. 124).

10 Pelsio...compertum : Pelsio nostro **TC** (p. 287).

11 alii...prodeunt : novi quotidie magno numero Scriptores proveniunt **TC** (p. 287).

12 confidenter : fidenter **TC** (p. 287).

13 Postrema...illi : Postrema eius die 16. Junii, ut et Heinsiana die 24. eiusdem mensis data, mihi utraque nuperrime reddita, quae fuerint, adiuncta apographa docent **TC**.

14 Hunc : Heinsium **TC**.

15 monui nuper : nuper monui **TC**.

16 constituisse : decrevisse **TC**.

17 optime : Sum enim optime **TC**.

18 promittens : promitto **TC**.

tantum non // aegre candorem meum et libertatem ingenuo dignam laturum, sed et me in istis, in quibus haereo, informare non neglecturum. Pag[ina] 3. in illis verbis, 'eumque ad ortum vergentem et retrogradum contra scilicet seriem signorum,' dubium oriri posse, indicasse Tibi sufficiat. Alia, si opus fuerit, alias dabo. Nunc ad Historica et Chronologica venio, illa scilicet quae ad Cometas pertinent. In his animadverto non pauca examinanda inveniri. Certum est paucos Cometas ante Christum natum fuisse ab Auctoribus annotatos. Sed cum Alstedius sex tantum annotavit, Eckstormius in sua *Hist[oria] Comet[arum]* 28. minimum recensuit:[19] post Christum vero natum ad illum Anni 1618. 200. amplius,[20] cum ille nonnisi 154. putarit. At vero Lavaterus et Keckermanus ab Augusto ad ann[um] 1556 solummodo 117. Ricciolus ab A[nno] ante Christum natum 480 ad post Christum natum 1618. Cometas 154 numerant. Quanta hic discrepantia! Eckstormius ille ante Christum natum A[nno] quem Ricciolus ut aeram Cometicam ponit, 480. sextum Cometam luxisse perhibet, et aeram Cometicam ab A[nno] Mundi 1657. ante C[hristum] N[atum] 2312. (parum in annis Christi a Funccio et Chytraeo distans, cum Buntingo vero et Bucholzero fere conveniens) arcessit, insigne hoc coeli phaenomenon Iridi in signum faederis a Deo constitutae coaevum faciens.[21] Sed leviora ista sunt, fateor, quae tamen, velut in transitu, attingenda erant. Venio ad alia Te propius spectantia. Pag[ina] 33. ponis Cometam quadrimestrem A[nno] Chr[isti] 380. apparuisse. Patere, Vir Humanissime, ut discam a Te quis eum annotarit? Augebit enim censum Cometarum,[22] nec a *Fasciculi Temporum* auctore Rolevinckio, nec a Funccio, nec a Calvisio, nec a Bucholzero, nec a Sigonio, nec a Ricciolo, nec ab Alstedio nec ab Eckstormio (quos habeo, et evolvi, tres vero postremos in recensendis Cometis diligentissimos experior) annotatus. Ibidem Cometas semestres ponis 4. scilicet A[nni] Chr[isti] 60, 145, 337, 1240. Ricciolus tantum tres, inter quos unus tantum Tuus est, ille postremus. Quod ad Cometam illum A[nno] 60. attinet,[23] hunc intelligit Alstedius, ponens eum in ann[um] 62. aerae quam ille, ut et Helvicus, Funccius, aliique Chronologi putant, verae, quae vulgarem // biennio praecurrit.[24] In ann[um] 60. aerae scilicet vulg[aris] ponit eum et Eckstormius semestremque notat (non ut Funccius et Calvisius, qui nullam[25] durationem eius describant ille ad Ann[um] 62. aere illius verae, hic ad 60. aerae vulg[aris]) qui quidem Cometam anni 54. aerae vulg[aris] qui mortem Claudii Caesaris antecessit, longo tempore fulsisse ait, sed tempus durationis non determinat[26] semestrem[27] illum Anni 60 sub Nerone pronunciat in Plin[ii] l[ibro] 2. c[apite] 25. quem Plinius assiduum dicit. Pronunciat autem Alstedius ut

19 recensuit : enumeravit **TC**.
20 amplius : amplius iste **TC**.
21 Eckstormius ille ante...faciens *om.* TC.
22 Nempe *del. post* Cometarum **BO**.
23 fere *del. post* attinet **BO**.
24 praecurrit : superat **TC**.
25 non ut...nullam : Funccius et Calvisius nullam **TC**.
26 quod facit Milichius quoque *del. post* determinat **BO**.
27 qui quidem...semestrem : Milichius quoque semestrem **TC**.

eundem ita et illum Anni 66. (sicut et Funccius ante eum putat[28] secundum aeram veram) qui Ricciolo, aeram vulgarem sequenti, est 64. fuisse semestrem. Nihil de durationis tempore habent reliqui, puta Funccius, Eckstorm[ius] qui soli ex illis, quos adduxi, eum annotant. Venio ad Tuum Cometam Anni 145. semestrem. Nullus illorum auctorum ullam eius facit mentionem. Ita, postquam me de ortu et parente eius certiorem feceris, proli Cometicae faecundae eum, sicut Tu stellas incognitas proli Uraniae, ut corpora corporibus caelestibus comparare liceat, annumerare non dubitabo. Succedit Cometes Anni 337. inter Tuos semestres ordine tertius. In eundem annum ponit eum Alstedius, ita secundum aeram vulgarem ponendus erit in ann[um] 335. quod innuit *Fasc*[*iculus*] *Temp*[*orum*] et Bucholzer (nisi tamen uterque, indefinite putans, in anni 337. initium eum referant, sicut Sigonius facit) exprimit autem Eckstormius, sed tempus durationis non ponit. Ricciolus vero eum prorsus silentio praeteriit. Quartus Tibi Cometa semestris est ille Anni 1240. quem et Ricciolus talem asserit. Alstedius eum notat, at tempus apparitionis non definit. Funccius, Calvisius, Bucholzerus,[29] eum non commemorant.[30] Sed nec Eckstormius, nisi ille sit, quem in Ann[o] 1239 collocat. Illum Riccioli A[nno] 603. visum Cometam, quem semestrem dicit ex Joan[ne] Praetorio, nullus illorum Auctorum tamdiu fulsisse ait. Ita dum rationes vestras puto, invenio 6.[31] Cometas, qui senis mensibus durarunt, nempe A[nno] secundum aeram vulg[arem] a Christo nato[32] 60, 64, 145, 337, 603 et 1240. Plures si dentur, investigabo, Deo volente. Unicus restat Cometes, prorsus singularis, octimestris nempe a Te productus, in Annum 375. relatus. Hic puto erit Alstedio Anni 377. a Funccio, Calvisio, Bucholz[ero] et Ricciolo omissus, ab Eckstorm[io][33] in Ann[um] 375. positus. At is[34] de duratione nihil. Annon vero error est in numero? Nam A[nno] 975. octimestrem annotat Calvisius ex Cedreno et Eckstorm[ius] et Schoffero, omissum a reliquis. Haec velim boni et aequi consulas. Charta et hora finire me jubent epistolam. Vale et salve

a
Tuo

Rolitsio//

Post Scriptum.[35] Exspecto porro abs Te responsum de illo phaenomeno ad borealem cinguli Andromedae, quod Gravianae observationi a me Tibi missae pro Tua eximia harum rerum peritia et humanitate debes. Si quid interim de hoc phaenomeno

28 putat : aliique putant **TC**.
29 Bucholzerus : Bucholzerus, Sigonius **TC**.
30 commemorant **TC** : commemorat **BO**.
31 6. : 6. dum vero et aliorum rationes excutio, reperio 8. **TC**.
32 XX *del. post* nato **BO**.
33 Eckstorm[io] : Eckstormio, Sigonio **TC**.
34 is : isti **TC**.
35 Haec velim...Post Scriptum *om.* **TC**.

ibidem observato Franequera, vel Roma, vel etiam aliunde, quod non immerito, ut vicem officii, spero, accepero, eius Te mox faciam participem. De cometa die 30. Maii s[tylo] n[ovo] Lipsiae viso scripsit mihi Amicus. De illo die 22. eiusdem mensis ad Ratisbonam viso, jam ante feci Te certiorem. An vero coniiciendam certo sit, observatum tunc fuisse circa Favonium in extremitate Tauri s[eu] in capite, vesperi circa horam 11. cum hora 10. adhuc fuerit lucidum nimis crepusculum, libenter ex Te discam. Sed vide quam multiplicentur illustria ista phaenomena. Cracovia die 8. Junii a Reverendo Viro Domino Wojenscio Canonico Cracov[iensi] Doctissimi illius in re Medica et, ni fallor, Mathematica Viri filio, ideoque literis humanioribus instructo, perscriptum est[36] Cometam ibi ingentem, barbatum, circa horam noctis 11. saepius visum fuisse. En denique Ratisbona accepi de cometa triduo ante excessum Seren[issimi] Archiducis Francisci Sigismundi supra Oenipontem viso. Quid de his omnibus habeas judicesve velim vel tribus verbis exponas.[37] Scriptum illud Gallicum, quod abs Te expectavi, utpote quo adhuc opus habeo, jam jam accepi, una cum literis die 8. Julii scriptis, mihi longe gratissimis: ad quas pauca, mantissae instar,[38] repono. Quod ad phaenomenon illud ad borealem cinguli Andromedae visum attinet, ego nihil determino. Nec enim tantum mihi arrogo, ut de his judicium ferre valeam. Volui tamen quaestionem hanc Tibi proponere, ut Tuam sententiam expiscarer. Sed hanc // Tu exponere mihi noluisti, utpote in Praeclarum opus Tuum Cometographicum relatam. Quod uti minime miror, imo aequum et conveniens judico, ita fateor me libenter visurum, si vel verbo dicas sententiam Tuam, de qua dum mentionem, simul et salivam nobis moves, tam de illo phaenomeno, quam de reliquis. Qua tamen de re statues id quod aequum et decorum est. Scito enim me antiqua amicitiae regula, ut ab amico justa, possibilia et grata postulentur, in amicitia colenda, nihil habere antiquius. Exemplaria *Prodromi* aliorumque librorum Tuorum cum desiderio exspecto, et pretium Tibi mox, Deo volente, gratus pro iis refundam. Dum iterum iterumque epistolam Tuam relego, perspicio Te propediem opusculum de posteriori Cometa editurum. Coniicio Te *Cometographiae* Tuae *Prodromum* alterum respectu posterioris Cometae praemissurum esse. Quod si ita est, gratulabor multum Te id quoque meis et aliorum Amicorum precibus dedisse. Votum Tuum pro successu operis mei conceptum mutuo voto gratus penso: et quia Tu quoque meis conatibus stimulos utilitatis publicae et laudis addis, omnem a me, quantum fieri poterit, movebo moram, ne in vota et monita vestra (paria enim faciunt tecum Ill[ustrissimus] Rautensteinius, Nobiliss[imus] Heinsius aliique Amici) publicaque commoda iniurius evadam. Quod superest monitum Tuum, tanquam a candida manu profectum, grato animo accipio, et si quae eiusmodi praedictiones occurrant in meo libro, iis abstinebo. Sed praedictiones nullae ei insunt Astrologicae: sed forte sic placuit Amico alicui ea vocare, quae Politice et Ethice dicimus consecutura ex praemissis, ut in Polonia v[erbi] g[ratia] ruinam Reip[ublicae] jacturamque totius libertatis nisi bona

36 perscriptum est : perscriptum esse accepi **TC**.
37 exponas : expone, et Vale bene ac prospere **TC** (In the TC, the letter dated 30 June ends here).
38 mantissae instar : e vestigio **TC**.

consilia obtineant. // Haec quoque si nolis modeste inseri in literis ad Ill[ustrissimum] Raut[ensteinium] et occasione literarum et judicii eius, verbo indicabis. Emolliri tamen quaedam poterunt, vel potius placidissime dici: quanquam ubique studio omnem servavi modestiam. De tuo illo libello an sit futurus *Prodromus* Posterioris Cometae et an figuras et quot sis ei additurus, velim quamprimum ex Te cognoscere. Vale et constanter fave.

Dab[am] die 14. Julii 1665.

Iterum Post Scriptum. Tandem his diebus perlegimus cum Eruditiss[imo] Henr[ico] Syvers, eximio Astronomiae et Tui cultore *Prodromum* Tuum et contulimus de iis in quibus haesimus, vel quae aperte erronea irrepsisse visa sunt. Haec pro jure amicitiae Tibi mitto, de Tua aequitate certus. Responsum exspectabimus quod si, ut optamus, obtinuerimus, erit haec correctura *Prodromi* Tui, Prodromo eius a me composito adiungenda. Fasciculum Tuum accepi hodie. Curabo Tibi 11 1/2 Imper[iales] brevi volente Deo numerari, ut conflentur 21 1/2 Imp[eriales]. En Tibi phaenomenon heri ab aliquot nostrum observatum. Iterum Vale.[39]

Greetings, most noble, worthy, and renowned Sir! I have not answered your most learned and kind letter sent on 16 June yet, but not out of negligence. I should be exceptionally careful (I say it not to boast) not to appear rude and impolite by neglecting duties towards friends, especially towards those distinguished ones (and you are such a person for me and for all the righteous ones due to your merits and kindness), and by answering letters too slowly. But you are well aware that sometimes (this fact

39 Quod superest…Iterum Vale : Quod reliquum est certus sum de Tua, sicut nuper dixi, aequanimitate et singulari humanitate. Huic et Clarissimus Bullialdus, aliique virtutibus et meritis Tuis, in epistola libri Ptolemaei *De Judicandi Facultate et Animi Principatu* ab se editi, quem jam jam accipio, et ad Tecum colendam sinceram amicitiam magis magisque stimulor, dignum testimonium praebuit. Itaque mitto Tibi, ea quae in lectione Tui *Prodromi Cometici* notavi cum Eruditissimo illo Viro Henrico Syvero, Tuorum meritorum reique Atronomicae strenuo cultore, Tibi proponenda. Quaedam aperte indicant errores esse calami, vel per solitam Typographi incuriam irrepsisse. Alia nos suspensos tenent, ita ut Te consulere necesse habeamus. Ad haec responsum abs Te exspectabimus cum desiderio. Quod si obtinemus ex voto, a Te correcta poterunt, si ita fieri velis, inseri vel adiungi Prodromo *Prodromi* Tui *Cometici*, a me gratia Amicorum Italorum compositi, ut ibi illius operis Tui praeclari compendium veluti cum annotatiunculis et correctione prodeat. Valere Vir Amplissime, Te jam juberem, cum epistola haec jam satis verbosa, vereor ne et molesta, praesertim Tibi gravioribus studiis occupato, evaserit. Sed unum superest Tibi perscribendum, me heri paulo post X. vespertinam cum aliquot e domesticis meis, inprimis Martino Ruaro, Doctissimi illius et Clarissimi Viri Urbis Vestrae olim incolae, cui cognominis est, filio, docto juvene, liberorum meorum Ephoro, observasse ad occasum tale ignitum phoenomenon. Apparuerunt primum duae faculae in linea recta in medio nube, instar cinguli, divisae. Hic paulo post a latere sinistrae partis inferioris assiluit tertia facula. Ex his compactum corpus lucidissimum, in medio tamen nube, velut cingulo divisum, paulo post ad sinistram, scilicet versus novum templum inclinare, et minui magis magisque coepit, et ita disparuit ante horam undecimam, luce clariori post se inter spissas nubes relicta. His Te tandem, Vir Clarissime bene valere et feliciter agere cupio. **TC.**

saddens me) it does not depend on ourselves, and we cannot always fulfil one duty or another to a friend, even to a very distinguished one. And such a misfortune has been happening to me for the last two weeks, since the time I received your letter. I do not want to say that both of us should excuse ourselves and apologise to each other. There are two things (concerning both of us) which caused the lack of my immediate answer; I was not satisfied with what I sent to you, with a sincere greeting, through the most renowned Pels,[40] our mutual friend. The first one is that so far I have been preoccupied with diligently finishing and sending (as soon as possible) the forerunner of your *Prodromus cometicus* to the most learned and renowned men, Riccioli[41] and Kircher,[42] together with some excerpts of your letters and those of other friends, which I could suitably enclose to your text. I finished and sent it through a messenger, thanks to God. In that way, I wanted to act kindly towards them and especially towards you. It is for you to decide if I succeeded. I can assure you that I wanted to show both to you and to them how much I esteem your work. The second thing is that I waited for the replies from some friends (whom I informed about this thing, as you requested) to know what they want to suggest to you (rather than to condemn or judge you) about that famous text of yours (as you wanted them to do). They promised me to do so a long time ago. But since they could not do it among their many occupations, I did not want to delay my duty any longer, not to seem negligent towards such a great friend. And there are also some other causes, apart from the answer I owe you, which force me to write. First of all, I should tell you what the most illustrious Master Rautenstein,[43] the emissary from Neuburg, and the most noble Master Heinsius,[44] the ambassador of the Dutch Republic, wrote to me about you. I do not doubt that you learned from the most noble Master Pels what the latter wrote to me on 18 June about receiving your *Prodromus* which was most pleasant for him; he asked me to convey it to you. His last letter, sent on 2 July, which I received three days ago, reads as follows: 'I will pass to Father Curtz[45] the letter from Hevelius, sent on 16 June to you, generous Sir, and because he mentioned me so honourably in his letter to you, generous Sir, I am indebted to him, so I will write to that most renowned Sir when I receive that book written by Father Curtz.'[46] What I received a few days ago from Heinsius from Stockholm, written on 24 June, reads as follows: 'Recently I received Hevelius's *Prodromus* from Danzig. I am still not sure whether the author of this gift was Hevelius or Pels. Not only did I eagerly and carefully read this work, but I gave it to read to Isbrants,[47] the envoy of the Dutch Republic, when the most famous Chancellor of the King of Sweden had given me a copy 14 days ago.

40 There is no trace of Lubieniecki's letter to Hevelius sent via Pels (see note to Letter 5).
41 See section 2.4.15 of the Introduction.
42 See section 2.4.8 of the Introduction.
43 See note to Letter 5.
44 See section 2.4.6 of the Introduction.
45 See section 2.4.5 of the Introduction.
46 TC p. 124.
47 Johan Isbrants (*fl.* 17th century), Dutch diplomat, emissary to Sweden in the years 1665–67.

Are you asking me what I think about this work? I was quite convinced about the author's talent and diligence, but everything I found there turned greater than I hoped for and expected. In the Dutch Republic, as far as I know, every day new writers step forward to present the same argument in order to make themselves famous, and they dare to dance after that Roscius[48] as successfully and surely as they can.'[49] These are his words. I have recently informed him that I understand from your letter that you decided to send him a copy of your *Prodromus*. Now I go to other things which I have read and noted in your *Prodromus* and about which I need to ask you, ensured of your kindness and civility. I hope that you, mindful of the laws of learned men, will not only tolerate my sincerity and freedom the mind deserves, but you will also rush to inform me about those things about which I am puzzled. On page 3, it is sufficient to tell you that one can have doubts about these words: 'and its movement is eastwards and retrograde, viz. opposite to the order of signs.' I will send you more another time, if there is need. Now I go to historical and chronological topics, viz. those concerning comets. Among them, I see many details which need examination. It is certain that the authors noted a few comets before the birth of Christ. But while Alsted[50] noted only 6 of them, Eckstorm[51] in his *Historia cometarum* listed at least 28, and after the birth of Christ up until the comet of 1618[52] he mentioned over 200, while the former author only 154. And Lavater[53] and Keckermann[54] list only 117 comets since Augustus until 1556,[55] and Riccioli[56] 154 comets since the year 480[57] before the birth of Christ until the year 1618 after the birth of Christ. What a discrepancy! Eckstorm states that in the year 480, which Riccioli takes as a beginning of the cometary age,[58] the sixth comet shone and he begins the cometary age in the year 1657 of

48 See note to Letter 16A.

49 TC p. 287.

50 Johann Heinrich Alsted (1588–1638), German Calvinist theologian and scholar. Lubieniecki refers to his *Thesaurus chronologiae* (Herbornae Nassoviorum: [s.e.], 1650), p. 503.

51 Heinrich Eckstorm (1557–1622), German Protestant theologian and chronicler. Lubieniecki refers to his *Historiae eclipsium, cometarum et pareliorum* (Helmaestadi: typis heredum J. Lucii, 1621), pp. 202–12.

52 See note to Letter 1F.

53 Ludwig Lavater (1527–86), Swiss Calvinist theologian. Lubieniecki refers to his *Cometarum omnium fere catalogus* (Tiguri: per Andream Genesum, [1556]). This books consists only of the catalogue of comets since Augustus until 1556.

54 Bartholomaeus Keckermann (see note to Letter 7A), *Systema physicum* (Hanoviae: apud Guilielmum Antonium, 1610), p. 690.

55 C/1556 D1, see Alexandre Guy Pingré, *Cométographie ou Traité historique et théorique des comètes*, vol. 1 (Paris: Imprimerie royale, 1783), pp. 502–07; Gary W. Kronk, *Cometography. A Catalogue of Comets. Volume 1: Ancient – 1799* (Cambridge: Cambridge University Press, 1999), pp. 309–11.

56 Giovanni Battista Riccioli, *Almagestum novum*, pars posterior tomi primi (Bononiae: ex Typographia Haeredis Victorii Benatii, 1651), pp. 3–23.

57 The comet of 480 BC is said to have appeared during the battle of Salamis, see Pingré, *Cométographie*, p. 255; Kronk, *Cometography*, pp. 2–3.

58 When Lubieniecki says that an author begins the cometary age in a certain year, he means that the first comet described by that author was dated to that year.

the world, viz. in the year 2312 before the birth of Christ (differing slightly from Funck[59] and Chyträus[60] in the years of the life of Christ, while Bünting[61] and Buchholzer[62] agree with him), and he makes that celestial phenomenon contemporary with the rainbow shown by God as a sign of the covenant. Those things are not very important, but they had to be touched upon by the way. I move on to another topic, more closely related to you. On page 33, you say that there was a four-month comet in the year 380[63] after the birth of Christ. Most kind Sir, let me ask you who noted that comet. Because it will increase the list of comets, as neither *Fasciculus temporum* by Rolevinck,[64] nor Funck, nor Calvisius,[65] nor Buchholzer, nor Sigonius,[66] nor Riccioli, nor Alsted, nor Eckstorm (whose books I own and have read, and I find the last three of them very diligent in listing comets) mentioned it. In the same place, you say about four six-month comets, viz. in the years 60,[67] 145,[68] 337,[69] and 1240[70] after the birth of Christ. Riccioli says only about three of them and only one is listed by you (the last one). As for the comet of the year 60, Alsted acknowledges it by placing it in the year 62 of the Real Age (as Helvicus,[71] Funck, and other chronologists call it, which is two years forward to the common age). Also Eckstorm places it in the year 60 of the common age and notes it was a six-month comet (Funck and Calvisius do otherwise, stating nothing about its duration; the former places it in the year 62 of the

59 Johann Funck (1518–66), German Protestant theologian. Lubieniecki refers to his *Chronologia* (Basileae: excudebat Jacobus Parcus sumptibus Joannis Oporini, 1554), pp. 8, 264. According to Funck and Chyträus, the world was created in the year 3962 before the birth of Christ and the deluge (the time of the first comet, as Eckstorm says) was in the year 1656 of the world, 2306 years before the birth of Christ.

60 David Chyträus (1530–1600), German Protestant theologian and historian. Lubieniecki refers to his *In Genesin enarratio* (Vitebergae: excudebat Johannes Crato, 1561), sig. Br–Bv.

61 Heinrich Bünting (1545–1606), German protestant theologian. Lubieniecki refers to his *Chronologia* (Servestae: impressum typis Bonaventurae Fabri, 1590), sig. 18r.

62 Abraham Buchholzer (1529–84), German historian and Protestant theologian. Lubieniecki refers to his *Index chronologicus* (Gorlicii: excusus typis et sumptibus Iohannis Rhambae, 1599), p. 2.

63 The sources about the comet of 380 are scarce, see Pingré, *Cométographie*, p. 303.

64 Werner Rolevinck (1425–1502), German Carthusian monk and historian. Lubieniecki refers to his *Fasciculus temporum* (Coloniae Aggripinae: [s.e.], 1474).

65 Sethus Calvisius (Seth Kalwitz, 1556–1615), German historian, astronomer and music theorist. Lubieniecki refers to his *Opus chronologicum* (Francofurti: impensis Johannis Thymii, 1620).

66 Carolus Sigonius (*c.* 1524–84), Italian humanist and classical scholar. Lubieniecki refers to his *Historiarum de occidentali imperio libri XX* (Francofurti: apud Heredes Andreae Wechelii, 1593).

67 The comet of 60 (X/60 P1) was mentioned by a number of Roman and Far Eastern sources, see Pingré, *Cométographie*, p. 285; Kronk, *Cometography*, pp. 31–32.

68 The sources about the comet of 145 are rather scarce, see Pingré, *Cométographie*, p. 292.

69 The sources place this comet rather in the year 336 but tend to connect it to the death of Emperor Constantine the Great (*c.* 272–337, Roman Emperor since 306), see Pingré, *Cométographie*, p. 301; Kronk, *Cometography*, pp. 63–64.

70 C/1240 B1, see Pingré, *Cométographie*, pp. 403–04; Kronk, *Cometography*, pp. 215–17.

71 Christophorus Helvicus (Christoph Helvig, 1581–1617), German theologian and historian. Lubieniecki refers to his *Theatrum historicum, sive chronologiae systema novum* (Gissae Hessorum: excudebat Nicolaus Hampelius, 1609).

Real Age, the latter in the year 60 of the common age). He says that the comet in the year 54[72] of the common age, which foreshadowed the death of emperor Claudius, shone for a long time, but he does not determine the time of its duration. He states that the comet in the year 60, during the reign of Nero, was a six-month one, according to Pliny's Book 2, Chapter 25. Pliny calls it 'constant.' And Alsted states that this one, like the one of the year 66 (he and Funck before him date it according to the Real Age), which Riccioli places in the year 64[73] after the common age, was a six-month comet. Nothing about its duration is said by others, viz. Funck and Eckstorm, who noted it only relying on what I mentioned above. I move to your six-month comet of the year 145. None of those authors mention it. So when you inform me about its appearance and origin, I will doubtlessly add it to the abundant cometary offspring, as you add the unknown stars to the offspring of Urania, if one can compare bodies to celestial bodies. What follows is the comet of the year 337, the third among your six-month comets. Alsted places it in the same year, so according to the common age it should be placed in the year 335. It agrees with *Fasciculus temporum* and Buchholzer (unless both of them, without giving much detail, place it at the beginning of the year 337, as Sigonius does), and Eckstorm discusses it in detail but does not give the time of its duration. But Riccioli has passed it in silence. Your fourth six-month comet is the one of the year 1240 about which Riccioli states the same. Alsted notes it but does not define the time of its appearance. Funck, Clavisius, and Buchholzer do not mention it. Neither does Eckstorm, unless it is the one he places in the year 1239. None of the authors agree that the Riccioli's comet (seen in the year 603[74] and which he says to be a six-month one, quoting Johannes Praetorius[75]) shone for so long. Having considered your arguments, I find six comets which were seen for six months, viz. in the years after the birth of Christ according to the common age: 60, 64, 145, 337, 603, and 1240. If there are more, I will investigate it, God willing. There is only one comet left, very peculiar, viz. the eight-month one, which you listed and placed in the year 375.[76] I think it is the one which Alsted places in the year 377, Funck, Calvisius, Buchholzer, and Riccioli omit, and Eckstorm places in the year 375. But he says nothing about its duration. However, is not there an error in number? For Calvisius (after

72 The comet of 54 (X/54 L1) was said to have foretold the death of Emperor Claudius (10 BC–54, Roman Emperor since 41), see Pingré, *Cométographie*, p. 284; Kronk, *Cometography*, pp. 28–30.

73 The comet of 64, see Pingré, *Cométographie*, pp. 286–87; Kronk, *Cometography*, p. 33.

74 Sources place the comet rather in 602, see Pingré, *Cométographie*, p. 326.

75 Johannes Praetorius (1537–1616), German mathematician and astronomer from Bohemia. Lubieniecki refers to his *De cometis, qui antea visi sunt, et de eo, qui novissimus mense Novembri apparuit* (Noribergae: in Officina Typographica Catharinae Gerlachin et Haeredum Johannis Montani, 1578), sig. [A4v].

76 Roman sources inform about the comet seen before the death of Emperor Valentinian I (321–75, Roman Emperor since 364), see Pingré, *Cométographie*, p. 302.

Cedrenus,[77] Eckstorm, and Schefferus[78]) notes one eight-month comet in the year 975,[79] which all the others omit. I would like you to answer me honestly and sufficiently. Paper and late hour force me to end the letter. Take a farewell and wishes

from
your

Rolitsius

Postscript. I am waiting for your answer about that phenomenon near the northern star in the Belt of Andromeda which (due to your extraordinary astronomical knowledge and kindness) you owe to de Grau's[80] observation which I sent to you. Meanwhile, if I receive anything from Franeker, Rome or somewhere else about that phenomenon observed in the same place (because I have some grounds to expect the return of favour), I will soon inform you. A friend wrote to me about a comet seen on 30 May, new style, in Leipzig.[81] About the one seen on the twenty-second of the same month in Regensburg, I informed you earlier. I would like to learn from you if one can suppose it was seen then in the east in the extreme part or in the Head of Taurus, in the evening (about 11 o'clock), because about 10 o'clock the darkening sky was still quite bright. But behold how often these celestial phenomena occur! It was written from Kraków on 8 June by a reverend man, Master Wojeński,[82] a canon there and the son of that man very learned in medicine and mathematics, himself very learned in liberal arts, that a large bearded comet was spotted by them there about 11 o'clock at night. And finally, I received some information from Regensburg about the comet seen over Innsbruck three days before the death of Archduke Sigismund Francis.[83] I would like you to tell me (even briefly) if you know about all of these occurrences, and what you think about them. I have just received a much-awaited French text from you, as I still need it. I received it together with your very pleasant letter, written on 8 July, which I answer briefly by the bye. As far as that phenomenon next to the northern star in the Belt of Andromeda is concerned, I can say nothing certain, because I lack the confidence to give my own statement about it. Nevertheless, I wanted to ask you this question in order to learn your opinion. But you did not want to provide me with the answer, as if you kept it for your famous *Cometographia*. It does not surprise me;

77 Georgius Cedrenus (Kedrenos, *fl.* 11th century), Byzantine historian, author of chronicle *Σύνοψις ἱστοριῶν* which was published with Latin transl. by Wilhem Xylander as *Georgii Cedreni annales* (Basileae: per Ioannem Oporinum, 1566). On the eight-month comets see p. 563 of that edition.

78 Probably Johannes Schefferus (see note to Letter 17). I could not identify the book to which Lubieniecki refers here.

79 X/975 P1, see Pingré, *Cométographie*, pp. 357–58; Kronk, *Cometography*, pp. 159–61.

80 See section 2.3.2 of the Introduction and note to Letter 7.

81 There are no records of a comet seen in May 1665. The earlier comet (C/1665 E1) disappeared in April.

82 Probably Stanisław Wojeński (d. 1685), since 1680 the bishop of Kamieniec Podolski.

83 Sigismund Francis (1630–65, Archduke since 1662), Archduke of Further Austria and Tyrol.

I rather find it just and proper, and so I admit I would willingly see how you express your opinion about that phenomenon and other things, even if briefly; you have only mentioned it and you have whetted my appetite. You will decide what is just and good about it. But know that I regard nothing to be more important than the principle of friendship which allows asking of a friend what is just, feasible, and pleasant, in order to cultivate the friendship itself. I am eagerly waiting for the copies of the *Prodromus* and for your other books; I will soon send you money for them, God willing. As I am re-reading your letter, I see you will soon publish a book about the latest comet. I guess you will send forward a new *Prodromus* of your *Cometographia* as a result of the late comet. If it is so, I will heartily congratulate you that you released it in response to the requests from me and other friends. I take your wishes for the success of my work with gratitude, and I return the wishes; and because you also add incentives (public utility and praise) for my efforts, I will do everything that is possible not to do any harm to your wishes and pieces of advice (the most illustrious Rautenstein, the most noble Heinsius,[84] and other friends make them together with you), and to the public good. As far as the rest is concerned, I thank you for your advice, as it comes from a just hand, and if there are any predictions of that sort in my book, I will abstain from them. But there are no astrological predictions in it. But perhaps a friend already evoked those things which we call political and ethical consequences based on some premises; for example, the ruin of the state and the fall of every freedom in Poland, against good advice. If you do not want me to humbly include it in the letter to the most learned Rautenstein, say only a word. Some things could be placated, or phrased more gently, but so far I have taken care to keep decency everywhere. I would like to learn about that book of yours as soon as possible; whether it is going to be a *Prodromus* of the latest comet, and if you are going to enclose any illustrations and how many. Farewell, and constantly stay benevolent!

Sent on 14 July 1665.

Second postscript. Recently, I have finally meticulously read your *Prodromus* together with the most learned young man Heinrich Sivers,[85] a great admirer of astronomy and yourself, and we have compared those places in which we had difficulties, or where clearly some mistakes seemed to have sneaked in. I send you them because of our friendship, sure about your kindness. We hope for an answer, and if we receive it, as we wish, those corrections of your *Prodromus* will be enclosed to its forerunner written by me. Today I received a package from you. I will take care to pay you 11 1/2 Reichstalers, God willing, so it will be 21 1/2 Reichstalers in total. I enclose the information about the phenomenon seen here yesterday. Farewell again!

84 See section 2.4.6 of the Introduction.
85 See note to Letter 20.

27A.

Excerpt of Rautenstein's letter to Lubieniecki, 6 July 1665

BO: C1, v. 7, 1031[86]
TC: 127

Ratisbonae 6. Julii 1665.

Mitto triste nuncium Oeniponte nuperrime ad nos perlatum de obitu die 25. Junii Archiducis Sigismundi Francisci, quem triduo ante Cometa supra urbem visus praesagivisse fertur. Hevelianas literas P[atri] Curtio communicavi.[87] Quid de Hevelii *Prodromo* judicem, jam ante perscripsi, meque literis meis ipsi gratis pro eodem acturum, quamprimum a P[atre] Curtio responsum et librum[88] obtinuero.

In Regensburg, on 6 July 1665.

I am sending the sad news conveyed to us recently from Innsbruck about the death of Archduke Sigismund Francis[89] on 25 June; it is said to have been foretold three days before by a comet seen over the city. I sent the letter from Hevelius to Father Curtz. I had written before what I think about Hevelius's *Prodromus*, and I will send him a letter with thanks as soon as I receive an answer and that book from Father Curtz.

27B.

Excerpts of Rautenstein's letter to Lubieniecki, 9 July 1665

BO: C1, v. 7, 1031
TC: 128

Ratisbona die 9. Julii.

De Cometa illo Dieteldorfii a me 22. Maii viso, nihil amplius memorare possum, nisi quod putem orientem tunc nobis a tergo, Occidentem circa faciem et Cometam

86 Each part of the enclosure is printed separately here.
87 Mitto...communicavi *om.* TC.
88 librum : librum illum TC.
89 See note to Letter 27.

ad laevam, id est ad Meridiem visum fuisse. Qualis vero fuit ille qui Oeniponti ante obitum ultimi Achiducis conspectus fertur, a Mathematicis eorum locorum discendum erit. Videmus Cometas, legimus Prophetas, deprehendimus seriem consiliorum et fatalem quasi omnium rerum confusionem. Et tunc vix sunt, quo ad meliora se inflecti et a perversis abduci patiuntur, ita ut nihil verius sit, quam illud vetus: 'Vitia erunt donec homines': homines vero quo propriores fini mundi, eo peiores. 'Optimi illi qui minimis urgentur', aut qui parvi moriuntur, antequam inquinentur pravitate seculi et procreent sobolem vitiosiorem.

De opere G[enerositatis] V[estrae][90] repeto pariter quae scripsi, et id suo tempore videre gestio, utinam cum commodo et utilitate G[enerositatis] V[estrae] cui Famae quidem et reputationis multum exinde accedet, nescio an etiam fortunae? Verumque autem opto, utpote qui G[enerositati] V[estrae] nihil nisi optima quaeque ex animo voveo. Quae inter eandem et Amicos Romanos, Bononienses, Gedanenses, Herbipolenses aliosque quam plurimos occasione cometae scripta hinc inde et agitata sunt, proculdubio doctissima et singularia orbi erudito non poterunt non placere. //

From Regensburg, 9 July.

As for the comet, I was in Dietldorf[91] on 22 May; I remember nothing more except that I think that then we had the east behind us, the west in front of us, and the comet was seen on the left, that is in the south. But mathematicians of those places should be asked what was the one which is said to have been seen in Innsbruck before the death of the last Archduke.[92] We observe the comets, we read the prophets, we perceive various pieces of advice and a somewhat fatal confusion of all things. And then there are hardly any people who allow themselves to be inclined to what is good and to be diverted from what is wretched, so there is nothing more accurate than the old saying: 'There will be vices as long as there are men'[93] because the closer the men are to the end of the world, the worse they are. 'They are the best men who are encumbered with the least'[94] or who die as children before they defile themselves with the wretchedness of the world and conceive a more sinful offspring.

About your work, generous Sir, I repeat what I wrote, and I expect to see it in due time, hopefully when it is convenient and useful for you, generous Sir, for whom it will bring much fame and reputate, and maybe even fortune. And I wish it as the one

90 G[enerositatis] V[estrae] *om.* TC.

91 Town in Bavaria, north of Regensburg.

92 See note to Letter 27.

93 Tacitus, *Historiae*, 4, 74, transl. by Alfred John Church and William Jackson Brodribb, in Tacitus, *Complete Works of Tacitus* (New York: Random House Inc., 1873).

94 Horace, *Satyrae*, 1, 2, 68–68, transl. by C. Smart, Theodore Alois Buckley, in Horace, *The Works of Horace* (New York: Harper & Brothers, 1863). Quoted with minor changes.

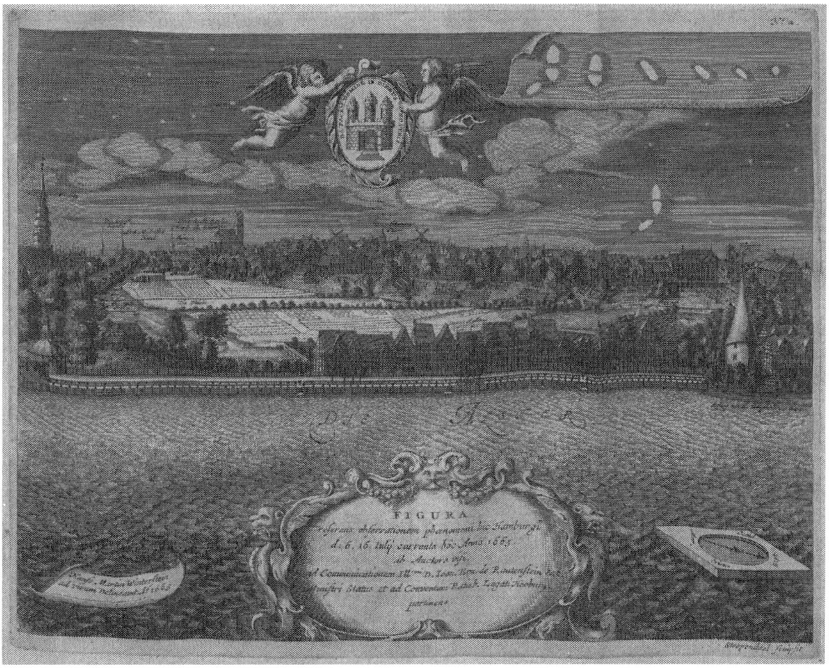

Figure 12. The extraordinary phenomenon over Hamburg in Lubieniecki's *Theatri cometici pars prior.* Biblioteka Narodowa, Warsaw, SD XVII.4.2660.

who wishes you, generous Sir, all the best from the heart. What was written in various places and inquired about by you and the friends from Rome, Bologna, Danzig, Würzburg, and many others on the occasion of the comet cannot be unpleasant for the learned world; beyond doubt, these things are learned and unique.

27C.

Lubieniecki's description of a celestial phenomenon seen in Hamburg, 7/[17] July 1665

BO: C1, v. 7, 1031

Hamburgi die 7. Julii Juliani 1665

Heri paulo post horam 10. vespertinam aliquot nostrum observavimus ad Occasum tale ignitum phaenomenon: Apparuerunt primum duae faculae in linea recta, in me-

dio nube instar cinguli divisae: his paulo post a latere sinistrae partis inferioris assiliit tertia facula. Ex his compactum corpus lucidissimum, in medio tamen nube velut cingulo divisum, paulo post ad sinistram scilicet versus novum templum inclinare, minuique magis magisque caepit, et ita disparuit, luce clariori post se relicta. Mitto Figuram eius.

In Hamburg, on 7 July, Julian style, 1665

Yesterday, soon after 10 o'clock in the evening, some of us saw in the west such a fiery phenomenon: at first, two torches appeared in a straight line, divided in half by a cloud so that it looked like a belt. Soon afterwards, a third torch jumped to the left side of the lower part of that belt. They created one very bright body, divided in half by a cloud, like with a belt, and soon after it began to lean towards left, viz. towards the new church, and to diminish, and finally disappeared, leaving only some bright light behind. I send you an illustration.

27D.

List of errors in the *Prodromus cometicus*[95]

BO: C1, v. 7, unnumbered (between C1, v. 7, 1033 and C1, v. 7, 1034)

Pag[ina] 8. lin[ea] 10. Legitur 'alto', pro 'altus': sed levissimum hoc

pag[ina] 10. lin[ea]. 10. Author sine dubio Ascens[ionem] rectam vult esse 40°. 20', quam Typographus ex errore Typothetae 4°, 20'. fecit.

pag[ina] 16[96] lin[ea] 30. Progressus Cometae spatio 24. horarum ponitur 5°, 30'. cum rectius ex mente Authoris videatur notari debere 5', 30".

pag[ina] eadem lin[ea] 37. Longitudo Cometae 16°, 20'. Arietis decem gradibus justo minor ex contextu cernitur.

26°[97]

95 On the page, there are few barely legible (or entirely illegible) notes in pencil, probably written by Hevelius (at the end of the enclosure, the date '1665, d[ie] 14 Julii' is definitively written in his hand). In the apparatus, I indicate what was written in pencil, or note that the annotation was illegible. See also Letter 32A.

96 10. *del.* **BO** (correction in pencil).

97 Note in pencil, preceded by an undeciphered sign.

pag[ina] 19. lin[ea] 11. Longitudo Cometae non apposito Dodecatemorii Arietis indice indeterminata offertur.[98]

pag[ina] 21. In Tabula priori, a 28.^{vi} Decemb[ris] hora 2.da mat[utina] ad 29.ni horam 9.^{am} 30'. (hoc est, intra spatium 23. horarum et 30') ponitur motus Cometae proprius 23°, 40'. Cum altera Tabula pro duobus integris diebus (28. scil[icet] et 29. Dec[embris]) intra spatium 48. horarum exhibeat motum proprium 23°, 45'. Puto igitur ex errore irrepsisse 23° pro 21° et c[etera].

recte[99]

pag[ina] eadem. Tab[ula] altera notat in columna 3^{ia} (a sinistris) pro 'Motu diurno' 'Min[utae] Sec[undae]' quae rectius 'Grad[us] Min[utae]' notari debent.[100]

p[agina] 26. lin[ea] 27. Praepositio 'ad' pro 'a' posita sensum obscurat.

p[agina] 28. lin[ea] 22. Occurris 'Pyperbolica' pro 'Hyperbolica', quod levissimum. Sed linea antepenult[ima] 'atque' pro 'adque' sensum ambiguum facit.

recte[101]

p[agina] 29. lin[ea] 1. Particula redditiva 'Quantum' videtur innuere, quod // lin[ea] 3. inter 'affirmare' et 'Cometam' forte omissum sit 'statuo' vel aliud quoddam verbum, pro integritate sensus orationis redditivae.

p[agina] eadem lin[ea] 2. Ambiguitatem admittit Author, dum motum Cometae inversum quidem, s[ecundum] s[ignorum] seriem tamen tendentem, quem exinde pag[ina] 18. lin[ea] 22. et 19. lin[ea] 12. directum dixit, vocat retrogradum contra notionem retrogradationis pag[ina] 4. lin[ea] 8. a fine et pag[ina] 18. lin[ea] 23. notatam.

pag[ina] 30. lin[ea] 20. Valor numeri 21 1/2 dubius est, propter notam eius neglectam. Ex Tabula tamen pag[ina] 21. suppleri potest.

pag[ina] 31. lin[ea] 12. pro 'statuant', ponendum 'statuunt'. levis error.

1665, d[ie] 14 Julii[102]

98 Followed by an undeciphered note in pencil.
99 Note in pencil.
100 Followed by an undeciphered note in pencil.
101 Note in pencil, together with undeciphered signs.
102 Note in pencil.

On page 8, line 10. There is 'alto' instead of 'altus,' but this is not important.

on page 10, line 10. Without a doubt, the author wanted the right ascension to be 40° 20', which the printer, due to a typesetter's mistake, gave as 4° 20'.

on page 16, line 30. The movement of the comet within 24 hours is stated to be 5° 30', while it seems that according to the author's judgement it should be written as 5' 30".

on the same page, line 37. The longitude of the comet given as 16° 20' of the Aries seems 10 degrees too small, as the context suggests.

26°[103]

on page 19, line 11. The longitude of the comet is given without the symbol of the sign of Aries.

on page 21, In the first table, from 28 December, 2 o'clock in the morning, to the twenty-ninth, 9.30 o'clock (this is within 23 hours and 30 minutes), the proper motion of the comet is given to be 23° 40', while the second table for the two whole days (namely 28 and 29 December) within 48 hours shows the proper motion to be 23° 45'. Therefore, I think the error has sneaked in, leaving 23° instead of 21° and so on.

correctly[104]

On the same page. In the second table, in the third column (from the left), titled 'Motu diurno,' there should be written 'Gradus, Minutae' instead of 'Minutae, Secundae.'

on page 26, line 27. Preposition 'ad' instead of 'a' makes the passage unclear.

on page 28, line 22. There is 'Pyperbolica' instead of 'Hyperbolica' which is irrelevant. But in the antepenultimate line 'atque' instead of 'adque' makes the sense ambiguous.

correctly[105]

on page 29, line 1. A redditive word 'Quantum' seems to suggest that in line 3, between 'affirmare' and 'Cometam,' 'statuo' or another similar word was perhaps omitted, as the meaning of the redditive phrase suggests.

103 Note in pencil.
104 Note in pencil.
105 Note in pencil.

on the same page, line 2. The author admits an ambiguity while he calls the movement of the comet inverse but tending according to the order of signs — which he called 'direct' on page 18, line 22, and page 19, line 12 — 'retrograde' contrary to the definition of retrograde given on page 4, line 8 from the bottom, and page 18, line 23.

on page 30, line 20. Value of 21 1/2 is dubious because it lacks the note. It can be supplemented from the table on page 21.

on page 31, line 21. Intead of 'statuant,' it should be 'statuunt.' An irrelevant mistake.

14 July 1665[106]

106 Note in pencil.

28.

25 August 1665, Lubieniecki to Hevelius

BO: C1, v. 7, 1032
TC: 401–02
Olhoff: 102–03

Salutem Plurimam Nobilissime, Amplissime, Clarissime Vir. Quod a multo tempore nec scribas nec rescribas mihi, Tuis gravioribus laboribus adscribo. Non dubito enim Te omni studio et virium nisu edendae Tuae *Cometographiae*,[1] quod maximum praestantissimi Tui ingenii monumentum et opus incomparabile orbis literatus magno cum desiderio et taedio exspectat, incumbere.[2] Crede mihi omnium animi, aures oculique in illud velut Astronomici ingenii miraculum directi sunt, et quamprimum illud sibi exhiberi, avide praestolantur. Neque vero putes me palpum Tibi obtrudere. Provoco enim ad totius non tantum Senatus Astronomici, in quo sine aemulo et rivali dominaris, sed et Reip[ublicae] literariae judicium. Fidem dictis faciunt illa quae ut antehac Ill[ustrissimo] Rautensteinio, Nobilissimo Heinsio, aliisque Viris Amplissimis, ita nuperrime a Celeberrimis Societatis Jesu Philosophis Ricciolo (de cuius valetudine nihil ab illo tempore accepi)[3] Kirchero et Schotto, tum Doctissimo Grutero perscripta ad Te misi. En quae Cl[arissimus] Bullialdus tot testimoniis maximae virtuti Tuae merito datis non contentus postremis literis ad me die 14. Augusti datis inseruit: 'Nullus apud nos praeter Agarratum Observationes Cometae, quas aliquando editurus est, per instrumenta habuit: nullius propterea operam his, quae quaerimus indagandis sufficientem puto. R[everendus] P[ater] Grandamicus libellum etiam edidit, sed quae praestanda forent, non[4] praestat. Ill[ustrissimi] Hevelii Observationes, quae // sunt certissimae, solae adhibendae erunt.' Haec ille.[5] En vero etiam gravissimum et amplissimum, sed meritis Tuis par debitumque testimonium a Cl[arissimo] Kirstenio datum, qui Tibi per me hasce literas, iisque inclusum *Cometam* suis versibus adornatum[6] ac nomini Tuo Clarissimo dicatum, mittit. Gratulor Tibi has laudes summas Copernicianis, Tychonianis et Keplerianis proximas in rei Astronomicae scientia, ut ab aliis Viris talium peritissimis, ita et ab hoc Viro Doctissimo inque Astronomicis versatissimo tributas. Is vero etiam gratum Tibi animum hoc publico monumento de *Prodromo Cometico*, quem ei per me miseras, testari se voluisse mihi dixit, nec pauca quae de Te vel Tui occasione de argumento Cometico inter nos locuti fueramus, inseruit. Caeterum post missas ad me hasce literas mihi heri adfuit,

1 edendae...*Cometographiae* : ad edendam Tuam *Cometographiam* **TC Olhoff**.
2 incumbere **TC Olhoff** *om.* **BO**.
3 (de...accepi) *om.* **Olhoff**.
4 nondum **TC** (p. 533).
5 Nullus apud nos...ille *om.* **TC Olhoff**.
6 *Cometam*...adornatum : *Cometam* suum versibus doctissimis adornatum **TC Olhoff**.

et plura exemplaria *Cometae* istius sui Tibi aliisque Amicis promittit, quae ego per Amicum notum, Deo volente, diligenter curabo. Quae Cl[arissimus] Vir Erasmius Bartholinus nuper de Cometis commentatus est, jam, ut puto, videris. Honorificam quidem Tui mentionem ibidem facit. Sed in sententiam de gemino Cometa in Corvo et Lepore inclinat. De qua judicium Tuum, si fas est, libenter intelligam, una cum aliis Amicis hac in urbe degentibus. Cl[arissimo] Buthnero, permultis impeditus negociis necdum rescribo: sed pl[urimam] salutem ei adscribo. Dabo operam, ut meo erga illum officio defungi quamprimum possim, certus humanitatis et aequitatis eius. // Heinsianas includo ad Dominum Pelsium mittendas: qui quia tacet, et nos tacebimus. Nam si ei intutum est scribere, intutum erit et literas recipere.[7] Vale Vir Amplissime et Humanissime, rem feliciter age, ac mihi constanter fave, qui sum

Cl[arissimo] Nomini Tuo
devotiss[imus]

T[ibi] N[otus]
manu propria

Hamb[urgi] 25. Aug[usti] 1665.

Greetings, most noble, worthy, and renowned Sir! Because you do not write nor answer me, I justify it with your serious works. I do not doubt that you devote every effort and toil with all your might to publish your *Cometographia* — this magnificent monument of your outstanding talent and exceptional work is expected eagerly and impatiently by the learned world. Believe me, everyone's minds, ears, and eyes are turned towards it as towards a wonder of astronomical talent, and everyone avidly anticipates to see it as soon as possible. And do not think I flatter you. Because I adduce not only the opinion of the whole astronomical senate (where you are an unparalleled leader) but also that of the Republic of Letters. The testimony is given by what I sent you: last time from the most illustrious Rautenstein,[8] the most noble Heinsius,[9] and other great men, and recently from the most renowned Jesuit scholars: Riccioli[10] (about whose health I have not received anything since that time), Kircher[11] and

7 qui quia...recipere : easdemque commendo **TC Olhoff**.
8 See note to Letter 5.
9 See section 2.4.6 of the Introduction.
10 See section 2.4.15 of the Introduction.
11 See section 2.4.8 of the Introduction.

Schott,[12] and the most learned Gruterus[13] as well. This is what the most renowned Boulliau[14] (displeased with giving not enough praises of your most excellent talent) wrote to me in his last letter, sent on 14 August: 'There is no one here, except for Agarrat,[15] who made observations of the comet with instruments. He will publish them soon. Therefore, I find no one's work sufficient enough to investigate what we are searching for. The Reverend Father Grandami has also written a booklet,[16] but he does not prove what should be proven. Only the observations of the most renowned Hevelius (which are the most reliable) will be used.'[17] These are his words. And this is a solemn and great (but equal to your merits and earned with them) testimony given by the most renowned Kirsten, who sends through me this letter and his *Cometa*,[18] full of beautiful verses and dedicated to you. I congratulate you on these high praises of your astronomical knowledge, similar to those of Copernicus, Tycho, and Kepler, which are ascribed to you by this man, very learned and experienced in astronomy, and by other most distinguished men as well. He also told me that with that public testimony he wanted to express his thanks to you for the *Prodromus cometicus* (which you had sent to him through me); and he included there a lot of what had been said about you, or related to you in the context of this cometary discussion. Moreover, having sent me this letter, he visited me yesterday and promised more copies of his *Cometa* which I will take care to send to you through a friend you know, if God allows it. What the most renowned Erasmus Bartholin has recently stated about comets,[19] you will see, I think. He also makes there an honorific mention of you. But he leans towards the opinion about two comets — one in Corvus and one in Lepus. Other friends and I, who live in this city, would willingly like to learn your opinion about it, if you do not mind. As many obligations impede me, I cannot reply to the

12 Gaspar Schott (1608–1666), German Jesuit scholar, student, and assistant of Athanasius Kircher. He studied in Würzburg. In the years 1631–52 he taught at the Jesuit school in Palermo. Between 1652–65, he worked in Rome with Kircher. In 1666, he was again in Würzburg. Schott worked on mathematics and mechanics. He was Lubieniecki's correspondent; in the TC pp. 761–96, there are 14 letters they exchanged (6 from Schott, 8 from Lubieniecki).

13 Isaac Gruterus (1610–1680), Dutch philologist, lawyer, and civil servant. He was a teacher and provost in classical schools in Middelburg and Nijmegen. He was Lubienicki's correspondent; in the TC pp. 863–80, TC3 pp. 69–73, 83–84, there are 28 letters they exchanged (14 from Gruterus, 14 from Lubieniecki).

14 See section 2.4.2 of the Introduction.

15 Antoine Agarrat (*fl.* 17th century), French astronomer and Pierre Gassendi's assistant. I could not identify his supposed publication of cometary observations. See also CJH3 p. 270, note to Letter 66, des Noyers to Hevelius, 4 February 1655.

16 Jacques Grandami (1588–1672), French Jesuit astronomer. See also 'Subsidia de cometis', CJH2 p. 463. The publication in question is his *Le Parallèle des deux comètes qui ont paru les années 1664 et 1665* (a Paris: chez Sebastien Cramoisy, et Sebastien Mabre-Cramoisy, 1665).

17 TC p. 533.

18 Michael Kirsten (see section 2.4.9 of the Introduction), *Cometa, poema ad Hevelium* (Hamburgi: apud Zachariam Hertelium, 1665).

19 Erasmus Bartholin (see section 2.4.1 of the Introduction), *De cometis anni MDCLXIV et MDLXV opusculum* (Hafniae: apud Danielem Paulli, 1665).

most renowned Büthner,[20] but I pass many greetings for him. Sure of his kindness and justice, I will take care to fulfil my duty towards him as soon as I can. I enclose a letter from Heinsius to Pels.[21] Because he is silent, so will we. If it is unsafe for him to write, it will also be unsafe for him to receive letters. Farewell, greatest and most kind Sir; be prosperous and always benevolent towards me, the one who is

renowned Sir,

Yours devoted,

the one whom you know
by my own hand

In Hamburg, on 25 August 1665.

20 See section 2.4.3 of the Introduction.
21 See note to Letter 5.

29.

28 August 1665, Lubieniecki to Hevelius

BO: C1, v. 7, 1033
TC: 402–03

Enclosure:
29A. Excerpt of the letter from Busmann to Lubieniecki, 21 August 1665

Salutem Plurimam Nobilissime, Amplissime, Clarissime Vir. Quamvis nondum quidquam responsi ad aliquot meas abs Te acceperim (quod uti occupationibus Tuis gravissimis et assiduis adscribo, ideoque haudquaquam miror, ita Te mihi captato temporis momento commodo brevi responsurum spero) et proxime scripserim, tamen cum ex eo tempore Cl[arissimi] Viri Domini Busmani[1] Hannoverensis Medici et Physici[2] celeberrimi literas ante octiduum scriptas acceperim, volui earum copiam Tibi facere.[3] Cum enim[4] de Te loquantur, imo Te alloquantur, facies pro eximia Tua humanitate rerumque naturalium scientia, ubi per graviora negotia Tua licuerit, ut quid ad haec respondere debeam, ex Te intelligam. Cumprimis vero ex Te cognoscere aveo, nec ego tantum, sed et plures Amici, hac in urbe degentes, an sperare debeamus posterioris Cometae *Prodromum*, quem nobis Tuae literae promittebant m[ense] Maio ad me scriptae, dummodo id votis et studiis nostris congrueret. Quod ita esse, aliquot literis meis Tibi testatum esse memini. Nam cum magno desiderio nihil satis unquam festinetur,[5] perquam libenter fieri vellemus, ut longae morae, antequam ingens illud et mirandum *Cometographiae* opus[6] ultimam manum experiatur et in lucem prodeat, subeundae et tolerandae taedium *Prodromus*, sicut nuper factum esse gratulamur, promta suavitate detergat. Itaque priusquam Tibi gravissimae occupationes ad priorum literarum mearum capita respondere permittant, velim verbo innuas, an et quam cito hunc Praecursorem exspectare debeamus. Fasciculum exemplorum *Cometae* Kirsteniani jam promovi Lubecam redemptum.[7] Hodie inter 12. et primam horam visa est ad Austrum stella splendida, quae forte erit Venus sine dubio jam pleno lumine fulgens: quam ideo sub luce Solis in distantia 60.

1 apud *del. post* Busmani **BO**.

2 Medicumi (um *del.*) et Physicumi (um *del.*) **BO**.

3 tamen…facere : tamen et nunc scribo. Nam ex eo tempore Clarissimi Viri Domini Busmanni, Hannoverensis Medici et Physici celeberrimi literas, ante octo dies scriptas accepi. Itaque volui earum copiam Tibi facere **TC**.

4 enim : autem **TC**.

5 magno…festinetur : magnum desiderium accelerari omnem festinationem velit **TC**.

6 opus **TC** *om.* **BO**.

7 redemptum *om.* **TC**.

fere graduum videri mirum non est. // Sed de his Tua sententia nobis[8] audienda, ac prius expetenda foret si Tuae occupationes fieri id vellent.[9] Vale et constanter fave

Tui studiosissimo

T[ibi] N[oto][10]

Hamb[urgi] 28. Aug[usti] 1665.

Greetings, most noble, worthy, and renowned Sir! I justify your lack of answer with your serious and constant works. Therefore, I am not surprised, and I hope that you will answer me soon, at the first given opportunity. Despite the fact that you still have not answered a few letters I sent, and that I recently wrote to you, I decided to send you a copy of the letter which was written eight days ago and which I have recently received from the most renowned man, Master Busmann,[11] a physician and physicist from Hannover. Since this letter talks about you (and even addresses you), out of your exceptional kindness and knowledge of nature, you will make sure, that I will learn from you how I should answer it, if you only can help me among your onerous duties. Firstly, I would like to learn from you (and not only me but many other friends who live in this city as well) if we should hope for a *Prodromus* of the second comet which you promised in your letter sent to me in May; it would match our wishes and expectations. I remember I attested it in a few letters. Since there is never enough hurry to satisfy a great desire, we would like your *Prodromus* to relieve with its natural sweetness the pains of enduring and tolerating this long delay, before that great work of the *Cometographia* is finished and published. Therefore, before your serious duties allow you to answer the main topics of my letters, I would like you to say just a word if and how soon we should expect this forerunner. I have already sent to Lübeck a package of Kirsten's *Cometa*[12] to be sold. Today, between 12 and 1 o'clock, a bright star was seen in the south (probably Venus), undoubtedly shining with its full brightness. It is no wonder that it was seen in daylight, about 60 degrees from the Sun. But we should listen to your opinion about this. And, first

8 nobis *om.* **TC**.

9 si...vellent : si id occupationes Tuae permitterent **TC**.

10 constanter...N[oto] : age prospere **TC**.

11 Christian Busmann (1610–81), German physician. He studied in Wittenberg and Helmstedt. He was the city physician of Hannover. In the TC pp. 641–85, there are 21 letters he exchanged with Lubieniecki (10 from Busmann, 11 from Lubieniecki).

12 Michael Kirsten (see section 2.4.9 of the Introduction), *Cometa, poema ad Hevelium* (Hamburgi: apud Zachariam Hertelium, 1665).

of all, we should ask if your duties allow you to answer. Farewell, and be benevolent towards

Yours devoted,

the one whom you know

In Hamburg, on 28 August 1665.

29A.

Excerpt of the letter from Busmann to Lubieniecki, 21 August 1665

BO: C1, v. 7, 1033 (follows the preceding one)
TC: 665–66

Excerptum ex litteris Cl[arissimi] Domini Busmanni
Hannovera die 21. Aug[usti] 1665 datis.

Non potui intermittere, quin significarem, me in absentia Domini Mulleri, Principis Nostri Consiliarii, Observationes, quas *Prodromum Cometicum* inscripsit, Joh[annis] Hevelii accepisse. Perlegi statim, et accuratissime delineatum phaenomenon novum animadverti, ut instar omnium Vir tam praeclare in caelestibus edoctus merito haberi debeat. Propter Telescopia enim magna et rara ea annotavit quae aliis abscondita manserunt. Percharum autem mihi est, quod tam insignis Astronomus eandem mecum assumpserit in plano adumbrationem. Persuasum hactenus fuit Mathematicis praeclarissimis, quoscunque Cometas circulum percurrere exacte maximum, quibus contrariari nefas duxi. Hunc lubens largior Hevelio, Cometam nuperum, de quo sermo est, sub notabili incurvatione cursum suum ultimis diebus reduxisse, et plus duobus tribusve gradibus a circulari orbita deviasse. An vero sub simili semper incurvatione ille reflexus vel recursus accidit, etiam Hevelius ipse, ex nullis iis omnibus bene multis Cometis, quos pervestigavit, id ipsum ullo modo penetrare potuit. Semper itaque addiscendum nobis est, nec ullus tam doctus, quin usus et aetas plura edoceat. Porro Cometas, ex mente Hevelii, ex halitibus fumidisque evaporationibus tam Solis quam reliquorum Planetarum omnium conflatos et compositos esse, quaestio est maximi ponderis et profundissimae contemplationis, inter Copernicanos quidem pro certo; at vero reliquis naturalium rerum indagatoribus, in aethere generationem et corruptio-//nem negantibus, pro magno paradoxo habita. Novissimum Cometen in Pegaso, novarum turbarum praenuncium in Polonia fuisse, res loquitur et historia porro loquetur.

An excerpt of the letter of the most renowned Master Busmann,
given at Hannover, on 21 August 1665.

I could not wait to tell you that when Master Mullerus, the advisor of our Duke, was absent, I received the observations of Johannes Hevelius titled *Prodromus cometicus*. I read it at once, and I saw the new phenomenon described precisely; that man (so exceptionally learned in celestial phenomena) should be undoubtedly worth as much as all others put together. Because he described, using his large and unique telescopes,[13] what for others was left hidden. I rejoice that such a famous astronomer adopted the same description as I did. So far, there was a firm opinion of the most renowned mathematicians that all comets move precisely in the great circle, and I found it improper to oppose. I readily agree with Hevelius that the latest comet, about which we are talking, in its last days changed its route at a notable curve and deviated 3 or 4 degrees from the circular orbit. Hevelius could not state (basing on not a few, but many comets he investigated) if this bending, or backward movement, always happens at a similar curve. Therefore, we should never cease to learn, as no one is so educated that experience and age cannot teach him more. Moreover, it is a matter of great importance and deep contemplation (certainly for Copernicans) whether, according to Hevelius's opinion, the comets are incited and formed from exhalations and smoky evaporations[14] — both from the Sun and all other planets. But for other investigators of nature (those who reject generation and corruption in Aether), it is considered to be pure nonsense. The facts prove, and the history will prove in the future that the latest comet in Pegasus was a harbinger of new unrests in Poland.

13 See note to Letter 3.

14 See section 1.2.1 of the Introduction.

30.

5 September 1665, Hevelius to Lubieniecki

BO: C1, v. 7, 1034
TC: p. 403

Illustrissimo Viro
Domino De Rolis
Johannes Hevelius Salutem

Veniam dabis pro summa Tua erga me humanitate et benevolentia variis occupationibus, gravissimisque laboribus, quos ratione recentiorum[1] Cometarum aggressus sum, cumprimis vero maximo meo moerori atque dolori, quem nuper ob praematurum et insperatum charissimi atque unici filioli mei obitum percepi, quod tardius quam merito debebam ad tuas plurimas mihi multo jucundissimas respondeam; easque ob causas etiam brevibus me ad singula tua quaesita expediam. Primo; phaenomenum illud, quod Romani observatores aliqui, tum Clariss[imus] Gravius ad borealem cinguli Andromedae viderunt minime[2] Cometa, sed stella tantummodo est nebulosa, quae Tubo tamen conspecta similem fere speciem Cometae exhibet; nec nova aliqua modo prognata est, sed a plurimis jam annis ibidem extitit: quippe a Simone Mario anno 1614 jam observata est, ut pluribus in descriptione mea posterioris Cometae edenda dicturus sum. Egregie itaque illusi[3] sunt Viri illi Eximii ab illa stella. Quod si vero negent, exeant, aspiciant Andromedam, eodem loco ubi antea illud ipsum invenient phaenomenum: nempe uno gradu paullo plus dissitum a boreali in cingulo, catenam versus Andromedae. Quaeris secundo an alterum *Prodromum* posterioris Cometae sim *Cometographiae* praemissurus? Constitui quidem ita, si Deo ita visum fuerit cum quibusdam iconismis; sed praepedior, cum papyrum // ad imprimendum Pyratae nuper mihi abstulerint. Hinc exspectandum erit, donec aliam mihi comparavero ex Batavia;[4] id quod autem vix intra mensem unum aut alterum fieri poterit. Interea observationes meas omnes tam prioris quam posterioris[5] Cometae, quas in *Cometographiam* hactenus[6] reservaveram, calculo subiiciam, quo simul eas tam nudas quam enodatas edere possim. Ad quod negotium Clariss[imus] Dominus Auzotius me impellit epistola quadam ad Clar[issimim] Petitum perscripta ac Parisiis recens[7] edita: qua partim meas observationes in dubium vocare suspectasque reddere satagit;

1 nuperum *del.* **BO**.
2 fuit *del. post* minime **BO**.
3 illusit (t *del.*) **BO**.
4 aliam...Batavia : aliam ex Batavia mihi comparavero **TC**.
5 omnes...posterioris : utriusque **TC**.
6 hactenus *om.* **TC**.
7 nuper *del.* **BO**.

sed annuente Divino Numine demonstraturus sum me Instrumentis meis omnia recte ac accuratissime observasse; atque ita Eruditis rem postmodum diiudicandam committam, quis nostrum ab ipso initio ad 14[8] Febr[uarii] rectius Cometam observaverit, et quisnam circa motum diurnum aberraverit. Tertio; de illa quae annotasti circa historiam Cometarum plene quidem hac vice respondere nequeo; sed brevibus saltem quaedam attingam. Reliqua Deo dante plenius ex Historia nostra Cometarum *Cometographiae* quoque inserenda suo tempore intelliges. Cometam illum qui anno Christi 380 apparuit Rockenbachius annotavit. De illo autem anno 145, haec te scire velim, quod contra mentem acciderit,[9] quod illum inter illos qui semestri spatio affulserunt reposuerim; quippe alium arripere volui. Nam hic non nisi per sex noctes adspectabilis fuit, autore Rockenbachio: sic ut in illo aberraverim. Quod autem facile fieri interdum potest, // dum enim ad plurima inter festinandum attenti sumus, accidit non[10] raro, ut in uno aut altero exorbitemus. Errare siquidem humanum, et nos homines admodum ad errores sumus proclives; vix puto ullum adeo felicem futurum, qui opus quoddam absque omni errore edere possit; imo dum nonnunquam errores annotamus alterius,[11] errorem committimus; id quod Tibi ipsi, amice honoratissime obtigit. Nam statim linea quarta eiusdem pagellae, ubi menda indicare voluisti, ipsemet exorbitasti.[12] Siquidem pag[ina] 10 lin[ea] 30 scribis: '<u>Progressus Cometae spatio 24 hor[arum] ponitur 5° 30', cum rectius ex mente authoris videatur notari debere 5' 30";</u>' quae tamen verba non pag[ina] 10 sed 16 inveniuntur.[13] Adhaec mi amice, quis attentus et harum rerum bene gnarus lector non prima fronte[14] perspicit[15] haec ita ex errore Typographi descendisse: cum consequentia et antecedentia loca illa bene illustrent. Etenim motus diurnus die 2 et 3 Febr[uarii] non extitit aliquot graduum sed saltem minutorum; sic etiam pag[ina] 16 lin[ea] 37 ubi expressa est longitudo Cometae 16° 20', statim subsequitur, '<u>vel saltem 26° 19'.</u>' Agnosco interea et Typographos et Correctores circa gradus et minuta dum signa diversa arripiunt ' pro ° et ° pro ', facile aberrare posse; id quod etiam in Tabula altera pag[ina] 21 factum est; sed quis adeo praeceps, qui primo intuitu illud non agnoscit. In Tabula vero priori circa motum diurnum 23° 30' nullus commissus est error; sed tu iterum ipsemet aberrasti dum ab 28 die Decemb[ris] hor[a] matut[ina] ad diem 29, horam 9 vespertinam, tantummodo 23 horas numeras, cum sint 43 horae: deinde tempora et horarum spatia // cuiusvis diei bene distinguenda sunt, a[16] quo termino hic motus adeo diversus et inaequalis numerari incipitur; quae clarius explicari possent, si res esset huius loci.[17] Pag[ina]

8 14 : 12 **TC**.
9 acciderit : cecederit **TC**.
10 nonnunq (nunq *del.*) **BO**.
11 annotamus alterius : alterius annotamus **TC**.
12 errorem commisisti *del.* **BO**.
13 Sed *del. post* inveniuntur **BO**.
14 statim *del.* **BO**.
15 ex *del. post* perspicit **BO**.
16 a : et a **TC**.
17 res...loci : vacaret **TC**.

29 lin[ea] 2 nulla ambiguitas: nam phaenomenum etiam retrogreditur, quando iter reflectit, etiamsi tunc procedat secundum s[eriem] signorum. Pag[ina] 18 lin[ea] 22 etiam recte loquutus sum: in directum nempe dicunt Astronomi quod s[ecundum] s[eriem] signorum] progreditur. Sic etiam pag[ina] 19 lin[ea] 12 et pag[ina] 4 lin[ea] 8, omnia recte se habent; sed necesse ut recte accipiantur, ac attente legantur. Reliqua ubi legi debet 'altus' pro 'alto'; 'hyperbolica' pro 'Pyperbolica',[18] 'statuunt' pro 'statuant' leviora sunt, quam ut de iis verba facere necessum sit. At pag[ina] 28 lin[ea] 22 non pro 'atque' legi debet 'adque'; sed pro 'collum', 'collo' legatur; sic et pag[ina] 3 pro 'ortum vergentem' legatur 'occasum'; quam[19] plagam reliqua verba subsequentia satis clare indicant, tum lapsum esse calami; verum de his satis. De caetero, rogo titulum historiae Cometarum Eckstormii mihi indices, vel potius si prostat eum autorem mihi emas pecuniam libenter reddam; mea enim interest eum legere. Imperiales illos 11 1/2 nondum accepi nec intellexi a quonam eos accipere debeam. Denique gratias habeo debitas pro communicatis quibusdam Eruditorum Virorum literis, quos omnes velim quam officiosissime resalutas. Has dum finio literas, posteriores Tuae die 25 Augusti datas mihi traduntur, testes pristini amoris atque singularis affectus tui, quem quo modo demereri queam, sane nescio. Interea gratias habeo quas possum maximas et Tibi et Clarissi[mo] Domino Kirstenio, inprimis quod *Cometam* suum, acutissimi ingenii sui foetum nomini meo inscribere, meque multis elogiis exornare, praeter omne meritum haud fuerit // dedignatus, ut suo tempore fusius declarabo, animo aliquanto a tristitia liberiore. Interea officia mea ipsi deferas, meoque nomine prolixissimas ipsi agas gratias etiam atque etiam rogo. Clar[issimi] Erasmi Bartholini *Cometam* vidi, miror sane in illa quaestione illum adhuc haesitare; non dubito tamen quamprimum observationes meas viderit, quin unicum tantum Cometum in Corvo et Lepore habiturus sit. Vale Vir Generosissime[20] et condona tristissimo animo quod[21] adeo confuse ad tua respondeam.[22]

Dabam Gedani anno 1665 die 5 Septemb[ris] st[ylo] n[ovo][23]

Postscriptum.

Dum obsignare volebam hasce, Tuae die 28 Augusti accipio, ad quas pariter brevissime respondeo. Observationes meas[24] recentiorum Cometarum cum dissertatione et descriptione praesertim posterioris Cometae, quam primum potero in lucem editurus sum. Opusculum enim maximi est laboris. Stellam quam videtis Venus est, quae quavis serena die nunc in maxima elongatione a Sole apparet. De deviatione Cometa-

18 'hyperbolica' pro 'Pyperbolica' **TC** : 'Pyperbolica' pro 'hyberbolica' **BO**.
19 quem ad quam (quem ad *del.*) **BO**.
20 Generosissime **TC** : Illustriss[ime] **BO**.
21 quod *om.* **TC**.
22 respondeam : respondenti **TC**.
23 Dabam...n[ovo] *om.* **TC**.
24 nec *del. post* meas **BO**.

rum satis fuse jam dictum est in *Prodromo* nostro p[agina] 28, 29 et 30 et adhuc plenius dicturus in ipsa *Cometographia*. Ad reliqua Cl[arissimi] et Erud[itissimi][25] Busmanni hac vice respondere nequeo, se omnia illa in *Cometographiam* reiicio. Novissimum vero Cometam praenuncium fuisse novarum turbarum in Polonia, valde dubito. Quibus iterum iterumque valeas, omnesque amicos nobis bene cupientes salutes peto.

To the most illustrious Sir,
Master De Rolis,
Johannes Hevelius sends greetings

Knowing of your enormous kindness and benevolence towards me, I ask you for forgiveness, as my various duties and serious works caused by the late comets consumed me. And first of all, I am affected by grief and pain because of the premature and unexpected death of my dearest only son.[26] Therefore, I answer your many pleasant letters to me later than I should. And due to these reasons, I will answer your questions one by one, however briefly. Firstly, that phenomenon which was seen by some observers in Rome and by the most renowned de Grau[27] near the northern star in the Belt of Andromeda[28] was not a comet but only a nebulous star which, through a telescope, looks quite similar to a comet. And it is not a new star which has recently come into existence, but it has been in that place for many years. It was seen by Simon Marius in 1614, as I will discuss in detail in my description of the second comet, which will soon be published. Therefore, those exceptional men were greatly deceived by that star. And if they do not agree, let them go out and see Andromeda; they will find that phenomenon where it was before, viz. one degree from the northern star of the Belt towards the Chain of Andromeda. Secondly, you ask if I plan to publish a *Prodromus* of the second comet before my *Cometographia*. I have decided (if God allows it) to do so with some illustrations, but I was impeded as pirates stole my printing paper. Therefore, I will have to wait until I get some new paper from the Netherlands, which will not happen sooner than in a month or two. Meanwhile, I will compute all my observations, both of the first and the second comet, which I have kept so far for the *Cometographia* so that I can publish them both with and without commentary. The most renowned Master Auzout forces me to do so because of a letter to the most renowned Petit, recently published in Paris.[29] In this letter he partly questions my observations and tries to make them look suspicious. But with the help of God's will, I will prove that I saw everything precisely and correctly with my instruments, and so I will leave this question for the learned men to decide which of us observed the co-

25 et Erud[itissimi] *om.* **TC**.

26 Johannes Adeodatus Hevelius (1664–5), see Chantal Grell, 'Hevelius en son temps', CJH1 p. 139.

27 See note to Letter 7.

28 See section 2.3.2 of the Introduction.

29 Adrien Auzout, *Lettre de monsieur Auzout du 17 Juin à monsieur Petit* (Paris: [s.e.], 1665). See section 2.2.4 of the Introduction.

met more accurately (since the beginning until 14 February) and who made a mistake about its daily movement. Thirdly, this time I cannot reply fully to what you have noted about the comets, but at least I will discuss some of those things in brief. You will learn the rest more thoroughly from our history of comets which, in due time, will be included in the *Cometographia*. That comet which appeared in the year 380[30] after the birth of Christ was noted by Rockenbach.[31] About the one of the year 145,[32] I would like you to know that it happened against my intention that I put it among those which shone for six months, because I wanted to choose a different one. The comet of the year 145 was visible for six nights only, according to Rockenbach, so it was my mistake. But such mistakes may easily occur because when we are busy with many things, it often happens that we err in one thing or another. To err is human, and we, as people, are very prone to mistakes. I think there will be no one so lucky to publish a book without any error. But sometimes when we note others' mistakes we make a mistake ourselves, as it happened to you, most honoured friend. Because at the beginning, in the fourth line of that page where you wanted to indicate mistakes, you made one yourself. About page 10, line 30, you write: 'The movement of the comet within 24 hours is stated to be 5° 30' while it seems that according to the author's opinion it should be given as 5' 30"' while these words can be found not on page 10, but page 16. Moreover, my friend, what kind of a reader (surely diligent and experienced in these matters) — would not see at first glance that it happened due to a typesetter's mistake as the text before and after it makes it clear? And daily movement on 2 and 3 February did not exceed a few degrees but only minutes. And also on page 16, line 37, where the longitude of the comet is given as 16° 20', just below it is written 'or 26° 19' only.' I realise that typesetters and proof-readers can easily make a mistake in degrees and minutes, since there occur various symbols, ' instead of ° and ° instead of '. It happened in the second table on page 21. But who is so hasty that he does not comprehend it at a glance? And in the second table, there is no mistake about 23° 30' of daily motion, but you erred yourself since you count only 23 hours from the morning hour of 28 December until 9 o'clock in the evening of 29 December, while there were 43 hours in fact. Moreover, the time and number of hours of a day should be distinguished from the point when this movement (so odd and irregular) begins. It could be explained more clearly if it was a proper place for it. On page 29, line 2, there is no ambiguity, because the phenomenon also has a retrograde movement when it inclines its course, even if it moves then according to the order of signs. On page 18, line 22, I was also right because astronomers say that a thing moves directly when it moves according to the order of signs. And also on page 19, line 12, and on page 4, line 8, everything is accurate, but it should be understood correctly and read carefully. Other places (where one should read 'altus' instead of 'alto,' 'hyperbolica' instead of 'Pyperbolica,'

30 See note to Letter 27.

31 Abraham Rockenbach (1536–1611), German philologist and mathematician, professor at the University of Frankfurt an der Oder. Hevelius refers to his *De cometis tractatus novus methodicus* (Witebergae: ex Officina Cratoniana, sumptibus Henningii Grosii, 1602).

32 See note to Letter 27.

'statuunt' instead of 'statuant') are of lesser importance, so there is no need to discuss them. And on page 28, line 22, 'adque' should not be read instead of 'atque,' but 'collo' should be read instead of 'collum.' And on page 3, instead of 'moving eastwards' should be 'westwards.' The text which follows suggests that direction quite clearly, so it was a scribal error. But enough about it. As far as the rest is concerned, I ask you to tell me the title of the history of comets by Eckstorm,[33] or rather (if it is available) to buy this book for me. I will readily return you the money because I would like to read it very much. I have not received those 11 1/2 Reichstalers yet, nor have I got to know from whom I should receive them. Finally, I send much-deserved thanks for the letters you sent me from the learned men. I would like you to pass my warm greetings to them. As I am finishing this letter, a more recent one from you, sent on 25 August, is being delivered to me. It testifies your earlier benevolence and exceptional affection towards me, leaving me clueless as for what I could do in return. Meanwhile, I am as thankful as I can to you and to the most renowned Master Kirsten,[34] especially because he found me worthy (more than I deserve) of dedicating his *Cometa* to me, the offspring of his most acute talent, and of granting me so many praises. I will write more in due time, when my mind is freer of sadness. Meanwhile, I ardently ask you to pass my regards to him and to thank him very much on my behalf. I have seen the *Comet*[35] of the most renowned Erasmus Bartholin, and I wonder why he still sticks to that opinion, but I do not doubt that as soon as he sees my observations, he will agree that there was only one comet in Corvus and in Lepus. Farewell, most generous Sir, and forgive the troubled mind that it answers your letters so disorderly.

Given at Danzig, on 5 September 1665, new style.

Postscript.

When I wanted to seal this letter, I received the one you sent on 28 August, which I answer briefly. I plan to publish (as soon as I can) my observations of the recent comets, especially of the last one, with the discussion and description. Such a booklet requires much work. The star you are seeing now is Venus which, in its greatest elongation, is visible during a calm day. About the irregular movement of comets, I have already said enough in our *Prodromus*, on pages 28, 29, and 30, and I will say more in the *Cometographia*. I cannot answer other questions from the most renowned and learned Busmann[36] this time, but I leave them all for my *Cometographia*. I doubt the latest comet was a sign of the recent unrests in Poland. I wish you good health and ask you to greet all friends who are benevolent towards us.

33 See note to Letter 27.

34 See section 2.4.9 of the Introduction.

35 Erasmus Bartholin (see section 2.4.1 of the Introduction), *De cometis anni MDCLXIV et MDLXV opusculum* (Hafniae: apud Danielem Paulli, 1665).

36 See note to Letter 29.

31.

11 September 1665, Lubieniecki to Hevelius

BO: C1, v. 7, 1037
TC: 405–06

Seal with Lubieniecki's arms (Rola coat of arms), address: 'A Mons[ieur] / Mons[ieur] Jean Hevelius / Senateur et Grand Astro-/nome / à Dantzic.' ('To *Monsieur* Johannes Hevelius, Councillor and great astronomer, in Danzig.')

Salutem Plurimam Nobilissime, Amplissime, Clarissime Vir.

Minime opus erat, ut solicite silentium Tuum apud me excusares, qui bene novi quantos labores in edenda Tua *Cometographia*, assiduo[1] subeas. Fateor tamen mihi longe gratissimum responsum Tuum exstitisse, hoc solo ingratum, quod tristem de excessu carissimi et unici filioli Tui nuncium attulerint. Doleo sane et ego (dolebunt et omnes Amici Tui, hic et alibi agentes, postquam id acceperint) immatura filioli Tui morte, matura tamen, utpote vitae datoris et rerum arbitri decreto sancita. Huius voluntati subsumus, et submittamus[2] nos per omnia oportet, ut gratiam eius inexhaustam ad tollenda, et reparanda damna modesta patientia provocemus.[3] Atque hoc Tibi ex amicitiae et officii homini Christiano incumbentis[4] ratione sincere precor. Doleo et damnum illud a piratis maria hac tempestate infesta habentibus illatum, tanto plus quod et publicum cum privato coniunctum sit desideratissimi operis, (talem enim esse Cometae posterioris *Prodromum* hic et ubique, certus esto) editionem retardaturum. Observationem Tuarum utriusque Cometae editionem, quam seorsum nobis dare promittis, pariter cum ingenti desiderio exspectamus. Varia Te jam experiri hominum judicia non est quod mireris. Nam

– quot capitum vivunt, totidem studiorum
Millia.

1 assiduo : assidue **TC**.

2 submittamus : eidem submittamus **TC**.

3 provocemus : provocemus. Merito huc referam illa Senecae: 'Sapiens ad omnem incursum minitus et intentus non si paupertas, non si luctus, non si ignominia, non si dolor impetum faciat, pedem referet. Interritus et contra illa ibit et inter illa.' Ep[istula] 59. tum et illa: 'Non affligitur sapiens liberorum amissione, non amicorum. Eodem enim animo fert illorum mortem quam suam exspectat' Ep[istula] 74. At quanto maiori virtute, universum quoque mundum vincente, induit nos Christiana disciplina, non tantum solatiorum hic in terra, sed et aeternorum gaudiorum in caelo consequendorum monstrans iter. **TC**.

4 incumbentis : convenientis **TC**.

Atque hoc omnium seculorum testatur memoria. Ideoque Vir bonus et sapiens,[5] invidia quoque et iniquitate aliorum, // sed nec ullo animi metu ambiguo, ab honesto proposito absterreri, et vel dimoveri se patitur, quin potius destinatum forti et constanti carpit iter pede

– – – multa laborum
Praemia laturus.

Hoc etiam ad Te, Magne Heveli, refer, et in Te virtutis cultoribus[6] exhibe. Intrepido mentis corporisque oculo adversa praesentia vel imminentia et minantia specta, et omen quidem Deus prohibeat; caeterum in ea et Tu natus es tempora, in quibus animum firmare expedit constantibus exemplis. Sed quod apud Te hic dico, ad me quoque invidiae malevolorum expositum et varia itidem judicia mox experturum, pertinere duco. Quis vero aequus nescit vel fateri nolet cuique viro honesto liberam non solum mentem, sed et linguam, virtutis praesertim studia commendantem, esse debere? Quod mihi adeo modeste et placide ad ea, quae occasione *Prodromi* Tui proposueram, responderis gratissimum est. Fateris enim quaedam inter tot gravissimos labores incogitanti Tibi excidisse, alia typographorum et correctorum incuria subrepsisse[7] quale et in me, vel meo scriba (id enim recte non memini, uter nostrum illa exararit) notas. Ita nos 'labi et errare, non solum posse sed et debere,' ut dicit et docet alicubi Magnus in Rep[ublica] literaria Vir J[ustus] Lipsius,

Scimus et hanc veniam petimusque damusque vicissim.

Velle Te *Cometographiae* Tuae Historiam Cometarum addere, valde rei literariae gratulor, quamvis idem argumentum sim tractaturus, ut[8] nosti. Nisi tot Amicis fidem in eo addixissem, lampada Tibi porrigerem. Sed nosti in juvanda re literaria tentandum esse omnibus pro virili. Nam recte ille dixit, 'Non omnia possumus // omnes.' Eckstormius *Historias Eclipsium, Cometarum et Pareliorum* Helmaest[ii] A[nni] 1621 edidit. Libenter Tibi hunc librum comparabo, si potero. Cum Cl[arissimo] Kirstenio Tuam epistolam hodie communicabo.[9] Quod ad observationem Romanam et Franequeranam phaenomeni ad cingulum Andromedae Boreal[em] attinet, gratias Tibi amplissimas ago de solido responso. Haec pauca, quae repono ad Tuas hodie mihi

5 sapiens : sapiens, quem ex Seneca paulo ante descripsi, nequaquam **TC**.
6 cultoribus : cultoribus, bono exemplo ad optima quaeque et laudabilia excitandis **TC**.
7 excidisse *del.* **BO**.
8 ut **TC** *om.* **BO**.
9 communicabo : communicavi **TC**.

redditas, e vestigio et manu itidem festina, velim boni et aequi consulas. Vale[10] et fave constanter

Tuo

observantissimo
et studiosissimo

Rolitsio
manu propria

Hamb[urgi] 11. Sept[embris] 1665.

Greetings, most noble, worthy, and renowned Sir!

There was no need for your afflicting excuses for your silence because I know how many endeavours you constantly undertake to publish your *Cometographia*. I admit your answer was very pleasant for me, and the only unpleasant thing was that it brought this sad information about the death of your dearest and only son.[11] I also grieve (and all your friends, wherever they live, will grieve when they learn about it) at the premature death of your son — but not premature, since it was decreed by the One who gives life and rules over all things. We are subject to His will, and we should obey Him in everything. In this way, we may call His inexhaustible grace to remove and expiate our sins with deserved penitence. And I sincerely pray for you, prompted by friendship and duty, as every Christian should. I also grieve at a loss caused by the pirates who nowadays dominate the seas, especially because it will delay the publication of your work which is much expected both publicly and privately (I am sure there is such a common expectation for the *Prodromus* of the latest comet). We await with equal eagerness the publication of your observations of both comets — you have promised to give it to us separately. There is no wonder that you will experience various judgements from people. Because

> as many thousands of people, so many different inclinations there are.[12]

And the memory of all centuries testifies this. Therefore, a good and wise man should not let himself be repelled or driven away from an honest plan by jealousy and ine-

10 Vale : Vale, feliciter age TC.
11 See note to Letter 30.
12 Horace, *Satyrae*, 2, 1, 27–28, transl. by C. Smart, Theodore Alois Buckley, in Horace, *The Works of Horace Translated Literally into English Prose*, vol. 2 (Philadelphia: Joseph Whetham, 1836). Quoted with minor changes.

quity of others, nor by any uncertain fear of his spirit. He should rather walk surely and confidently towards what he intended,

> certain of bearing off ample rewards for his pains.[13]

Apply it to yourself, great Hevelius, and show it to those who admire the virtue in you. With a fearless eye of mind and body, face what is present and imminent and watch what is dangerous, and may God keep the omen away. After all, you were born in the times when it is useful to strengthen the spirit with constant exampla. But what I say here about you also applies (I believe) to me, as I am exposed to jealousy of people of ill will, and I will soon likewise experience various judgements. Who of the righteous does not know or does not want to admit that an honest man should have not only a free mind but also a tongue to encourage the studies of virtue? It is most agreeable that you have answered so sensibly and kindly the questions which I asked about your *Prodromus*. You admit some of the errors I noted had been made unknowingly, as you were compelled to complete many serious tasks; others have sneaked into it due to the negligence of typesetters or proof-readers, and you ascribe some of them to my secretary or me as well (I do not recall which one of us wrote it). And so, as Justus Lipsius (a great man in the Republic of Letters) says and teaches somewhere, 'that we not only can but we also should fall and err'[14]

> we are conscious of this, and this privilege we demand and allow in turn.[15]

I congratulate the Republic of Letters on the fact that you want to add the history of comets to your *Cometographia*; however I also want to discuss this subject, as you know. If I had not made so many promises about it to my friends, I would have passed the baton to you.[16] But you know that in promoting sciences and arts everyone should try as hard as they can. Because these words are right: 'we cannot all do everything.'[17] Eckstorm published *Historiae eclipsium, cometarum et pareliorum* in Helmstedt in 1621.[18] I will readily buy you this book if I can. Today I will pass your letter to the most renowned Kirsten.[19] As for the observations from Rome and

13 Horace, *Satyrae*, 2, 1, 11–12, transl. by C. Smart, Theodore Alois Buckley, in Horace, *The Works of Horace Translated Literally into English Prose*. Quoted with minor changes.

14 Justus Lipsius (1547–1606), Flemish philosopher, classical scholar, and political thinker. Lubieniecki quoted his *Politicorum sive civilis doctrinae libri sex* (Amsterdami: apud Guilielmum Bleau, 1632), p. 18 (with minor changes).

15 Horace, *De arte poetica*, 11, transl. by C. Smart, in Horace, *The Works of Horace Translated Literally into English Prose*.

16 Literally 'passed the torch' (*lampada porrigerem*), an allusion to the torch race at some of ancient Greek games (*lampadedromia*, λαμπαδηδρομία).

17 Virgil, *Eclogae*, 8, 63, transl. by H. Rushton Fairclough, in Virgil, *Eclogues. Georgics. Aeneid I–VI* (Cambridge, Mass., London: Harvard University Press, 2004).

18 See note to Letter 27.

19 See section 2.4.9 of the Introduction.

Franeker about the phenomenon near the northern star in the Belt of Andromeda, I am grateful for your detailed answer. I would like you to accept my short response to your letter (which I received today) honestly and sufficiently. I was replying immediately and in haste. Farewell, and always be benevolent towards

Yours obliged and devoted,

Rolitsius
by my own hand

In Hamburg, on 11 September 1665.

32.

15 September 1665, Lubieniecki to Hevelius

BO: C1, v. 7, 1038
TC: 406

Enclosure:
32A. List of errors in the *Prodromus cometicus*

Salutem Plurimam Vir Nobilissime, Amplissime, Clarissime.

Insigni mihique tot experimentis probatae humanitati Tuae confisus proximis literis subiicio maiori cum cura descriptum Observationum nostrarum ad *Prodromum* Tuum exemplar. Addidit mihi hic animum et modestia Tua, et erga me benevolentia, ac rei literariae juvandae studium singulare, cuius nuperrimae Tuae literae manifesta praebent documenta. Quibus de caussis volo, si Tu id fieri velis, Prodromo *Prodromi* Tui a me Italorum Amicorum caussa composito, addere pauca illa, quae in illo libro Tuo censes esse corrigenda. Hoc enim non tantum de longe lateque celebrata clarissimi nominis Tui fama vel minimum decerpere non poterit, sed ut magnam eruditionis, ita maiorem modestiae et studii publicae utilitatis laudem Tibi conciliabit. Adderem enim illa pauca quae Tibi tot studiis coelestibus dies noctesque distracto et festinanti exciderunt vel per incuriam Typographi[1] vel Correctoris irrepserunt,[2] Te consentiente et volente correcta esse: reliqua, si quae occurrerint, ab erudito et aequo Lectore, facile corrigenda. Qualia et illa sunt ad marginem in fine pag[inae] 52. et initio 53. posita, 'Ex eo quod cometae sunt corpora discea' et c[etera] quae ad proximum paragraphum 'Ideoque,' pertinent et hic posita, illic omittenda sunt. Non amplius recordor quid tunc p[agina] 4. lin[ea] 8 annotarim. Non enim hanc annotationem in meis tabulis reperio. Tu si id tanti esse judices, me monebis. Cometas novos literae Neoburgenses et Amsterodamenses ferunt. Illic enim visa fuisse insolita stella fertur, quam talem esse Curtius negat. Forte quidam Jovem vel Venerem, dum radii eius per nubeculam oblongam et e regione astri visusque nostri subiecta distenduntur, pro Cometis ducunt, // quo eos et aliquot Cometarum visorum experientia et de pluribus sparsa fama facile inducere potest. Eckstormii librum Tui caussa heri quaesivi. Sed frustra fui. Videbo an alia aliqua ratione eum Tibi comparare queam. Meretur enim istius Viri industria Tuae adiungi eidemque famulari. Licet vero ille singulari diligentia annotet Cometas non tamen illum quadrimestrem Anni 380. annotavit. De quo quod me certiorem reddideris, gratias Tibi ago debitas. Velim autem etiam respondeas de illo octimestri Cometa, quem in Ann[um] 375. ponis, cuius ego

1 Typographiorum (orum *del.*) **BO**.
2 vel...irrepserunt *om.* **TC**.

hanc durationem non possum apud Auctores, qui ad manum sunt, invenire. At dixi invenisse me alium octimestrem A[nn]o 975. Dic itaque verbo quae de illo Tua sit sententia. Cl[arissimus] Kirstenius intra octiduum ad nuptias secundas, utinam secundo omine! transibit. Cl[arissumum] Mullerum nudius tertius vidi et compellavi. Humanissime Te salutat. Illam stellam, quae ante paucos dies coelo sereno visa est, Cor Leonis esse putabat. At dixi illam Tibi esse Venerem, quod etiam Eruditus ille Vir Juvenis Syversius initio affirmavit. Vale Vir Amplissime et Humanissime. Cl[arissimum] Büthnerum saluta, silentium diuturnum inter tot labores excusa, vicem officii tamen ei promitte. 'Vivite felices memores et vivite nostri.'

Clarissimo Nomini Tuo
addictissimus

Rolitsius
manu propria

Hamb[urgi] 15. Sept[embris] Gregor[iani] 1665.

Greetings, most noble, worthy, and renowned Sir!

Assured of your kindness (great and proven so many times) I add to this letter a copy of our observations about your *Prodromus*. I wrote it with great diligence. Your modesty and benevolence towards me and your efforts in promoting the sciences and arts (testified by your last letter) encourage me to do so. Because of these reasons (if you wish) I want to add something which you think should be corrected. I want it to be included in the forerunner of your *Prodromus*, written by me for the Italian friends. It will not cause a detriment to the widely celebrated fame of your most renowned name; instead, it praises you both for the erudition and your great modesty, as well as for your efforts for the public good. Because I would add only those few mistakes which have escaped your notice (as a result of your constant preoccupation with the celestial studies, or out of haste), or which have sneaked into your work due to the negligence of typesetters or proof-readers, and which should be amended according to your judgement and wishes. Others, if there are any, will be easily corrected by a learned and just reader. There is a note in the margin at the end of page 52 and at the beginning of page 53, reading: 'From the fact that comets are discoid bodies' and so on, which concerns the next paragraph, 'And therefore,' and should be placed next to it and omitted in its current place. I do not recall what I noted about page 4, line 8. I cannot find this note in my tables. If you think it is important, you will tell me. The letters from Neuburg and Amsterdam bring new comets. It is said that an extraordinary star was seen there, but Curtz[3] denies it. Perhaps they consi-

3 See section 2.4.5 of the Introduction.

der Jupiter or Venus to be comets — when their rays pass through a long cloud and extend to our sight from the stellar region. The experience of a few comets we saw and the prevailing opinion about many others can easily suggest such a thing. Yesterday, I looked for Eckstorm's book[4] for you, but in vain. I will see if I can get it for you in some other way, because that man's talent is worth merging with yours and serving yours. Even though he lists comets with extraordinary diligence, he did not note the four-month one of the year 380.[5] I am very thankful for your information about it. But I would like you to answer me about that eight-month comet which you dated back to the year 375,[6] whose duration I cannot find in the books I have here. And as I said, I found another eight-month one, in the year 975.[7] Therefore, tell me briefly what you think about it. In eight days, the most renowned Kirsten[8] is getting married for the second time; may it be a good time! Three days ago, I saw the most renowned Müller[9] and I talked to him. He sends you his kind greetings. He thought that the star seen a few days ago when the sky was clear was the Heart of Leo. And I said it was Venus because that learned young man, Sivers,[10] initially claimed so. Farewell, greatest and kindest Sir! Greet the most renowned Büthner,[11] apologize to him for my long silence caused by so many works, but promise him that I will fulfil my duty. 'Live happily and remember us.'[12]

Renowned Sir,

Yours devoted,

Rolitsius
by my own hand

In Hamburg, on 15 September, Gregorian style, 1665.

4 See note to Letter 27.
5 See note to Letter 27.
6 See note to Letter 27.
7 See note to Letter 27.
8 See section 2.4.9 of the Introduction.
9 See section 2.4.11 of the Introduction.
10 See note to Letter 20.
11 See section 2.4.3 of the Introduction.
12 Tibullus, *Elegiae*, 5, 3, 31.

32A.

List of errors in the *Prodromus cometicus*[13]

BO: C1, v. 7, 1038 (follows the preceding one)

Pagina 3. pro voce Ortum, ponendam puto vocem Occasum.

Pag[ina] 8. lin[ea] 10. Legitur 'alto,' pro 'altus': sed levissimum hoc

Pag[ina] 10. lin[ea]. 10. Author sine dubio Ascens[ionem] rectam vult esse 40°. 20', quam Typographus ex errore Typothetae 4°. 20'. fecit.

Pag[ina] 16 lin[ea] 30. Progressus Cometae spatio 24. horarum ponitur 5°, 30'. cum rectius ex mente Authoris videatur notari debere 5', 30".

Pag[ina] eadem lin[ea] 37. Longit[udo] Cometae 16°. 20'. Arietis decem gradibus justo minor ex contextu cernitur.

Pag[ina] 19. lin[ea] 11. Longit[udo] Cometae non apposito Dodecatemorii Arietis indice indeterminata offertur.

Pag[ina] 21. In Tabula priori, a 28.^{vi} Decemb[ris] hora 2.^{a} mat[utina] ad 29.^{ni} horam 9.^{am} 30' (hoc est, intra spatium 43. horarum et 30'.) ponitur motus Cometae proprius 23°. 40'. Cum altera Tabula pro duobus integris diebus (28. scil[icet] et 29. Dec[embris]) intra spatium 48. horarum exhibeat motum proprium 23°. 45'. Puto igitur ex errore irrepsisse 23° pro 21. et c[etera].

Pag[ina] eadem. Tabula altera notat in columna 3.^{a} (a sinistris) pro 'motu diurno' 'Min[utae] sec[undae]' quae rectius 'Grad[us] Min[utae]' notari debent.

Pag[ina] 26. lin[ea] 27. Praepositio 'ad' pro 'a' posita sensum obscurat. //

Pag[ina] 28. lin[ea] 22. Occurrit 'Pyperbolica' pro 'Hyperbolica,' quod levissimum. Sed linea antepenult[ima] 'atque' pro 'adque' sensum ambiguum facit.

Pag[ina] 29. lin[ea] 1. Particula redditiva 'Quantum' videtur innuere, quod lin[ea] 3. inter 'affirmare' et 'Cometam' forte omissum sit 'statuo' vel aliud quoddam verbum, pro integritate sensus orationis redditivae.

13 See also Letter 27D.

Pag[ina] eadem lin[ea] 2. Ambiguitatem admittit Author, dum motum Cometae inversum quidem, s[ecundum] s[ignorum] seriem tamen tendentem, quem exinde pag[ina] 18. lin[ea] 22. et 19. lin[ea] 12. directum dixit, vocat retrogradum contra notionem retrogradationis pag[ina] 4. lin[ea] 8. a fine et pag[ina] 18. lin[ea] 23. notatam.

Pag[ina] 30. lin[ea] 20. Valor numeri 21 1/2 dubius est, propter notam eius neglectam. Ex Tabula tamen pag[ina] 21. suppleri potest.

Pag[ina] 31. lin[ea] 12. pro 'statuant,' ponendum 'statuunt': levis error.

On page 3 I think it should be 'Occasum' instead of 'Ortum.'

On page 8, line 10. There is 'alto' instead of 'altus,' but it is not important.

On page 10, line 10. Without a doubt, the author wanted the right ascension to be 40° 20', which the printer, due to a typesetter's mistake, gave as 4° 20'.

on page 16, line 30. The movement of the comet within 24 hours is stated to be 5° 30', while it seems that according to the author's judgement it should be written as 5' 30".

On the same page, line 37. The longitude of the comet given as 16° 20' of Aries seems 10 degrees too small, as the context suggests.

On page 19, line 11. The longitude of the comet is given without the symbol of the sign of Aries.

On page 21. In the first table, from 28 December, 2 o'clock in the morning, to the 29th, 9.30 o'clock (this is within 43 hours and 30 minutes), the proper motion of the comet is given to be 23° 40', while the second table for the two whole days (namely 28 and 29 December) within 48 hours shows the proper motion to be 23° 45'. Therefore, I think there have erroneously sneaked in 23° instead of 21° and so on.

On the same page. In the second table, in the third column (from the left), titled 'Motu diurno,' there should be written 'Gradus, Minutae' instead of 'Minutae, Secundae.'

On page 26, line 27. Preposition 'ad' instead of 'a' makes the passage unclear.

On page 28, line 22. There is 'Pyperbolica' instead of 'Hyperbolica' which is not important. But in the antepenultimate line, 'atque' instead of 'adque' makes the sense ambiguous.

on page 29, line 1. A redditive word 'Quantum' seems to suggest that in line 3, between 'affirmare' and 'Cometam,' 'statuo' or another similar word was perhaps omitted, as the meaning of the redditive phrase suggests.

On the same page, line 2. The author admits an ambiguity while he calls the movement of the comet inverse but tending according to the order of signs — which he called 'direct' on page 18, line 22, and page 19, line 12 — 'retrograde' contrary to the definition of retrograde given on page 4, line 8 from the bottom, and page 18, line 23.

On page 30, line 20. The value of 21 1/2 is dubious because it lacks the note. It can be supplemented from the table on page 21.

On page 31, line 21. Instead of 'statuant,' it should be 'statuunt.' An irrelevant mistake.

33.

29 September 1665,[1] Hevelius to Lubieniecki

BO: C1, v. 7, 1039
TC: 407

Domino
De Rolis
Hamburgi

Illustrissime Vir,

Cum iterum iterumque[2] verba facias de erroribus typographicis *Prodromi Cometici* corrigendis, operae pretium esse duco, ut quid sentiam libere dicam. Quaeso sinas mihi ipsi mea corrigere[3] in opusculo meo prima occasione edendo; ubi omnia illa quae erronea sunt in catalogum referam, non autem omnia ea quae Tu corrigere putas. Nam in quibusdam nullo modo aberravi; Siquis bene distingueret inter motum antecedentium et subsequentium horarum facile rem perspiceret. E[xempli] g[ratia] Pag[ina] 21. in Tabula priori ubi de motu proprio loquutus sum; etiam si ibidem rem non omnino accurate tractaverim, nihilominus non tantopere aberravi ut Tu quidem existimas, id quod ex opusculo meo edendo clarius intelliges. Pag[ina] 29. lin[ea] 2. nullo modo ambiguitatem commisi. Sed recte et studio ita loquutus sum, dommodo recte accipiant. Item Pag[ina] eadem l[inea] 1. ad verbulum quantum nihil omissum est. Idcirco rogo quam officiose ut illa omnia mihi relinquas. Nescio profecto num unquam repererim alterius errores in alieno operi; atque ita convenientius erit ut quilibet sua explicet et corrigat.[4] Interea cum compendium ex *Prodromo* confeceris poteris in ipso opusculo, si meis plane verbis loquaris re ipsa illa corrigere quae revera corrigenda sunt, ut ut talia sint quae quilibet harum rerum gnarus prima fronte ipsemet videt, tum ex antecedentibus et consequentibus agnoscit errores esse typographicos. E[xempli] g[ratia] 'Occasum' pro 'ortu';[5] 'altus' pro 'alto'; 44.° 20.' pro 4.° 1.'; 5' 30" pro 5.° 30'; 26.° 20.' Arietis pro 16.° 20.' Arietis; In frontispicio tabulae secundae[6] ° ' pro ' "; 'Hyperbolica' pro 'Pyperbolica'; 'statuunt' pro 'statuant'. Haec inquam per me Tibi integrum erit in tuo compendio re ipsa corrigere; sed rogo ne illa in peculiarem ut dixi catalogum errorum referas. Interea condones Vir honorande, quod tam saepius hisce rebus Tibi // adeo sum molestus; cum sciam aliis occupationibus te satis

1 In the TC, the letter is dated 30 September.
2 iterumque *om.* **TC**.
3 operae...corrigere : volui sententiam meam sincere dicere, me scilicet omnia corrigere velle **TC**.
4 non autem omnia ea...corrigat *om.* **TC**.
5 pone *del. post* ortu **BO**.
6 post secundae (post *del.* **BO**) : post **TC**.

superque esse districtum.[7] Quaecunque de novis cometis narrant, nullo sane nituntur fundamento. Cum praeter binos etiam a te conspectos, nulli amplius hoc anno apparuerint. Cometa ille qui per octo menses fulsit, non anno 375. sed 975. extitit, id quod etiam corrigendum erit. Plura vide apud Lycosth[enem] et Rochenb[achium]. Amicos omnes inprimis Clariss[imum] Dominum Kirstenium et Cl[arissimum] Dominum Mullerum quam officiose saluta, quibus, quamprimum tantum a laboribus susceptis longe laborissimis paullulum respiravero, responsurus sum. Vale et fave

Illustrissimae Dominationi Tuae
Addictissimo

J[ohanni] Hevelio

Anno 1665 die 29 Septembr[is]

To Master
De Rolis
in Hamburg

Most illustrious Sir,

As you are constantly referring to the correction of typographical errors in my *Prodromus cometicus*, I think it is worthwhile for me to speak freely about it. I ask you to let me correct them myself, in a work I will publish at the earliest opportunity. I will list all the mistakes, but not all those which you think should be corrected. Because in some of them there is no error. If anyone knew the difference between the movement of antecedent and subsequent hours correctly, they would understand it easily. For example, on page 21, in the first table where I speak about the proper motion, even if I had not discussed everything precisely, I was not entirely wrong, as you think, and you will surely learn it from the book I will publish. On page 29, line 2, there is no ambiguity. Instead, I stated it accurately and purposefully — if only it were correctly understood. Also, on the same page, in line 1, concerning the word 'quantum,' nothing was omitted. Therefore, I ask you very kindly to leave all of these to me. I do not know if I have ever noted anyone's mistake in his book, and it will be more convenient if everyone explains and corrects what belongs to him. Meanwhile, because you made an abstract of the *Prodromus*, you can amend in this little work (if you indeed speak with my words) what really should be corrected, so that anybody, especially an expert in this matter, can see these mistakes at once, and, from the preceding and subsequent text, he can understand that they are typographical errors. For example: 'occasum' instead of 'ortum,' 'altus' instead of 'alto,' 44° 20' instead of

7 sed rogo...districtum *om.* TC.

4° 1', 5' 30" instead of 5° 30", 26° 20' of Aries instead of 16° 20' of Aries, ° ' instead of ' " in the label of the table, 'Hyperbolica' instead of 'Pyperbolica,' 'statuunt' instead of 'statuant'. I sincerely advise you to correct them in your abstract, but I ask you, as I said, not to make an errata to my work. Meanwhile, forgive me, honourable Sir, that I disturb you with these things so often, as I know you are busy enough with other obligations. There are no grounds for what they say about the new comets. Because apart from those two which you saw, no more appeared this year. The comet which shone for eight months was not in the year 375[8] but in 975.[9] It also should be corrected. You will find more in Lycosthenes[10] and Rockenbach.[11] Greet kindly all friends, especially the most renowned Masters Kirsten[12] and Müller,[13] to whom I will answer as soon as I take some rest from very toilsome tasks which I have undertaken. Farewell, and be benevolent towards,

illustrious Sir,

Your devoted,

Johannes Hevelius

On 29 September 1665

8 See note to Letter 27.

9 See note to Letter 27.

10 Conrad Lycosthenes (Wolffhart, 1518–61), German humanist. Hevelius refers to his *Prodigiorum ac ostentorum chronicon* (Basileae: per Henricum Petri, 1557), p. 365.

11 Abraham Rockenbach (see note to Letter 30), *De cometis tractatus novus methodicus* (Witebergae: ex Officina Cratoniana, sumptibus Henningii Grosii, 1602), p. 173.

12 See section 2.4.9 of the Introduction.

13 See section 2.4.11 of the Introduction.

34.

6 October 1665, Lubieniecki to Hevelius

BO: C1, v. 7, 1070
TC: 407

Salutem Plurimam Nobilissime, Amplissime, Humanissime Vir.

Rem gratam mihi literis Tuis postremis die 30. Sept[embris] datis, hodie mihi redditis,[1] fecisti, quod mentem Tuam de erroribus, qui in *Prodromum Cometicum* irrepserunt, declarasti, Te nimirum velle eos in opusculo Tuo brevi edendo summatim recensere. Ego certe nunquam rem tam delicatam tetigissem, ni ipsa Tua humanitas modesta bonique publici studiosa eam mihi in manus porrexisset, et ita quidem, ut Amicorum judicia exquirerem. Cum vero etiam *Prodromum* Tuum Te consentiente in compendium redegerim, tanto libentius de his Te consului. Itaque qui didici nil contra aequitatem et amici voluntatem facere, non plus in eo quam permittis, faciam. Librum vero illum Tuum cum desiderio, una cum omnibus Amicis beneque Tibi cupientibus exspecto. Rudbeckianam de Cometis commentationem cum XI. figuris, quarum duae desunt, per Nobiliss[imum] et Humaniss[imum] Heinsium nuper accepi. Ibi quoque, ut et ubique meritam laudem feres. Cl[arissimo] Büthnero pl[urimam] salutem adscribo, utcunque per Te meo in illum officio perfungens. Quamprimum me aliis laboribus expedivero, ad binas eius humanissimas literas rite respondebo. Nec Rochenbachium pro me, nec Eckstormium pro Te, utrumque sollicite quaesitum, invenire hic possum. Vale et constanter fave

Tuo

Rolitzio
manu propria

Hamb[urgi] 6. Octobr[is]
Gregor[iani] 1665.

Greetings, most noble, worthy, and kindest Sir!

You gave me much pleasure with your last letter, written on 30 September, which was delivered to me today, because you expressed your opinion about the mistakes which sneaked into your *Prodromus cometicus*, viz. that you want to discuss them at once in

1 mihi redditis : vero huc perlatis **TC**.

your booklet which will be published soon. Of course, I would have never touched upon such a delicate matter, if your kindness, modesty, and mindfulness for the public good had not encouraged me to do so. And I would not have asked my friends for their opinions. And as I prepared an abstract of your *Prodromus*, for which you agreed, I felt more prompted to ask you about the mistakes. Therefore, because I was told not to do anything against justice and my friend's will, I will not do more than you let me. I wait eagerly for that book of yours, together with all the friends who wish you all the best. From the most noble and kindest Heinsius,[2] I have recently received Rudbeck's discussion of comets with eleven illustrations,[3] two of which are lacking. In this work, you are praised as deservedly as everywhere else. I send many greetings for the most renowned Büthner,[4] as I fulfil my duty towards him through you. As soon as I am free from other works, I will answer two very kind letters from him. I cannot find here either Rockenbach[5] for me or Eckstorm[6] for you, although I search thoroughly.

Yours,

Rolitsius
by my own hand

In Hamburg, on 6 October 1665, Gregorian style.

2 See section 2.4.6 of the Introduction.

3 Lubieniecki refers here probably to Rudbeck's (see section 2.4.16 of the Introduction) letter to Heinsius, published in the TC pp. 348–59, in the *Communicatio Ubsalensis*. Despite his efforts, Lubieniecki seemingly received from Rudbeck no direct answer to his letters.

4 See section 2.4.3 of the Introduction.

5 See note to Letter 30.

6 See note to Letter 27.

35.

17 November 1665, Lubieniecki to Hevelius

BO: C1, v. 7, 1071
TC: 407–08

Salutem Plurimam Nobilissime, Amplissime, Humanissime Vir.

Cum uterque nostrum assiduis laboribus distringamur, aequam veniam mutuo dabimus silentio. Ego me in Tuis rationibus, ut puto, omni aere alieno liberavi, vicem responsi abs Te exspectans, tum et nuncium de *Prodromo Cometico* secundo, sive altero illo opere, quod ad vindicandam Tuam sententiam Cl[arissimo] Auzutio opponere paras. Huius amicitiam Cl[arissimus] Bullialdus mihi conciliavit, ita ut mihi illas suas Ephemerides Cometicas miserit. Ego hostimenti loco misi ei Tuum *Prodromum*, quem habuit acceptissimum. Confidenter in suis ad me literis de sententia sua loquitur, Tuumque opus cum taedio[1] se exspectare, testatur. Integrum erit utrique vestrum opinionem et existimationem vestram cordate, ut veritas obtineat, illaesa amicitia tueri. Ego, si quis apud Vos meis precibus locus dari potest, velim hanc controversiam amice, Cl[arissimi] Hugenii et Bullialdi vel etiam aliorum communium Amicorum reique Mathematicae apprime peritorum opera[2], componi. Qua in re ego quod jusseris faciam, ut et omnia, quae Tibi grata fore cognovero. Scio enim hanc esse amicitiae legem, usibus amici consulere et votis obsequi. Cl[arissimus] Isaacus Gruterus nuper me certiorem fecit, *Prodromum Tuum* e typographeo Janssoniano cum quibusdam nescio cuius annotationibus prodiisse. Scripsi ad illum, ut me plenius de hac re edoceat. Novum cometen m[ense] Sept[embri] apparuisse negat Curtius, Schottus, Busmanus, quod ex eorum literis habeo. Sed Rudbeckius // affirmat, locumque ei nuper exeunte Oct[obri] Juliano in Andromeda assignavit, qua de re Heinsianae literae, nuper ad Te per nautam missae Te docebunt. Fac, Vir Amicissime, sciam quid de hoc apud animum Tuum statueris, et an opus Tuum contra Auzutium perfectum sit, quod videre quamprimum gestio. Certus porro esto, me Tuis magnis et praeclaris meritis testimonium dignum ubique dare, quotiescunque res fert. Vale bene quam diutissime publico bono, mihique constanter fave.[3] Dab[am] Hamburgi die 17 Nov[embris] Gregor[iani] 1665.

1 cum taedio : avidissime **TC**.
2 opera **TC** *om.* **BO**.
3 bene…fave : diu et feliciter **TC**.

Clarissimo Nomini Tuo
omnibus studiis et officiis
deditissimus

Rolitsius
manu propria

Greetings, most noble, worthy, and kindest Sir!

As we are both busy and constantly preoccupied, we will forgive each other for the silence between us. I have paid, I think, all my debts towards you, and I await your answer and some information about the second *Prodromus cometicus* or that other work which you prepare to defend your opinion against the most renowned Auzout.[4] The most renowned Boulliau[5] contacted me with Auzout, who sent me his ephemerides of comets. In return, I sent him your *Prodromus* which he accepted with pleasure. In his letter to me, he is confident about his opinion, and he testifies that he expects your book impatiently. It would be good if both of you watched each others' opinions closely in order to find the truth and not to harm your friendship. If there is any room for my requests, I would like this controversy to end in a friendly manner, with the help of the most renowned Huygens[6] and Boulliau, or also other mutual friends experienced in mathematics. I will do in this matter whatever you say, and I will do everything that is kind to you. I know that this is the principle of friendship to care about a friend's business and to listen to his wishes. The most renowned Isaac Gruterus[7] has recently informed me that your *Prodromus cometicus* was published by Janssonius's publishing house[8] with someone's commentary. I have asked him to tell me more about it. Curtz,[9] Schott,[10] and Busmann[11] deny that there was a comet in September, as I gather from their letters. But Rudbeck[12] confirms the comet and places it in Andromeda[13] at the end of October, Julian style. You will learn about it from Heinsius's[14] letter, recently sent to you through a sailor. Tell me, my friend, your opinion about it and whether your text against Auzout is ready; I would like to

4 See section 2.2.4 of the Introduction.

5 See section 2.4.2 of the Introduction.

6 See section 2.4.7 of the Introduction.

7 See note to Letter 28.

8 Johannes Janssonius (1588–1664), Dutch publisher and cartographer who ran his printing house in Amsterdam. After his death, the business was inherited by his son-in-law, Johannes van Waesbergen (d. 1681). See section 2.2.3 of the Introduction.

9 See section 2.4.5 of the Introduction.

10 See note to Letter 28.

11 See note to Letter 29.

12 See section 2.4.16 of the Introduction.

13 See section 2.3.2 of the Introduction.

14 See section 2.4.6 of the Introduction.

see it as soon as possible. Be sure that I testify your great and well-known merits everywhere, any time I can. Farewell and be in good health for very long for the public good, and always be benevolent towards me. Given at Hamburg, on 17 November 1665, Gregorian style.

Renowned Sir,

Yours most humbly obliged,

Rolitsius
by my own hand

36.

28 November 1665, Hevelius to Lubieniecki

BO: C1, v. 7, 1073
TC: 408–09

Illustri Viro
Domino Rolissio
Johannes Hevelius Salutem

Cum nullum praegnans argumentum hactenus habuerim a quo tuas posteriores die 6 Octob[ris] datas obtinuerim, lubens hucusque scriptionem hanc meam differre volui, non ignorans, Te aliis multis gravissimis negotiis nullo non tempore esse districtum, sic ut nefas esse duxerim otiosis literis Te interpellare, imprimis[1] cum et ego non minus occupatissimus fuerim opusculo meo, quod sub manibus versatur. Nam ob fusissimum calculum maximum mihi facessivit laborem; quem autem Deo favente, fere jam exantlatus sum.[2] Spero, si Altissimo ita porro visum fuerit, post festum Nativitatis Domini[3] typis tractatum istum commissurum. Quo in statu tuum versetur opus libenter itidem cognoscerem: videmus quantum temporis requiratur ad quaedam typis evulganda, ne dicam conscribenda, praesertim si innumeris numeris omnia prius elicienda et conquirenda sunt. Nuper restantes illos 11 1/2 Imperiales tuo nomine accepi, pro quibus gratias habeo debitas.[4] Haec dum scribo literae tuae die 17 Novemb[ris] scriptae mihi traduntur, ad quas nunc simul respondebo. Gaudeo quod amicitiam contraxeris cum Clariss[imo] domino Auzotio, quem profecto ob vividum suum ingenium atque raram rerum Astronomicarum atque Opticarum experientiam, inter maximos quosque huius seculi Viros repono. Atque idcirco non est quod verearis, etiamsi aliqua magni momenti quaestio inter nos versetur ratione prioris Cometae, amicitam // exinde posse obnubilari, sed potius Tibi persuadeas velim affectum reciprocum ea ratione posse magis magisque confirmari, si quod sentiam candide et aperte pari modo ut ipse Dominus[5] Auzotius fecit, humaniter tamen et decenter, veritatis unice gratia, in medium proferam. Operae enim pretium est ut et ego audiar, resque ab utraque parte discutiatur examinetur. Atque tum facile controversia illa componetur; non vereor ipsi Domino Auzotio aliisque Viris Eruditis rem totam diiudicandam committem: cumprimis manifestum fiet, an hallucinatus fuerim in motu diurno circa exitum Januarii et initium Februarii? ut pag[ina] 8 meminit suarum literarum, tum circa magnitudinem faciemque an aliquid alieni exhibuerim?

1 et imprimis (et *del.* **BO**) : praeprimis **TC**.
2 exantlatus sum : exantlavi **TC**.
3 Nativitatis Domini : Epiphan[iae] **TC**.
4 Nuper...debitas *om.* **TC**.
5 Dominus *om.* **TC**.

Quorsum autem a die 12 et 14 Februarii, quo nudis oculis atque debitis Organis Sextantibus scilicet et Quadrantibus observari a nobis desiit, utrum ad primam, an vero ad secundam Arietis stellam viam porrexerit, tam ex Domini Auzotii quam meis observationibus inter se exquisite postmodum collatis unicuique satis evidenter suo tempore patebit. Si *Prodromus* meus *Cometicus* denuo Amsterodami prodierit, rogo ut prima occasione mihi unicum exemplar compares, atque transmittas, quo videam, quid commentaria illa sibi velint, et quisnam illius rei auctor sit. Miror sane quod me inscio id tentaverint. Cl[arissimi] Rudbeckii commentarium pariter nondum vidi, vellem tamen eum quantocyus legere. Novus autem iste Cometa quem dictus Auctor in Andromedam refert, non nisi est stella illa nebulosa in eius Cingulo; quae tamen inter Cometas nullo jure refertur. Heinsianas literas, quas promittis // necdum accepi,[6] quid causae sit, ignoro. Vale Vir Illustrissime[7] diuque feliciter vive. Dabam Gedani Anno 1665 die 28 Novemb[ris].

To the illustrious Sir,
Master Rolitsius,
Johannes Hevelius sends greetings

Because since the time I received your letter sent on 6 October until now I have had no urgent subject to write about, I have willingly decided to delay my answer. I was also aware that you had been constantly preoccupied with many important tasks, so I have thought it would be improper to distract you with some careless letters, especially as I have been very busy myself, working on my booklet which is almost ready. I had much work with the long calculations, but (God willing) I have almost finished everything. If this is the decision of the Almighty, I hope that after Christmas I will send this text to the printer. I would eagerly like to learn about the progress of your work; we know how much time it takes to publish (not to mention to write) something, especially if everything needs to be discovered and examined with countless calculations beforehand. I recently received those remaining 11 1/2 Reichstalers from you; I am very grateful for them. While I am writing this, your letter, sent on 17 November, is being delivered to me, and I am going to answer it at once. I rejoice because you have started corresponding with the most renowned Master Auzout,[8] whom I place among the greatest men of our age due to his talent and rare experience in astronomy and optics. Therefore, even if a serious dispute about the first comet arose between us, you should not be afraid that it can harm our friendship. Instead, I would like to convince you that it may make our mutual amity grow stronger, if I present sincerely and clearly, but also humanely and decently, what I think — in the way Auzout did it. It would be only for the sake of truth. Because I should be

6 accepii (*secundum* i *del.*) **BO**.
7 Vir Illustrissime *om.* **TC**.
8 See section 2.2.4 of the Introduction.

heard out as well, and the case should be discussed and examined from both points of view. And then that dispute will be quickly composed. I will surely let Master Auzout and the other learned men adjudicate. First of all, it will become clear whether I said some nonsense about the daily motion at the end of January and the beginning of February, as he notes on page 8 of his *Lettre*, and if I presented anything wrong about the magnitude and appearance of the comet. In due time, on the basis of the observations conducted by Master Auzout and myself, after they are compared, it will become evident for everyone where the comet went after 12 and 14 February (the day when it ceased to be seen by us with the naked eye and proper instruments, viz. sextants and quadrants) either towards the first, or the second star of Aries. If my *Prodromus cometicus* is published again in Amsterdam, I ask you to buy a copy and send it to me at the earliest opportunity, so that I can investigate those commentaries and see who their author is.[9] I am wondering why the publisher did it without telling me anything. Also, I have not seen the commentary by the most renowned Rudbeck[10] yet, but I would like to read it as soon as possible.[11] That new comet, placed by him in Andromeda,[12] is nothing else but that nebulous star in her belt, which cannot be placed among comets. I have not received the letter from Heinsius[13] yet (the one you promised); I do not know why. Farewell, most renowned Sir, and live long and happily! Given at Danzig, on 28 November 1665.

9 See section 2.2.3 of the Introduction.
10 See section 2.4.16 of the Introduction.
11 See note to Letter 34.
12 See section 2.3.2 of the Introduction.
13 See section 2.4.6 of the Introduction.

37.

28 November / [8 December] 1665, Lubieniecki to Hevelius

BO: C1, v. 7, 1072
TC: 409–10

Salutem Plurimam Vir Nobilissime, Amplissime et Clarissime.

Minime miror silentium Tuum. Tanto vero magis hoc excusatione dignum est, quod et argumentum scribendi defuerit, et Tibi tot molestae et longae computandae et componendae rationes fuerint. Gratulor Tibi librum Te Tuum, omnium votis expetitum, nobis et publico, fere pro strena daturum. Id ego, si vel maxime vellem, praestare non potero, ob insignem Operis molem. Hanc ut argumenti diversitas et communicantium non illiberalis copia, nec parca industria, bono publico admisit,[1] ita et, apud aequos rerum aestimatores et bonarum artium amantes, ut spero, excusabit, vel etiam commendabit. Tu, qua es humanitate, fac ut Tuum librum, quamprimum perfectus fuerit, accipiamus. Promtam pariter officii vicem a me praestolare. Gratum mihi quoque fuit ex Te intelligere, vera me Cl[arissimis] Auzutio, Hugenio aliisque Amicis de Tua modestia et aequitate pollicitum esse. Macte eximia tam rarae exulcerato hoc seculo virtutis gloria et ingenio, Magne[2] Heveli, excellentissime caeli interpres. Mea si quid valet hic opera, ea Tuis votis et studiis omni ex parte dicata est. De exemplo *Prodromi* Tui a Janssonio, ut accepi, editi, libenter Tibi gratificabor, postquam eius copia mihi facta // fuerit. Scripsi tunc ea de re ad Amicum. Sed nil responsi ab eo usque adhuc accepi. Crede mihi non parum nobis in commercio literarum aliisque liberalibus institutis belli Monasteriensis iniquitas incommodavit. De qua et ego queri debeo. Scribo itaque de eadem re ad alium Amicum, ut voto potiare Tuo, si illi rumori constat fides, saepe alias affecta et afflicta. Interim ego strenae loco destinavi tria *Prodromi* a Te editi exempla, Almae Academiae Hafniensi, et Clarissimis Bartholinis Thomae et Erasmio: quorum ille in *Consilio* suo *Medico de Cometis* sua sponte,[3] et supra meritum quidem meum, de me bene meritus est, ut nosti. In his enim terris, ubi nullus suspicioni vel invidiae locus est, gentilitium non a stemmate mutuatum gero nomen, quod ex Nobilissimo Pelsio jamdiu cognoris. Nuncius de novo Cometa non confirmatur aliunde. Nec ego ex eo tempore quidquam a Cl[arissimo] Rudbeckio per Nobiliss[imum] Heinsium, imo nec ab illo ipso (quod miror, utpote eius eximiae benevolentiae et in colendis literarum commerciis amicitiaeque juribus diligentiae, quam Rautensteinianae supparem merito dixero, contrarium) magis 15. // abhinc diebus accepi. Tu vero quod nondum illas, quarum ego praenuncius fui,

1 admisit : attulit **TC**.
2 Magne *om.* **TC**.
3 de me *del. post* sponte **BO**.

acceperis, miror. Sed memini me Tibi scribere, Cl[arissimum] Heinsium suas ad Te deferendas nautae cuidam tradidisse. Nosti autem hoc hominum genus ferri navi, quae saepe ventis ludibrium debet, ac ferre fidem, quae itidem saepius par levibus ventis promissa in ludibria vertit. Ego Heinsium ea de re hodie monebo. Tu vero an ille nauta ludificans vel ludificatus dici mereatur, quamprimum nuncia. Rudbeckii scriptum sat curiosum et longum ad Heinsium pertinet, et a me illius solertiae et benevolentiae debetur. Itaque cum nullum in mora periculum esse existimem, de illo commentario Tecum communicando ad eum referam.[4] Alioqui libenter occasionem istam bene de Te et publico merendi, si res mihi integra foret, arriperem. Spero enim Te non pauca illic inventurum, quae singularem Tuam provocabunt sagacitatem et caelestem industriam, magno rei Mathematicae totiusque literariae emolumento. Ut autem, quantopere praeclari nominis Tui famae consultum eam, tester cuncta illa *Prodromi* Tui exempla cum debita commendatione Amicis et Fautoribus dono fero, // ita ut etiam unum in Novo Chiloniensi Parnasso, meae in illum juxta et Te observantiae et benevolentiae monumentum, deponere, ad me retulerim et apud me decreverim. Stellam illam in cingulo Andromedae, quam indicas, non dubito eandem esse cum illa, quae Romae et Franequerae observata pro Cometa habita est. Postquam hucusque literas istas scripsi, accipio literas a Domino Busmanno die 24. Nov[embris] S[tylo] V[etere] Hannoverae datas, longe eruditissimas mihique gratissimas de argumento haud levi, quod ipsi explicandum proposui, de aetate scilicet illius Magni Astronomi Hipparchi, ad cuius gloriae societatem Te, Magne HEVELI, consentiens orbis literarii judicium[5] admisit; tum de aetate Mithridatis Magni, nec non de Cometis circa mortem Ludovici Pii. Alioqui de nuperis Cometis nihil epistola eius habet. Accipio et Heinsianas pariter longe acceptissimas die 15. 25. Nov[embris] datas, errore quodam, qui in curandis literis nostris saepe committitur, serius justo mihi redditas. Sed nec ille Cometici argumenti quidquam continent, quod alias ad Te perscribere non negligerem. Cl[arissimo] Büthnero, cui salutem adscribo, si velit aliquid ex Tuo Rochembachio mei caussa excerpere, dabis, quae Tua est humanitas, huius libri copiam. Vale bene et rem age feliciter, mihi vero favere perge, qui sum

Clarissimo Nomini Tuo
omni studio et officio
addictissimus

Rolitsius

Post Scriptum. Literas Cl[arissimo] Kirstenio tradidi: qui *Cometae* sui exempla aliquot huc illuc mittenda mihi donavit.[6]

4 referam : hodie retuli **TC**.

5 Te *del. post* judicium **BO**.

6 Vale bene...donavit : Literas Tuas Clarissimo Kirstenio statim reddi curavi. Quam gratae illi Tuae fuerint, vel ex eo apparet, quod aliquod exempla Poematis sui Cometici Tibi inscripti mihi miserit, Amicis dono mittenda. Vale et mihi constanter fave. **TC**.

Hamb[urgi] 28 Nov[embris] Juliani 1665.

Greetings, most noble, worthy, and famous Sir!

Your silence does not surprise me. The fact that there were no subjects to write about and that you had lots of tedious and time-consuming calculations to make is a decent excuse. I congratulate you that your book, which everyone expect, will be given to me and the public as a New Year gift. I cannot do likewise (however much I would like) because of the great amount of work. This multifarious discussion, a great volume of correspondence, and my numerous efforts were meant only for the public good. I hope that they will also excuse (or even recommend) me in the eyes of some righteous experts and all who love good arts. Out of your kindness, make sure that we can receive your book as soon as it is ready. Expect from me a due return for it. It was pleasant for me to learn from you that what I said about your modesty and equity to the most renowned Auzout,[7] Huygens,[8] and other friends was true. May you be successful, great Hevelius, most excellent interpreter of the sky, with your exceptional glory from your talent and virtue — which is so rare in this rotten century! If my works have any meaning, they are entirely devoted to your wishes and efforts. I will readily give you a copy of your *Prodromus* (which, as I found out, was published by Janssonius[9]) as soon as I get one. I wrote about it to a friend. But so far I have received no answer. Believe me, the inequity of the Münster war[10] caused serious problems with epistolary communication and other liberal endeavours. I should also wail because of it. Therefore, I write to another friend to keep my promise and to ask if that rumour can be trusted — because trust can be often weakened and harmed. Meanwhile, I have intended three copies of your *Prodromus* as New Year gifts for the university in Copenhagen, and for the most renowned Thomas[11] and Erasmus Bartholin;[12] the former decided to do me good service (greater than any of my services for him) with his *De cometa consilium medicum*, as you know. In this country, where there is no room for suspicion or jealousy, I bear my family name, not the one borrowed from my coat of arms, as you already know from the most noble Pels.[13] The information about the new comet is not confirmed from elsewhere. Nor have I re-

7 See section 2.2.4 of the Introduction.

8 See section 2.4.7 of the Introduction.

9 See note to Letter 35 and section 2.2.3 of the Introduction.

10 Lubieniecki refers here to the war between England and the Netherlands in 1665–67. The Prince-Bishop of Münster, Christoph Bernhard von Galen (1606–78, Bishop in 1650–72), sided with the English King Charles II and attacked the Netherlands, but in 1666 he was forced to surrender by Louis XIV of France and Frederick William of Brandenburg. See also CJH3 p. 236, note to Letter 45, des Noyers to Hevelius, 1 or 2 April 1653.

11 See note to Letter 4A.

12 See section 2.4.1 of the Introduction.

13 See note to Letter 5.

ceived anything since that time, for over 15 days, from the most renowned Rudbeck[14] through the most noble Heinsius,[15] and not even from him personally (this fact surprises me, as it is contrary to his exceptional benevolence and diligence in keeping the correspondence and the principles of friendship; I can honestly say that in this he even matches Rautenstein).[16] I wonder why you have not received what I had written to you yet. But I remember writing to you that the most renowned Heinsius had given his letter for you to a sailor. I know that this kind of people is carried by ship (which often becomes a toy for the winds), and they carry the trust (which, equal to light winds, often turns into a toy). Today, I will tell Heinsius about it. Inform me as soon as possible whether we should say that this sailor trifles or is being trifled with. Rudbeck's text is quite interesting and long; it concerns Heinsius, and I owe it to his craftiness and benevolence. Therefore, because I think there is no danger in delay, I will write him about the commentary which I will send to you. Nevertheless, I would take that opportunity to render a service for you and for the public, if I only can. But I hope that you will find there many things which will incite your singular sagacity and celestial industry for the great progress of mathematics and all sciences. In order to testify how much I care about the glory of your renowned name, I send all the copies of your *Prodromus* as a gift to friends and benefactors with a proper recommendation. I have considered and decided to give one of them to the new Parnassus in Kiel[17] as a proof of my reverence and benevolence towards them and to you. I do not doubt that the star in the Belt of Andromeda[18] which you mentioned is the same as the one seen in Rome and in Franeker, which was considered to be a comet. After I had written this, I received a very learned and pleasant letter from Master Busmann,[19] sent on 24 November, old style, from Hannover.[20] It touches a difficult subject about which I asked him, viz. the age of that great astronomer Hipparchus[21] (to whose glorious company the unanimous opinion of the learned world has admitted you, great Hevelius) and also the age of Mithridates the Great,[22] and the comets at the death of Louis the Pious as well.[23] But his letter says nothing about the latest comets. I also received an equally anticipated letter from Heinsius, sent on 15/25 November, which was delivered to me much later than it should be, due to an error which often happens in our correspondence. But it also does not contain anything about the cometary discussion; I will write to you about it another time. If the most renowned Büthner[24]

14 See section 2.4.16 of the Introduction.
15 See section 2.4.6 of the Introduction.
16 See note to Letter 5.
17 Lubieniecki probably refers to the university in Kiel which was founded in 1665.
18 See section 2.3.2 of the Introduction.
19 See note to Letter 29.
20 TC pp. 678–81.
21 Hipparchus lived between *c.* 190 and *c.* 120 BC.
22 Mithridates VI the Great (135–63 BC), king of Pontus in Asia Minor since 120 BC.
23 See Letter 42A.
24 See section 2.4.3 of the Introduction.

(whom I salute) wants to check anything in your copy of Rockenbach's book for me,[25] let him use it, if you would be so kind. Farewell, and prosper, and still be benevolent towards me, the one who is

renowned Sir,

Yours most humbly devoted,

Rolitsius

Postscript. I have passed the letter to the most famous Kirsten.[26] He gave me a few copies of his *Cometa*[27] for me to send them to a few places.

In Hamburg, on 28 November, Julian style, 1665.

25 In a letter to Büthner, dated 16 November 1665, TC p. 825, Lubieniecki asks the addressee to buy (or to make) a copy of some cometary catalogues, including Rockenbach's (see note to Letter 30). He adds that if Büthner is unable to obtain this book, Hevelius has a copy.

26 See section 2.4.9 of the Introduction.

27 Michael Kirsten, *Cometa, poema ad Hevelium* (Hamburgi: apud Zachariam Hertelium, 1665).

38.

19/29 January 1666, Lubieniecki to Hevelius

BO: C1, v. 7, 1093
TC: 410–11

Salutem Plurimam Nobilissime, Amplissime, Humanissime Vir.

Exierunt duo menses, ex quo nil ad Te literarum dedi. Tum quod exspectarem responsionem ad meas ultimas tum quod nimium essem occupatus, idque assiduo; tum denique quod nil magnae rei scribendum haberem. Ita nec novi anni felix auspicium et continuatam in plures annos valetudinis et successuum prosperitatem Tibi de more precatus sum[1] epistola ad Te scripta. Feci tamen id corde et literis ad communes Amicos Nobiliss[imum] Pelsium et Doctiss[imum] Büthnerum scriptis. Facio et nunc antequam Januarius Gregorianus praetereat. Tu, qua es eximia humanitate et aequitate, cuncta quae a manu candida veniunt boni et aequi consule. Magnum profecto desiderium tenet nos Tui *Prodromi* secundi et syntagmatis rationum et observationum Cometicarum. Sperabamus enim Te nobis ista, imo et *Cometographiam* ipsam, pro strena daturum fuisse. At nihil horum comparet. Verumenimvero ipse demum experior, quam gravia impedimenta obiiciantur aliquod publico committere volentibus. Tanto magis si quid arduum et ingens sit: quale tuum opus est. Sed nec meum leve et facile est: idque horis tantum succisivis (nam meliores vitae tuendae cura et officii ratio sibi vendicat) peragendum. Quo loco sint opera Tua, fac ut sciam. Amici apud Batavos in meo opere urgendo pergunt. Figurarum aeri cum debita cura incidendarum haud exigua copia, nec parva libri moles justum tempus requirit. De Commentario Rudbeckiano nondum Tibi gratificari potui. Cum enim ante tres hebdomadas per Ornatissimum Juvenem // Dominum Charisium[2] Nobilissimi Domini Charisii[3] Legati apud Faederatos Belgas pro Serenissimo Danorum Rege Residentis filium apographum eius in Bataviam mittere decreverim, nolui autographum simul casibus itinerum, hac praesertim tempestate, exponere. Sed cum is iter de die in diem differre necesse habeat, ego id officii erga Te differendum diutius non ratus, scriptum hoc et ipsam Rudbeckii manum Tibi mitto. Votis enim Tuis Humanissimus Heinsius consensit. Speramus non parum utilitatis inde redundaturum in publicum, postquam hanc commentationem curiosam expenderis. Heinsius optaret figuras Te Rudbeckianas simul habere posse. Sed has mense IIXbri[4] in Hollandiam misi. Quamprimum autem inde eas habere potero, etiam non perfecto opere nec aliis fi-

1 per *del. post* sum **BO**.
2 Charisium : Carisium **TC**.
3 Charisii : Petri Carisii **TC**.
4 IIXbri : Novembri **TC**.

guris sculptis, libenter eas ad Te mittam, ut studiis Tuis et usibus publicis consulam. Hyemalem Cometam fama vulgare non destitit. Imo vero ei locum in plaga Orientali, cursum Septentrionem versus, tempus horam 5.^{tam} matutinam assignavit. Sed artis periti nihil de eo hic vel alibi, quod sciam, observarunt. Fecit me nuper Doctiss[imus][5] Büthnerus certiorem voluisse Te ei mea gratia Rochenbachii Tui facere copiam.[6] Sed illum ex suo exemplari ea jam tunc descripsisse,[7] quae ab eo postulaveram. Ego nihilominus gratias Tibi de hac promta benevolentia ago. Vale quam diutissime et quam felicissime. Dabam Hamburgi die 19/29 Januarii Anno 1666.

Clarissimo Nomini Tuo
omnibus studiis et officiis
addictissimus

S[tanislaus] L[ubieniecius] Rolitsius

Greetings, most noble, worthy, and kindest Sir!

Two months have passed since I sent a letter to you (partly because I was awaiting an answer to my last letter, partly because I was very busy all the time, and partly because I had nothing important to write about. Therefore, I have not had a chance to wish you (in a letter to you) neither all the best in the New Year nor good health and much success in the future, as I should have. But I have done it cordially in the letters written to our mutual friends: to the most renowned Pels[8] and the most learned Büthner.[9] And I am doing it now, before the Gregorian January ends. Accept honestly and sufficiently (out of your exceptional humanity and kindness) all that comes from this honest hand. We are eagerly waiting for the second *Prodromus* of yours and for a summary of the opinions and observations of the comets. Because we were hoping you would send this book to us, or even the *Cometographia* itself, as a New Year gift. But nothing like this happened. However, I myself experience serious problems which impede those who want to publish anything. It occurs especially when the task is arduous and enormous; and it is the case with your book. But mine is not small and simple either, and it requires spare hours to be completed (because taking care of life and other duties demand those best ones). Write to me about the progress of your works. Friends in the Netherlands keep asking me about my book. The copies of engraved illustrations (which have to be prepared with care) and a substantial volume of the book require some time. So far, I could not send you the commentary

5 Doctiss[imus] : Humanissimus **TC**.
6 facere copiam : copiam facere **TC**.
7 descripsisse : describi curasse **TC**.
8 See note to Letter 5.
9 See section 2.4.3 of the Introduction.

from Rudbeck.[10] When three weeks ago I decided to send a copy of it to the Netherlands through the most capable young man (Charisius,[11] the son of the most noble Master Charisius,[12] the ambassador of the most serene King of Denmark[13] in the Dutch Republic) I did not want to expose the original text to the misfortunes of roads, especially nowadays. But because he had to necessarily delay his journey, I decided not to delay my obligation towards you, and I am sending you the original manuscript from Rudbeck.[14] The kindest Heinsius[15] agreed with your wishes. We hope that it will be beneficial for the public when you evaluate this interesting commentary. Heinsius suggested that you should receive Rudbeck's illustrations along with it. But I had sent them to the Netherlands in October. I will willingly send them to you when I get them back (even if my work is not finished and other images are not engraved) for the sake of your studies and for the public use. There are still popular rumours about the winter comet. They say it was in the east, it moved northwards and was seen at 5 o'clock in the morning. But those who are experienced in astronomy have not seen anything anywhere, as far as I know. The most learned Büthner recently wrote to me that you wanted to give him your copy of Rockenbach[16] for me. But he has already duplicated from his copy what I had asked for. Nevertheless, I am very thankful for such benevolence. Farewell, and be in good health and prosperous! Given at Hamburg, on 19/29 January 1666.

Renowned Sir,

Yours most humbly devoted,

Stanisław Lubieniecki Rolitsius

10 See section 2.4.16 of the Introduction.
11 Ambrosius Charisius (1644–1700), Danish lawyer and politician.
12 Peder Charisius (1608–85), Danish diplomat and ambassador in the Netherlands in 1651–69.
13 Frederick III (1609–70, King since 1648).
14 There is no extant enclosure with Rudbeck's manuscript.
15 See section 2.4.6 of the Introduction.
16 See note to Letter 30.

39.

13 February 1666, Hevelius to Lubieniecki

BO: C1, v. 7, 1094
TC: 411

Domino Rolisio
Hamburg[i]

Illustrissime Vir.

Quanto rarius amicos nunc literis compello, tanto diligentius ea, quae prae manibus habeo, curo; sed ut ut omne tempus, quod reliquum tantum ab aliis occupationibus tam publicis quam privatis fuit illis negotiis ac studiis hactenus tribuerim, neutiquam tamen citius descriptionem meam Posterioris Cometae, cum *Mantissa Prodromi* ad umbilicum omnino producere potui: quippe non solum opusculum hocce sub manibus praeter opinionem excrevit, sed etiam maximi fuit laboris tantum prolixissimum expedire calculum, ut suo tempore videbis. Jam vero, Gratia Divina, eo usque sum progressus, ut post paucos dies, dummodo gratissimas quasdam occupationes, quae se se novissime ob recens natam mihi Filiolam, obtulerunt, expedivero, praelo illud sum commissurus; nihilominus tamen, menses aliquot excurrent priusquam typis, uti optime nosti, penitus excribi, atque Icones manu mea aeri incidi possint. Interea iterum iterumque Literatos omnes rogo vehementer, ut judicium de illis controversiis inter Dominum Auzoutium et me exortis eousque suspendant, quo eo accuratius postmodum rem istam decernere possint. Pro Commentario Rudbeckiano magnas Tibi habeo gratias; oculo fugitivo tantum adhuc illum perlustravi; video tamen ingeniose ac accurate conscripta esse omnia, multasque praegnantes quaestiones movisse: de quibus autem, ut mihi videor, pene omnibus in *Cometographia*, aliisque // scriptis meis, si non omnino ex composito, implicite tamen passim nobis jam sermo fuit. De parallaxibus vero investigandis beneficio altitudinum Meridianarum in diversis Terrae Regionibus, plane aliter sentio eas nimirum multo certius aliis rationibus, etiam ab uno observatore explorari posse; de quibus autem hac vice plura dicere nequeo. Nullum Cometam post illum Mense Aprili in caelo apparuisse pro certo habeas velim, quicquid etiam sentiunt alii. Puto tamen nonnullos phaenomenum illud ad cingulum Andromedae, quod hactenus et adhucdum ex oriente singulis diebus Septentrionem versus motu sc[ilicet] diurno proficiscitur arripuisse. A clariss[imo] Domino Bartholino,[1] praevia salute explores velim; an exemplar illud *Prodromi* per Dominum nostrum Pelzium ipsi jam olim transmissum acceperit? tum, quo in statu versentur observationes Tychonis, quas avidissime expectat Orbis Astronomicus. Literas Celeberrimi Heinsii tandem accepi,

1 Bartholino quem meo nomine (quem meo nomine *del.* **BO**) : Erasmo Bartholino **TC.**

ad quas etiam proximo tabellario responsurus. De illis phaenomenis nuper Romae in Jove, umbra sc[ilicet] Jovialium, atque peculiari macula in ipso corpore Jovis obvia, ex qua motus huius Planetae circa axem periodicus accurate stabilitur, a Joseph Campani, nec non Eximio Cassino detectis,[2] ac quibusdam Epistolis satis fuse descriptis, Romaeque Mens[ibus] Oct[obri] et Novemb[ri] editis, nec dubito quin jam perceperis. Ad quam maculam videndam, Cl[arissimus] Hugenius itidem tubum praeparat longissimum. Vale Literarum egregie[3] Cultor, omnesque amicos saluta. Inprimis vero Cl[arissimum] Mullerum roges velim ne meum diuturnum silentium sequius interpretetur[4].

Tuae Illustrissimae Dominationis
observantissimus

J[ohannes] Hevelius

Gedani 1666
die 13 Febr[uarii]

To Master Rolitsius
in Hamburg

Most illustrious Sir!

The less often I write letters to my friends, the more attention I pay to what I have to do here. But despite the fact that so far I have assigned all the time which is free from my other public and private duties for these studies and work, I could not finish my description of the second comet with the *Mantissa Prodromi cometici* earlier; not only because this booklet has grown more extensive than I presumed, but it was also tedious to solve those long calculations, as you will see in due time. Thanks to God's grace, I have come to that point that in a few days (when I finish enjoyable duties which have recently occurred with the birth of my daughter[5]) I intend to send it to the printer. Nevertheless, it will take a few months, as you well know, before the fonts are correctly set and I engrave the images. Meanwhile, I keep on asking all the learned men to suspend their judgement about that controversy which arose between Master Auzout and me until they can see the case more precisely.[6] I am very grateful for Rudbeck's[7] commentary. So far I have only leafed it through, but I see that everything is described cleverly and

2 des *del.* **BO**.
3 egregie **TC** : egregius **BO**.
4 interpretetur **TC** : interpretur **BO**.
5 Catherine Elisabeth Lange (née Hevelius, 1666–1745), daughter of Johannes Hevelius, since 1696 married to Ernst Lange, a Councillor of Danzig, see Chantal Grell, 'Hevelius en son temps', CJH1 p. 139.
6 See section 2.2.4 of the Introduction.
7 See section 2.4.16 of the Introduction.

accurately, and it touches upon many important issues. I think I have discussed almost all of them in several places in my *Cometographia* and other writings, even indirectly, if not openly. I hold an opposite view on finding parallaxes with altitudes in the meridian in various parts of the Earth, viz. that they can be discovered with more certainty in a different way, even by one observer; I cannot say more about it this time. I would like to assure you that no comet appeared after April, no matter what other people say. But I think that many people were interested in that phenomenon near the Belt of Andromeda[8] which so far has been moving every day, viz. with its daily motion, from the east towards north. I would like you to ask the most renowned Bartholin[9] (sending him my greetings) if he received that copy of the *Prodromus* I had sent him some time ago through our Master Pels,[10] and to tell me how his work on Tycho's observations proceeds; the astronomical world eagerly awaits them. I have finally received the letter from the most renowned Heinsius,[11] which I will answer through the next messenger. I do not doubt that you already know about those phenomena near Jupiter recently discovered in Rome, viz. about the shadow of its companions and a particular spot seen on the very body of Jupiter which allows establishing the periodical movement of this planet around its axis. They were discovered by Giuseppe Campani[12] and the exceptional Cassini[13] and described quite exhaustively in his letters published in Rome in October and November.[14] The most renowned Huygens[15] prepares a very long telescope to see that spot. Farewell, great worshipper of arts, and greet all our friends. I would like you to ask the most renowned Müller[16] not to take my long silence as an insult.

Illustrious Sir,

Yours obliged,

Johannes Hevelius

In Danzig, on 13 February 1666.

8 See section 2.3.2 of the Introduction.

9 See section 2.4.1 of the Introduction.

10 See note to Letter 5.

11 See section 2.4.6 of the Introduction.

12 Giuseppe Campani (1635–1715), Italian astronomer, optician, and instrument maker. He provided large telescopes for the Paris Observatory. He conducted observations of the rings of Saturn and of Jupiter's satellites. See also CJH3 p. 127, note to Letter 7, des Noyers to Hevelius, 24 July 1647, and p. 239, note to Letter 239, des Noyers to Hevelius, 6 August 1683.

13 See note to Letter 8.

14 Giovanni Domenico Cassini, *Lettera astronomica sopra l'ombre de pianetini Medicei in Giove* (Roma: appresso Fabio de Falco, 1665); Giovanni Domenico Cassini, *Lettere astronomiche sopra la varietà delle macchie osservate in Giove, e loro diurne rivoluzioni* (Roma: [s.e.], 1665).

15 See section 2.4.7 of the Introduction.

16 See section 2.4.11 of the Introduction.

40.

26 February 1666, Lubieniecki to Hevelius

BO: C1, v. 7, 1097
TC: 412

Salutem Plurimam Nobilissime, Amplissime, Clarissime Vir.

Nullum est dubium Tuum silentium esse diligentiae, quam perficiendo operi tam bene inchoato impendis, documentum. Gratulamur Te nobis brevi Cometae posterioris *Prodromum* daturum: cum Tibi huius consilii auctores extiterimus; Tum vero etiam Te nuper filiola auctum esse: simul autem optamus, ut et filiolae isti a matre rubenti, et puerperae Tuae, Tibique ipsi omnia ex voto eveniant. Si scire possum, quot figuras observationum Cometae posterioris nobis in *Prodromo* sis daturus[1], libenter is ex Te cognoscam. Scripsi ad Amicos, Viros rei Astronomicae gnaros, Te optare, ut judicia sua de hac, quae Tibi cum Domino Auzutio intercedit, controversia, tantisper suspendant, usquequo liber Tuus prodierit. Nec dubito eos hac aequitate erga Te usuros. Rudbeckiana Tibi placuisse gaudeo. Figurae eius jam aeri apud Batavos inciduntur. Cui laboris aliquot menses dandi erunt: sed et reliquis, nempe recognitioni communicatorum inter me et amicos: tum prosequendae historiae Cometicae. Novum Cometam Cl[arissimus] Rudbeckius nondum asseruit. Necdum quidquam Heinsiana industria ab eo obtinuit, quod ad fidem dicto praestandam faciat. Tu vero quod sententiam tuam toties affirmes, a // me nomine Amicorum hic degentium gratias debitas accipe. Illas epistolas Italicas, quae Jovis orbem exploratum dant, non vidi. De Campani tamen observatione et organo praestantissimo quod jam antea aliquid cognoverim, Kircherianae benevolentiae debeo: sicut Schottianae Doctissimum Ricciolum vivere, et edita *Geographia* et *Astronomia reformata*, *Chronologiam* quoque *reformatam* nobis dare velle, imo brevi daturum esse. Ad Cl[arissimum] Erasmium Bartholinum proxime, Deo volente, scribam de tabulis[2] Tychonianis, quas Tu alter Tycho jure meritoque urges. Vale diu et feliciter. Dab[am] Hamburgi die 26. Febr[uarii] 1666.

Tuus
omni studio
et officio

S[tanislaus] L[ubieniecius] Rolitsius
manu propria

1 daturus **TC** : daturum **BO**.
2 de tabulis : scribam et monebo de illis observationibus **TC**.

Greetings, most noble, worthy, and renowned Sir!

There is no doubt that your silence is a proof of your diligence which you devote to finishing your work which has begun very well. We congratulate you on the soon publication of the *Prodromus* of the latest comet because we authored this idea. We also congratulate you on the recent birth of your daughter;[3] we also wish you everything you could dream to happen to that recently born daughter of yours and to your wife,[4] and to you. If I may, I would like to ask you how many illustrations of your observations of the latest comet you want to include in the *Prodromus*. I wrote to our friends who are experienced in astronomy that you would like them to suspend their judgements about your controversy with Master Auzout until you publish your book.[5] I do not doubt they will be just for you. I am glad that you like what Rudbeck[6] wrote. His images are being engraved right now in the Netherlands. This work will take a few months; other things, viz. preparing my correspondence with friends and the accompanying history of comets, will take as much time. The most renowned Rudbeck has not confirmed the new comet yet. Neither Heinsius's[7] efforts have managed to retrieve any confirmation of his statement from him. I send you many thanks on behalf of the friends who live here because you corroborated your opinion so many times. I have not seen those Italian letters which talk about the new discoveries about Jupiter. But I have learned something about Campani's[8] observation and his exquisite instrument, and I owe it to Kircher's[9] benevolence; I have also learned from Schott[10] that Riccioli[11] is alive and after publishing his *Geographia*[12] and *Astronomia reformata*,[13] he wants to give us also the *Chronologia reformata*,[14] and that he will do it soon. I will soon write to the most renowned Bartholin,[15] God willing,

3 See note to Letter 39.

4 Elisabeth Hevelius (née Koopmann, 1647–93), the second wife of Johannes Hevelius (since 1663). She helped her husband in conducting observations and published his *Prodromus astronomiae* (Gedani: typis Johannis Zachariae Stolli, 1690) and *Firmamentum Sobiescianum* (Gedani: typis Johannis Zachariae Stolli, 1690) after Hevelius's death, Chantal Grell, 'Hevelius en son temps', CJH1 pp. 136–38.

5 See section 2.2.4 of the Introduction.

6 See section 2.4.16 of the Introduction.

7 See section 2.4.6 of the Introduction.

8 See note to Letter 39.

9 See section 2.4.8 of the Introduction.

10 See note to Letter 28.

11 See section 2.4.15 of the Introduction.

12 Giovanni Battista Riccioli, *Geographiae et hydrographiae reformatae libri duodecim* (Bononiae: ex typographia Haeredis Victorii Benatii, 1661).

13 Giovanni Battista Riccioli, *Astronomiae reformatae tomi duo* (Bononiae: ex typographia Haeredis Victorii Benatii, 1665).

14 Giovanni Battista Riccioli, *Chronologia reformata* (Bononiae: ex typographia Haeredis Dominici Benatii, 1669).

15 See section 2.4.1 of the Introduction.

about the publication of Tycho's tables, on which you (as a new Tycho) rightly insist. Farewell, and be prosperous.

Yours most devoted,

Stanisław Lubieniecki Rolitsius
by my own hand

41.

2 April 1666, Lubieniecki to Hevelius

BO: C1, v. 7, 1098
TC: 412–13

Enclosure:
41A. Observations of Mira Ceti by Heinrich Sivers

Seal with Lubieniecki's coat of arms (Rola coat of arms), address: 'A Monsieur / Monsieur Jean Hevelius / Conseiller de la ville de / Dantzic' ('To *Monsieur* Johannes Hevelius, Councillor of the city of Danzig')

Salutem Plurimam Vir Nobilissime, Amplissime et Humanissime.

Cum paulo ante et nuperrime denuo per litera Nobilissimi Heinsii edoctus sim Cl[arissimum] Rudbeckium Tuae tandem accessisse de illo phaenomeno in Andromeda observato, ab eo pro novo Cometa habito, sententiae mei officii esse duxi, Te hac de re facere certiorem. Qua occasione ex Te, nisi sit molestum, libenter discam sitne id quippiam Cometae cognatum vel affine nec ne? Tum an nulla ignita impressio sive traiectio, quas meteora esse et in aere fieri constat, Cometa dici possit, si aliquantum duret? Nam Cometas quosdam sublunares et aereos dari multi eorum, qui eos coelo intulerunt ac Tibi noti sunt, inter quos et communis noster amicus Ricciolus est, concedunt. Atque Te ipsum eo propendere animadverto, si quid hic video. Adhaec Keckermannus olim vester et Petrus Creuster Astronomus hora tantum 1 1/4 durasse Cometam Anno 1527. aiunt. Joannes // Schulerus autem Prof[essor] Bredanus, qui vivit ac nuper admodum *Cometologiam* suam, quam Te vidisse arbitror, doctam sane, sed tamen Cometis iniquam et secundum Astrologica placita factam edidit, Tuae accedens aliorumque sententiae, Cometas scil[icet] ex corporum caelestium exhalationibus sive effluviis generari, eosdem Meteora aetherea appellat. Ad haec, Clarissime Heveli, duobus tribusve verbis velim respondeas. En vero Tibi ne sine symbola Te conveniam, observationes[1] stellae illius in Ceto mirae, quam Tu praeclaro opere illustrasti. Has in urbe ista Eruditus et industria perinsignis Vir ille Juvenis Henricus Sivers sicut antehac ita et nuper admodum mense Febr[uario] factas mecum communicavit, eo fine ut ad Te has mitterem. Cum vero proximo die Martis tardius eas acceperim, propterea nec meas istas literas ad Te destinatas mittere tunc poteram. Noctuas quidem Athenas, qui haec ad Te mitto. Sed dabis id meo in Te candori, in bonas literas studio. Adultarum et concubiarum, imo et inso-//mnium magis[2] quam

1 observationes **TC** : observatione **BO**.
2 magis : magis, ut cum Tacito loquar **TC**.

pervigilium Tuarum noctium observationes magno praestantissimorum instrumentorum apparatu subnixae, pluria eiusdem generis nobis dabunt. Sed Tu sat bene scio, alienos labores acceptos habes, nec Tuae vel scientiae vel diligentiae imitamenta aspernaris, ne qui Te suspiciunt despexisse videare. Vale, Vir Humanissime, ac favere perge ei qui est

Tuus
omni studio et officio

S[tanislaus] L[ubieniecius] Rolitsius
manu propria

Hamb[urgi] 2. Apr[ilis] 1666.

Greetings, most noble, worthy, and kindest Sir!

As I recently learned from a letter from the most noble Heinsius[3] that Rudbeck[4] had finally received your opinion about that phenomenon in Andromeda,[5] which he considered to be a comet, I have decided I should inform you about it. On this occasion — if I may — I would like to learn from you whether it was something related and similar to a comet or not, and whether a fiery impression or a falling star (which are considered to be meteorological phenomena and happen in the air) could be called comets if they last for some time. As you know, many of those who put comets in the sky (including our mutual friend Riccioli[6]) agree that there are some sublunary comets in the air. If I am right, I think you lean towards this opinion. Moreover, your Keckermann[7] and an astronomer Peter Creutzer[8] say that the comet in 1527[9] lasted for only 1 1/4 hours. Johannes Schuler, a professor from Breda, who is alive and has recently published his *Cometologia* (which I presume you have already seen; it is quite a learned book, but unfavourable for comets and based on astrological opinions), is close to the opinion you share with others, viz. he says that comets originate from ex-

3 See section 2.4.6 of the Introduction.

4 See section 2.4.16 of the Introduction.

5 See section 2.3.2 of the Introduction.

6 See section 2.4.15 of the Introduction.

7 Bartholomaeus Keckermann (see note to Letter 7A), *Systema physicum* (Hanoviae: apud Guilielmum Antonium, 1610), p. 680.

8 Peter Creutzer (*fl.* 1st half of the 16th century) discussed the supposed comets of 1527 in a pamphlet titled: *Außlegung uber den erschröcklichen Cometen, so im Westrich und umbligenden grentzen erschinen, am XI. tag Weinmonats, des M.D.XXVII. jars* (Nürnberg: durch Georg Wachter, [1528]).

9 The phenomenon of 1527 was often described as a comet, see Alexandre Guy Pingré, *Cométographie ou Traité historique et théorique des comètes*, vol. 1 (Paris: Imprimerie royale, 1783), p. 485.

halations or effluxes of celestial bodies, and he calls them aethereal meteora.[10] I would like you, most renowned Hevelius, to answer this briefly. Not to turn to you without a gift, I send you the observation of Mira Ceti, which you described in your brilliant work.[11] A very learned and industrious young man, Heinrich Sivers,[12] made those observations in this city before and recently (in February), and handed them over to me in order to send them to you. Because I received them much later (on the last day of March) I could not send my letter to you then. I carry owls to Athens when I send you this.[13] But take it as a symbol of my honesty towards you, and as an effort made for the noble sciences. Your studies, conducted early and late at nights (which must have been sleepless rather than watchful),[14] supported by your excellent astronomical inventory, will give us many more observations of that kind. But I know well that you welcome the works of others and that you will not reject such imitations of your knowledge and diligence, and you will not scorn them. Farewell, kindest Sir, and always be benevolent towards

Yours most devoted,

Stanisław Lubieniecki Rolitsius
by my own hand

In Hamburg, on 2 April 1666.

10 Johannes Schuler (d. 1676), German philosopher. Lubieniecki refers to his *Cometologia, sive de Cometis Disquisitio Philosophica* (Hagae-Comitis: ex typographia Adriani Ulaco, 1665), p. 8.

11 The observations of the variable star Mira Ceti (o Ceti) were published by Hevelius in his *Mercurius in Sole visus* (Gedani: autoris typis et sumptibus, 1662). See Chantal Grell, 'Hevelius en son temps', CJH1 p. 107; CJH3 pp. 311–18, Letter 93, Hevelius to des Noyers, 18 December 1660, pp. 319–20, Letter 94, des Noyers to Hevelius, 23 January 1661, and pp. 480–84, Letter 182, Hevelius to des Noyers, 30 April 1677.

12 See note to Letter 20.

13 See note to Letter 12.

14 In the TC, Lubieniecki adds that he alludes to Tacitus, *Annales*, 1, 65.

41A.

Observations of Mira Ceti by Heinrich Sivers

BO: **C1, v. 7, 1098**
TC: 929–30

Caeterum mitto tandem Observationes meas de stella illa mira in Ceto, quas ut saepius desideratas recusare non possum, licet tantum obiter a me institutae fuerint, idque sine apparatu alias necessario, quem deesse novisti.[15]

Anno 1647 Mense Februario Clariss[imus] Dominus Joach[imus] Jungius mihi inter alios ad se accersito stellam ostendit in loco caeli G, quem concursus rectae DF per stellas D, E[16] et F evidentiores in lino piscium boreo, ultra nodum F productae, cum recta (sive basi trianguli Menkar) AC per C continuata notat: eamque diligenter observare jussit, utpote, accremento et decremento obnoxiam. Feci ergo, et phaenomena eius, quotquot serenitas caeli offerebat, sedulus retuli usque ad occasum heliacum; observationes autem retinuit Jungius. Post, ubi cetus iterum emersit, stella evanuerat, nec per reliquum anni tempus visui se unquam obtulit.

Anno 1648. Observationem ulteriorem aggressurus vocabar in Holsatiam, ad instituendum Nobilissimi Domini Henningi Wohndfleten filium Valdemarum.[17] Abhinc vero per integros 7 annos, (quorum plerosque revocatus ex Holsatia impendi studiis Fratrum nobilissimorum de stirpe Generosa Sangerhusiana dirigendis)[18] occasio nulla data est quicquam de stella ista notandi. Observasse quidem subinde memini; sed, quum id plerumque in itineribus factum sit, de tempore vel apparitionis vel occultationis certus non sum. Tandem in Patriam reversus.

Anno 1655, Per Octobr[em]: Novembr[em]: et Decembrem item per

Anno 1656, Januarium caelum contemplatus sum; stellam vero in loco G nullum inveni. Regionem igitur istam caeli rimatus ope Telescopii trium circiter pedum, stellulas[19] istas quas diagramma[20] inter F et G item G et C exhibet, offendi.

Anno 1657 et 1658 Animus variis negotiis distractus contemplationibus non vacavit

15 Caeterum...novisti *om.* TC.
16 E *om.* TC.
17 ad...Valdemarum *om.* TC.
18 quorum...dirigendis *om.* TC.
19 stellasulas (*primum* as *del.*) **BO**.
20 diagramma : diagramma superius **TC**.

Anno 1659 Mensibus Septembri exeunte, Octobri,[21] Novembri et Decembri comparuit iterum stella saepius laudata in G, primum magnitudine aequalis fere ipsi A; deinde sensim deficient: sed incommoditas temporis non permisit observare, utrum etiam stellae circa hunc locum antea observatae eodem modo sese haberent.

Anno 1660: 17[mo] Septembris iterum apparuit haec stella aeque magna ac anno praecedenti. Verum // nec tum morbo correptus, observationem instituere valui. Ubi autem convalui, et Anno 1661, 13[tio] Januarii observaturus fui situm stellarum circumvicinarum, illa in G plane evanuerat.

Anno 1661: 16[to] Augusti hor[a]: 11, comparuit stella mira, magnitudine fere aequalis stellae trianguli viciniori[22] C. Abhinc caelum fuit nubilum usque ad Septembrem, per quem fere integrum eandem stellae istius, visui redditae, magnitudinem observavi. Deinde decreverit sensim, usque adeo ut circa initium Novembris libero visu amplius conspicua non esset.[23]

Anno 1662: 28[vo] Augusti hor[a]: 20,[24] 30' stella in G iterum visa aequabat magnitudinem eiusdem stellae vicinae C. Posthac caelum continue nubilum eius contemplationem subtrahebat usque ad 2 Octobr[is], quo vix conspiciebatur. ita enim jam decreverat.

Anno 1663: 30[mo] Augusti nondum comparuit. Dehinc caelum partim nubilum: partim Lunae crescentis splendore occupatum speculationibus incommodum fuit.

21, 22, 23, et 24 Sept[embris] singulis diebus a 9[na] in 12[mam] usque observavi locum ope Telescopii pedalis (maior et melior enim tum ad manus non fuit) et offendi stellam perparvam intensi quidem splendoris, quae tamen visu non armato conspici nequivit.

12 Octobr[is] et aliis subsequentibus diebus, quibus Nubes coelum non obvelabant eodem modo observatio instituta est, nec tamen vel hoc anno vel proxime sequentis[25]

Anni 1664[ti26] tam autumno quam hyeme quicquam aliud deprehendi, nisi stellulam illam.

Anno 1665, 10 Julii ab h[ora][27] ad 1 usque observavi locum caeli, ubi stella illa apparere solita est; sed nec tum voti compos, stellulam iterum et quidem per tubum

21 et *del. post* Octobri **BO**.
22 viciniorii (*ultimum* i *del.*) **BO**.
23 fuerit *del.* **BO**.
24 20 : 10 **TC**.
25 sequentis : sequenti **TC**.
26 Anni 1664[ti] : Anno 1664 **TC**.
27 h[ora] *om.* **TC**.

opticum conspexi reliquis stellulis non minus locum suum servantibus. Idem quoque per reliquam huius anni partem mihi contigit. Sed[28]

Anno 1666: 11mo Februarii hor[a]: 7. tandem mihi stella desiderata[29] visa est liberis[30] oculis, loco consueto, magnitudinem exhibens fere tertiam et aequalem ferme vicine C in triangulo Menkar.

16. Eiusdem, denuo visa aequabat[31] magnitudinem et lucem duarum viciniorum B et C trianguli Menkar, eratque jam 3^{tiae} classis.

23. Eiusdem hor[a] 7. tertium apparuit in parte caeli occidua circa Horizontem. Magnitudo adhuc eadem apparebat; nec accrementum facile animadverti poterat, cum radiis Solis crepuscularibus involuta teneretur, quibus etiam deinde magis magisque tecta (accedente simul splendore Lunae abhinc crescentis) ad occasum heliacum vergens conspectum sui porro negavit.

Haec sunt, quae huc usque observare licuit. Vale et fave constanter

Tui

H[enrico] S[ivers][32]

Furthermore, I am sending you my observations of Mira Ceti; since you have asked, I cannot refuse to do it, even though I made them cursorily and without the needed instruments which you know I lack.

In February 1647, the most famous Master Joachim Jungius[33] called for me and showed me (among other things) the star in the sky in point G, which was indicated by the straight line DF, running through visible stars D, E, and F in the Northern

28 Sed *om.* TC.

29 desiderata : desiderata optato TC.

30 liberis : et quidem liberis TC.

31 aequabat : aequabat jam TC.

32 Haec...S[ivers] : Nec post ortum Ceti heliacum per residuas anni partes visus est, licet saepius suo loco quaesitus. Anno 1667. 17. Jan[uarii] ad duas circiter horas nubilis, quae continuo per tres hebdomadas caelum obsederant, hoca octava iterum comparuit stella ista vaga, idque liberis oculis, magnitudinem exhibens quartam. Abhinc observationibus Luna obfuit. 1. Febr[uarii] hor[a] 7. 30'. visui reddita fere accedebat ad tertiam magnitudinem. 14. eiusd[em] hor[a] 7. tertium visa fulgorem tertiae magnitudinis, qualem convenit stellae B. spargebat TC.

33 Joachim Jungius (1587–1657), German logician, mathematician, and philosopher. Since 1629, he was a professor at the *gymnasium academicum* in Hamburg. For more information about him see Christoph Meinel, *In physicis futurum saeculum respicio. Joachim Jungius und die Naturwissenschaftliche Revolution des 17. Jahrhunderts* (Göttingen: Vandenhoeck & Ruprecht, 1984); Christoph Meinel, *Die Bi-*

Cord of Pisces beyond the Knot F, together with the straight line AC (or the base of the Menkar triangle), prolonged beyond C. He told me to watch it diligently, because it was increasing and decreasing. I did so, and I noted its phenomena sedulously, whenever the sky was clear enough, until its heliacal setting, but Jungius kept the observations. Later, when Cetus arose again, the star had disappeared, and it was not visible for the remaining part of the year.

In 1648, when I was about to start further observations, I was appointed to Holstein as a teacher for Valdemarus, the son of the most noble Master Henningius Wohnsflet. Since then, for the whole seven years (for many of which I was revoked from Holstein to supervise the studies of the most noble brothers of a family from Sangerhausen), I had no chance to see anything regarding this star. I remembered to carry out the observations, but because I made them during my journeys, I am not sure about the time of its appearance or of its occultation. Finally, I came back home.

In 1655, in October, November, and December, and also

In 1656, in January, I watched the sky, but I found no star in point G. When I investigated that part of the heavens with a three-foot-long telescope, I found those stars which are between F and G and also between G and C in the picture.

In 1657 and 1658, my mind was distracted by many other duties, and it had no time for such contemplations.

In 1659, at the end of September, in October, November, and December, the aforementioned star appeared again in G, and initially its magnitude was nearly equal to A; later, it was gradually extinguishing, but the inconvenient time did not allow me to see whether other stars around that place changed or not.

In 1660, on 17 September, this star reappeared, as bright as the year before. But I was sick at that time, and I could not make any observations. When I recovered and, on 17 January 1661, I wanted to see the place of the neighbouring stars, the one in G was entirely invisible.

In 1661, on 16 August, at 11 o'clock, Mira Ceti appeared, nearly equal in its magnitude to the star in the triangle which is close to C. Later, the sky was obscured by some clouds until September, when (for the entire month) I was watching the magnitude of that star, as it was visible again. Then it diminished so at the beginning of November it could not be seen with the naked eye.

bliothek des Joachim Jungius. Ein Beitrag zur Historia litteraria der frühen Neuzeit (Göttingen: Vandenhoeck & Ruprecht, 1992).

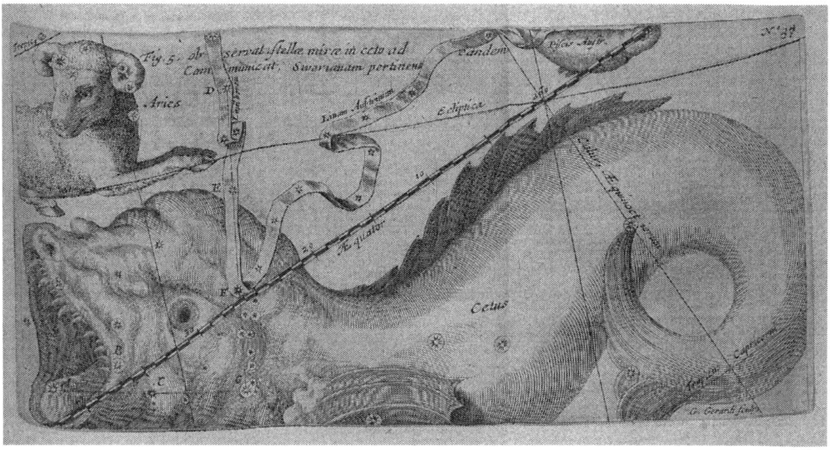

Figure 13. Sivers's observations of Mira Ceti. Illustration from Lubieniecki's *Theatri cometici pars prior*. Biblioteka Narodowa, Warsaw, SD XVII.4.2660.

In 1662, on 28 August, at 8.30 in the evening, the star in G was seen again, and it was equal to the magnitude of the neighbouring star C. Later the clouds, continually obscuring the sky, made the observation impossible until 2 October when it was barely visible. It had already diminished.

In 1663, it did not appear until 30 August. Later, the sky was not favourable for observations because it was partially obscured by the clouds and partially illuminated with the moonlight.

On 21, 22, 23 and 24 September, every day since 9 till 12 o'clock, I watched that place with a one-foot-long telescope (because then I did not have a bigger and better one) and I saw a very small star shining with a bright light which could not be seen with the naked eye.

On 12 October and for the next few days, when the clouds did not cover the sky, I conducted the observations in the same way, but neither this year nor the next,

1664 (neither in autumn, nor in winter) I saw anything else but that little star.

In 1665, on 10 July, since 11 until 1 o'clock, I watched the place in the sky where that star used to be, but my wish was not fulfilled, and I did not see that little star again, even with a telescope, while the other stars kept their places. I also failed to do so for the remaining part of the year. But

In 1666, on 11 February, at 7 o'clock, I saw the star I wished for with the naked eye, in its proper place, of about the third magnitude and nearly equal to the neighbouring C in the Menkar triangle.

On the sixteenth of the same month, the star, visible again, had the equal brightness and light to two neighbouring B and C in the Menkar triangle, and it was of the third magnitude.

On the twenty-third of the same month, at 7 o'clock, the star appeared for the third time in the western part of the sky, near the horizon. So far, it has shown the same magnitude, and no increment could be seen; however, it was veiled with the rays of the sunset, in which it was later enshrouded (as well as in the light of the crescent Moon) when it moved towards its heliacal setting and became invisible again.

This is what I could see so far. Farewell, and be benevolent towards

Yours,

Heinrich Sivers

42.

16 April 1666, Lubieniecki to Hevelius

BO: C1, v. 7, 1109
TC: 413

Enclosures:
42A. Excerpt of the letter from Lubieniecki to Boulliau, 18 April 1666
42B. Excerpts of the letters from Gruterus to Lubieniecki, 12 March and 1 April 1666

Salutem Plurimam Nobilissime, Amplissime, Clarissime.

Non diu Te nunc morabor, sed paucis potius me absolvam.[1] Nam et Te assiduo occupatum esse scio, et me esse fateor.[2] Ideo neque nunc Tuas curas et operas turbarem nisi haec quae ad Te pertinent, a Cl[arissimo] Viro Isaaco Grutero, haud ita pridem accepissem. Sunt vero mihi ista[3] reddita simul, licet diversis temporibus scripta. Si quid typographo illi indicari velis, me mone. Ego dabo operam, ut id per Amicum fiat. En Tibi quae nuper ad Cl[arissimum] Bullialdum Amicum communem de Cometis[4] dedi in tabellis, ut res planior reddatur, proposita. De his velim sententiam Tuam, si fieri potest, paucis dicas. Cum autem de illo Cometa[5] A[nno] C[hristi] 837. in Virgine viso Auctor quidem vitae Ludovici Pii non more Planetarum in Orientem perrexisse dicat Aventinus vero ei contradicat, velim verbo[6] hic quoque sententiam Tuam indices. Vale et age feliciter ac fave constanter. Dabam Hamburgi die 6. 16. April[is] 1666.

Tuus
omni studio, cultu et
officio

S[tanislaus] L[ubieniecius] Rolitsius
manu propria

1 absolvam : expediam **TC**.
2 assiduo…fateor : scio et me fateor assiduo esse occupatum **TC**.
3 mihi ista : ista mihi **TC**.
4 Cometis : Cometis circa mortem Lodovici Pii visis **TC**.
5 de…Cometa : illum Cometam **TC**.
6 indices *del. post* verbum **BO**.

Greetings, most noble, worthy, and renowned Sir!

I will not wait for your answer any longer, but instead, I will write something more to you. I know that you are continually preoccupied, and I admit that I am too. I would not disturb your affairs and works now had I not received some letters which concern you from the most renowned Isaac Gruterus.[7] They were both delivered to me at once, even though they had been written at different times. Tell me if you would like me to send them to the printer. I will take care to do so through a friend. I enclose what I sent about the comets to the renowned Boulliau,[8] our mutual friend; I added a table there in order to make it clearer. I would like you to tell me your opinion about them, if you only can. Because about that comet seen in Virgo in the year 837[9] after the birth of Christ, the author of the life of Louis the Pious says that it moved eastwards (contrary to the planets[10]), while Aventinus[11] disagrees with him. I would like you to make a brief comment about it. Farewell, and be prosperous and stay constantly benevolent! Given at Hamburg, on 6/16 April 1666.

Your most obliged admirer,

Stanisław Lubieniecki Rolitsius
by my own hand

42A.

Excerpt of the letter from Lubieniecki to Boulliau, 18 April 1666

BO: C1, v. 7, 1109 (follows the preceding one)
TC: 544

Excerptum ex Literis ad Cl[arissimum] Bullialdum datis

Cum novem Cometae intra septennium circa excessum Ludovici Pii annotari videantur, en Tibi rem in tabella propositam, et conatum simul conciliationis, quae quatuor Cometas ex illis novem fecit.

7 See note to Letter 28 and section 2.2.3 of the Introduction.

8 See section 2.4.2 of the Introduction. The date of the enclosure, given in the TC, suggests that it was sent after the letter to Hevelius.

9 See note to Letter 42A.

10 *Vita Hludovici imperatoris*, in *Monumenta Germaniae historica. Scriptorum tomus II*, ed. by Georg Heinrich Pertz (Hannoverae: impensis Bibliopolii Aulici Hahniani, 1829), p. 642.

11 Johannes Aventinus (Johann Georg Turmair, 1477–1534), German historian and philologist from Bavaria. Lubieniecki refers to his *Annalium Boiorum sive veteris Germaniae libri VII*, ed. by Nicolaus Cisner, (Francofurti: impensis Ludovici Regis, 1627), p. 236.

Cometa A[nno] C[hristi]
1. 837. in Libra *Annal*[*es*] *Fuld*[*enses*] in Virgine, Calvis[ius] Vid[e] Eckstorm[ium]
2. 838. in Libra Rock[enbachius] in Scorpio, Calvis[ius] Vid[e] Aventin[um] et Eckstorm[ium]
3. 839. in Virgine (corr[ectum] 837.) Auct[or] Vitae Lud[ovici] in Ariete Vincent[ius] Func[cius] Rock[enbachius] Eckstorm[ius]
4. 840. in Scorpio (corr[ectum] 838.) Auct[or] Vitae Lud[ovici] Vid[e] Alsted[ium]
5. 841. in Aquario *Annal*[*es*] *Fuld*[*enses*] Vid[e] Alsted[ium]
6. 842. in Aquario Func[cius] Rock[enbachius] Eckst[ormius] Vid[e] Alst[edium] tum *Chron*[*icon*] *Fontanell*[*ense*][12] et Nithard[um]
7. 8. 9. Anno 843. in Libra, Ariete et Aquario Ricciol[us] ex Cardano.

Cometa A[nno] C[hristi]
1. 837. (corr[ectum] 836) qui autumno Solem antecessit. Eckstorm[ius]
2. 838. (corr[ectum] 837) qui vere Solem secutus est. Eckst[ormius] Avent[inus] Rockenb[achius] apparuit in Virgine, duravit 25. diebus Anonym[us] Calv[isius] confer Eckst[ormium] *Ann*[*ales*] *Fuld*[*enses*] et Rockenb[achium]
3. 838. in Scorpio Anonym[us] Calv[isius] 839. in Ariete Vincent[ius] Func[cius] Rock[enbachius] Eckst[ormius]
4. 841. exeunte et 842. ineunte in Aquario.

Hic verbo dicendum est, quod probabilis coniectura suggerit, Cometam in Libra primo 837. visum ad Virginem perrexisse, et in Tauro disparuisse. Pariter illum 838. a Scorpio ad Arietem: utrumque contra Signorum seriem cursum tenuisse. Ita tres Cometas Alstedii annis[13] 840, 841 et 842 signis Scorpii, Arietis et // Aquarii (nam illum in Virgine Alstedius in annum 814. retulit parum, fateor, provide) annis 838, 839 et 841. simulque 842. adscribendos: illos vero Cardani in annum 837. 838 et 841. simulque 842. collocandos puto.

An excerpt of the letter sent to the most renowned Boulliau

Because nine comets seem to have been described during the seven years around the death of Louis the Pious,[14] I have created a table and taken some effort to reconcile the data in order to make four comets out of those nine.[15]

12 Fontanell[ense] **TC** : Fortunatell[ense] **BO**.

13 annis **TC** : annii (*secundum* i *del.*) **BO**.

14 Louis I the Pious (778–840), son and successor of Charlemagne, since 813 co-emperor with his father, Emperor since 814.

15 An extraordinary number of comets appeared in the late thirties and early forties of the ninth century. In the 837, the Comet Halley appeared (1P/837 F1). Significant comets were also seen in the 838 (X/838 V1), 839 (X/839 B1), and 841 (X/841 Y1), see Alexandre Guy Pingré, *Cométographie ou Traité historique et théorique des comètes*, vol. 1 (Paris: Imprimerie royale, 1783), pp. 340–47; Gary W. Kronk,

Comet in the year after the birth of Christ
1. 837 in Libra — *Annales Fuldenses*;[16] in Virgo — Calvisius,[17] see Eckstorm,[18]
2. 838 in Libra — Rockenbach;[19] in Scorpio — Calvisius,[20] see Aventinus[21] and Eckstorm,[22]
3. 839 in Virgo (corrected to 837) — the author of the life of Louis the Pious;[23] in Aries — Vincent,[24] Funck,[25] Rockenbach,[26] Eckstorm,[27]
4. 840 in Scorpio (corrected to 838) — author of the life of Louis the Pious,[28] see Alsted,[29]
5. 841 in Aquarius — *Annales Fuldenses*,[30] see Alsted,[31]
6. 842 in Aquarius — Funck,[32] Rockenbach,[33] Eckstorm,[34] see Alsted[35] and *Chronicon Fontanellense*,[36] and Nithard[37] as well

Cometography. A Catalogue of Comets. Volume 1: Ancient – 1799 (Cambridge: Cambridge University Press, 1999), pp. 125–32. Reports collected by Lubieniecki refer to these phenomena.

16 *Annales Fuldenses*, in *Monumenta Germaniae historica. Scriptorum tomus I*, ed. by Georg Heinrich Pertz (Hannoverae: impensis Bibliopolii Aulici Hahniani, 1826), p. 361.

17 Sethus Calvisius (see note to Letter 27), *Opus chronologicum* (Francofurti: impensis Johannis Thymii, 1620), p. 550.

18 Heinrich Eckstorm (see note to Letter 27), *Historiae eclipsium, cometarum et pareliorum* (Helmaestadi: typis heredum J. Lucii, 1621), p. 233.

19 Abraham Rockenbach (see note to Letter 30), *De cometis tractatus novus methodicus* (Witebergae: ex Officina Cratoniana, sumptibus Henningii Grosii, 1602), p. 170.

20 Calvisius, *Opus chronologicum*, p. 550.

21 Johannes Aventinus (see note to Letter 42), *Annalium Boiorum sive veteris Germaniae libri VII*, ed. by Nicolaus Cisner, (Francofurti: impensis Ludovici Regis, 1627), p. 236.

22 Eckstorm, *Historiae eclipsium, cometarum et pareliorum*, p. 233.

23 *Vita Hludovici imperatoris*, in *Monumenta Germaniae historica. Scriptorum tomus II*, ed. by Georg Heinrich Pertz (Hannoverae: impensis Bibliopolii Aulici Hahniani, 1829), p. 642.

24 Vincent of Beauvais (*c.* 1184/1194–*c.* 1264), medieval encyclopaedist of the Dominican Order. Lubieniecki refers here to his *Speculum historiale*, see Vincent of Beauvais, *Speculi maioris* [...] *tomus quartus, qui Speculum historiale iscribitur*, (Duaci: ex officina typographica Baltazaris Belleri, 1624), p. 974.

25 Johann Funck (see note to Letter 27), *Chronologia* (Basileae: excudebat Jacobus Parcus sumptibus Joannis Oporini, 1554), p. 446.

26 Rockenbach, *De cometis tractatus novus methodicus*, p. 170.

27 Eckstorm, *Historiae eclipsium, cometarum et pareliorum*, p. 233.

28 *Vita Hludovici imperatoris*, in *Monumenta Germaniae historica. Scriptorum tomus II*, p. 644.

29 Johann Heinrich Alsted (see note to Letter 27), *Thesaurus chronologiae* (Herbornae Nassoviorum: [s.e.], 1650), pp. 506–07.

30 *Annales Fuldenses*, in *Monumenta Germaniae historica. Scriptorum tomus I*, p. 363.

31 Alsted, *Thesaurus chronologiae*, pp. 506–07.

32 Funck, *Chronologia*, p. 446.

33 Rockenbach, *De cometis tractatus novus methodicus*, p. 170.

34 Eckstorm, *Historiae eclipsium, cometarum et pareliorum*, pp. 233–04.

35 Alsted, *Thesaurus chronologiae*, pp. 506–07.

36 *Fragmentum chronici Fontanellensis*, in *Monumenta Germaniae historica. Scriptorum tomus II*, p. 301.

37 Nithard (*c.* 795–844), Frankish historian. Lubieniecki refers to his chronicle, see *Nithardi historiarum libri IIII*, in *Monumenta Germaniae historica. Scriptorum tomus II*, pp. 666–67.

7. 8. 9. 843 in Libra, Aries, and Aquarius — Riccioli[38] based on Cardano.[39]

Comet in the year after the birth of Christ
1. 837 (corrected to 836) which moved before the Sun in autumn — Eckstorm,[40]
2. 838 (corrected to 837) which followed the Sun in spring — Eckstorm,[41] Aventinus,[42] Rockenbach;[43] it appeared in Virgo, lasted for 25 days — anonymous,[44] Calvisius,[45] compare to Eckstorm, *Annales Fuldenses*,[46] and Rockenbach,
3. 838 in Scorpio — Anonymous,[47] Calvisius;[48] 839 in Aries — Vincent,[49] Funck,[50] Rockenbach,[51] Eckstorm,[52]
4. at the end of 841 and at the beginning of 842 in Aquarius.

It should be added that a probable guess suggests that the comet seen for the first time in Libra in 837 moved towards Virgo and disappeared in Taurus. The one of the year 838 also moved from Scorpio towards Aries. Both moved against the order of signs. So, three comets of Alsted[53] — in 840, 841, and 842 in the signs of Scorpio, Aries, and Aquarius respectively (because the one in Virgo Alsted placed in 814 carelessly, I admit)[54] should be ascribed to the years 838, 839, and 841–42. And I think those of Cardano[55] should be placed in the years 837, 838, and 841–42.

38 Giovanni Battista Riccioli (see section 2.4.15 of the Introduction), *Almagestum novum*, pars posterior tomi primi (Bononiae: ex Typographia Haeredis Victorii Benatii, 1651), p. 6.

39 Girolamo Cardano (1501–76), Italian mathematician, astrologer and physician. After Riccioli, Lubieniecki refers here to his *In Cl. Ptolemaei de astrorum iudiciis, aut (ut vulgo appellant) Quadripartitae Constructionis Lib. IIII*, (Basileae: ex officina Henricpetrina, 1578), p. 276.

40 Eckstorm, *Historiae eclipsium, cometarum et pareliorum*, p. 233.

41 Eckstorm, *Historiae eclipsium, cometarum et pareliorum*, p. 233.

42 Aventinus, *Annalium Boiorum sive veteris Germaniae libri VII*, p. 236.

43 Rockenbach, *De cometis tractatus novus methodicus*, p. 170.

44 Lubieniecki refers here to the anonymous author of the life of Louis the Pious, see *Vita Hludovici imperatoris*, in *Monumenta Germaniae historica. Scriptorum tomus II*, p. 642.

45 Calvisius, *Opus chronologicum*, p. 550.

46 *Annales Fuldenses*, in *Monumenta Germaniae historica. Scriptorum tomus I*, p. 361.

47 Lubieniecki refers here to the anonymous author of the life of Louis the Pious, see *Vita Hludovici imperatoris*, in *Monumenta Germaniae historica. Scriptorum tomus II*, p. 644.

48 Calvisius, *Opus chronologicum*, p. 550.

49 Vincent of Beauvais, *Speculi maioris* [...] *tomus quartus, qui Speculum historiale iscribitur*, p. 974.

50 Funck, *Chronologia*, p. 446.

51 Rockenbach, *De cometis tractatus novus methodicus*, p. 170.

52 Eckstorm, *Historiae eclipsium, cometarum et pareliorum*, p. 233.

53 Alsted, *Thesaurus chronologiae*, pp. 506–07.

54 Alsted, *Thesaurus chronologiae*, p. 506.

55 Cardano, *In Cl. Ptolemaei de astrorum iudiciis, aut (ut vulgo appellant) Quadripartitae Constructionis Lib. IIII*, p. 276.

42B.

Excerpts of the letters from Gruterus to Lubieniecki, 12 March and 1 April 1666

BO: C1, v. 7, 1109 (follows the preceding one)
TC: 873

Excerpta ex Literis a Cl[arissimo] Viro Isaaco Grutero Roterodami datis.

1. die 12. Martii 1666.

Typographus mihi author, secundum *Prodromi* Heveliani editionem e suo promitti praelo, post longi temporis intervallum rogatus an constet sibi, nihil reponit; ut verear ne haec jactantius effutierint eius primae et rudes de ista re cogitationes, quod vero convenientius. Vale.

2. Kal[endas] April[es]

Illudebat exspectationi meae Typographus de Hevelianae cometologiae *Prodromo* rogatus, an operas eius exerceret liber et quando secundae editionis spem faceret: cum pateretur me duntaxtat ancipitem luctari cum promissi umbra, quo poscebat[56] me grandi in congressu ante menses complusculos. Visum ergo per alios, sed dissimulato meam hic rem agi; exploratius cognoscere singula, quae tui coniectabam esse voti. Reponunt, perstare eiusdem pollicitationis[57] tenacem, sed frigus aliquod suffundi, ut certior ab eo quam antea non discedas: et putent insanire[58] typographum: quem suo permittere genio statui. Graevianae exemplar orationis non reperio in nostris officinis: et author vir mihi longe // amicissimus, hoc nomine interpellatus, silet: distractus, opinor laboribus Academicis, et limandis quae in Hesiodum paravit, mox prelo subiicienda. Ut doleam fraudari te opella mea; quam et hic et alibi velim experiaris animo meo parem, id est, alacrem, et nullo tardandam frigusculo.

56 poscebat **TC** : pascebat **BO**.
57 pollicitationis **TC** : pollicitationes **BO**.
58 insanire : felle nimio insanire **TC**.

Excerpts of the letters from the most renowned Isaac Gruterus, given at Rotterdam.

1. 12 March 1666.

The printer,[59] when I asked him whether he knew if the second edition of Hevelius's *Prodromus* had been published at his printing press, did not answer me, so I presume that his first and crude statements about that thing were nothing but boasting — which would be better on the whole. Farewell!

2. 1 April

The printer mocked my expectations when I asked him about the *Prodromus* of Hevelius's cometology, whether this book was prepared at his workshop and when the second edition would be published; meanwhile it is clear that he is fighting against me with a shadow of the promise he gave me in a great gathering of people many months ago. I decided to learn more details about it through others, keeping my involvement secret, which would agree with your wishes, as I suppose. Those people answer that he is willing to keep his promise, but he is careless, so you will not be more assured after meeting him than you were before; and they think the printer is mad, and I have decided to leave him alone to his mind. I cannot find a copy of Graevius's *Oratio* in our bookstores, and the author, who is my good old friend, is silent when I ask about it; I think he is distracted by his scholarly works and by finishing what he prepared about Hesiod, which will be published soon.[60] Therefore, I grieve that you were deceived by my actions, even though I would like you always to experience them as matching my intentions, viz. as ardent and not hindered by any negligence.

59 See note to Letter 35 and section 2.2.3 of the Introduction.

60 Johann Georg Graevius (1632–1703), German classical scholar, professor of rhetoric at the University of Duisburg (1656–62) and at the University of Utrecht (since 1662), professor of history and politics (since 1667) at the University of Utrecht. In this passage, Gruterus refers to his *Oratio de cometis contra vulgi opinionem cometas esse malos nuncios* (Traiecti ad Rhenum: typis et officina Gisberti a Zyll, 1665) and *Ησιόδου Ἀσκραίου τὰ εὑρισκόμενα. Hesiodi Ascraei quae exstant*, ed. by Johann Georg Graevius (Amstelodami: apud Danielem Elzevirium, 1667).

43.

17 April 1666, Hevelius to Lubieniecki

BO: C1, v. 7, 1099
TC: 413–14

Domino Rolissio
Hamburg[i]

Illustrissime Vir.[1]

Descriptio mea posterioris Cometae, nec non *Mantissa Prodromi Cometici*, jam sub praelo serio fervet;[2] quam primum typis penitus exscripta erit, quod ultimo Maii fieri posse, DEO annuente, spero, vobis quantocyus transmittam. Hucusque computando, ad scribendo hocce opusculo satis mihi fuit negotii; nunc iconismos aeri incidendos, quorum quinque sunt, aggredior. Phaenomenum illud ad Cingulum Andromedae pro nullo Cometa, sed tantum pro stella quapiam nebulosa haberi potest, quanquam diversam faciem ab aliis nebulosis exhibet; de qua in opusculo meo aliquanto plenius dicendum erit. Nulla aerea impressio, sive traiectio, etiamsi aliquantisper duret, tum caudam, instar Pareselenarum ac Pareliorum et c[etera]: referat, potest dici Cometa. Non ignoro, quidem plurimos inveniri, utpote Licciolum aliosque, qui etiam nonnullos aereos Cometas nonnunquam dari, statuunt; sed ego plane aliter sentio; nullum videlicet genuinum Cometam in aere generari ac toto durationis tempore ibidem subsistere. Saepius quidem lubens[3] largior, plurimos per aerem nostrum sublunarem traiicere; sed brevi tantum temporis spatio in eo commorari posse: prout plurimis exemplis Cometa 1472, 1618, 1652 aliisque demonstrari posset; sed de his in *Cometographia* plenius. Sic ut minime eo propendeam dari Cometas etiam aereos sed mere aethereos, ex materia utplurimum aetherea prognatos; atque sic non aeterna, ut plurimi hoc nostro aevo sentiunt, sed temporanea esse corpora[4] ac nulla ratione, certa periodo // redeuntia. An phaenomenum istud, cuius noster[5] Keckermannus et Pet[rus] Creüster meminit, Cometa fuerit genuinus?[6] valde dubito; cum nullis probis observationibus et rationibus unquam poterit probari. Dictus Keckermannus alioquin omnes Cometas aereos esse voluit: purus enim putus fuit Philosophus Peripateticus, uti probe nosti. Schüleri

1 Domine *del.* **BO**.
2 fervet : servet **TC**.
3 lubens *om.* **TC**.
4) *del. post* corpora **BO**.
5 noster *om.* **TC**.
6 Cometa...genuinus : genuinus fuerit Cometa **TC**.

Cometologiam non vidi; si Hamburgi prostat,[7] ut eum emas, ac per Lubecam data occasione mihi transmittas humanissime peto, pecuniam lubentissime quantocyus restituam. Pro observationibus istis mirae stellae in Collo Ceti Cl[arissimi] Henrici Syverti,[8] quem meo nomine salutes rogo, gratias habeo debitas: optime observationibus meis in *Historiola* mea huius stellae exhibitis respondent. Ad quam hactenus ad finem Januarii usque probe et diligenter attendi; sed a duobus annis et amplius prorsus latuit; Mense autem Febr[uario] et Martio, non memini me ad locum illum, nescio qua incuria, ut ut saepius plurimas observationes isto tempore peregerim, attendisse; sic ut illam minime adhuc viderim; fieri tamen potuit, ut eo tempore a Cl[arissimo] Syverto[9] indicato, prodierit; de quo Ipsi gratulor eam conspexisse. Notandum[10] igitur erit,[11] an Mense Julio denuo apparitura sit? sed vix puto eousque eam posse perdurare; poterit tamen post Solstitium Brumale redire. An *Astronomia reformata* Riccioli, jam edita sit, et an Hamburgi prostet, scire valde desidero. Si Tibi obtinget occasio ad illum scribendi, meo nomine illum humanissime salutes; ego aliquoties, ante aliquot annos ad illum scripsi; sed nullum plane responsum obtinui, qui factum sit, nescio. Vale[12] prosperrime. Gedani 1666 17 Apr[ilis].

Tuae Illustrissimae Dominationi
ex animo bene cupiens

J[ohannes] Hevelius

To Master Rolitsius
in Hamburg

Most illustrious Sir!

My *Descriptio* of the second comet and the *Mantissa Prodromi cometici* are being printed right now; when they are ready (I hope it will happen at the end of May, with God's help) I will send them to you as soon as possible. So far, I have had enough work with computing and writing this little book; now I am starting to engrave the illustrations; there are five of them. That phenomenon near the Belt of Andromeda[13] cannot be considered a comet, but only a nebulous star, however it looks different from other nebulous ones; I will say something more about it in my

7 prostat : prostet **TC**.
8 Syverti : Syversi **TC**.
9 Syverto : Syversio **TC**.
10 Notandum : Attendendum **TC**.
11 erit : erit nobis omnibus **TC**.
12 P *del. post* Vale **BO**.
13 See section 2.3.2 of the Introduction.

booklet. No sighting in the air or a falling star can be called a comet, even if it is visible for some time and shows a tail, similar to paraselenes and parhelia, and so on. I know there are many people who, like Riccioli[14] and others, state that sometimes there are comets in the air, but my opinion is entirely different, viz. that no real comet comes to existence in the air and stays there for the whole time of its duration. I will rather willingly admit that more often comets cross our sublunary air, but they can remain there only for a short time; it could be proven with many examples, like the comets of 1472,[15] 1618,[16] 1652[17] and others, but there will be more about it in the *Cometographia*. Hence, I would not lean to the opinion that there are also comets in the air, but rather that they are only in Aether, and that they originate generally from aethereal matter; therefore, they are not eternal, as many people in our century think, but they are temporal bodies and by no means do they return in any specified period of time. I doubt whether that phenomenon, mentioned by our Keckermann[18] and Peter Creutzer,[19] was really a comet, because it could never be proven by reliable observations or statements. Moreover, Keckermann thought all the comets were in the air because he was a pure and devoted peripatetic philosopher, as you surely know. I have not seen Schuler's *Cometologia*;[20] if it is available in Hamburg, I kindly ask you to buy a copy for me and to send it through Lübeck at your earliest convenience; I will send you the money back as soon as possible. I am grateful for the observations of Mira Ceti made by the most renowned Heinrich Sivers,[21] whom I ask you to greet on my behalf; they are consistent with my observations published in the *Historiola*.[22] So far I had watched it carefully and diligently until the end of January, but for the last two years it has not been visible; in February and March, I forgot to watch that spot due to some carelessness, despite the fact that I conducted many other observations at that time, so I have not seen that star so far. However, it could happen that at the time indicated by the most renowned Sivers it was visible; I congratulate him on spotting it. It should be noticed if it reappears in July, but I do not think it will last until then; nonetheless, it can return after the autumnal solstice. I would like to know if Riccioli's *Astronomia reformata*[23] has already been published and if it is available in Hamburg. If you have a chance to

14 See section 2.4.15 of the Introduction.

15 C/1471 Y1, see Alexandre Guy Pingré, *Cométographie ou Traité historique et théorique des comètes*, vol. 1 (Paris: Imprimerie royale, 1783), pp. 471–77; Gary W. Kronk, *Cometography. A Catalogue of Comets. Volume 1: Ancient – 1799* (Cambridge: Cambridge University Press, 1999), pp. 285–89.

16 See note to Letter 1F.

17 See note to Letter 2B.

18 See notes to Letters 7A and 41.

19 See note to Letter 41.

20 See note to Letter 41.

21 See note to Letter 20.

22 Hevelius refers to his description of Mira Ceti published in his *Mercurius in Sole visus*, see note to Letter 41.

23 Giovanni Battista Riccioli, *Astronomiae reformatae tomi duo* (Bononiae: ex typographia Haeredis Victorii Benatii, 1665).

write to him, greet him kindly on my behalf; some years ago, I wrote a few letters to him, but I got no answer — I do not know why. Farewell, and be prosperous! In Danzig, on 17 April 1666.

Illustrious Sir,

Your affectionate friend,

Johannes Hevelius

44.

30 April 1666, Lubieniecki to Hevelius

BO: C1, v. 7, 1110
TC: 414

Seal with Lubieniecki's coat of arms (Rola coat of arms), address: 'A Mons[ieu]r / Mons[ieu]r Jean Hevelius Cons[eille]r / de la ville / à Dantzic' ('To *Monsieur* Johannes Hevelius, city Councillor, in Danzig')

Salutem Plurimam Nobilissime, Amplissime et Humanissime Vir.

Non diffiteor Cometam illum anni 1527. qui vix ultra horam duravit, in dubium vocari posse. Quia tamen a Clarissimis Astronomis, ipsoque Ricciolo et apud eum Jo[anne] Vogelino et Corn[elio] Gemma, in censum Cometarum relatus est, satis tuto eum Cometis annumerari posse arbitror. In controversiam veniunt, ut animadverto, hic sicut ubique, quae insolentia sunt. Nempe tam de annuis, quam dialibus Cometis dubitatur. Illi extraordinariis phaenomenis, hi meteoris accensentur. Sed quid de stella Anno 1033. auctore Calvisio a Septentrione versus Meridiem traiiciente et a die 28. Febr[uarii] ad 15. Martii durante, sentias, dic verbo. Idem fac ad illam nuper propositam Auctoris vitae Ludovici Pii et Aventini de eodem Cometa[1] contradictionem. Apud Cl[arissimum] Kirstenium hersterna die amicissimam Tui mentionem in lectione Tuarum literarum feci. Is Te humanissima salute impertit. Non licuit illi tam beato esse, // ut stellam illam miram Ceti hoc anno observaret. Invigilabit tamen et, ut sic dicam, insidiabitur ei brevi, ut sperat, rediturae. Cum paulo ante literas Tuas Humanissimo Sivero communicassem, rem me gratissimam ei fecisse perspexi. Plurimam ille reddit salutem Tibi. Neque vero ei per Januarium stella illa apparuerat, licet aliquoties caelum pro desiderio Auditorum lustraverit. Neque tamen videtur ei tunc primum illam prodiisse, utpote[2] jam valde magnam, sed fortassis per aliquot dies praecedentes vel etiam in fine Januarii jam apparituram fuisse, ni caelum nubilum prohibuisset. *Astronomia Reformata* Riccioli jam prodiit in lucem, sed hic nondum prostat. Imo vero eiusmodi libri perraro in has oras perferuntur. Ubi ad eum scribere mihi contigerit, officio salutationis Tuo nomine apud eum rite defungar, imo et literas Tuas recte curabo ac responsum obtinebo, ut spero. *Cometologiam* Schuleri heri ad Te misi per Generosiss[imum] Cyprianum Crenscium[3] Rautensteinianum er Peltzianum Amicum, // meum vero etiam popularem, qui, qua est humanitate in amicos officiosa, librum illum Domino Peltzio tradet. Pretium pro eo nullum exigo, nisi

1 Cometa : Cometa anni 837 **TC**.
2 utpote : utpote quae fuerit **TC**.
3 Crenscium : Crensium **TC**.

mihi alicuius Libelli Tui exemplum forte mittere velis. Spero nos aliquando volente Deo commutationem exemplarium aequam facturos, postquam Tuum meumque opus fuerit absolutum. Vale Vir Clarissime, et ad opera Tua praeclara edenda prospere incumbe. Dabam Hamburgi die 20. 30 April[is] 1666.

Tuus
omni studio et officio

S[tanislaus] L[ubieniecius] Rolitsius
manu propria

Post Scriptum.[4] Addo verbum. Pax inter Praesulem Monaster[iensem] et Batavos die 9. 19. huius mensis feliciter conclusa et mutuis Delegatorum subscriptionibus firmata est. Speratur vero et Anglicanam compositum et Rhenanam confirmatum iri. Quis non videat Laeta quoque Cometis e vestigio succedere?[5] Iterum V[ale].

Greetings, most noble, worthy, and kindest Sir!

I agree that the comet of 1527,[6] which lasted for a little more than an hour, can be questionable. But because it was mentioned in the cometary lists by most famous astronomers, even by Riccioli[7] himself and — as he informs us — by Johannes Vögelin[8] and Cornelius Gemma,[9] I think it can be considered a comet. There are usually discussions, as I observe, about unusual comets. Surely there are doubts both about year-long comets, as well as those day-long ones. The former are thought to be extraordinary phenomena, the latter — meteorological ones. But tell me briefly what

4 Dabam...Post Scriptum *om.* **TC**.

5 Succedere : Vos quidem id etiam experimini, qui vobis, ut ab aliis accepi, de tanta omnis generis frumenti secundo Vistula et Bugo ad Vos ex Polonia et Russia devecti copia, quanta ab integris decem annis non fuerit visa, gratulamini. Qua de re et ego vobis et patriae gratulor, cui pacem justam et fidam semper opto ex animo. **TC**.

6 See note to Letter 41.

7 Giovanni Battista Riccioli (see section 2.4.15 of the Introduction), *Almagestum novum*, pars posterior tomi primi (Bononiae: ex Typographia Haeredis Victorii Benatii, 1651), p. 9.

8 Johannes Vögelin (1500–49), German mathematician and astronomer, professor in Vienna. He was the author of *Significatio cometae qui anno M.D.XXVII apparuit cum passionibus eius*, printed in Tadeáš Hájek, *Dialexis de novae et prius incognitae stellae inusitatae magnitudinis et splendidissimi luminis apparitione, et de eiusdem stellae vero loco constituendo*, (Francofurti ad Moenum: [s.e.], 1574), pp. 150–67.

9 Cornelius Gemma (1535–78), astronomer and physician. Lubieniecki refers (after Riccioli) to his *De prodigiosa specie, naturaque cometae, qui nobis effulsit altior lunae sedibus, insolita prorsus figura, ac magnitudine, anno 1577 plus septimanis 10. apodeixis tum physica tum mathematica* (Antverpiae: ex officina Christophori Plantini, 1578), pp. 39–40.

you think about the star of 1033[10] which, according to Calvisius,[11] moved from the north southwards and lasted from 28 February until 15 March. Tell me also about that contradiction between the author of the life of Louis the Pious and Aventinus about the comet I have recently outlined to you.[12] Yesterday, at Kirsten's[13] house, when we were reading your letters, I made a friendly mention of you. He sends you his kindest greetings. He did not manage to see Mira Ceti[14] this year. But he will spy upon it and (let me say so) he will set an ambush for it, because he believes it will return soon. Recently, when I gave your letters to the kindest Sivers,[15] I saw I made him very happy. He sent you many greetings. He also did not see that star in January, however he checked the sky a few times when the public asked him. Neither does he think that the star appeared then (as it is quite big), but perhaps it could have appeared a few days before, or even at the end of January, if the cloudy sky had not hindered the observation. Riccioli's *Astronomia reformata* has already been published but is not available here yet. Books of this kind seldom come here. When I have a chance to write to him, I will fulfil my duty to greet him on your behalf; I will even send him your letter, and I will get the answer, I hope. Yesterday I sent you Schuler's *Cometologia*[16] through the most generous Cyprianus Crenscius, a friend of Rautenstein[17] and Pels,[18] my kinsman; out of his great kindness towards friends, he will deliver this book to Master Pels. I do not want money for it if you are willing to send me a copy of one of your books. I hope one day, God willing, we will make a fair exchange of books when our works are finished. Farewell, most renowned Sir, and work prosperously on publishing your famous books! Given at Hamburg, on 20/30 April 1666.

Yours obligingly,

Stanisław Lubieniecki Rolitsius
by my own hand

10 X/1033 E1, see Alexandre Guy Pingré, *Cométographie ou Traité historique et théorique des comètes*, vol. 1 (Paris: Imprimerie royale, 1783), pp. 369–70; Gary W. Kronk, *Cometography. A Catalogue of Comets. Volume 1: Ancient – 1799* (Cambridge: Cambridge University Press, 1999), pp. 171–72.

11 Sethus Calvisius (see note to Letter 27), *Opus chronologicum* (Francofurti: impensis Johannis Thymii, 1620), p. 618.

12 See Letter 42.

13 See section 2.4.9 of the Introduction.

14 See note to Letter 41.

15 See note to Letter 20.

16 See note to Letter 41.

17 See note to Letter 5.

18 See note to Letter 5.

Postscript. I add a word here. The peace between the bishop of Münster and the Dutch was fortunately concluded on the ninth/nineteenth of this month and confirmed with the signatures of the delegates.[19] One can hope that the English one[20] will be made and the Rhine one[21] will be confirmed, too. Who would fail to see that comets are also followed by good things? Farewell again!

19 See note to Letter 37.

20 Lubieniecki probably does not allude here to a particular treaty but wishes the early end of the Anglo-Dutch war.

21 Lubieniecki probably refers here to the treaty signed in 1666 in Cleves. It reaffirmed the decisions of the treaty of Xanten, signed in 1614, which divided the Duchies of Jülich, Cleves, and Berg between Brandenburg and Neuburg.

45.

22 May 1666, Hevelius to Lubieniecki

BO: C1, v. 7, 1111
TC: 947

Ad Dominum De Rolis
Hamburg[i]

Illustrissime Vir.

Ad binas Tuas posteriores die 16 et 30 Aprilis datas paucis respondeo. Cum animus meus plane aliis speculationibus Cometicis nunc occupatur. Tu Chronologica et Historica, ut percepi, tractas, ego vero nunc mere Astronomica, sic ut plane hac vice Tibi satisfacere, ob varias occupationes nequeam. De cometa vero 837 ita sentio: illum in Signo Virginis primum apparuisse, et quidem in ea parte qua pedes eius caudam Corvumque contingunt, hinc spatio 25 dierum motu retrogrado per Leonem, Cancrum et Geminos progressum esse, et tandem in Capite Tauri sub Aurigae pedibus evanuisse. Caeterum, stellam illam anno 1033, etiamsi Calvisius referat, a Septentrione versus Meridiem traiecisse et a 28 Febr[uarii] ad 15 Martii perdurasse, valde dubito Cometam fuisse, multo minus stellam aliquam novam. Cum nemo, qui Cometas descripsit (quantum sciam) Cometam aliquam illo anno[1] notaverit. Adhaec Cometam cum sonitu et fragore unquam traiectum esse, vix credo; forte diversos duos globulos ignotos[2] (quorum alter die 28 Febr[uarii], alter rursus 15 Martii visus est) vel faces ardentes, pro Cometis arripuerunt: ut saepius fieri solet ab Imperitis. Quot Cometae, quaeso, hoc solo unico anno apparuissent, si ex opinione vulgi res dirimenda esset. Ego puto, nonnullos, plurima meteora ac phasmata, imo etiam ipsos Planetas nonnunquam pro Cometis apprehendisse. Quamobrem hac in parte non facile // credo omnibus, nec ipsis Historicis, nisi inter se probe conveniant et circumstantias adiiciant fide dignas. Typographo illo Batavo nihil significandum habeo; faciat quicquid velit, mihi perinde erit. Pro *Cometologia* Schüleri gratias Tibi habeo magnas, daboque operam ut par pari suo tempore referam. Vale feliciter et porro fave

Tuae Illustrissimae Dominationis
studiosissimo

J[ohanni] Hevelio

1 ad illumo annumo (ad, um *et* um *del.* **BO**) : eo anno **TC**.
2 ignotos : ignitos **TC**.

Dabam Gedani, Anno 1666, die 27 Maii

To Master de Rolis
in Hamburg

Most illustrious Sir!

I answer two letters of yours, sent on 16 and 30 April, only briefly because my mind is entirely preoccupied with other cometary deliberations. You deal with chronological and historical subjects, and I with strictly astronomical ones, so I cannot fulfil your wishes now, due to many other obligations. My opinion about the comet of 837[3] is as follows: it appeared in the sign of Virgo, in its parts where its Feet touch the Tail[4] and Corvus; thence for 25 days it was moving with retrograde movement and passed through Leo, Cancer, and Gemini, eventually disappearing in the Head of Taurus near the Feet of Auriga. Moreover, I doubt much that the star of 1033[5] was a comet, even though Calvisius says that it moved from the north southwards and lasted from 28 February until 15 March;[6] and I doubt even more that it was a new star. Because no one who described comets noted a comet in that year (as far as I know). Moreover, I do not believe that any comet has ever moved with noise and roar; perhaps two different unknown balls (one of which was seen on 28 February, another on 15 March) or burning torches were considered comets, as often happens among those who are inexperienced. If we had to decide relying on the voice of the public, how many comets would have appeared in this year only? In my opinion, many meteorological phenomena and other sightings (and sometimes even planets) were considered to be comets. Therefore, I am not inclined to believe everyone (even historians) on this subject, unless they agree with one another and add some trustworthy circumstances. I have nothing to say to that Dutch printer;[7] he may do whatever he wants; I do not care.[8] I am grateful for Schuler's *Cometologia*,[9] and I will do my best to pay you evenly in due time. Farewell, and be benevolent towards

illustrious Sir,

Yours devoted,

Johannes Hevelius

Given at Danzig, on 27 May 1666

3 See note to Letter 42A.

4 The Tail of Hydra.

5 See note to Letter 44.

6 Sethus Calvisius (see note to Letter 27), *Opus chronologicum* (Francofurti: impensis Johannis Thymii, 1620).

7 See note to Letter 35.

8 See section 2.2.3 of the Introduction.

9 See note to Letter 41.

46.

1 June 1666, Lubieniecki to Hevelius

BO: C1, v. 8, 1123
TC: 947

Salutem Plurimam Nobilissime, Amplissime, Clarisssime Vir. Responsionis officium aequitas: brevitatem tempus imponit mihi, cui prope dolor imponit silentium. Hoc enim ipso inauspicato die, quo communem olim utriusque nostrum, adeoque et Nobiliss[imi] Peltzii, Cl[arissimi] Titii aliorumque Virorum Egregiorum in urbe vestra degentium amicum Illustrissimum Rautensteinium, peracta Suecica legatione ad suos revertentem domi meae excepturus eram, nil nisi domicilium praestantissimae illius mentis recepi. Epitaphium quod levandi doloris et exsolvendi officii causa extemplo ad aram amicitiae optimae notae Viro et Amico posui, Nobiliss[imus] Peltzius cui exemplum mitto, Tecum commune habebit. In iis, quae ad me de argumento Cometico scripsisti, Tibi assentio. Nam de Cometa anni 837. idem auctoritate[1] Virorum Praeclarorum nixus judicavi. Sed Tu, ut et Cl[arissimus] Bullialdus (a quo nuper literas 21. Maii datas accepi) assensu gravi meam aliorumque sententiam uterque firmastis. Stellam illam anno 1033, visam inter Cometas non retuli. Meteoron enim aereum fuisse arbitror. Observationes Tuas de cometa verno quamprimum mihi mitti cupio, una cum communibus Amicis, Tibi optime cupientibus. *Cometologiam* Schuleri Te accepisse, et grato quidem animo, valde mihi gratulor, grata Tibi officia praestare semper gestiens. Vale Vir Humanissime, et fave mihi meisque studiis, quae cum Tuis mere Astronomicis juncta, aliquid utile et jucundum simul Reip[ublicae] literariae, ut spero, prodent promentque.

Dab[am] Hamburgi die 1. Junii Gregor[ianii] 1666.

Clarissimo Nomini Tuo
omnibus studiis et officiis addictissimus

S[tanislaus] L[ubieniecius] Rolitsius
manu propria

1 auctoritate : idonea auctoritate **TC**.

Greetings, most noble, worthy, and renowned Sir! Justice forces me to answer, but time forces me to be brief, as grief almost forces me to keep silence. On that very miserable day when I was about to welcome in my home the most illustrious Rautenstein[2] (the friend of ours, as well as of the most noble Pels,[3] of the most renowned Titius,[4] and of other exceptional men who live in your city) after his return from his delegation to Sweden, I only received the information that his great mind became quiet. I send a copy of the funeral speech (which I have placed on the altar of friendship for the best man and friend, in order to ease the grief and fulfil the duty) to the most noble Pels, and he will share it with you. I agree with you in what you wrote to me about the cometary discussion. Because I think the same about the comet of 837,[5] relying on the opinions of renowned men. But both you and the most renowned Boulliau[6] (whose letter, sent on 21 May, I have recently received) have confirmed this opinion of mine and of others with your adamant consent. I did not place that star of 1033[7] among comets, because I think it was a meteorological phenomenon. I would like you to send your observations of the vernal comet to me and to our mutual friends who wish you all the best. I am glad that you welcomed the *Cometologia* by Schuler[8] with pleasure because I am always willing to serve you. Farewell, kindest Sir, and always be benevolent towards me, and towards my studies which (aligned with your purely astronomical ones) can produce and do something useful and pleasant for the Republic of Letters, I hope!

Given at Hamburg, on 1 June 1666, Gregorian.

Renowned Sir,

Yours most humbly devoted,

Stanisław Lubieniecki Rolitsius
by my own hand

2 See note to Letter 5.

3 See note to Letter 5.

4 Johann Peter Titius (Titz, 1619–89), pedagogist and teacher from Danzig. Since 1644, he was a teacher in St Mary's school in Danzig. Since 1652, he was a professor at the *gymnasium academicum* in Danzig — until 1653 a professor of Greek and oriental languages, then professor of rhetoric and poetry. Titius was Hevelius's correspondent and there are 7 letters he sent to him between 1656 and 1684, cf. Klaus-Dieter Herbst, 'Alphabetical list of the correspondents', CJH1 p. 273.

5 See note to Letter 42A.

6 See section 2.4.2 of the Introduction.

7 See note to Letter 44.

8 See note to Letter 41.

47.

23 June 1666, Hevelius to Lubieniecki

BO: C1, v. 8, 1123
TC: 947–48

Domino Rolisio
Hamburgum

Illustrissime Vir.

Tandem opusculum meum ab Amicis hactenus desideratissimum Tibi per Tabellarium ordinarium, ut Tibi quantocyus morem gererem, transmitto; de quo quid Vobis videatur, et an desiderio etiam vestro satisfaciant, suo tempore mihi aperiatis etiam atque etiam rogo. Nunc liberum esto omnibus harum rerum Eruditis, de omnibus illis controversiis judicium ferre; sed necesse est, ut omnia et singula non superficietenus, atque oculo fugitivo perlustrentur, ac pervolutentur, sed ordine, diligenter ac scrupulose perlegantur, perquirantur, examinentur ac ponderentur, tam ipsae observationes, quam fusissimus ac laborissimus calculus inde deductus; non dubito, quin quilibet aequus Censor, qui litem suam nondum fecit, et minime adversam aliquam in pectore gerit praeconceptam sententiam (quorum optime scio, nonnullos passim dari, qui me etiam inauditum indefensumque jam olim condemnarunt) sic indicet, me (absit gloria) in observandis his Cometis officio meo, pro virili functum esse, ac minime, quousque fidis Instrumentis, ab initio scilicet ad 12 13 et 14 Febr[uarium] Cometam insectatus sum in motu diurno aberrasse, ut quidem Cl[arissimus] Auzoutius pag[ina] 8 *Epist*[*olae*][1] me accusat; sed econtrario illum jugiter hallucinatum esse, sicuti prolixe satis a nobis demonstratum est. Quo autem Cometa a die 12, 13 et 14 Febr[uarii], ubi illum Instrumentis congruis amplius observare haud // potui, viam suam duxerit, an primam, an vero secundam Arietis versus salvis quorumcumque Observatorum observationibus, et remotis inconvenientiis et absurditatibus omnibus? Tibi aliisque diiudicandum relinquo; tum utrum Observationes illae omnes contra me in medium allatae, tam accurate inter se conveniant, ut Cl[arissimus] Auzoutius sibi quidem imaginatus est? et an tanto fundamento nitantur, ut omnes meas observationes penitus iis evertere possit. De his et reliquis, inquam, omnibus, judicium Amicorum cum exploraveris nobis communices iterum iterumque humanissime peto. Interea amicos omnes officiose saluta, quibus prima occasione aliquot Exemplaria per Lubecam transmittam, nec non quaedam in omnem eventum pro Bibliopolis, si Tibi ita videbitur, quodlibet exemplar uno Imperiali redimendum.[2]

1 *Epis*[*olae*] : *Epist*[*olae*] *ad Cl*[*arissimum*] *Petitum* TC.
2 nec non...redimendum *om.* TC.

Vale et pristino favore ac amore fac nos prosequaris. Dabam Dantisci Anno 1666 die 23 Junii.

Tuus
affectu amicissimo
deditus

J[ohannes] Hevelius
manu propria.

To Master Rolitsius
Hamburg

Most illustrious Sir!

Finally, I am sending you my little work (so anticipated by our friends) through an ordinary messenger, in order to fulfil my duty as soon as possible; I ask you to tell me, in due time, what you think about it and if it meets your expectations. Now each of the learned men may freely present his opinion about that controversy, but nothing should be checked and leafed through superficially and hastily, but instead, it should be read, inquired, examined and considered accurately, diligently and meticulously — both the observations and those long and toilsome calculations based on them. I have no doubts that every just judge who has not taken part in the dispute yet and who does not have any prejudiced opinion in mind (because I know there are many who have condemned me without listening to my defence) will confirm that (I say it not to boast) I properly fulfilled my duty of observation of those comets and that I made no mistake about the daily movement of the first comet when I was watching it with my reliable instruments since the beginning until 12, 13, and 14 February. The most renowned Auzout accuses me of this mistake in his *Lettre*, on page 8, but, on the contrary, he was the one who was constantly deluded, as will be proven.[3] I leave it to you and to others to decide (on the basis of some reliable observations by some observers, without any mistakes or absurdities) where the comet headed to on 12, 13, and 14 February when I could not see it anymore with appropriate instruments: to the first or to the second star of Aries; and whether all of those observations presented against me agree so much with one another, as the most renowned Auzout thinks, and whether they are so firmly based that they could ruin all my observations entirely. When you investigate the opinion of our friends about all of these, I kindly ask you to inform me about it. Meanwhile, kindly greet all the friends, for whom I will send some copies of the book through Lübeck at the first opportunity; and, just in case, I will send some

3 See section 2.2.4 of the Introduction.

for the booksellers to be sold for one Reichstaler each, if you decide so. Farewell, and be as benevolent and kind towards me as before! Given at Danzig, on 23 June 1666.

Your devoted friend,

Johannes Hevelius
by my own hand

48.

29 June 1666, Lubieniecki to Hevelius

BO: C1, v. 8, 1149
TC: 948
Olhoff: 113–14 (part)

Seal with Lubieniecki's coat of arms (Rola coat of arms)

Salutem Plurimam Nobilissime, Amplissime, Clarissime Vir. Tabellarius, qui hodie ad nos accessit ex terris vestris, attulit mihi continuationem *Prodromi* Tui *Cometici* cum adiuncta *Mantissa*, opus profecto eximium[1] et ut subtilissimi ingenii, ita et affectus erga me Tui insigne documentum. Scio me primum in hac urbe esse, qui hunc librum pretiosum et praeclarum, et munus Tuum, crede mihi, acceptissimum possideo. Video, si quid hic video, postquam librum fugitivo oculo inter multa negocia hodie expedienda percurri, Te Cl[arissimo] Auzoutio accuratissime, sed et humanissime ac modestissime respondere. Qua in re ut officium boni Viri praestas, ita Te vere bonarum literarum peramantem esse probas. Macte tantis Virtutum et eruditionis laudibus, macte ingenti in rem Mathematicam merito, Magne Heveli! Judicia Amicorum sincera nec minimum adulatoria Tibi per me, sicut antea,[2] patebunt. Ideo ut exspectati operis Tui dulcedine animum Magnifici et Ampliss[imi] ac Humanissimi Viri Domini Bartholdi Mülleri, bonarum artium studiosissimi et de hac civitate Consulis optime meriti, exsatiarem, illud jam ei perlustrandum misi. Reliqua, quae promittis exemplaria, cum desiderio exspecto. // Gratulor mihi non tam quod gratum officium, quam quod etiam publice utile Tibi praestiterim submissis observationibus Rudbeckianis, quod toties in libro Tuo fateris et demonstras. Quod si promtae voluntati facultas responderet, essem profecto strenuus Mathematicorum studiorum Promotor, qualem me esse praedicas. Nunc vix sperare possum, me, sicut ante Te volente feci, in compendium redacturum *Prodromum* hunc Tuum Reverendorum Virorum Kircheri et Riccioli gratia, licet utrique par hic quoque officium praestare cuperem. In isto, ut volebas, salutando, me officio defunctum rite esse, jam ante, nisi fallor, ad Te perscripsi. Indicavi etiam Te aliquot ad eum literas dedisse, nihil autem responsi accepisse. Si tamen ulla ratione fieri poterit, dabo operam sedulo, ut huius quoque operis Tui potiora caeteris Tuis operi meo jam insertis addantur. Ita, quantum in me erit, praestantissimus labor Tuus a me quoque omni posteritati et seculorum aeternitati, quamdiu scilicet literis debitus tribuetur honos, commendabitur. Nunc autem demum video, quam utilem Tibi praestitissem operam, si ipsas demonstrationes Rudbeckianas Tibi misissem. Sed necdum ipse eas ex Batavia accepi, missurus eas ad Te evestigio, postquam hae ad me perlatae

1 eximium **BO Olhoff** : praecipuum **TC**.
2 antea : antea, quamprimum ea expiscari potero **TC**.

// fuerint. Nobiliss[imum] Hugenium a Christianissimo Rege Lutetiam evocatum, honeste et liberaliter ex Cl[arissimo] Bullialdo paulo ante accepi. Cl[arissimus] Auzutius nil mihi ad meas posteriores respondit. Sed et Cl[arissimus] Bullialdus ab aliquo tempore silet, domestica quadam molestia lacessitus. Salutem adscribo Cl[arissimis] Viris Titio et Büthnero: sed et Nobiliss[imo] Peltzio, quem ab urbe abesse ex diuturno silentio colligo: tum quod nullam epitaphii Rautensteiniani, quod ad eum miseram Tecum communicandum, facias mentionem.[3] Vale Vir Humanissime et Amicissime, mihique constanter fave. Dabam Hamburgi die 29 Junii Gregor[iani] 1666.

Clarissimo Nomini Tuo
addictissimus

S[tanislaus] L[ubieniecius] Rolitsius
manu propria

Post Scriptum.[4] Proximo die Sabbathi cum adessem Seren[issimae] Reginae Suecorum Christinae, eique rationem de opere meo redderem, fecit Sua M[aiestas] mentionem Tui honorificam, Teque ut Excellentissimum Astronomum favore suo prosequitur.[5]

Greetings, most noble, worthy, and kindest Sir! A messenger from your land, who arrived here today, brought me the continuation of your *Prodromus cometicus* with the *Mantissa*, an outstanding work of the most subtle mind and a clear testimony of your benevolence towards me. I know that I am the first one in this city who owns this precious and excellent book, a very pleasant (believe me) gift from you. I see, if I can see anything after I have quickly leafed it through among many other tasks I have to complete today, that your answer to the most renowned Auzout is very precise, kind, and polite.[6] In this way, you fulfil the duty of honest men, and you prove that you really are the lover of arts and sciences. Good luck to you, great Hevelius, amidst so many praises of your virtues and learning; good luck with your vast merits for mathematics! I will inform you about the sincere (not flattering) opinions of our friends, as I did before. In order to satisfy the mind of the magnificent, greatest, and kindest Sir, Master Barthold Moller[7] (a very learned man and a distinguished Councillor of this city) with the sweetness of your much-expected book, I have sent it to him to read. I eagerly await the other copies you promised. I congratulate myself because I made something

3 mentionem : mentionem. Unum addo: Proximo die Sabbathi cum adessem Serenissimae Reginae Suecorum Christinae, eique, prout Princeps illa Magnanima litteras assiduo amat, fovetque, rationem de opere meo redderem, fecit Sua Majestas mentionem Tui honorificam, Teque ut Excellentissimum Astronomum gratia sua prosequi testata est. **TC**.

4 Judicia Amicorum...Post Scriptum : et c[etera] **Olhoff**.

5 Dabam...prosequitur *om*. **TC**.

6 See section 2.2.4 of the Introduction.

7 See note to Letter 15.

pleasant for you and useful for the public when I sent you Rudbeck's[8] observations, which you so often mention and use in your book. If my abilities were equal to my will, I would become a sure helper of mathematical studies, as you call me. Now I have a slight hope that I could prepare an abridged version of your *Prodromus* for the reverend Kircher[9] and Riccioli[10] — as I did before with your consent; however, I would like to fulfil my duty towards both of them. I wrote to you before, if I am not mistaken, that I kept my promise of greeting Riccioli, as you wished. I told you I had sent him a few letters, but I received no answer. If I only can, I will take care to add some things from this book of yours to what I included in my work previously. In this way, to the best of my ability, I will promote your outstanding work to the whole posterity and for endless centuries, viz. as long as there is any due respect for sciences. Now I see what a useful thing I would have done for you if I had sent you those images by Rudbeck. But I have not received them from the Netherlands; I will send them when they are delivered to me. Some time ago, I learned from the most renowned Boulliau[11] that the most noble Huygens[12] had been called to Paris by the Most Christian King. The most renowned Auzout did not answer my last letter. But the most renowned Boulliau has also been silent for some time because he is burdened with some cares. I send greetings for the most renowned Masters Titius[13] and Büthner,[14] and also for the most noble Pels[15] who is absent from the city, as I presume from his prolonged silence and the fact that you do not mention my funeral oration for Rautenstein[16] which I sent to him in order to share it with you. Farewell, kindest and friendly Sir, and always be benevolent towards me. Given at Hamburg, on 29 June 1666 Gregorian.

Renowned Sir,

Yours devoted,

Stanisław Lubieniecki Rolitsius
by my own hand

Postscript. Last Saturday, when I visited the most serene Christina, Queen of Sweden,[17] and described my book to her, her majesty made an honorific mention of you, and she favours you as the most excellent astronomer.

8 See section 2.4.16 of the Introduction.
9 See section 2.4.8 of the Introduction.
10 See section 2.4.15 of the Introduction.
11 See section 2.4.2 of the Introduction.
12 See section 2.4.7 of the Introduction.
13 See note to Letter 46.
14 See section 2.4.3 of the Introduction.
15 See note to Letter 5.
16 See note to Letter 5.
17 See note to Letter 12. Between 1666 and 1668, Christina lived in Hamburg.

49.

24 July 1666, Hevelius to Lubieniecki

BO: C1, v. 8, 1140
TC: 948

Domino Rolisio
Hamburgi

Vir Illustrissime.

Hac vice nullum aliud scribendi argumentum occurrit, quam quod significem, me nuper per Lubecam 23 exemplaria novissimi mei opusculi Tibi transmississe; ex quibus unicum Tibi reservabis, alterum Cl[arissimo] Viro Domino Kirstenio, et tertium Cl[arissimo] Mullero meo nomine officiose exhibebis; reliqua vero Bibliopolis, si ea desiderent tradas rogo, quodlibet exemplar uno Imperiali parata pecunia redimentum, cum praetium ordinarium constitutum alias sit 4 flor[enis][1] polon[icis].[2] Caeteris amicis fere omnibus, cum quibus aliqua mihi intercedit necessitudo opusculum hocce pariter jam obtuli, tam Italis, Gallis, Anglis, Polonis, Germanis quam[3] Batavis, nec non quibusdam in Suecia utpote Celeberrimo Heinsio et Rudbeckio. Cl[arissimo] Bartholino etiam per doct[issimum] M[agistrum] Reichelium prima occasione transmissurus sum; sic ut jam supra centum exempl[aria] amicis distribuerim. Nuper Eximius Cassinus observationes suas in Marte[4] habitas ratione quarundam Macularum mihi transmisit, ex quibus pariter clare deducit, ut[5] Jovem sic Martem quoque circum axem[6] spatio 24 hor[arum] et 40' gyrari.[7] Literae illae, in quibus ea omnia demonstrat Bononiae sunt impressae.[8] Vale et fave

1 fol *del.* **BO**.
2 reliqua...polonicis *om.* **TC**.
3 quam *om.* **TC**.
4 Martio **TC**.
5 in *del. post* ut **BO**.
6 ut...axem : ut in Jove clare deducit Martem **TC**.
7 gyrari : circa axem gyrari **TC**.
8 impressae : impressae: Saluta quam humanissime Cl[arissimos] Viros Kirstenium et Müllerum **TC**.

Tuae Illustrissimae Dominationis
studiosissimo

J[ohanni] Hevelio
manu propria

Dabam Dantisci
anno 1666, die 24
Julii. Raptim

To Master Rolitsius
in Hamburg

Most illustrious Sir!

This time, there is no other reason to write but to notify you that I have recently sent you 23 copies of my newest booklet through Lübeck. I would like you to keep one, and I kindly ask you to give one to the most renowned Master Kirsten[9] and one to the most renowned Müller[10] on my behalf; pass the rest to the booksellers, if they want them, for one Reichstaler each, paid in cash, while otherwise the usual price is set to 4 Polish florins. I have already sent copies to almost all other friends towards whom I was obliged: to the Italians, the Frenchmen, the Englishmen, the Poles, the Germans, the Dutchmen, and to some people in Sweden, such as the most renowned Heinsius[11] and Rudbeck.[12] I am planning to send a copy to the most renowned Bartholin[13] through the most learned Master Reichelt[14] at the earliest opportunity. In this manner, I have already distributed more than one hundred copies. The exceptional Cassini has recently sent me his observations of some spots on Mars; on their basis, he firmly concluded that Mars (like Jupiter) revolves around its axis in 24 hours

9 See section 2.4.9 of the Introduction.

10 See section 2.4.11 of the Introduction.

11 See section 2.4.6 of the Introduction.

12 See section 2.4.16 of the Introduction.

13 See section 2.4.1 of the Introduction.

14 Probably Julius Reichelt (1637–1717), German mathematician and astronomer, professor at the University of Strasbourg. Reichelt was Hevelius's correspondent and they exchanged 9 letters (7 from Reichelt, 2 from Hevelius) between 1666 and 1681, cf. Klaus-Dieter Herbst, 'Alphabetical list of the correspondents', CJH1 p. 270. See also CJH3 p. 546, note to Letter 210, des Noyers to Hevelius, 10 January 1681.

and 40 minutes. The letters in which he demonstrates all of this have been printed in Bologna.[15] Farewell, and be benevolent towards

illustrious Sir,

Yours devoted,

Johannes Hevelius
by my own hand

Given at Danzig
on 24 July 1666.
In haste

15 Giovanni Domenico Cassini (see note to Letter 8), *Martis circa axem proprium revolubilis observationes* (Bononiae: ex typographia HH. de Ducciis, 1666); Giovanni Domenico Cassini, *De aliis Romanis observationibus macularum Martis* (Bononiae: ex typographia HH. de Ducciis, 1666); Giovanni Domenico Cassini, *De periodo quotidiana revolutionis Martis* (Bononiae: typis HH. de Ducciis, 1666).

50.

10 November 1666, Lubieniecki to Hevelius

BO: C1, v. 8, 1152
TC: 948–49

Salutem Plurimam Vir Nobilissime, Amplissime, Clarissime.

Non miraberis, ut spero, tandiu dilatum responsi ad postremas Tuas literas 24. Julii datas officium, postquam nunc cognoris me ante quatridium demum Tuum illum fasciculum Lubeca accepisse. Alioqui, licet continuis nec profecto levibus laboribus in contexenda *Historia Cometarum* occupatus, honestis et necessariis erga Amicos officiis deesse non soleo. De oblato novo libri Tui exemplo debitas Tibi ago gratias. Exosculor in hoc munere Tuam benevolentiam officiosam, incredibilem vero et singularem rerum caelestium peritiam ac in texendis rationibus ac deducendis firmandisque argumentis solertiam etiam veneror, et magis magisque demiror. Cuius rei ut perenne monumentum operi meo insit, animus mihi est Tuas Cometae posterioris observationes, quas vides, aeque ut prioris, ei inserere: dummodo Tu assentias. Licet vero hac de re jam antea ad Te retulerim, volui tamen et observationes ipsas ex opere Tuo excerptas Tibi mittere, ut, si forte alicubi aliquid demendum, addendum vel mutandum sit, me pro solita aequitate moneas. Nunquam enim quidquam votis amici contrarium, in iis praesertim quae ad eum pertinent, facere in animum induco. Velim itaque has observationes cum judicio de iisdem Tuo ad me remittas. Exempla Tui libri Cl[arissimis] Viris Kirstenio et Mullero nomine Tuo obtuli. Summopere illi depraedicant Tuam humanitatem et rerum Astronomicarum scientiam. Operosas illas tabulas, tot // et tantis rationibus completas magis mirantur, quam rimantur, aliis praesertim distracti. Approbare tamen eos tua argumenta et observata, certum mihi et exploratum est, et tunc quidem jam, cum primum exemplum libri cum eis communicassem. Hunc ego libenter, ut prioris Cometae *Prodromum*, Reverendorum Virorum Kircheri et Riccioli, utpote Amicorum communium et a nobis dissitorum longe[1] gratia in compendium redigerem. Sed ita me multiplex historiae Cometalis labor distinet et premit, ut huic alteri labori operam commodare nulla ratione queam. Nunc quidem, dum Amici et typographi aliaque urgent. Ad finem enim historiae, cui decurrens anno decurso mensis iste initium dedit, propero. Priusquam vero haec aliaque quae me premunt, amoveam, *Prodromus* iste Tuus meo apud Italos quoque, amplius opus non habebit. Editorum a Cl[arissimo] Cassino nil prorsus vidi. Sed virum esse in Mathematicis exercitatissimum, ex multis cognovi: tum vero etiam ex Temet ipso. Ideo libenter observationes eius, ubi commodum erit, videbo. Potitus jamdiu earum humani-

1 dissitorum longe : longe dissitorum **TC**.

tate Kircheriana fuissem, nisi tanta locorum distantia impediisset. Rudbeckianas figuras nondum ex Hollandia accepi, licet assiduis monitis Amicos ibi agentes meaque curantes, ut eas ad me mitterent, urserim. Non parum vota nostra in multis funesti istius belli, cuius incendium die ab hinc octavo in Albi et oculis huius urbis aliquantum exarsit, incommoda mo-//rantur. Hoc utinam eodem modo, ut nuper in carissima patria, exstinguatur. Magnum hoc meae sententiae de indifferenti Cometarum significatione adferret momentum. Ad Nobiliss[imum] Pelsium proxime scripsi per tabellarium Berolinensem, quod ei cum pl[urima] salute indicari peto. Salveant et Cl[arissimi] Viri Pastorius et Titius, qui per Vos rogantur Viduae Rautensteinianae nomine, ut honori defuncti mariti eius, Amici omnium nostrum communis, epitaphia vel alioqui carmina Lugubria conscribant. Vale, Vir Humanissime, ac mihi fave constanter.[2] Dab[am] Hamb[urgi] 10. Sept[embris] Greg[oriani] 1666.

Tuus
omni studio et officio

S[tanislaus] L[ubieniecius] Rolitsius
manu propria

Post Scriptum. Exemplaria libri Tui distribuere nudius tertius inter Bibliopolas caepi.[3]

Greetings, most noble, worthy, and renowned Sir!

You will not be surprised why I delayed my answer to your last letter (sent on 24 July) for so long, when you learn that four days ago I finally received the package from you from Lübeck. As much as I am busy with constant and serious work on preparing my history of comets, I am not used to neglecting what is honest and required towards friends. I am grateful because you sent me a copy of your new book. I marvel at your great benevolence seen in this gift, and I also revere and greatly admire your incredible and singular experience in heavenly things and mastery in weaving opinions and deducing and confirming arguments. In order to put the perennial monument of that fact into my work, I would like to enclose to it, as you can see, the observations

2 Ad Nobiliss[imum]...constanter : Quod reliquum est, velim Vir Humanissime, Cl[arissimum] Titium vestrum, quem Virum ego magni meritissimo facio, salutes, eique nuperum postulatum per Nobiliss[imum] D[ominum] Peltzium Foederati Belgii nomine apud vos residentem, factum in memoriam revoces. Est autem illud pro Vidua communis olim nostrum omnium amici D[omini] Rautensteinii, quae per me mariti optimi et optatissimi Amicos rogat et obsecrat, ut marito ad amicitiae aram jura persolvant. Vale ac mihi constanter fave. TC.

3 Dab[am]...caepi *om.* TC.

of the last comet (as I did with the former one), if you only agree.[4] Even though I asked you about it last time, I have decided to send you the observations excerpted from your book, so you can (with your usual equity) tell me if anything needs to be removed, added or changed. Because I never want to do anything against the will of a friend, especially in a matter which concerns him. I would like you to send me these observations back together with your opinion about them. On your behalf, I handed the copies of your book to the most renowned Kirsten[5] and Müller.[6] They worship your kindness and astronomical knowledge. Because they are distracted by other occupations, they rather admire than study those labour-consuming tables concluded with so many calculations. However, immediately after I gave them the copies of the book, I learned that they approve of your opinions and observations. I would willingly make an excerpt from them, as I did with your *Prodromus cometicus*, for the reverend Kircher[7] and Riccioli[8] — our mutual friends who live far away from us. But my complex work on the history of comets dissuades me and forces me not to undertake another task. Especially now, when I am urged by my friends, printers, and other things. Because I hurry towards the end of that history of comets, which I began in the last month of the previous year. Before I finish all the things which overwhelm me, your *Prodromus* will not need my help in Italy any longer. I have not seen what the most renowned Cassini published.[9] But from many people (including you) I have learned that he is very experienced in mathematics. Therefore, I will willingly see his observations when I can. I would have received it a long time ago thanks to Kircher's kindness, if such a great distance between us had not impeded it. I have not received Rudbeck's[10] illustrations from the Netherlands yet,[11] however I always urge my friends who live there and take care of my business to send them to me. Our wishes will be delayed among many inconveniences of that terrible war, whose fire burned for some time eight days ago over Elbe and was seen in this city as well.[12] May it extinguish in the same way as it happened recently in our dearest homeland![13] It makes my sentence about the indifferent meaning of comets much more valid. I have recently written to the most noble Pels[14] through a messenger from Berlin; I would like you to tell Pels about it and to send him many greetings. May also the most

4 There is no enclosure with the extract from Hevelius's *Descriptio cometae*, but see Appendix 2.

5 See section 2.4.9 of the Introduction.

6 See section 2.4.11 of the Introduction.

7 See section 2.4.8 of the Introduction.

8 See section 2.4.15 of the Introduction.

9 See notes to Letters 8 and 49.

10 See section 2.4.16 of the Introduction.

11 See note to Letter 34.

12 Lubieniecki probably refers here to the war between Bremen and Sweden which began in August 1666.

13 Lubieniecki refers here to the Lubomirski rebellion, see section 1.1.3 of the Introduction.

14 See note to Letter 5.

renowned men, Pastorius[15] and Titius,[16] be greeted; Rautenstein's widow[17] wants to ask them (through you) to write an epitaph or some mourning poems to honour her late husband, a friend to all of us.[18] Farewell, kindest Sir, and always be benevolent towards me! Given at Hamburg, on 10 September 1666 Gregorian.

Yours most obliged,

Stanisław Lubieniecki Rolitsius
by my own hand

Postscript. Three days ago, I began to distribute copies of your book among the booksellers.

15 Joachim Pastorius (von Hirtenberg, 1611–81), Polish diplomat and historian. From 1655 to 1667, he was the professor of law and history at the *gymnasium academicum* in Danzig. He was the royal historian at the court of King Jan Kazimierz. Born in a Lutheran family, he sided with the Polish Brethren and the Reformed Church, in 1658 converted to Catholicism, and in 1675 he was ordained a priest. Pastorius was Hevelius's correspondent and there are 13 letters he sent to Hevelius between 1652 and 1679, cf. Klaus-Dieter Herbst, 'Alphabetical list of the correspondents', CJH1 p. 269. See also Chantal Grell, 'Introduction', CJH2 p. 106.

16 See note to Letter 46.

17 See note to Letter 5.

18 Titius's poem is in TC3 pp. 57–59. There is no mourning poem by Pastorius. See section 1.3.2 of the Introduction.

51.

20 October 1666, Hevelius to Lubieniecki

BO: C1, v. 8, 1153
TC: 951

Domino
Rolisio
Hamburgi

Vir Illustris

Ob varias curas, inprimis quod spatio quatuordecim fere dierum peregre abfuerim, aliquanto tardius ad tuas 10. Sept[embris] datas respondeo. Ex quibus autem cum perceperim Te decrevisse observationes meas posterioris Cometae etiam Tuo praeclaro operi adscribere, gratias Tibi habeo permagnas; sed, ut pace Tua dicam, in quibusdam fere nimis brevitati studes, et ideo res fit obscurior: quiquid tamen sit, Tuo arbitrio omnia committo, dummodo inscriptionem, sive titulum immutes. Ea enim quae Tu ipse ex opusculo meo in compendium redegisti, non sunt ipsae genuinae meae observationes, sed tantummodo est generalis et perbrevis illius Cometae descriptio. Quare praestat, mea opinione ut subsequens titulus illi praefigatur: Generalis Cometae qui apparuit Anno 1665, mense Aprili descriptio Dantisco Heveliana, ab Autore (sc[ilicet] Rolisio) in compendium redacta.[1] Nam, cum Tua verba sint, necesse est, ut etiam nomen Tuum praeponatur. De caetero, si Tabulam meam G aeri incidi curaveris, rogo operam des sedulam,[2] ut omnia et singula quam accuratissime delineentur, secus tum Lector, tum negotium ipsum nimium quantum turbabitur, quanquam, mea opinione, haud opus est, ut Tabula illa G Tuo operi adiiciatur. Sufficeret, si Lectores ad ipsum *Mantissam* remitteres. Nam // ex sola illa Tabula G, notisque tuis brevissimis, nemo sane arduam illam Quaestionem inter Cl[arissimum] Auzoutium et me exortam intelliget, nedum eam diiudicabit. Denique Tecum sentio, minime opus esse, integram meam *Mantissam* in compendium redigere: cum certo sciam Italos opusculum illud jam vidisse: poteris igitur commode isto labore supersedere. Imo etiamsi velles, nulla tamen ratione, nisi maximo meo praeiudicio, id fieri posset. Etenim, si[3] brevioribus ea omnia pertractandi potuissem utique et ego fecissem. Cl[arissimo] Domino Titio desiderium Tuum exposui, non dubito quia prima occasione Tibi satisfaciat.[4]

1 ut subsequens...redacta : ut talis titulus praefigatur, qui indicet hanc esse generalem Cometae, qui apparuit Anno 1665. mense Aprili descriptionem a Te in compendium redactam **TC**.

2 sedulam *om.* **TC**.

3 quicquam omitti potuisset, ipsemet id *del. post* si **BO**.

4 prima...satisfaciat : jam Tibi satisfecerit **TC**.

Vale Vir Illustris,[5] ac favore Tuo porro me prosequere. Dabam Gedani Anno 1666 die 20 Octob[ris].

Tuae Illustrissimae Dominationi
Addictissimus

J[ohannes] Hevelius
manu propria

To Master
Rolitsius
in Hamburg

Illustrious Sir!

Due to my many duties, and especially because I was abroad for nearly fourteen days, I answer your letter sent on 10 September quite late. I have learned from it that you had decided to add my observations of the last comet to your famous book; I am very grateful, but (to be honest) in some points you are too concise and, therefore, the thing becomes too obscure.[6] I leave it, however, to your decision, if only you change the heading or the title. Because what you excerpted from my work into this summary is not my real observations but only a general and concise description of that comet. Therefore, in my opinion, it would be better to label it with the following title: 'A general description of the comet which appeared in April 1665, written by Hevelius in Danzig, summarized by the author (viz. Rolitsius).' Since you wrote it, it should contain your name. As far as the rest is concerned, if you commission engraving my table G, I ask you to make sure to have all the details drawn with the utmost precision; otherwise, both the reader and the task itself will be quite perplexed. In my opinion, however, there is no need to enclose table G to your book. It would be enough, if you sent the readers to the *Mantissa*. Because no one will understand that discussion which arose between the most renowned Auzout and myself, nor will they be able to make their own opinion about it relying on table G and your short notes only.[7] Finally, I agree with you that there is no need to make a summary of my whole *Mantissa*, because I am sure that the Italians have already seen this booklet, so you can spare yourself that work. But if you wish to do so, it will be done with nothing but my great dissent. Because if I had been able to discuss all of it more concisely, I would have done so myself. I have passed your wish to the most renowned Titius;[8] I do not

5 Vir Illustris *om.* TC.
6 See Appendix 2.
7 See section 2.2.4 of the Introduction.
8 See note to Letter 46.

doubt he will fulfil it at the first given opportunity. Farewell, most renowned Sir, and always be benevolent towards me! Given at Danzig, on 20 October 1666.

Illustrious Sir,

Yours devoted,

Johannes Hevelius
by my own hand

52.

2 November 1666, Lubieniecki to Hevelius

BO: C1, v. 8, 1160
TC: 951

Enclosures:
52A. Observation of the comet on 22 December 1664 from Reyher
52B. Observation of the phenomenon in Andromeda from Reyher

Salutem Plurimam Nobilissime et Amplissime Vir. Merebatur ultima epistola Tua officiosae humanitatis et benevolentia[1] plena, ut ad eam statim rescriberem. Sed variae occupationes et labores mei continui adhuc mihi impedimento erant. Exspectabam praeterea exempla figurarum Rudbeckianarum ex Hollandia, quae libenter Tibi misissem. Sed haec nondum accepi, brevi tamen, volente Deo, accipiam, quod mihi hodie redditae literae Amsterodamenses promittunt. Interim en Tibi, quae hodie a Cl[arissimo] Samuele Reyhero Math[ematico] in nova Acad[emia] Chiloniensi Prof[essore] accepi. Ego coniicio eum illud ipsum phaenomenum in Andromeda observasse, quod observarant Romani Math[ematici] et Cl[arissimus] Gravius. Quod ad illa quae ex descriptione Tua Cometae verni excerpsi, attinet, consentio Tibi, titulum eis talem, ut mones, praefigendum, quo scil[icet] constet hanc generalem descriptionem Cometae a me, Tuo cum consensu, in compendium esse redactam. In tota hac re, ut existimationi mei operis, ita praecipue Tui Clarissimi nominis, sed et publicae utilitati, consulere, et priorum observationum Tuarum pariter Te consentiente in compendium redactarum conatum imitari volui. Scito autem et crede firmam mihi stare sententiam nil Tuis votis contrarium facere. Figura G. jam erit ut puto, perfecta.[2] Brevitas illa mea, cuius obscuritas // poterit aliquo auctario illustrari, laudes saltim Tuas serae posteritati dicet, et artis Mathematicae amantibus viam ad incomparabilia Tua opera monstrabit. De *Mantissa* Tua in compendium redigenda tanto magis omnem cogitationem deposui, quod Tu id fieri nolis: nempe quod tam ampla et ardua materia a me, praesertim in perbrevi commentario, ex dignitate pertractari nequeat. Sed, crede mihi, meas quoque id vix ferre rationes posse. Nam et alii labores, tempus omne sibi vendicantes, fieri id aegre paterentur, et

1 benevolentia : benevolentiae **TC**.
2 perfecta : perfecta, in cuius, ut et aliarum sculptura, debita adhibita est industria **TC**.

operis moles ac sumtus ultra opinionem excrevit.[3] Vale, Vir Humanissime, mihique constanter fave.

Tuus
omni studio et officio

S[tanislaus] L[ubieniecius] Rolitsius
manu propria

Hamburgi die 2. Nov[embris]
Greg[oriani] 1666.

Greetings, most noble, and worthy Sir! Your last letter, full of kindest humanity and benevolence, deserved my immediate answer. But my various duties and constant work were an impediment. I was also waiting for Rudbeck's[4] illustrations from the Netherlands,[5] which otherwise I would have sent to you willingly earlier. I have not received them yet, but soon, God willing, I will, as was promised in the letter from Amsterdam which arrived today. Meanwhile, I am sending you what I have received from the most renowned Samuel Reyher,[6] a professor of mathematics at the new academy in Kiel. I presume that he saw the same phenomenon in Andromeda, which had been observed by the mathematicians in Rome and by the most renowned de Grau.[7] As far as my excerpt of your description of the vernal comet is concerned, I agree with you that it should be labelled, as you remarked, with a title from which it is clear that the general description of the comet was summarised by me with your consent.[8] In all this matter, I wanted to care both for the appraisal of my work and the most renowned name of yours, but especially for the public benefit. And I wanted to repeat my efforts about your earlier observations which I had made into a summary with your consent. Be assured and believe me that I am resolved not to do anything against your will. The picture G will be completed soon, I think. That brevity of mine (whose obscurity will be explained by some additions) will certainly sing your praises in the ages to come, and it will show the way to your excellent works to all who love mathematics. I have given up all my plans of preparing a summary of your *Mantissa*,

3 excrevit : excrevit. Cl[arissimus] Titius docto Epicedio et Amici manibus et viduae meisque votis satisfecit. TC.

4 See section 2.4.16 of the Introduction.

5 See note to Letter 34.

6 Samuel Reyher (1635–1714), German lawyer and mathematician, professor of mathematics at the university in Kiel. He was Lubieniecki's correspondent; in the TC p. 881–sig. [Vvvvv2 bis v], pp. 965–66, there are 16 letters they exchanged (8 from Reyher, 8 from Lubieniecki). See section 2.3.2 of the Introduction.

7 See note to Letter 7 and section 2.3.2 of the Introduction.

8 See Appendix 2.

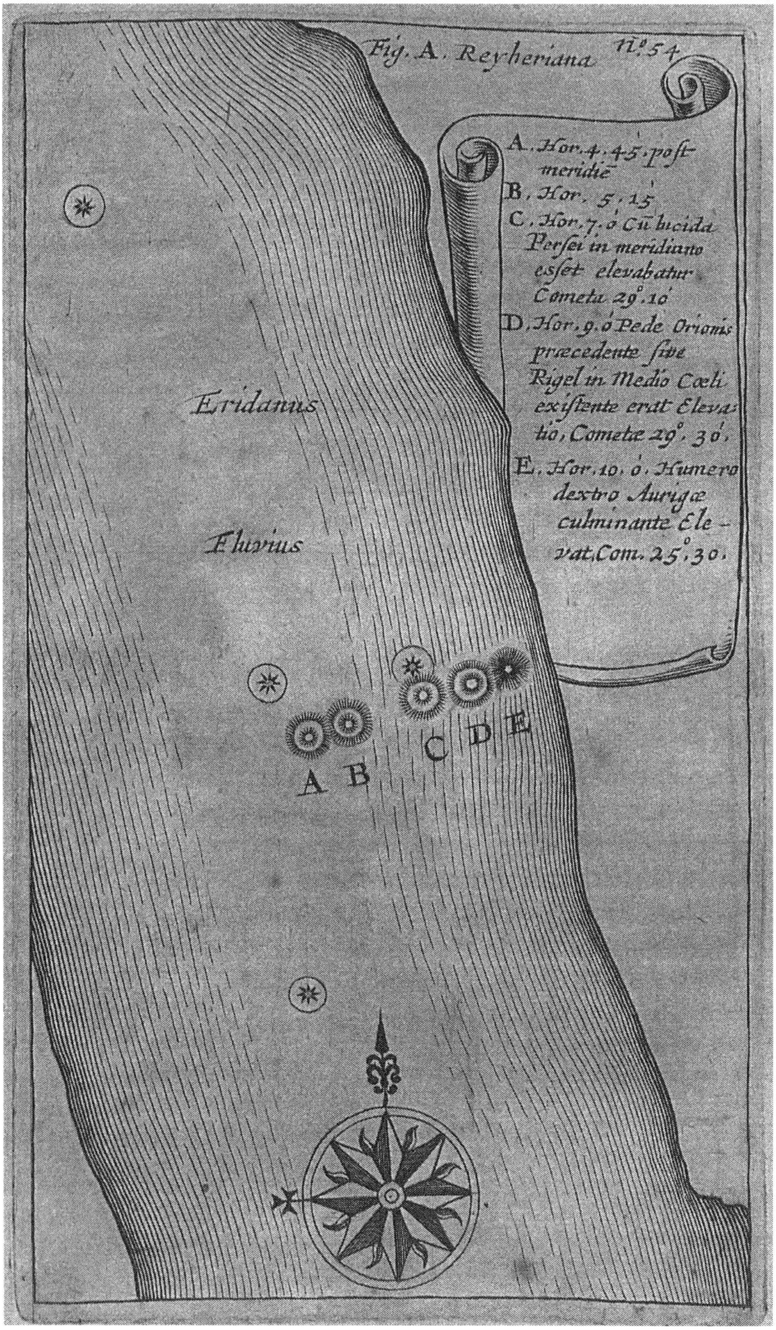

Figure 14. Reyher's observation of the comet in Eridanus. Illustration from Lubieniecki's *Theatri cometici pars prior*. Biblioteka Narodowa, Warsaw, SD XVII.4.2660.

as you do not want me to do so, and because such a broad and challenging matter cannot be discussed sufficiently by me — especially in a short text. But believe me, there are other reasons which have prevented it. My other works (which require all of my time) could suffer from it, and the size and costs of my work could grow beyond my expectations. Farewell, kindest Sir, and always be benevolent towards me!

Yours obligingly,

Stanisław Lubieniecki Rolitsius
by my own hand

In Hamburg, on 2 July 1666 Gregorian.

52A.

Observation of the comet on 22 December 1664 from Reyher

BO: C1, v. 8, 1160
TC: picture no. 54 (after p. 884)

A. Hor[a] 4.45' post meridiem
B. Hor[a] 5.15'.
C. Hor[a] 7.0'. cum Lucida Persei in Meridiano esset, elevabatur[9] Cometa 29°.10'.
D. Hor[a] 9.0' Pede Orionis praecedente, sive Rigel in Medio Coeli existente, erat Elevatio Cometae 29°.30'.
E. Hor[a] 10.0' Humero dextro Aurigae culminante, Elevat[io] Com[etae] 25°.30'.

A. At 4.45 in the afternoon.
B. At 5.15.
C. At 7 o'clock, when the bright star in Perseus was in the meridian, the elevation of the comet was 29° 10'.
D. At 9 o'clock, when the First Foot of Orion, or Rigel, was in the middle of the sky, the elevation of the comet was 29° 30'.
E. At 10 o'clock, when the Right Arm of Auriga was culminating, the elevation of the comet was 25° 30'.

9 elevabatur **TC** : elevabatus [*sic*] **BO**.

52B.

Observation of the phenomenon in Andromeda from Reyher

BO: C1, v. 8, 1160
TC: 885

Excerptum ex litteris Cl[arissimi] Viri Samuelis Reyheri Math[ematicae] in Acad[emia] Chilon[ensi] Prof[essoris]

Kilonii XIV. Kalen[das] Nov[embres] A[nno] Jul[iano] MDCLXVI.

Coronidis loco pauca addere debui de stella (num nova, nec ne, dubito) ante aliquot[10] Hebdomadas in linea recta, cum cingulo Andromedae a me primum observata, locus ex addito Schemate D. melius innotescet. Nudis oculis nubecula esse videtur. Per Tubum vero decempedalem, instar cometae caudam a Sole avertentis apparet; cum tamen reliquas nebulosas idem Tubus ex multis stellis minoribus compactas esse testetur. Quod autem nihil de hoc phaenomeno in lucem ediderim, unica causa fuit imperceptibilis hactenus situs mutatio, quae ob solitariam observationem cogitationes de eodem suspendit.

An extract from a letter by the most renowned Samuel Reyher, a professor of mathematics at the academy in Kiel

In Kiel, on the fourteenth day before the Kalends of November in 1666, Julian.

As a supplement, I should add something about the star (I do not know whether new or not) which I observed a few weeks ago in a straight line with the Belt of Andromeda — the place is shown better in the attached Picture D.[11] With the naked eye, it looks like a little cloud. But with a ten-foot-long telescope, it looks like a tail of a comet (turning away from the Sun), while the same telescope shows all other cloud-like objects as consisting of many little stars. The only reason why I have not published anything about this phenomenon is that the change of its place has been so far imperceptible; due to this fact, and because there was only one single observation, all the considerations about this phenomenon have been suspended.

10 aliquodt (d *del.*) **BO**.
11 See section 2.3.2 of the Introduction.

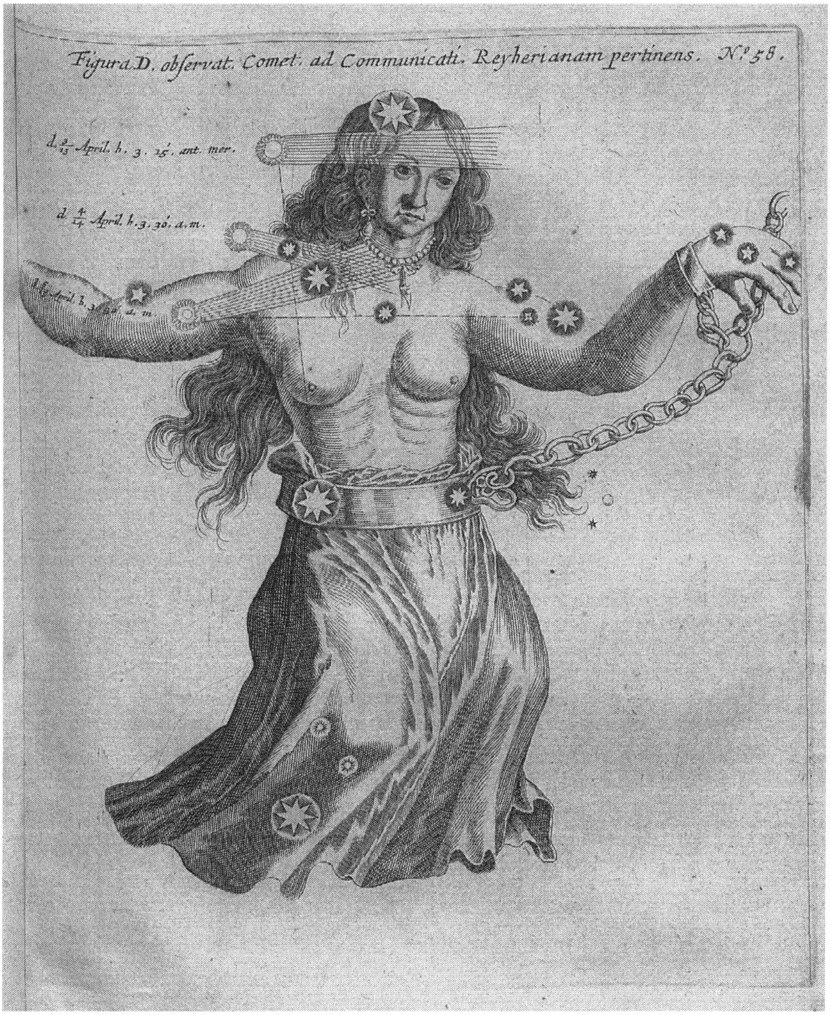

Figure 15. The constellation of Andromeda in Reyher's *communicatio* in Lubieniecki's *Theatri cometici pars prior*. Biblioteka Narodowa, Warsaw, SD XVII.4.2660.

53.

9 November 1666, Lubieniecki to Hevelius

BO: C1, v. 8, 1161
TC: 951

Salutem Plurimam Nobilissime, Amplissime, Clarissime Vir.

Nemo, crede mihi, lubentius me Tua studia praecelsa et egregios conatus juvare cupit. Nec defuissem officio in mittendis ad Te extemplo figuris Rudbeckianis, si eas citius accepissem, quod hodie demum factum. En ergo eas Tibi, quae dubio procul usui Tibi erunt in praeclaro opere Tuo Cometographico, cum et sine his ex solo textu non parum utilitatis rei Mathematicae attulisti.[1] En vero Tibi et observationes Britanicas aeri incisas. Innues verbo, quid sentias de opinione Hibernorum, qui hybernum nostrum Cometam geminum esse putarunt. Nil praeter haec nunciandum Tibi habeo, quam pacem Bremensem, quae nos gravissimi belli Germanici metu liberavit. Ita et multiplex pax, non tantum bellum, Cometas nostros adhuc excepit. Vale Vir Humanissime et mihi porro fave.[2] Hamb[urgi] 9. 9br[is] Gregor[iani] 1666.

Tuus
omni studio et officio

S[tanislaus] L[ubieniecius] Rolitsius
manu propria

Greetings, most noble, worthy, and renowned Sir!

Believe me, no one more willingly than I wants to help your exceptional studies and great efforts. And I would not have neglected my duty to send you Rudbeck's[3] illustrations, had I received them earlier; it finally happened today. I am sending them here.[4] They will be undoubtedly useful in your famous work on comets; but even without them — with the text only — you have prepared a thing beneficial for the mathematics. I am sending you also engraved observations from Britain.[5] Tell me briefly what you think about the opinion of the Irish that there were two comets in

1 attulisti : attuleris **TC**.
2 Vale…fave : B[ene] V[ale] **TC**.
3 See section 2.4.16 of the Introduction.
4 I could not find this enclosure. On the materials from Rudbeck, see note to Letter 34.
5 I could not find this enclosure.

winter.[6] I have nothing more to tell you, except that the peace of Bremen has freed us from the fear of war in Germany. Therefore, a lot of peace — and not only war — has followed our comets so far.[7] Farewell, kindest Sir, and always be benevolent towards me! In Hamburg, on 9 November 1666 Gregorian.

Yours obligingly,

Stanisław Lubieniecki Rolitsius
by my own hand

6 I could not determine to which Irish astronomers Lubieniecki refers here. It is probable, however, that there is a pun here, based on the similarity of the words 'Hibernorum' (Irish) and 'hybernum' (winter).

7 Lubieniecki is alluding to his general opinion about effects of comets, see section 3.3.1 of the Introduction.

54.

18 December 1666, Hevelius to Lubieniecki

BO: C1, v. 8, 1162
TC: 951

Illustrissimo Viro
Domino Rolisio

Recte auguraris Cl[arissimum] Reyherum stellam nebulosam Andromedae pro novo phaenomeno arripuisse. Miror quod hucusque nihil quicquam hac de re perceperit, cum tamen[1] inter Eruditos[2] de eo[3] phaenomeno jam varie[4] disceptatum fuerit. De reliquo gratias habeo ingentes pro transmissis figuris Rudbeckianis; materiam autem illam in opere meo Cometographico jam plene pertractatam esse suo tempore videbis. Literas hasce Cl[arissimo] Domino Müllero destinatas haud gravatim exhibeas rogo, cui simul observationem geminam Eclipsium hic Dantisci habitarum communicavi; si iis delectaris curabis ea describi; interea typos ipsos aeri incisos his adiungere volui.[5] Vale et me amare perge.[6]

Tuus
promptissima voluntate

J[ohannes] Hevelius

Gedani Anno 1666
die 8 Decemb[ris]

To the most illustrious Sir,
Master Rolitsius

Your guess that the most renowned Reyher[7] considered the nebulous star in Andromeda to be a new phenomenon is correct. I am surprised why he had not heard about

1 tot ac tot inter [illegible] *del. post* tamen **BO**.
2 ea *del. post* Eruditos **BO**.
3 fuerint habitae, tam etiam a nonnullis typis evulgatae *del. post* eo **BO**.
4 de...varie : jam varie de hoc phaenomeno **TC**.
5 Literas...volui *om.* **TC**.
6 et...perge *om.* **TC**.
7 See note to Letter 52.

it before, as there was a discussion about that phenomenon among learned men.[8] As far as the rest is concerned, I am grateful for Rudbeck's[9] illustrations you sent me. In due time, you will see that matter discussed thoroughly in my book about comets. I ask you to pass (without any delay) the letter I destined for the most renowned Master Müller,[10] to whom I have also sent two observations of eclipses made here in Danzig; if you like them, have them copied. Meanwhile, I wanted to attach the engraved text to them. Farewell, and always be benevolent towards me!

Your friend,

Johannes Hevelius

In Danzig, on 18 December 1666.

8 See section 2.3.2 of the Introduction.
9 See section 2.4.16 of the Introduction.
10 See section 2.4.11 of the Introduction. I have been unable to identify this letter.

55.

15 January 1667, Lubieniecki to Hevelius

BO: C1, v. 8, 1185
TC: 952

Salutem Plurimam Nobilissime, Amplissime, Clarissime Vir.

Exspecto quid amplius Cl[arissimus] Reyherus de illo phaenomeno in Andromeda nuper demum ab se observato, dicturus sit. Misit quidem mihi m[ense] Nov[embri] figuram istius phaenomeni, quam en Tibi, sed spero eum de hoc argumento plura daturum. Promisit enim plura se mihi missurum, quae nondum accepi. Figuras Rudbeckianas Te accepisse recte gratoque animo, multum gaudeo. Non dubito Te in praeclaro opere Tuo rem hanc accuratissime discussurum. Meum, ne Te quid rerum mearum lateat, properat ad finem. Sed novae et assiduae Amicorum preces tantum apud me valuerunt, ut consentirem in figuras omnium Cometarum Historiae eorum addendas. Cui fini M[agistrum] Conr[adum] Ysenium, eximium Matheseos cultorem, sub tectum meum accepi et huic labori adhibeo, nullam bene de Rep[ublica] literaria merendi occasionem negligere volens. Vale ac mihi porro[1] fave.

Hamb[urgi] 15. Jan[uarii] 1667.
quem, ut et plures alios
Tibi Tuisque prosperos precor.

Tuus
omni studio et officio

S[tanislaus] L[ubieniecius] Rolitsius
manu propria

Greetings, most noble, worthy, and renowned Sir!

I am waiting for what the most renowned Reyher[2] will say about that phenomenon in Andromeda which he has recently seen.[3] In November, he sent me a drawing of that phenomenon (which I enclose here),[4] but I hope he will say more about this subject.

1 ac…porro : et feliciter age bono publico, ac mihi, ut soles, **TC**.
2 See note to Letter 52.
3 See section 2.3.2 of the Introduction.
4 In the BO, this letter lacks the enclosure. It is probable, however, that is was the same as in Letter 52B.

Because he promised to send me more, but so far I have got nothing. I am glad that you received Rudbeck's[5] illustrations without any problems and that you liked them. I do not doubt that you will discuss that matter in your famous book. My work is not hidden from you and almost finished. But new and constant requests from my friends are so important for me that I agreed to add the illustrations of all the comets to their descriptions. To that end I hired Master Conradus Ysenius,[6] a great lover of mathematics, and I told him to do this work because I do not want to neglect any opportunity to have some merits for the Republic of Letters. Farewell, and always be benevolent towards me!

In Hamburg, on 15 January 1667.
I pray, may this year and many years to come
be prosperous for you and yours.

Yours obligingly,

Stanisław Lubieniecki Rolitsius
by my own hand

5 See section 2.4.16 of the Introduction.

6 Probably Swedish scholar Conrad Ysing (1637–71), later a professor of medicine in Pärnu. In the TC3, p. 21–23, there is a letter from him to Lubieniecki.

56.

8 February 1667, Hevelius to Lubieniecki

BO: C1, v. 8, 1186
TC: 952

Domino Rolisio
Hamburgi

Vir Illustris

Si Cl[arissimus] Reiherus sibi persuadet Cometam illum, quem in transmissa delineatione descripsit, peculiare phaenomenum esse, et ab aliis minime animadversum[1], longe aberrat. Siquidem hocce phaenomenon ille ipse Cometa est, quem nos omnes Mense Aprili anno 1665 observavimus, et cujus observationes et delineationem in recentiori nostro opusculo dedimus. Nihil itaque novi, nec in Nebulosa illa Cinguli Andromedae aliquid deprehendit; syrma quidem ei apposuit sed nec hucusque aliquid ostendit, prout cuilibet experiri integrum est. Gratulor opus Tuum ad finem properare, et ex voto omnia succedere. Figuras illas Cometarum quod attinet, scias me pariter eas jam pridem perfecisse, non solum ante aliquot jam annos illas manu mea delineavi et sculpsi, sed et omnes jam typis sunt exscriptae, exspectant solummodo posteriorem et residuam partem *Cometographiae*. Optandum esset ut Tuae et meae Icones inter se debite convenirent; de quo autem constabit[2] omnibus, cum utriusque autoris schemata in publicum proferentur: quod ut quantocyus feliciter fieri possit Faxit DEUS O[ptimus] M[aximus]. Nova stella[3] in collo Ceti nunc tandem denuo apparet, id quod et Vobis Hamburgi non latent[4] puto. Vale et me amare perge.

Tuus
optima quaequae vovens

J[ohannes] Hevelius
manu propria

Gedani Anno 1667
die 8 Febr[uarii]
st[ylo] n[ovo]

1 esse *del. post* animadversum **BO**.
2 [illegible] *del post* constabit **BO**.
3 Novam stellam (m *bis del.*) **BO**.
4 non latent : sine dubio cognitum **TC**.

To Master Rolitsius
in Hamburg

Illustrious Sir!

If the most renowned Reyher[5] thinks that the comet which he described in the account I received is some new phenomenon that was not observed by others, he is wrong. Because that phenomenon is the comet which we saw in April 1665 and whose observations and description I have recently published in my booklet. Moreover, he has not seen anything new about that nebulous star in the Belt of Andromeda; he added her a dress[6] but has not demonstrated anything, as anyone can see. I congratulate you that your book is almost finished and that everything proceeds according to your wishes. As far as the plates of the comets are concerned, you should know that not only did I make them before, but I also painted and engraved them with my own hand a few years ago; all of them have descriptions made with fonts and they only wait for the final and remaining part of my *Cometographia*. It should be desired that my and your illustrations are consistent with each other; everyone will know it when the diagrams of both authors are published; may the greatest and almighty God make it happen successfully as soon as possible. The new star in the Neck of Cetus[7] has finally reappeared; I think it is not hidden from you in Hamburg. Farwell, and always be benevolent towards me.

Your affectionate friend,

Johannes Hevelius
by my own hand

In Danzig, on 8 February 1667, new style

5 See note to Letter 52.

6 Hevelius refers to the full-figure representation of the constellation of Andromeda sent by Reyher to Lubieniecki, see 52B.

7 See section 2.3.2 of the Introduction and note to Letter 41.

57.

18 February 1667, Lubieniecki to Hevelius

BO: C1, v. 8, 1201
TC: 952

Enclosure:
57A. Letter from Oldenburg to Lubieniecki, 3 January 1667
57B. Letter from Lubieniecki to Oldenburg, 26 January / [5 February] 1667

Salutem Plurimam Vir Nobilissime, Amplissime, Clarissime.

Tuas gratissimas literas d[ie] 8. huius mensis datas ante triduum recte accepi. Ad quas tunc evestigio respondere non potui. Id nunc paucis facio. Occupationum enim et temporis utriusque nostrum, habenda nobis est ratio. Nisi quod brevitatem huius epistolae pensent adiuncta, quae haud ita pridem a Cl[arissimo] Oldenburgio[1] accepi et ad eum dedi. Accipe haec, Vir Humanissime, pro Tua aequitate et boni consule, utpote ab Amico candido et boni publici amante profecta. Scito autem me semper, ut par est, singularem magnorum Tuorum in Rep[ublica] literaria meritorum rationem habiturum, cunctaque officia, quae et Tibi grata et in mea potestate fuerint, libenter praestiturum. Revera enim Te ob illa merita singulari cultu et studio prosequor. Vide autem quam et studio rem Mathematicam et bonum publicum juvandi uterque conveniamus et tacita quadam animi consensione veluti conspiremus, licet ego Tibi hic palmam libenter meritoque concedam et ultro offeram. Non equidem laborem hunc magnum et cum haud exiguo sumtu, tum vero et retardatione operis mei coniunctum in delineandis et aeri incidendis Cometarum figuris, diversorum Amicorum hortatu, subivissem, si mihi de Tuo proposito ante aliquot menses, ex quo illi id urgent, constitisset. Nunc referre gradum salva promissi fide, exspectatione amicorum, tum vero existimatione et re mea privata, vix licet. Qua de re tamen cum Amicis // agam, ut si forte evadendi via detur, labori frustraneo me subducam et vel subtraham, ne post Homerum *Iliada*, post Virgilium *Aeneida*, post Ciceronem[2] *Philippicam* scribere velle videar. Utinam mihi Tua prius videre, quam mea edere contingeret saltim, quo ex Tuis mea corrigerem. In quibus si abs Te dissentiam fide clarorum et classicorum Auctorum nixus, occasionem inquirendi veri, amantibus eius, dabo. Victus vero, non tantum manus, sed et palmam Tibi dabo. Id enim cum opus est, sapientis esse Cicero alicubi pronunciavit. Quin et ab Hevelio in Mathematicis vinci, in parte gloriae ducam. Nam et Turno a Virgilio

1 Oldenburgio **TC** : Oldendorpio [*sic*] **BO**.
2 Ciceronem : Demosthenem et Ciceronem **TC**.

dictum gloriaeque ductum 'Aeneae magni dextra cadis.'[3] Quanquam procul a me absit haec temeritas, ut me Tibi hic adversarium sistam. Doceri et vinci abs Te laudi mihi et ab omnibus dabitur, et ipsemet ducam. Cl[arissimum] Reyherum de illo phaenomeno Tuaque sententia monebo. Necdum enim quidquam ex eo tempore de hoc argumento ab illo accepi, et responsum eius indies exspecto. *Monita* Cl[arissimi] Bullialdi de stella in collo Ceti et de nebulosa Andromedae publici juris facta, Te jam pridem accepisse non dubito. Observavit ille stellam illam exspectatione citius, die scil[icet] 20. Jan[uarii]. Nobis quoque eam hic conspicere licuit. Cl[arissimus] Mullerus, qui Te officiose // salutat et silentium diuturnum affecta jamdudum valetudine excusat, eam se ante medium Novembrem observasse dicit. Vale, Vir Cl[arissime] diu et feliciter.

Tuus
omni studio, cultu, officio

S[tanislaus] L[ubieniecius] Rolitsius
manu propria

Hamburgi 18 Febr[uarii] Greg[oriani] 1667.

Post Scriptum.[4] Adfui heri Illustrissimo Isbrandio Faeder[atorum] Belg[arum] Oratori ex Suecia redeunti, qui, ut Vir est humanissimus et bonarum literarum artiumque amantissimus, in meos Tuosque labores Cometicos diligenter inquisivit, Tuique cum laude mentionem fecit, virtutes, studia, et machinas Tuas Mathematicas, ut par est, depraedicans. Nil scio de Cl[arissimo] Büthnero. Responsum ab eo jam diu exspecto. Vivitne et valet bene?[5]

Greetings, most noble, worthy, and renowned Sir!

Three days ago, I received a very pleasant letter you had sent on the eighth of this month. I could not answer it at once then. Now I am doing it only briefly. Because each of us should take care of his duties and time. The brevity of this letter will be compensated by the enclosed letters which I have recently received from the most renowned Oldenburg[6] and which I sent to him. Take them, kindest Sir, and accept them honestly and sufficiently, as they come from an honest friend who cares for the public good. Know that I will always praise your exceptional merits in the Republic of Letters — as I should — and I will readily fulfil everything which is kind for you and possible for me to do. I truly have the utmost worship and respect for

3 Turno...cadis : olim jam quodammodo cum gloria coniunctum visum fuit 'Aeneae magni dextra cadere' **TC**.

4 Vale...Post Scriptum *om.* **TC**.

5 Nil...bene? : Vale, Vir Cl[arissime] diu et feliciter, ac mihi constanter fave. **TC**.

6 See section 2.4.12 of the Introduction.

you because of your merits. See how we both agree in supporting mathematics and the public good and how we are unanimous in silent concord; however, I willingly and rightfully yield and offer the laurel to you. Even though I was encouraged by my friends, I would not have taken up that vast and expensive work of painting and engraving the illustrations of comets (a task which also causes the delay of my book) had I known about your plan a few months earlier, when they induced me to do so. Now I cannot turn back due to the promise I made, because of my friends' expectations and for the sake of my own repute and finances. But I will discuss that matter with my friends to see if there is a way out of it — whether I can resign or withdraw from that pointless task so that it will not seem that I want to write the *Iliad* after Homer, the *Aeneid* after Virgil, the *Philippics* after Cicero. If I could at least see your work before I publish mine so that I could correct mine according to yours! If I disagree anywhere with you trusting too much the famous and classic authors, I will give the opportunity to seek the truth for those who love it. But if I am defeated, not only will I surrender, but I will also give the laurel to you. Cicero said somewhere that a wise man should do so when it is necessary.[7] To be defeated in mathematics by Hevelius is a reason to be proud, I think. Because in Virgil it was said to Turnus[8] (and declared a cause of his glory): 'you fall by the hand of great Aeneas.'[9] Nonetheless, may I stay away from such audacity as to become your opponent! Because everyone will admit (and I will agree) that to be defeated and taught by you is praise for me. I will inform the most renowned Reyher[10] about that phenomenon and about your opinion. I have not received anything from him since then, and I expect his answer soon. I do not doubt that you have already received the most renowned Boulliau's *Monita*[11] about the star in the Neck of Cetus[12] and that nebulous star in Andromeda.[13] He saw the star sooner than he expected, viz. on 20 January. It was also visible here. The most renowned Müller[14] — who sends his greetings to you and excuses his prolonged silence with his weak health — claims that he saw it before the half of November. Farewell, most renowned Sir, and be in good health and prosperous!

7 Probably an allusion to Cicero, *Ad familiares*, 4, 9, 2: 'in the first place, it has ever been considered the mark of a wise man to yield to circumstances, in other words, to bow to the inevitable', transl. by W. Glynn Williams, in Cicero, *The Letters to his Friends*, vol. 3 (London, Cambridge, Mass.: Harvard University Press, 1958).

8 Turnus, mythical King of the Rutuli in Italy. In Virgil's *Aeneid*, he was killed by Aeneas.

9 Virgil, *Aeneis*, 10, 830, transl. by H. Rushton Fairclough, in Virgil, *Aeneid VII–XII. Appendix Vergiliana* (Cambridge, Mass., London: Harvard University Press 2002).

10 See note to Letter 52.

11 Ismaël Boulliau (see section 2.4.2 of the Introduction), *Ad astronomos monita duo. Primum de stella nova, quae in collo ceti ante annos aliquot visa est. Alterum de nebulosa in Andromeda cinguli parte borea ante biennium iterum orta* (Parisiis: apud Sebastianum Mabre-Cramoisy, 1667).

12 See note to Letter 41.

13 See section 2.3.2 of the Introducton.

14 See section 2.4.11 of the Introduction.

Your obliged admirer,

Stanisław Lubieniecki Rolitsius
by my own hand

In Hamburg, 18 February 1667 Gregorian.

Postscript. Yesterday, I visited the most illustrious Isbrants,[15] the ambassador of the Dutch Republic, who is returning from Sweden. He is a man of great humanity and the lover of good sciences and arts; he inquired about my and your cometary works and he made an honorific mention of you, justly praising your virtues, studies, and mathematical instruments. I do not know anything about the most renowned Büthner.[16] I have been waiting for his answer for a long time. Is he alive and sound?

57A.

Letter from Oldenburg to Lubieniecki, 3 January 1667

BO: C1, v. 8, 1202[17]
TC: 909–10
CHO: 303–04[18]

Londino Hamburgum d[ie] 3 Januarii 1667.

Ferme jam desperaveram de Tuo ad meas responso, quando id ipsum d[ie] 18 Aug[usti] datum, ad manus meas pervenit. Miror interim, ubi id tam diu delituerit, cum nonnisi duabus abhinc Septimanas illud acceperim. Quicquid sit, mihi gratulor, Te vivere et valere, atque ornandis augendisque utilibus studiis animum jugiter addictum gerere. Author Tibi sim liceat, ut quamprimum opus Tuum Cometicum lucem viderit, Exemplar aliquod nobis transmittas. Adhaec, si absque magno Tuo incommodo fieri posset, rem nobis praestares gratissimam, si adderes *Sareptam* Mathesii, licet Germanice scriptam, nec non *Catalogum Mineralium* Swenckfeldii, et Jungii *Logicam* una cum Petri Lambecii *Historia litteraria*, primo scil[icet] eius Tomo. Vel remittemus Tibi argentum, vel libros, quos Tibi significare visum fuerit. Si quidem opus Tuum non tam cito manum Tuam ultimam senserit, fac quaeso, ut caeteri libri navigio quodam, prima quaqua occasione, ad me curentur. Certum Te esse velim, me

15 See note to Letter 27.
16 See section 2.4.3 of the Introduction.
17 Each part of the enclosure is printed separately here.
18 Printed on the basis of the TC.

paratissimum esse ad huiusmodi officia Tua pro virili, quoties id imperare volueris, pensanda.

Occupatur jam Societas nostra experimento transfundendi sanguinem unius animalis in Alterum, sive speciei eiusem, sive diversae. Num aliquam in indole mutationem id produxerit, an minus, brevi constabit, ni forte hyemis rigor, sanguinem sistens et coagulans, obfuerit.

Dominus Auzoutus in confutando Hevelio nunc laborat. Suspendit suum de hac controversia Cometica judicium (quod quidem rogitavit Hevelius) nostra Societas, donec quae Auzoutus dicenda habet, audiverit. Fortassis et observationes, hic loci a quibusdam Astronomis de eodem argumento factae, publici Juris fient, si quidem id opus esse judicatum fuerit, ad negotium hoc elucidandum.

In votis maxime habeo, ut Coloniae, Francofurti, Lipsiae et Augustae Vindelicorum, viros quosdam Philosophos mihi procures, qui commercium colere literarium nobiscum aveant. Ne dubites, paria nos reddituros iis, quae ipsis visum fuerit nobis impertiri, in Mechanicis, Anatomicis, Opticis, Astronomicis, Physicis, Chymicis et similibus. Theologica, Juridica, Politica, Metaphysica et hisce gemina Scholis et Templis relinquimus. Vale, et quantocius, si placet, responde.

From London to Hamburg, on 3 January 1667.

I had almost lost hope for your answer to my letter, when it got into my hands, sent on 18 August. I am wondering where it had been hiding for so long because I received it only two weeks ago. Whatever it was, I am glad that you are alive and in good health, and that your mind is always devoted to praising and promoting useful studies. I would like to ask you to send us a copy of your cometary book as soon as it is published. Moreover, if it is not any inconvenience for you, you would show us your great kindness if you added Mathesius's *Sarepta*[19], even written in German, as well as Schwenckfeld's *Catalogus mineralium*[20], and Jungius's *Logica*[21] together with Peter Lambeck's *Liber primus prodromi historiae literariae*.[22] We will send you back either

19 Johannes Mathesius (1504–65), German Lutheran theologian. The book in question is his *Sarepta, oder Bergpostill* (Nürnberg: durch Johann von Berg und Ulrich Newher, 1562 and Nürnberg, [s.e.], 1578).

20 Caspar Schwenckfeld (1563–1609), Silesian physician and natural philosopher. The book in question is his *Stirpium et fossilium Silesiae catalogus* (Lipsiae: impensis Davidis Alberti, 1600).

21 Joachim Jungius (see note to Letter 41A), *Logica Hamburgensis* (Hamburgi: sumptibus Bartholdi Offermans, 1638).

22 Peter Lambeck, *Liber primus prodromi historiae literariae* (Hamburgi: sumptibus autoris, typis Michelis Piperi, 1659). Peter Lambeck (1628–1680) was German polymath and humanist, since 1652 the professor of history at the *gymnasium academicum* in Hamburg, since 1662 the Imperial librarian,

the money or the books which you would like to get. If your work is not going to be finished any time soon, I ask you to ship the remaining books to me at the first given opportunity. I would like to assure you that I am ready to reciprocate such services as much as I can, whenever you ask.

Our Society is working on an experiment of transfusing the blood of one animal to another,[23] either of the same kind or another. Whether it causes any changes in their abilities or not, it will be soon determined, unless the frost of winter (which causes blood to curdle and congeal) prevents it.

Master Auzout is working now on his refutation of Hevelius.[24] Our Society suspends its judgement on that cometary controversy (as Hevelius asked) until we hear out what Auzout has to say. Perhaps the observations made here by some astronomers about that discussion will be published if there is a need to make a judgement in order to elucidate this case.

I would like you to inquire for me if there are any learned men in Cologne, Frankfurt, Leipzig, and Augsburg who would like to conduct correspondence with us. Be assured that we will send them in return something that they would like to ask us about in the subject of mechanics, anatomy, optics, astronomy, physics, chemistry, etc. We leave theology, jurisprudence, politics, metaphysics, and similar topics for schools and churches. Farewell, and reply as soon as possible, if you only wish.

57B.

Letter from Lubieniecki to Oldenburg, 26 January / [5 February] 1667

BO: C1, v. 8, 1202 (follows the preceding one)
TC: 910–12
CHO: 323–26[25]

Hamburgo Londinum XXVI. Jan[uarii] Juliani 1667.

Heri demum sero vespere literas Tuas III. huius mensis datas accepi. Ad quas en Tibi promtum responsum, quo in officii genere debitam solertiam cuique probo. Gratulor

counsellor, and historian in Vienna. In the TC, there is one letter sent by Lubieniecki to Lambeck (TC3 pp. 77–78).

23 The first blood transfusions were performed in February 1665 by Richard Lower (1631–91). Experiments with transfusion were widely described in *Philosophical Transactions*.

24 See section 2.2.4 of the Introduction.

25 Based on: Royal Society MS L 5, no. 8.

multum Reip[ublicae] literariae de praeclaris conatibus Inclytae Vestrae Societatis, commoda eius magno animo studioque promoventibus. Gratulor et Tibi, Viro eruditione celebri, quod, sicut hisce tuis literis indicas, // amplum munus consecutus sis, ut a Secretis tam Illustri Societati constituereris. Pergite, o boni, bene de genere humano mereri laudabilibus Philosophicis studiis, ita tamen ne studia pacis et virtutis (quod nec temere nec frustra moneo) negligatis. Nosti quid in hanc rem Seneca ad Lucilium hic et illic scribat. Nosti Oraculum sacrum, omni exceptione maius, Heroica quoque, licet alioqui Christiana, facinora ipsaque Angelica dona prae caritate nihili ducere. Atque haec quidem mea est sententia, qui talia studia Vobis non tantum ornamento, sed et emolumento fore scio. Aequitate Vos in Amplissimum Hevelium, Virum de re Astronomica optime, si quisquam alius nostro seculo, meritum, uti, Viris humanitate excultis convenit. Optaram ego tam Hevelio quam Auzouto amicus, Bullialdi, et Hugenii, utpote praestantissimorum Mathematicorum et utrique amicissimorum, opera controversiam illam dirimi potuisse. Sed illi, qua sunt modestia, tantum sibi tribuere, vel etiam alterutrum amicorum (quanquam reliquis omnibus, ex veteri quoque scito, magis amica sit veritas et Heveliana modestia et sinceritas, quam etiam literis ad me de hoc argumento scriptis testatus est, nota omnibus) offendere forte nolent. At vero Inclutae Vestrae Societatis auctoritas et accuratissimarum observationum fides, longe maius huic rei decidendae adferet momentum. Libros, quos a me postulas, et a me comparari in usus vestros poterunt, sine mora habebis. Nec in conciliando Vobis Philosophico literarum commercio in his oris exercendo quidquam a Vobis in mea industria desiderari patiar. Qua de re hodie amicum meum Noricum, melioris notae Virum, Vestrisque studiis juvandis idoneum, scripto adibo. Observationes Cometae nuperi hyberni et verni saltem a d[ie] 12/22 ad 21/31 Dec[embris] habitas velim mihi mittas. Cl[arissimus] Bullialdus misso nuperrime ad me scripto suo publici juris facto, nos de observata ab se d[ie] XX. Jan[uarii] Greg[oriani] stella illa mira in collo Ceti (quam et hac in Urbe Mathematici, alii citius, alii serius observarunt) tum de phaenomeno in Cingulo Andromedae Boreali, quod Anno 1665. M[ense] Febr[uario] a Cl[arissimo] Gravio Prof[essore] Math[eseos] Franequerano, mense vero Aprili ab Mathematicis Romanis, alias ab aliis observatum esse scio, monuit. Quid vero Vobis videtur de iride, quam urente intentissimo gelu ipsis huius Mensis Kalendis cum domesticis meis post horam primam pomeridianam[26] observavi et per integram amplius horam spectavi? Domus enim mea in aggere ad Alstriam fluvium urbem Veterem a Nova dividentem sita et ab utraque parte libero coelo gaudens, observandis sideribus et phaenomenis est percommoda. Cuius rei experimentum in observato d[ie] 6. 16. Julii 1665. phanomeno fig[ura] 2. in opere meo DEO volente, videbis. En vero tibi nostrum hocce phaenomenon nuperrime visum et cum cura a me et M[agistro] Conrado Ysenio, Mathematices cultore eximio, qui mihi σύνοικος καὶ σύσσιτος[27] est observatum. Apparebat vergere illud ad occasum brumalem et pene super angulum Austrinum (eorum nempe qui occasum spectant in turri

26 pomeridianam **BO TC** : meridianam **CHO**.
27 σύσσιτος **BO TC** : συάδιτος [*sic*] **CHO**.

nova S[ancti] Michaelis, quem situm et totam hanc urbis istius regionem vel potius mediam et maiorem quidem, ambitus ratione, partem in dicta tabula videbis) existens. Exhibebatur oculis instar orbis magnitudine circiter ipsius disci Solaris coloribus iridis variis, prout videri solent, distincti. Ab hoc tendebant radii tres, literam T quodam modo referentes, quorum duo portionem iridis referentes in una quasi recta linea vel arcu circulari jacebant, et quidem alter in occasum brumalem fere directus ab an-//gulo supra dicto austrino versus partem turris occiduam paululum declinabat ad Boream, alter vero tantundem ab angulo boreo versus partem orientalem in Austrum fere. Maxime vero conspicuus fuit orbis ille intermedius, obscuriores autem ab eo tendentes radii, velut a Cometis cauda, nisi quod radii isti in diversas mundi plagas porrigerentur, et latitudo eorum non modo non excesserit, sed ne quidem omnino aequaverit orbis diametrum. Hi radii circa finem in mucronem fastigati erant, et in cuspidem, velut disparituri, desinebant. Radius tertius in occasum fere aestivum tendens, ab orbe medio globum magis compactum referente, reliquis velut perpendicularis pene ad angulum boreum versus partem occidentalem turris coeteris videbatur pallidior et lucidior, minus coloribus distinctus, praesertim in productione paulo longius ab orbe intermedio. Sol momento primo apparitionis meridianum pertransierat ad mensuram oculorum judicio tertiae circiter parti quadrantis aequalem in circulo, quem sol eo die describere videbatur. Orbis hic paulo proprior horizonti quam Soli, sicque ab eo circiter unius signi paulo plus portione distare, centrum vero Solis, orbis illius et oculi, triangulum Scalenum, cuius latus maximum referebat oculi distantia[28] a phaenomeno, minimum oculi a Sole distantia,[29] efficere videbatur. Duravit hoc jucundum sane phaenomenon sesquihorae circiter spatio, ex quo quidem a nobis animadversum est, satis conspicuum et distinctis coloribus oculos ad se trahens. Sed ultra duas horas cum dimidia ab initio apparitionis, vel potius observationis nostrae (poterat enim longe prius, nobis cibum capientibus apparuisse) vestigia illius satis notabilia fuere. Denique, quod diligenter, nec sine admiratione et delectatione observavimus, prout Sol paulatim ad occasum tendebat, ita phaenomenon nostrum eadem fere proportione secedere videbatur in occasum et boream, sive in occasum aestivum, ut tandem, cum eius vestigia tantum incipiebant apparere, circiter 10. vel 12. gradibus a Turri distare et fere medium locum inter horizontem ac turrim occupare videretur. Caelum toto illo pomeridiano tempore fuit sudum et considerando phaenomeno aptum, nubeculis, claris tamen, hic et illic dispunctum, quarum quae phaenomeno proprius accedebant, semper longiores se exhibentes, in latitudine vero decrescentes rubicundo purpureoque tingebantur colore. Atque haec de isto phaenomeno sufficiant. Quid jam judicandum erit de communi sententia, iridem generari[30] ex refractione radiorum Solis ex adverso et in linea recta cum oculo nostro et centro iridis positi, in nube rorida, // et ita quidem, ut non appareat, nisi dimidia circuli forma? Haec enim cuncta phaenomenon nostrum refutat. Hyeme illud apparuisse

28 distantiam (m *del.*) **BO** : distantia **TC** distantiam **CHO**.
29 distantiam (m *del.*) **BO** : distantia **TC** distantiam **CHO**.
30 generari **BO CHO** : fieri **TC**.

non miror. Scio nempe etiam crescentibus, non tantum decrescentibus diebus circa aequinoctia conspici. Nec noctu eam adulto nuper auctumno, ut ab amico accepi apparuisse miror, quod inter ἀδύνατα Plinius posuit. Nam annorum 1523, 1527. et 1537. aliorumque experientia satis eum refutavit, et cum Aristotele dicentem plures simul duabus non videri, cum tres saepius visae fuerint. Eundem plures simul quam tres Soles ad sua tempora non visos, aliud nunc doceret experientia. Milichius enim ad h[unc] l[ocum] Plinii l[ibrum] scil[icet] 2. c[aput] 31. adnotat Regem Poloniae sex Soles vidisse. Verum hic Eckstormius in *Histor*[*ia*] *Parel*[*iorum*] tres legit. Sed diversae Milchii editiones sex Soles legunt. Atque hoc adnotat Cartesius lib[ro] de Meteoris c[apite] X. ubi etiam caussam reddit de quinque Solibus Romae 1629. (Eckstormio *Histor*[*ia*] *Parel*[*iorum*] Anno 1621. finiente) d[ie] 20 Martii visis.[31] Habebat autem nostrum phaenomenon ut iridis magnam, ita et aliquam parelii (qui etiam iridem sociam habere solet)[32] similitudinem. Eodem certe modo productum fuit, quo illi produci solent, ex reflectione radiorum Solis in nube densa et aequabili a latere posita. Nam hic plane Phaenomenon a latere Solis sinistro fuit, et cum oculo nostro triangulum constituit. Hic qui cum linea recta, in qua communis sententia Solem, oculum et Iridem ponit, conveniet? Dicendum arbitror aut Iridem fuisse extraordinariam, aut tantum quidpiam analogum iridi. Atque hoc potius eligo quam illud, ne sine necessitate communi sententiae reclamemus. Opus meum ad finem properat. Aliquot tamen Menses exibunt, priusquam figurae omnium Cometarum, quorum quidem historia accurata habetur, depingantur, et aeri incidantur. Has enim me precibus Amicorum dare oportuit. Vale.[33]

From Hamburg to London, on 26 January 1667 Julian.

Yesterday, late in the evening, I finally received your letter, sent on the third of this month. Now I am sending you an eager reply to testify my abilities in that kind of service. I heartily congratulate the Republic of Letters on the notable efforts of your famous Society which helps the Republic with such a great mind and service. I also congratulate you, a man of great learning, because (as you said in your letter) you were entrusted with an important duty when you were appointed the Secretary of such a famous Society. Good men, keep serving the humankind well with your laudable philosophical studies, but also do not neglect efforts for peace and virtue (I do not warn you hastily or in vain)! You know that Seneca often wrote about it to Lucilius.[34]

31 Sed diversae...visis **BO TC** : Cartesius, nisi memoria fallit, in Opticis quinque Soles simul memoriam prodidit. **CHO**.

32 solent (n *del.*) **BO** : solet **TC** solent **CHO**.

33 Vale **BO TC** : Vale. Dabam Hamburgi XXVI Jan[uarii] Juliani MDCLXVII. Post Scriptum. Indica quamprimum ad quem tuto libri, quos postulas, transmitti possint. Iterum Vale. **CHO**.

34 It seems that Lubieniecki does not allude to any particular passage in Seneca's *Ad Lucilium*. Similar opinions can be found in many places throughout this work, for example 28, 7: 'I disagree with those who strike out into the midst of the billows and, welcoming a stormy existence, wrestle daily in har-

You know this holy prophecy[35] (which has no exceptions) that heroic deeds (even those entirely Christian ones) and angelic gifts themselves mean nothing when compared to love. And this is my opinion too, because I know that those studies will not only be the cause of your glory but will be useful as well. As cultured men, you should act justly towards the greatest Hevelius, a man who did more for astronomy than anyone else in our century. As a friend to both Hevelius and Auzout, I would like that controversy to be solved with the help of Boulliau[36] and Huygens[37] — great mathematicians and friends to both of them.[38] But out of their modesty, they did not want it to be assigned to them; or perhaps they did not want to offend any of their friends (everyone else believes, however, as the ancient saying goes, that truth and Hevelius's modesty and sincerity are greater friends[39] — which he testified in his letter to me concerning that discussion, and which are known to everyone). But the authority of your famous Society and the faith in your perfect observations will make your decision crucial. You will soon have the books which you asked for and which I can buy for you. I will also not allow myself to neglect your wishes about obtaining some scholarly epistolary contacts in this country for you. Today, I will write and ask

dihood of soul with life's problems. The wise man will endure all that, but will not choose it he will prefer to be at peace rather than at war', transl. by Richard M. Gummere, in Seneca, *Ad Lucilium epistulae morales*, vol. 1 (Cambridge, Mass., London: Harvard University Press, 1979); 66, 5: 'The first day we investigated this problem: how can goods be equal in they are of three kinds? For certain of them, according to our philosophical tenets, are primary, such as joy, peace, and the welfare of one's country. Others are of the second order, moulded in an unhappy material, such as the endurance of suffering, and self-control during severe illness. We shall pray outright for the goods of the first class; for the second class we shall pray only if the need shall arise. There is still a third variety, as, for example, a modest gait, a calm and honest countenance, and a bearing that suits the man of wisdom', transl. by Richard M. Gummere, in Seneca, *Ad Lucilium epistulae morales*, vol. 2 (London, New York: William Heinemann, G. P. Putnam's Sons, 1920); 89, 8: 'Certain of our school, however, although philosophy meant to them "the study of virtue," and though virtue was the object sought and philosophy the seeker, have maintained nevertheless that the two cannot be sundered. For philosophy cannot exist without virtue, nor virtue without philosophy. Philosophy is the study of virtue, by means, however, of virtue itself; but neither can virtue exist without the study of itself, nor can the study of virtue exist without virtue itself. For it is not like trying to hit a target at long range, where the shooter and the object to be shot at are in different places. Nor, as roads which lead into a city, are the approaches to virtue situated outside virtue herself; the path by which one reaches virtue leads by way of virtue herself; philosophy and virtue cling closely together', transl. by Richard M. Gummere, in Seneca, *Ad Lucilium epistulae morales*, vol. 2.

35 Probably an allusion to 1 Corinthians 13. 1–2: 'Though I speak with the tongues of men and of angels, and have not charity, I am become as sounding brass, or a tinkling cymbal. And though I have the gift of prophecy, and understand all mysteries, and all knowledge; and though I have all faith, so that I could remove mountains, and have not charity, I am nothing.'

36 See section 2.4.2 of the Introduction.

37 See section 2.4.7 of the Introduction.

38 See section 2.2.4 of the Introduction.

39 An allusion to a well-known phrase: 'Amicus Plato, sed magis amica veritas' ('Plato is a friend, by the truth is a greater friend'), attributed to Aristotle. Considered to be a paraphrase of a passage from the *Nicomachean Ethics*, 1096a11–15, it was probably coined in anonymous 12th-century *De vita Aristotelis*, see *Aristotelis qui ferebantur librorum fragmenta*, ed. by Valentinus Rose (Lipsiae: in aedibus B. G. Teubneri, 1886), p. 447.

my friend from Noricum[40] who is a man well-known and an appropriate person for promoting your studies. I would like you to send me your observations of the former comet from winter and spring, at least from 12/22 until 21/31 December. In a recently published text which he sent to me,[41] the most renowned Boulliau informed us that on 20 January Gregorian he saw Mira Ceti[42] (which was seen by all the mathematicians in this city, earlier or later); he also wrote about that northern phenomenon in the Belt of Andromeda which I know was seen in February 1665 by de Grau,[43] a professor of mathematics in Franeker, and in April by the mathematicians in Rome, each time separately.[44] And what is your opinion about the rainbow which I saw on the first day of this month, when it was very cold, after 1 o'clock in the afternoon, and which I was watching for an entire hour together with the members of my household? Since our house is located in the dyke on the river Alster (which separates the Old Town from the New Town), it is very convenient for observing stars and other phenomena, because fortunately it has an open sky view from both sides. You will see an example of it in the observation of the phenomenon of 6/16 July 1665 in Illustration 2 in my book, God willing.[45] And this is that phenomenon, recently seen and described carefully by me and Master Conradus Ysenius[46] — a great worshipper of mathematics — who lives and dines with me. It seemed to lean towards the winter place of the sunset and was nearly over the southern quarter (viz. the quarter of those who watch the sunset in the new tower of Saint Michel;[47] this place and all the surrounding region of the city (or rather its central and main part) will be shown in a panoramic view in the illustration I mentioned). It looked like a circle of the size of the solar disc itself, divided into various colours of the rainbow, as they are usually seen. There were three rays stretching from it, similar in shape to the letter T, two of which constituted a part of the rainbow and were in a straight line or on an arc; one — directed almost towards the winter sunset, from the aforementioned southern quarter towards the western part of the tower — was leaning a little northwards, and another — directed equally from the northern quarter towards the eastern part — a little southwards. The most visible was that circle in the middle — the rays stretching from it were darker, like a tail of a comet; but those rays extended in various parts of the world and their width not only did not exceed but even did not equal the diameter of the circle. Those rays at their ends were sharpened like blades and ended like thorns,

40 Probably Johann Ernst Noricus (1634–78), German lawyer and philosopher, since 1665 counsellor at the court in Merseburg, as identified in the CHO p. 408. In the TC3 pp. 32–40, there are 6 letters he exchanged with Lubieniecki (3 from Noricus, 3 from Lubieniecki).

41 Ismaël Boulliau, *Ad astronomos monita duo. Primum de stella nova, quae in collo ceti ante annos aliquot visa est. Alterum de nebulosa in Andromeda cinguli parte borea ante biennium iterum orta* (Parisiis: apud Sebastianum Mabre-Cramoisy, 1667).

42 See note to Letter 41.

43 See note to Letter 7.

44 See section 2.3.2 of the Introduction.

45 See Letter 27C.

46 See note to Letter 55.

47 Saint Michel's church in Hamburg was built between 1647 and 1669.

as if they were about to disappear. The third ray, pointed towards the summer sunset, emerging from the centre of the circle — which looked like a denser sphere — was almost perpendicular to two others, and it was directed towards the northern quarter, to the western part of the tower; it seemed more pallid and brighter than others, with blurred colours, especially because it extended a little longer from the central circle. At the moment when it appeared, the Sun had already passed the meridian by a third part of a quarter of the orbit it seemed to move that day — according to the measurement made with the naked eye. The circle was a little bit closer to the horizon than to the Sun, and it seemed to be a little bigger than the length of one sign. The centre of the Sun, the circle, and the eyes seemed to make a scalene triangle; its longest edge was the distance from the eyes to the phenomenon; the shortest was the distance from the eyes to the Sun. That pleasant phenomenon lasted for about one and half hours since the moment when it was spotted; it was visible enough and attracted attention with its various colours. But even after two and a half hours after it had appeared, or rather after we had seen it (because it could have appeared a long time before, when we were eating), its traces were fairly visible. Eventually, when we were watching it diligently and with fascination and pleasure, the Sun was slowly tending towards the sunset; and our phenomenon — in almost the same proportion — seemed to go down towards its set and northwards, as if it was moving towards the summer sunset; finally, when only its traces were visible, it seemed to be about 10 or 12 degrees from the tower, placed nearly in the middle between the horizon and the tower. For the whole afternoon, the sky was clear and favourable for the observation of that phenomenon; in some places the sky was spotted with bright little clouds only — when they got closer to the phenomenon they always seemed longer but decreasing in width and coloured in red and violet. Enough of that phenomenon. What can be said about the common knowledge that rainbows arise from the refraction of the rays of the Sun which is on the opposite side and is placed in a straight line with our eyes and the centre of the rainbow in the humid cloud; and that the rainbows appear in a shape of a semicircle only? Nothing of these apply to our phenomenon. I do not wonder at the fact that it appeared in winter. Because I know they are seen also when days become longer (not only shorter) around equinoxes. I am not surprised by its nightly appearance in autumn, as I learned from a friend, but Pliny placed it among impossible things.[48] Because the experience of the years 1523, 1527, 1537, and others is a sufficient refutation of it, as well as of the one who says with Aristotle that more than two are never seen at the same time, while often three are seen at once. And even though he says there were never more than three suns seen at the same time, now the experience contradicts that view. Because Milich, in his commentary, says about the pas-

48 Pliny Major, *Natural History*, 2, 31, 99: 'Again, several suns are seen at once, neither above nor below the real sun but at an angle with it, never alongside of nor opposite to the earth, and not at night, but either at sunrise of at sunset', transl. by H. Rackham, in Pliny, *Natural History in Ten Volumes*, vol. 1 (Cambridge, London: Cambridge University Press, 1967).

sage in Pliny, Chapter 31, that the king of Poland saw six suns.[49] But Eckstorm in his *Historia pareliorum* says about three.[50] Yet various editions of Milich say about six suns.[51] And Descartes discusses it in his *Les Météores*, Chapter 10, where he mentions five suns seen in Rome in 1629 (while Eckstorm's *Historia pareliorum* ends in 1621), on 20 March.[52] Our phenomenon was both very similar to a rainbow and a little bit to a parhelion (which usually has something in common with a rainbow). It certainly arose in the same way as they usually do, viz. from the reflection of sunrays in a dense cloud placed next to the Sun. The phenomenon was clearly on the left side of the Sun and formed a triangle with our eyes. How does it agree with the straight line in which the common opinion places the Sun, the eye and the rainbow? I think it should be stated that it was an extraordinary rainbow or at least something similar to a rainbow. And I choose rather this one than the former one — not to oppose the common view when it is unnecessary. My book is reaching the end. But there will be a few months before the illustrations of all the comets — which are described in their precise history — are drawn and engraved. I should do it because my friends requested it from me. Farewell!

49 Jakob Milich (1501–59), German physician and mathematician. Lubieniecki refers to his edition of Pliny: *Liber secundus C. Plinii de mundi historia*, ed. by Jakob Milich (Halae Suevorum, ex officina Petri Brubachii, 1538), sig. 102v. Milich gives neither the year of the observation nor the name of the King.

50 Heinrich Eckstorm (see note to Letter 27), *Historiae eclipsium, cometarum et pareliorum* (Helmaestadi: typis heredum J. Lucii, 1621), p. 291.

51 Milich's book was published again in 1543, 1563, and 1573.

52 René Descartes, *Discours de la méthode pour bien conduire sa raison et chercher la vérité dans les sciences, plus la dioptrique, les météores et la géométrie qui sont des essais de cette méthode* (Leyde: à l'imprimerie de Ian Maire, 1637), p. 288; see also modern edition: *Oeuvres de Descartes*, vol. 6, ed. by Charles Adam and Paul Tannery (Paris: Léopold Cerf, 1902), p. 361.

58.

19 March 1667, Hevelius to Lubieniecki

BO: C1, v. 8, 1203
TC: 953

Domino
Domino Rolisio
Hamburgum

Illustrissime Vir.

Beasti me sane, quod inprimis copiam literarum Cl[arissimi] Domini Oldenburgii mecum communicasti. Ex quibus intellexi Ill[ustrissimam] Regiam Societatem, suum de nostra controversia judicium suspendisse, donec Auzoutium si quae dicenda adhuc haberet, audierit, prout Hevelius (ut scribit) rogitavit. Verum enimvero Cl[arissimus] Oldenburgius mentem meam plane secus intellexit. Ego quidem Astronomos Anglicos rogavi, qui rem examinare tum susceperant, statim, quamprimum *Epistolam* Auzoutii legissem, ut judicium suspenderent, donec responsionem meam parassem, ne inauditus condamnarer; factum tamen nihilominus est, ut sententiam suam, priusquam *Mantissa Prodromi* mei prodiit, super ea re pronuntiaverint, quemadmodum ex diario tam Gallico quam Anglico Eruditorum clare patet. Minime vero petii, ut Dominus Auzoutius praerogativa aliqua gauderet, ac secunda vice audiretur super *Mantissam* meam.[1] Si hoc fieri debeat, profecto et mihi indulgendum erit, ut pari jure fruar, paribusque armis dimicem. Nam cui obtingit bis audiri prae altero, fortunam ne dicam, Judicem longe habet benigniorem. Si igitur Cl[arissimus] Auzoutius denuo audiendus erit, jure oportet ut replicam meam non minus exspectent; atque sic controversia non amputabitur, sed in infinitum protrahetur. Sufficeret, meo quali quali judicio, si protinus // super ea re, quae prima vice ab utraque parte prodiit,[2] Ill[ustrissima] R[egia] Societas sententiam ferret; sicque nemini primatum concederet. Tu, quaeso, ipse indices anne petitum meum in summa consistat aequitate? Verum quacunque etiam ratione res mecum suscipiatur, bono ero animo: Solatio est, observationes meas ab ipso initio ad 12, 13 et 14 Febr[uarii] fixo stare talo, a quibus totum dependet negotium. Stellam in collo Ceti ego itidem a 23 Januar[ium] deprehendi; die vero 13 eiusdem nondum apparuit:[3] unde certum est[4] (cum toto anno

1 audiretur…meam : super *Mantissam* meam audiretur **TC**.
2 prodieruntit (erunt *del.*) **BO**.
3 sic ut *del. post* apparuit **BO**.
4 sit *del.* **BO**.

diligentissime ad eam attenderim) Mens[e] sc[ilicet] Novemb[ri] nondum fulsisse.[5] Vale et fave[6]

Tuo
affectu officio et studio

J[ohanni] Hevelio

Gedani Anno 1667
die 19 Martii

To Master
Master Rolitsius
to Hamburg

Most illustrious Sir!

You were very kind to me when you sent me a copy of Oldenburg's[7] letter. I learned from it that the most renowned Royal Society had suspended its judgement on our controversy until they listen to what Auzout has to say[8] — because Hevelius (as Oldenburg writes) asked for it. But the most renowned Oldenburg has misunderstood me completely. As soon as I read the *Lettre* by Auzout, I asked the English astronomers (who had already begun to examine that case) to suspend their judgement until I prepare my answer, in order not to be condemned without being heard out; nevertheless, it happened that they had stated their opinion before the *Mantissa* of my *Prodromus* was published, as it is clear from both the French and English learned journals.[9] I asked neither for inquiring Master Auzout, nor for hearing him out again about my *Mantissa*. If it has to be so, I should undoubtedly be allowed to use an equal privilege and to fight with equal arms. Because whoever is heard out twice before his opponent, he will have much more benevolent fate, not to say the judge. Therefore, if the most renowned Auzout is heard out again, they should reasonably expect my answer too; and thus the controversy will not be finished but prolonged forever. It would suffice, in my opinion, if the most renowned Royal Society made their judgement

5 fulsisse : fuisse **TC**.

6 et fave *om.* **TC**.

7 See section 2.4.12 of the Introduction.

8 See section 2.2.4 of the Introduction.

9 *Journal des sçavants* and *Philosophical Transactions*. For the list of Hevelius's publications in these journals see Maria Otto, 'Prace Jana Heweliusza', in: *Jan Heweliusz*, ed. by Maria Pelczar and Jarosław Włodarczyk (Radom: Instytut Technologii Eksploatacji – Państwowy Instytut Badawczy w Radomiu, 2011), pp. 245–47; Chantal Grell, 'Hevelius dans l'histoire des sciences: bibliographies', CJH1 pp. 575–79; CJH3 p. 616, note to Letter 237, des Noyers to Hevelius, 16 July 1683.

on the basis of what was said about that thing by both sides at the first opportunity; then no one would be privileged. Please, tell me if my request is entirely justified. In whatever way the matter is discussed, I am hopeful; what consoles me is the fact that my observations from the beginning until 12, 13, and 14 February — on which my whole work depends — have solid grounds. I have been seeing again the star in the Neck of Cetus[10] since 23 January, but on the thirteenth of the same month it was not visible yet. Therefore, it is certain (because I have been watching it diligently for the whole year) it did not shine in November. Farewell, and be benevolent towards

Yours obligingly,

Johannes Hevelius

In Danzig, on 19 March 1667

10 See note to Letter 41.

59.

22 March 1667, Lubieniecki to Hevelius

BO: C1, v. 8, 1205

Enclosure:
59A. Excerpt of the letter from Oldenburg to Lubieniecki, 19 February 1667

Salutem Plurimam Vir Nobiliss[ime] Ampliss[ime] Clariss[ime]

Ante biduum misit ad me literas hasce, quas vides, Cl[arissimus] Mullerus. Hodie accepi Oldendorpianas Londini 19. Febr[uarii] datas, quibus inter alia inerant haec quae adiungo. Dum ad finem operis propero, valde sum occupatus. Ideo et hic solito sum brevior. Exemplaria Tua nondum ulla ratione distrahere potui. Duo vendidi, duo donavi; pro quibus trium pretium Tibi dabo. Dolendum est homines hic parum esse Mathematicarum rerum curiosos. Sed hic morbus et alibi regnat. Vale Vir Humaniss[ime] et feliciter age.

Tuus
omni studio et officio

S[tanislaus] L[ubieniecius] Rolitsius
manu propria

Hamb[urgi] d[ie] 22. Martii
1667.

Greetings, most noble, worthy, and renowned Sir!

Two days ago, the most renowned Müller[1] sent me this letter which you see here.[2] Today, I have received a letter from Oldenburg,[3] sent in London on 19 February which — among other things — contained what I enclose. I am very busy as I am approaching the end of my work. Therefore, my letter is shorter than usual. I could not circulate all the copies of your book[4] in any way. I sold two and gave two as a gift; I will send you the money for three of them. Unfortunately, there are not many

1 See section 2.4.11 of the Introduction.
2 I could not find this enclosure.
3 See section 2.4.12 of the Introduction.
4 Lubieniecki is probably referring to Hevelius's *Descriptio cometae*.

people here who are interested in mathematics. But there is a plague here, as in other places.[5] Farewell, kindest Sir, and be prosperous!

Yours obligingly,

Stanisław Lubieniecki Rolitsius
by my own hand

In Hamburg, on 22 March 1667.

59A.

Excerpt of the letter from Oldenburg to Lubieniecki, 19 February 1667[6]

BO: C1, v. 8, 1205 (follows the preceding one)
TC: 912
CHO: 345[7]

Summopere laetor, Tractatum tuum Cometicum ad finem decurrere. Dominus Hevelius brevi quoque Colophonem addet *Cometographiae* suae; quamprimum scil[icet] molesta illa // de loco Cometae d[ie] 18. Febr[uarii] controversia diremta fuerit. Pars utraque supremum de re tota judicium Societ[ati] Regiae deferunt. Est quidam ex illius sociis, qui brevi, ni admodum fallor, observata sua Cometica est editurus. Id factum cum fuerit verosimile est Soc[ietatem] praedictam candide sententiam suam de re tota quantum quidem per observationes licuerit, pronunciaturam. Molimur jam Speculae contemplandis Astris idoneae aedificationem, estque spes magna, dummodo pax nobis redeat multa in caelis insignia porro a Nobis observatum iri. Non omittemus putem stellam in Collo Ceti, jam jam monente Bullialdo visendam contemplari: nec deerunt quibus curae erit Orbis Terrae paralaxim rimari: quae res rite deprehensa agitatam illam de terrae motu vel quiete controversiam plane opinor dirimet.

5 Outbreaks of plague were very frequent in the early modern period. A brief list of outbreaks in the 16th and 17th centuries is given in W. J. Simpson, *A treatise on Plague, dealing with the Historical, Epidemiological, Clinical, Therapeutic and Preventive aspects of the Disease* (Cambridge: at the University Press, 1905), pp. 26–27.

6 The same text is in Letter 61A.

7 Based on: Royal Society MS O 1, no. 40 and on the TC.

I am glad that your cometary treatise will soon be finished. Master Hevelius will also complete his *Cometographia* soon, viz. as soon as that controversy about the place of the comet on 18 February is solved.[8] Both sides have passed the final decision about it to the Royal Society. There is one of its fellows who will soon publish his cometary observations, if I am not mistaken.[9] When it happens, the Society will probably announce its opinion on that matter clearly, as far as the observations allow it. Now we are struggling with the construction of an observation post[10] which would be suitable for watching stars, and there is much hope that when the peace returns,[11] we will see many notable things in the sky. I think we will not omit the observation of the star in the Neck of Cetus,[12] because Boulliau[13] informs us that it will be visible soon; there will also be those who will engage in the search for the parallax of the Earth's orbit. In my opinion, when this issue is appropriately examined, it will undoubtedly end that ongoing controversy whether the Earth moves or not.

8 See section 2.2.4 of the Introduction.

9 In the CHO, p. 347, this person is identified as probably Robert Hooke (1635–1703), English natural philosopher. His remarks about the comets of 1664 and 1665, however, were published with his discussion of the comet of 1677, titled 'Cometa, or Remarks about comets', in Robert Hooke, *Lectures and Collections* (London: printed for John Martyn, printer to the Royal Society, at the Bell in St Paul's Church-yard, 1678), included in Robert Hooke, *Lectiones Cutlerianae* (London: printed for John Martyn, printer to the Royal Society, at the Bell in S. Pauls Church-yard, 1679). See also Voula Saridakis, 'The Hevelius-Hooke Controversy in Context: Transforming Astronomical Practice in the Late Seventeenth Century' in *Johannes Hevelius and His World. Astronomer, Cartographer, Philosopher and Correspondent*, ed. by Richard L. Kremer and Jarosław Włodarczyk (Warsaw: Institute for the History of Science, Polish Academy of Sciences, Copernicus Center for Interdisciplinary Studies, 2013) (*Studia Copernicana*, vol. 44), pp. 103–35; Chantal Grell, 'Hevelius en son temps', CJH1 pp. 109–17; Chantal Grell, 'Introduction', CJH2 p. 91; CJH3 pp. 465–66, note to Letter 174, Hevelius to des Noyers, 26 September 1675.

10 In the CHO, pp. 345–47, this object is translated as 'mirror' and the note suggests that it could mean 'lens'. In the text of the letter, however, the word in question is not *speculum* ('mirror') but *specula* ('look-out post').

11 i.e. when the war between England and the Netherlands (1665–67) ends.

12 See note to Letter 41.

13 See section 2.4.2 of the Introduction.

60.

12 April 1667, Lubieniecki to Hevelius

BO: C1, v. 8, 1206
TC: 953

Salutem Plurimam Nobilissime, Amplissime, Clarissime Vir.

Debueram Tibi ex officii ratione ad gratissimas literas Tuas d[ie] 19. Martii exaratas respondisse jam pridem: post acceptas praesertim a Cl[arissimo] Oldenburgio literas d[ie] 19. Febr[uarii] datas, ad me tamen demum d[ie] 22. Martii redditas. Sed, crede mihi, ita per bimestre integrum diversorum hospitum non tantum praetereuntium urbem hanc, sed et in ea aliquandiu agentium (quos inter Cl[arissimus] Heinsius Tibi quoque amicissimus fuit) frequenti praesentia et conversatione, tum puerperio uxoris, filiam alteram feliciter enixae per Dei gratiam,[1] aliisque negotiis et impedimentis distractus fui, ut nequaquam officia amicis, ac Tibi cumprimis, debita praestare ex voto potuerim. Dabis igitur tantillae morae veniam aequus et facilis, qualem Te in colenda amicitia semper expertus sum. En Tibi excerptum ex literis Oldenburgianis: unde quae sit Societatis illius celeberrimae mens, percipies. Doleo interim sic Tecum agi, ut scribis. Nemo enim est qui judicem vel arbitrum aequum jus utrique parti relinquere nec in alterutram partem propendere debere, ignoret vel neget. Itaque propediem ea de re Cl[arissimum] Oldenburgium pro jure amicitiae et studio aequitatis admonebo. Fateor enim cum tot Viris vastissima eruditione Clarissimis, Heinsio, Bullialdo, Kirchero, Curtio, Ricciolio aliisque, quibuscum literarum commercium colo, Herculeis Te laboribus optime de re Mathematica meritum, ideoque omni favore et cultu prosequendum esse. Ego si quid hic possum, experiar omnem industriam meam, praesertim si abs Te monitus fuero, Tibique gratum[2] id officiosae benevolentiae fore[3] intellexero. Nam ut amicitiae, sic et veri-// tati ante omnia deesse nunquam volo. Quo in statu sit Tua *Cometographia*, scire aveo. Opus meum, nisi intra aliquot menses, Deo juvante absolvetur ob molem suam et figurarum copia. Addere quaedam operi de Significatione Cometarum animus est. Horum compendium in istis XXX. rationibus, quibus velut fundamentis tota mea sententia innitur, videbis. De his judicium Tuum et Doctissimi ac Amicissimi Büthneri, cui pl[urimam] saltutem hic adscribo, scire desidero. Vale ac mihi porro fave, qui sum

1 enixae...gatiam : per DEI gratiam enixae **TC**.
2 gratum *om.* **TC**.
3 fore : fore gratum **TC**.

Tuus
omni studio, cultu, officio

S[tanislaus] L[ubieniecius] Rolitsius
manu propria

Hamburgi 12. April[is] Greg[oriani] MDCLXVII.

Greetings, most noble, worthy, and renowned Sir!

I should have answered earlier your most pleasant letter written on 19 March, especially when I received the message from the most renowned Oldenburg,[4] sent on 19 February, which had finally arrived on 22 March.[5] Believe me, for the entire two months I was very distracted due to often meetings and conversations with various guests who did not only pass through this city, but also stayed here for some time (among whom there was the most noble Heinsius,[6] a great friend of yours); and also because my wife[7] gave a fortunate birth to my second daughter,[8] thanks to God; there were other obligations and obstacles, so I could not fulfil proper duties towards my friends, especially towards you. You will forgive me for this delay because you are just and calm as I have always seen you in our friendship. I enclose Oldenburg's letter from which you will learn the opinion of that renowned Society.[9] Meanwhile, I grieve that he treated you in the way you write. Because there is no judge or mediator who would not know or would deny that equal rights should be granted to both sides and that he should not lean towards one of them. Therefore, I will soon admonish the most renowned Oldenburg about this case in a friendly and just manner. Because together with many famous men of great learning — Heinsius, Boulliau,[10] Kircher,[11] Curtz,[12] Riccioli,[13] and others with whom I correspond, I admit that you have many merits for mathematics with your Herculean works and, therefore, you should be treated with all benevolence and respect. If I can do anything, I will use all the means I have, especially if you tell me to do so; and I shall know that it will be a pleasant sign of my benevolence towards you. Because I never want any detriment for the friendship and, first of all, for the truth. I would like to know the progress of your work on the

4 See section 2.4.12 of the Introduction.
5 See Letters 59A and 61A.
6 See section 2.4.6 of the Introduction.
7 See section 1.1.2 of the Introduction.
8 Gryzella Konstancja Lubieniecka, died in 1675.
9 Lubieniecki sent the enclosure with Letter 61.
10 See section 2.4.2 of the Introduction.
11 See section 2.4.8 of the Introduction.
12 See section 2.4.5 of the Introduction.
13 See section 2.4.15 of the Introduction.

Cometographia. My book will be finished in a couple of months, with God's help, because of its volume and the number of illustrations. I plan to add there something about the meaning of comets. You will see its summary in these 30 arguments;[14] my opinion is based on them like on a fundament. I would like to know what you and the most learned and friendly Büthner[15] think about it; pass on many greetings to him. Farewell, and always be benevolent towards me, the one who is

Yours obligingly,

Stanisław Lubieniecki Rolitsius
by my own hand

In Hamburg, on 12 April Gregorian 1667.

14 See Appendix 3 and section 3.3.3 of the Introdution.
15 See section 2.4.3 of the Introduction.

61.

15 April 1667, Lubieniecki to Hevelius

BO: C1, v. 8, 1207

Seal with Lubieniecki's coat of arms (Rola coat of arms), address: 'A Monsieur / Mons[ieu]r Jean Hevelius / Conseiller de la ville / à Dantzic' ('To *Monsieur* Johannes Hevelius, city Councillor, in Danzig')

Enclosure:
61A. Excerpt of the letter from Oldenburg to Lubieniecki, 16 February 1667

Salutem Plurimam Vir Nobilissime, Amplissime, Humanissime.

Proxime inter molem negotiorum vix licuit opere tumultuario aliquid literarum ad Te dare, excidit vero memoria haec ex Oldenburgianis adiungere. Hoc nunc facio, utque festinationis erroribus tunc commissis veniam des rogo. Cl[arissimo] Büthnero, cuius gratissimas literas hodie accepi, sal[utem] pl[urimam] adscribo. Vale et fave

Tuo

Rolitsio
manu propria

Hamb[urgi] 15. Apr[ilis]
1667.

Post Scriptum. Velim indices mihi nomina horum Virorum, quos sibi Oldenburgius nominari postulat. Iterum V[ale].

Greetings, most noble, worthy, and kindest Sir!

Recently, among a great number of duties, working in haste, I could barely send a letter to you, and I forgot to enclose the message from Oldenburg.[1] I am doing it now, and I ask you to forgive that mistake caused by my haste. I pass many greetings for

1 See section 2.4.12 of the Introduction.

the most renowned Büthner,[2] whose pleasant letter I received today. Farewell, and be benevolent towards

Yours,

Rolitsius
by my own hand

In Hamburg, on 15 April 1667.

Postscript. Tell me, please, the names of those men about whom Oldenburg would like to be informed. Farewell again!

61A.

Excerpt of the letter from Oldenburg to Lubieniecki, 19 February 1667[3]

BO: C1, v. 8, 1207 (follows the preceding one)
TC: 912
CHO: 345[4]

Excerptum ex literis Cl[arissimi] Henrici Oldenburgii Londino d[ie] 19. Feburarii Anno 1667. datis.

Summopere laetor, Tractatum tuum Cometicum ad finem decurrere. Dominus Hevelius brevi quoque colophonem addet *Cometographiae* suae; quam primum scil[icet] molesta illa de loco Cometae d[ie] 18. Febr[uarii] controversia dirempta fuerit. Pars utraque supremum de re tota judicium Societ[ati] Regiae deferunt. Est quidam ex illius sociis, qui brevi, ni admodum fallor, observata sua Cometica est editurus. Id factum cum fuerit vero simile est Soc[ietatem] praedictam candide sententiam suam de re tota, quantum quidem per observationes licuerit, pronunciaturam. Molimur jam speculae contemplandis Astris idoneae, aedificationem, estque[5] spes magna, dummodo pax nobis redeat, multa in coelis insignia porro a nobis observatum iri. Non omittemus, putem Stellam in Collo Ceti, jam jam monente Bullialdo visendam contemplari: nec deerunt, quibus curae erit, Orbis terrae paralaxim rimari; quae

2 See section 2.4.3 of the Introduction.
3 The same text is in Letter 59A.
4 Based on: Royal Society MS O 1, no. 40 and TC.
5 estque **TC CHO** : atque **BO**.

res rite deprehensa, agitatam illam de terrae motu vel quiete controversiam, plane, opinor, dirimet.

Lubens scirem, quinam hoc tempore in Germania // et Polonia Viri in solida Philosophia et Mathesi prae ceteris excellunt. Scisciteris quaeso, Vir Praestantiss[ime], amicos Tuos per Germaniam utramque et eorum, quos edoctus fueris, nomina quamprimum commode poteris, significes rogo. Vale.

An excerpt of the letter from the most renowned Henry Oldenburg, sent in London, on 19 February 1667.[6]

I am glad that your cometary treatise will soon be finished. Master Hevelius will also complete his *Cometographia* soon, viz. as soon as that controversy about the place of the comet on 18 February is solved. Both sides have passed the final decision about it to the Royal Society. There is one of its fellows who will soon publish his cometary observations, if I am not mistaken. When it happens, the Society will probably announce its opinion about that matter clearly, as far as the observations allow it. Now we are struggling with the construction of an observation post which would be suitable for watching stars, and there is much hope that when the peace returns, we will see many notable things in the heavens. I think we will not neglect the observation of the star in the Neck of Cetus, because Boulliau informs us that it will be visible soon; there will also be those who will engage in the search for the parallax of the Earth's orbit. In my opinion, when this thing is appropriately examined, it will undoubtedly end that stiff controversy whether the Earth moves or not.

I would like to know which men now excel above others in the solid philosophy and mathematics in Germany and Poland. I ask you, most distinguished Sir, to inquire about your friends in the whole of Germany and to tell me, at the first given opportunity, the names of those about whom you will learn. Farewell!

6 For notes see Letter 59A.

62.

6 May 1667, Lubieniecki to Hevelius

BO: C1, v. 8, 1208
TC: 953

Enclosure:
62A. Letter form Reyher to Lubieniecki, 18 April 1667

Salutem Plurimam Nobilissime, Amplissime, Clarissime Vir. Mitto Tibi Cl[arissimi] Reyheri, cum quo illa, quae ad me scripsisti, ex amicitiae, aequitatisque jure, communicavi, responsum, quod ante paucos dies accepi.[1] Hoc mihi Tuam de eius observationibus sententiam, quam me non celabis, promittit. Nil praeterea, quod nunc scribam, habeo, quam ut Cl[arissimum] Buthnerum, et alios amicos salutes. Dolor Vestrae et literariae Reip[ublicae] ex amisso Viro Doctissimo et Consultissimo Vincentio Fabricio me quoque tangit. Eadem urbis huius fortuna, quae non ita pridem primarium Consulem, Virum Virtutibus et meritis inter Magnos reputandum Bartholdum Möllerum amisit. Ita Praeclarorum quoque Virorum fatis cedentium exemplis, mortalitatis et officii admonemur. Vale, Vir Praestantissime et Humanissime, diu et prospere, ac mihi fave constanter. Hamburgi VI. Maii Greg. MDCLXVII.

Tuus

S[tanislaus] L[ubieniecius] Rolitsius
manu propria

Greetings, most noble, worthy, and renowned Sir! I am sending you the answer from the most renowned Reyher[2] which I received a few days ago; following the principle of friendship and equity, I passed to him what you had written to me. This answer promises me your opinion about his observations — you will not hide it from me. I have nothing more to write about here but to ask you to greet the most renowned Büthner[3] and other friends. Just as you and the Republic of Letters, I was also stricken with grief caused by the loss of the most learned and wisest man, Vincentius Fabricius.[4] The same is the fate of this city which has recently lost the best Councillor, the

1 accepi : accepi d[ie] 18. Aprilis datum **TC**.

2 See note to Letter 52.

3 See section 2.4.3 of the Introduction.

4 Vincentius Fabricius (Schmidt, 1612–67), town clerk and Councillor from Danzig. He was also a poet (one of prefatory poems to Hevelius's *Selenographia* was written by him) and the author of political and commemorative speeches and legal texts. He exchanged 6 letters with Hevelius (4 sent by

man worthy of a place among the greatest men due to his virtues and merits, Barthold Moller.[5] And so we are admonished by these examples of mortality and duty of the most renowned men who yield to fate. Farewell, most excellent and kindest Sir, be in good health and prosperous, and always benevolent towards me! In Hamburg, on 6 May 1667 Gregorian.

Yours,

Stanisław Lubieniecki Rolitsius
by my own hand

62A.

Letter from Reyher to Lubieniecki, 18 April 1667

BO: C1, v. 8, 1208 (follows the preceding one)
TC: [Vvvvv2 bis r][6]

Kilonio Hamburgum d[ie] XIIX. Aprilis MDCLXVII.

Literas Tuas nudius tertius datas jam accipio jam quoque ad praecipuum caput respondeo, reliqua ob temporis brevitatem ad proximam occasionem differre coactus. Hevelii illustris illius nostrorum temporum Atlantis verba non penitus intelligo, quantum autem coniicere possum, Magnus ille Vir ex solo schemate judicium de mea sententia tulit, quia putat me Cometam Aprilem Anni 1665. pro insolito phoenomeno et a me unice animadverso venditare, quod nunquam in mentem venit, nec etiam venire potuit. Observavi enim illum Cometam Lugduni Batavorum, ubi Cl[arissimus] Golius aliique sescenti eandem stellam deprehenderunt. Nec etiam aliud in mea descriptione dixi. Ad alterum quod attinet, nempe de nebulosa Andromedae, itidem aliud testantur verba mea, quibus expresse testatus sum me dubitare, utrum nova sit, nec ne? et propter solitariam observationem judicium // meum me suspendere; non quod ab aliis observata non esset sed quod alios observasse non satis mihi constiterat. Tunc ergo solicitus eram de aliorum accuratioribus observationibus. Alias enim publico scripto sententiam meam propalassem, quam merito a Tua Generositate certior de aliorum sententiis factus, retraxi. Quid autem per illa verba velit: 'Sed nec hucusque aliquid ostendit,' fateor me non intelligere. Utrumque enim Phaenomenon

Fabricius, 2 by Hevelius) between 1647 and 1661, cf. Klaus-Dieter Herbst, 'Alphabetical list of the correspondents', CJH1 p. 259. See also CJH2 p. 140, note to Letter 1, Hevelius to Dupuy, 10 April 1654 and CJH3 pp. 307–08, note to Letter 90, des Noyers to Hevelius, 30 October 1660.

5 See note to Letter 15.

6 In the TC, after p. 888, there are two pages without page or folio numbers.

et Cometicum et Nebulosum ostendi, quamvis non eo modo, prout saepe laudatus Hevelius meas observatiunculas interpretatus est. Sed jam vale et Favere ne desine.

From Kiel to Hamburg, on 18 April 1667.

I have just received your letter sent three days ago, and I am answering it in haste, being forced to postpone other things to the next opportunity, due to lack of time. I do not fully understand the words of Hevelius — that famous Atlas of our time — but as far as I can guess, this great man's judgement about my statement was based on my drawing only; he thinks I claim that the comet of April 1665 was an unusual phenomenon and it was seen once only by me; I have never claimed such a thing, and I could never claim it.[7] Because I saw it in Leiden when the most renowned Golius[8] and six hundred other men watched that star. And I did not say anything else in my description. As far as other things (viz. the nebulous star in Andromeda) are concerned, my words again testify something different; because I testified explicitly that I doubt whether it is a new one or not and that I suspend my judgement because of this single observation — not because it was not seen by anyone else, but because I do not know enough about the observations conducted by others. I was then uneasy about more precise observations made by others. Otherwise, I would have published my opinion in a booklet which I had rightly retracted when I was informed by you, generous Sir, about others' opinions. But what Hevelius meant by these words: 'but has not demonstrated anything,' I admit I do not understand.[9] Because I have demonstrated both phenomena — the cometary and the nebulous one — but not in the way in which the praised Hevelius understood my petty observations. Farewell for now, and stay benevolent!

7 See section 2.3.2 of the Introduction.

8 Probably Jacob Golius (1596–1667), orientalist and mathematician, professor of mathematics and Arabic at the University of Leiden.

9 See Letter 56.

63.

31 May 1667, Lubieniecki to Hevelius

BO: C1, v. 8, 1214
TC: 953–54

Enclosures:
63A. Excerpt of the description of the comet of 1661
63B. Excerpt of the letter from Reyher to Lubieniecki, 6 May 1667

Salutem Plurimam Vir Ampliss[ime] et Clariss[ime].

Dum responsi vicem abs Te ad meas binas 12 et 26. Apr[ilis] exspecto, et ad finem operis propero, volui etiamnum paucis Te compellare, ut ad Te referrem, quae de Te initio historiae de Cometae Anno 1661. visi ortu et situ, motu et eventibus[1] ponere decrevi. En et excerptum ex Literis Cl[arissimi] Reyheri prid[ie] Nonas Maias datis. Tribus verbis respondebis mihi ad haec et superiora. Ita enim nunc nobis occupatis agendum est. Vale et fave

Tuo

S[tanislao] L[ubieniecio] Rolitsio

Hamb[urgi] 31. Maii
1667.

Post Scriptum. Excerpta ex Sim[one] Mario non mitto. Hunc nunc credo habes. Saluta amicos. Iterum vale.[2]

Greetings, most worthy, and renowned Sir!

While I am waiting for your answer to my two letters of 12 and 26 April, and as I am approaching the end of my book, I wanted to address you and report what I have decided to write about you at the beginning of the history of the origin, place, movement, and the circumstances of the comet seen in 1661.[3] I also enclose an excerpt

1 de Cometae...eventibus : Cometae anni 1661 visi **TC**.
2 et fave...vale *om.* **TC**.
3 See note to Letter 2B.

from the most renowned Reyher's[4] letter sent a day before the Nones of May. Answer briefly to these and to the earlier ones. Because this is the way we should act when we are busy. Farewell, and be benevolent towards

Yours,

Stanisław Lubieniecki Rolitsius

In Hamburg, on 31 May 1667.

Postscript. I am not sending the excerpts of Simon Marius.[5] I believe you already have them. Greet our friends. Farewell again!

63A.

Excerpt of the description of the comet of 1661

BO: C1, v. 8, 1214[6] (follows the preceding one)
TC2: 438

Omnium de hoc Cometa (1661)[7] observationes, quas adhuc videre mihi contigit, inimitabilis Heveliana solertia superat. Affirmat enim se eum feliciter usque ad 28. Martii cum Cl[arissimo] Bullialdo observasse, tubi praestantissimi beneficio, cum alios ultra 15. vel 20. Febr[uarii] suas observationes huius phaenomeni non extendisse arbitretur. *Mantissa Prodr[omi] Comet[ici]* p[agina] 181. Superius hunc nostrum medio Martii uno die 12', ultimo vero tantum 6'. confecisse, et quidem sub evidenti incurvatione orbitae, dicit, seque id in *Cometographia* demonstraturum promittit. p[agina] 174. Inferius vero eundem, ut et illum 1652. Cometam a nupero hyberno diversissima facie fulsisse pariter se demonstraturum pollicetur p[agina] 186. Cl[arissimus] Frider[icus] Mullerus ait Cl[arissimum] M[agistrum] Abdiam Trew in suis Cometae huius observatis c[apite] 3. monere Aristotelicos, Tychoni, Cometae parallaxin Lunari minorem esse observationibus Astronomicis contra Aristotelem probanti, tanquam perito artifici in sua arte credendum esse. *Comm[entario] de Cometa 1664* l[ibro] 2. c[apite] 2. Cum tamen Claremontius, et ex ipso alii, Tychonis observa-//ta in dubium vocent, illumque potius ex hypothesi nondum probata suas rationes deduxisse, quam evidenti Mathematica demonstratione id ipsum evicisse,

4 See note to Letter 52.
5 See section 2.3.2 of the Introduction.
6 Each part of the enclosure is printed separately here.
7 (1661) *om.* **TC2.**

opinentur, exspectanda nobis erit a Magno Hevelio tantae controversiae decisio. Is enim vel Solus, vel certe eximia diligentia doctrinam de Cometarum parallaxi pertractat.

All observations of that comet (of 1661),[8] which I have had so far a chance to see, are surpassed by Hevelius's unique talent. He claims that he watched it successfully until 28 March with the most renowned Boulliau,[9] using his perfect telescope, while (as he believes) others did not extend their observations of that phenomenon beyond 15 or 20 February (*Mantissa Prodromi cometici*, on page 181). At first, he says that our comet in the middle of March moved 12' in one day and at the end of March — 6' only, and that its orbit was visibly inclined; and he promises (on page 174) that he will show it in his *Cometographia*. Then, he says it was the same as the one of 1652.[10] He also promises (on page 186) to show that the comet was shining with a different face from the recent one seen in winter. The most renowned Friedrich Müller says (in the commentary of the comet of 1664,[11] in Book 2, Chapter 2) that the most renowned Master Abdias Trew (in his observations of that comet,[12] in Chapter 3) admonished the Aristotelians[13] that one should more believe Tycho, who with his astronomical observations proves against Aristotle that the cometary parallax is smaller than the lunar one, and he is an experienced artist in his art. But Because Claremontius[14] and others following him doubt in Tycho's observations and think that he rather deduced his arguments from a hypothesis which is not proven yet than that he confirmed it with a precise mathematical demonstration, we should wait for the great Hevelius's judgement in this controversy. Because he (either alone or with his exceptional diligence) discusses the subject, of the cometary parallaxes.

8 See note to Letter 2B.

9 See section 2.4.2 of the Introduction.

10 See note to Letter 2B.

11 Friedrich Müller (1630–67), German astronomer. The book in question is his *Cometologiae libri tres* (Giessae Hassorum: imprimebat Fridericus Kargerus, 1665).

12 Abdias Trew (1597–1669), German Protestant theologian, mathematician, and astronomer. The book in question is probably his *Gründtlicher Bericht von deme im Ende verwichenen Januarii Alten und Anfang deß Februarii Neuen Calenders erschienenen Cometen* (Nürnberg: verlegt von Johann Taubern, 1661).

13 Labels such as 'Aristotelians' and 'Peripatetics' were used to denote opponents of post-Copernican discoveries of astronomy in general, and they did not have to refer to the followers of Aristotle's teachings and writings.

14 Probably Scipione Chiaramonti (1565–1652), Italian philosopher and professor at the University of Pisa. He opposed Galileo and all contemporary astronomical discoveries. He supported the Ptolemaic system of the universe and he criticized both the Tychonian and Copernican models.

63B.

Excerpt of the letter from Reyher to Lubieniecki, 6 May 1667

BO: C1, v. 8, 1214 (follows the preceding one)
TC: [Vvvvv2 bis r–v][15]

Excerptum ex literis a Cl[arissimo] Reyhero
Kilonii prid[ie] Non[as] Maias 1667. datis.

De phaenomeno Andromedae nihil habeo quod moneam, nisi quod ex Bibliotheca Generosissimi[16] a Qualen, summi astronomiae cultoris, *Mundum Jovialem* Simonis Marii acceperim, et monente Ricciolio Non nulla in praefatione invenerim, quae excerpenda curavi.

An excerpt of the letter from the most renowned Reyher,
given at Kiel a day before the Nones of May 1667.

I have nothing to say about the phenomenon in Andromeda except that I have received the *Mundus Jovialis*[17] by Simon Marius from the library of the most generous von Qualen,[18] a great worshipper of astronomy, and, prompted by Riccioli,[19] I found something in its preface which I have excerpted.

15 See Letter 62A.

16 Generosissimi **TC** : Generossimi [*sic*] **BO**.

17 Simon Marius, *Mundus Jovialis anno MDCIX derectus ope perspicilli Belgici* ([Nürnberg]: sumptibus et typis Johannis Lauri, 1614). See section 2.3.2 of the Introduction.

18 Heinrich von Qualen (1594–1667), a learned nobleman, see Hans Schöder, 'Genealogische und biographische Nachrichten von der Familie Qualen', *Nordalbingische Studien. Neues Archiv der Schleswig-Holstein-Lauenburgischen Gesellschaft für Vaterländische Geschichte*, 3 (1846), 103–45 (pp. 123–24).

19 See section 2.4.15 of the Introduction.

64.

7 June 1667, Hevelius to Lubieniecki

BO: C1, v. 8, 1209
TC: 954

Domino Rolisio
Hamburgi

Illustrissime Vir.

Cum nihil notatu dignum Tibi perscribendum habuerim, variisque occupationibus curisque districtus fuerim, studio responsum meum ad ternas tuas, imo quaternas, ni fallor, hucusque distuli. Gratias habeo multas, quod sententiam tuam de effectibus et significationibus cometarum, quae sane perplacet, mecum communicare volueris; quam etiam jussu tuo nuper Domino Büthnero transmisi. Ego vix paucula quaedam ea de materia in *Cometographia* attingam, sufficiat Te, aliosque plurimos rem eandem jam plenius pertractasse. De caetero in *Cometographia* mea alacriter pergo, spero exeunte anno, si papyrum ex Batavia tantummodo impetrare possim, plane ad finem eam me perducturum. Tuum vero opus, uti percipio, citius in publicum prodibit,[1] quod ut fiat, faxit DEUS O[ptimus] M[aximus]. Lubenter nomina eorum Virorum Tibi significarem, quos sibi Cl[arissimus] Oldenburgius (quem humanissime salvere velim) postulat, sed nullos plane Mathesi et Philosophiae solidae operam dantes, in Polonia Lithuania ac Livonia novi. Quinam vero in Germania adsint, Tu melius explorare poteris: ego namque nec ibidem ullum habeo cum quo commercium exerceo literarium, praeter Balthasarum Moretum P[atrem] Soc[ietatis] Jes[u] Vratislaviae commorantem, vosque Hamburgenses. Haud ita pridem a Celeberrimo R[everendo] P[atre] Alberto Curtio perbreves quidem sed humanitate plenas accepi, quibus me certiorem reddit, se aliquot exemplaria *Historiae* suae *Coelestis* ad me transmisisse; sed nulla prorsus hactenus accepi, qui factum sit, nescio. Hasce itaque literas, nisi grave est, rogo Ei perferendas cures, // ad quaevis officia Tui gratia subeunda rursus paratissimus. Vale et porro fave

Tuae Illustriss[imae] D[ominationi]

J[ohanni] Hevelio

Gedani Anno 1667
die 7 Junii.

1 prodibit *om.* TC.

To Master Rolitsius
in Hamburg

Most illustrious Sir!

Since I had nothing noteworthy to write to you about and I was distracted by various duties and troubles, I have so far delayed my answer to three or four (if I am not mistaken) letters from you. I am grateful because you decided to give me your opinion about the effects and meanings of comets; I like this text very much, and I passed it to Master Büthner,[2] as you asked. I almost do not touch upon that matter in the *Cometographia*; it is enough that you and many others have already discussed that thing in a more detailed way. As far as the rest is concerned, I am working eagerly on my *Cometographia*; I hope that I will have completed it by the end of the year, if only I can get some paper from the Netherlands.[3] But your book, as I notice, will be published sooner — may the greatest and omnipotent God make it so. I would tell you willingly the names of those men about whom the most renowned Oldenburg[4] (to whom I would like you to pass my kindest greeting) asks, but I know almost no one who works on mathematics and solid philosophy in Poland, Lithuania, or Livonia. And those who are in Germany can be better investigated by you, because I have no one there with whom I keep epistolary contacts, except for Balthasar Moretus[5] (a Father from the Society of Jesus who lives in Breslau) and you, in Hamburg. Not so long ago, I received a very short but a very kind letter from the Reverend Father Albert Curtz in which he informed me that he had sent a few copies of his *Historia coelestis*[6] to me; but so far I have received nothing, and I do not know the reason why. I ask you to pass this letter to him, if it is not too coarse, and I am ready to return the favour in any form you wish. Farewell, and always be benevolent towards

illustrious Sir,

Yours,[7]

Johannes Hevelius

In Danzig, on 7 June 1667.

2 See section 2.4.3 of the Introduction.

3 Cf. Letter 30.

4 See section 2.4.12 of the Introduction.

5 Hevelius probably meant Theodorus Moretus (1602–67), a Jesuit mathematician with whom he exchanged 20 letters (13 from Moretus, 7 from Hevelius) between 1660 and 1667, cf. Klaus-Dieter Herbst, 'Alphabetical list of the correspondents', CJH1 p. 267.

6 Albert Curtz (see section 2.4.5 of the Introduction), *Historia coelestis* (Augustae Vindelicorum: apud Simonem Utzschneiderum 1666).

7 The letter in the BO is a draft and, therefore, it lacks the full signature.

65.

26 July 1667, Hevelius to Lubieniecki

BO: C1, v. 8, 1215
TC: 954

Domino Rolisio
Hamburgi

Illustrissime Vir.

Tandem observationes Tychonicas omnes, studio atque opera Lucii Baretti editas, a Rev[erendo] P[atre] Alberto Curtio obtinui; de quibus multum multumque Rei literariae gratulor, et ut Autori ea de caussa quoque gratias agere possim debitas, has conscripsi literas; quas peto humanissime ut data occasione Ei transmittas: ad paria officia rursus sum paratissimus. In dicto opere deperditus est Liber duodecimus Observationum Tychonicarum anni 1593. Qui, si Hafniae asservatur, rogandus erit Cl[arissimus] Bartholinus, ut eundem quantocyus proferat, nec non totum exemplar impressum cum manuscripto Danico conferre curet: quo si in quibusdam discrepantia quaedam intercederet, ea de re Astronomos quamprimum admoneret; profecto non solum in meum, sed in omnium Coelestium rerum Inspectatorum magnum cederet emolumentum. Vale. Dabam Gedani Anno 1667. die 26 Julii.

Tuae Illustrissimae Dominationis
studiosissimus

J[ohannes] Hevelius
manu propria

To Master Rolitsius
in Hamburg

Most illustrious Sir!

I have finally received all the Tychonic observations, published with great diligence and effort by Lucius Barettus, from the Reverend Father Albert Curtz;[1] I heartily congratulate the Republic of Letters on that book, and I have written this letter in

1 Albert Curtz (see section 2.4.5 of the Introduction), *Historia coelestis* (Augustae Vindelicorum: apud Simonem Utzschneiderum 1666). 'Lucius Barettus' is an anagram of 'Albertus Curtius.'

order to send my well-deserved thanks to the author for that publication; I kindly ask you to pass it to him when you have a chance; I am ready to provide similar services in return. In the aforementioned book, the twelfth book of Tycho's observations (from the year 1593) is missing. If it is kept in Copenhagen, the most renowned Bartholin[2] should be asked to publish it as soon as possible and to take care to compare the entire published copy with the Danish manuscript;[3] if there happen to be any discrepancies in anything, he should quickly notify the astronomers about it; he would certainly render a beneficial service not only to me but to all those who observe celestial phenomena. Farewell! Given at Danzig, on 26 July 1667.

illustrious sir,

Yours devoted,

Johannes Hevelius
by my own hand

2 See section 2.4.1 of the Introduction.

3 Since 1664, under the auspices of King Frederick III, Erasmus Bartholin was working (with no results) on the publication of Tycho Brahe's observations.

66.

26 July 1667, Lubieniecki to Hevelius

BO: C1, v. 8, 1217
TC: 954

Enclosure:
66A. Excerpt of the history of the comet of 807

Salutem Plurimam Vir Nobilissime, Amplissime, Clarissime.

Expectabam adhuc Reverendi Curtii responsum ad Tuas a me rite curatas: quod huic meo adiungere volebam. Sed cum nil responsi ab eo compareat, nec mihi quidquam de Viri Doctissimi ab excessu Illustrissimi Rautensteinii, communis Amici constet rebus studiisve, nolui diutius officium meum differre. Multis autem laboribus occupatum non minus et ipse occupatus, paucis compello. Exspecto abs Te responsum quod tribus verbis absolves, de parallaxibus Cometarum, quarum in nuperis ad Te meis d[ie] 31. Maii datis, feci mentionem. Nunc autem velim tribus quoque verbis respondeas de Mercurio Anno 778. (notante Angelocratore Chronol[ogia] l[ibro] 6) in Sole viso. En vero Tibi ea quae de illa celebri observatione Anno 807. facta in historia Cometae CLXI. commentatus sum,[1] quaeque[2] operi ad calcem addenda decrevi. Vale et mihi constanter fave.

Tuus
omni studio et officio

S[tanislaus] L[ubieniecius] Rolitsius
manu propria

Post Scriptum. Spero Tibi meo nomine ab Isaaco illo Beuningio 5. Imper[iales] numeratum iri pro 5. exempl[aribus] *Prodromi* posterioris, ex quibus duo tantum vendidi, alia donavi. Quid sit de reliquis statuendum, velim me reddas certiorem. Iterum V[ale].[3]

Hamb[urgi] 26. Julii Greg[oriani] 1667.

1 commentatus sum : dixi et deinceps commentatus sum **TC**.
2 quaeque : ac **TC**.
3 et mihi...Vale *om.* **TC**.

Greetings, most noble, worthy, and renowned Sir!

I have been waiting for the reverend Curtz's[4] answer to what you asked me to take care of; I wanted to add it to this letter of mine. But because there is no answer from him, and I know nothing about his life or studies after the death of the most illustrious Rautenstein[5] (our mutual friend), I did not want to delay my duty any longer. Therefore, I write to you briefly, as someone who is busy with his many works to another busy person. I await your answer, even a very concise one, about cometary parallaxes which I mentioned in my last letter to you, sent on 31 May. Now, I would like you to tell me briefly about Mercury seen in the Sun in the year 778 (as Angelocrator says in the sixth book of his chronology[6]). I am sending you what I wrote about that famous observation of the year 807 for the history of the 161st comet; I have decided to include it all in my book. Farewell, and always be benevolent towards me!

Yours obligingly,

Stanisław Lubieniecki Rolitsius
by my own hand

Postscript. I hope that Isaacus Beuningius will pay you, on my behalf, 5 Reichstalers for the 5 copies of your second *Prodromus*, from which I sold only two and gave the rest away. I would like you to tell me what to do with the remaining copies. Farewell again!

In Hamburg, on 26 July 1667 Gregorian.

66A.

Excerpt of the history of the comet of 807

BO: C1, v. 8, 1217
TC2: 132 (main text) and 447 (additions)

Excerptum ex Historia Cometae CLXI.

Hic etiam manifesto videmus laeta quoque succedere Cometarum apparitionibus, sicut et Eclipsibus et aliis vulgo habitis prodigiis, quae potius naturalibus causis tri-

4 See section 2.4.5 of the Introduction.

5 See note to Letter 5.

6 Daniel Angelocrator (1569–1635), German Calvinist minister and chronologist. Lubieniecki refers to his *Rationis temporum et rerum totius orbis memorabilium ab initio mundi usque ad Rodolphum II Romanorum imperatorem libri novem* (Francofurti: typis Johannis Lancelloti, impensis Jonae Rhodii, 1611), p. 318.

buenda sunt. Nam anno 807. Jovis stella per Lunam ult[imo] Januarii transire, et Sol ac Luna mense Februarii deliquium pati visa est. Mense Martio Mercurius in Sole visus, si credimus Adelmo et Sabellico, [a] imo illo et Adone tum Reginone, Buntingo ipsoque Eckstormio teste, ter illo anno Luna defecit.

Addenda.

[a] Imo illo et Adone [et cetera][7] Enn[eade] 8. l[ibro] 8. Visa eadem stella in solis medio Anno 778 notante[8] Angelocratore.[9]

Ibidem [sectione] ult[ima][10] ter illo anno Luna defecit. Vide de his Anonymum vitae Karoli M[agni] auctorem apud Pithaeum p[agina] 269 et Bunting[um] in *Chronol[ogia]* praesertim p[agina] 385 ubi probat Mercurium tunc in Sole longe distante videri non potuisse, sed fuisse illud novum quoddam phaenomenon. Unde incomparabilis Hevelius cum solertissimo Schickardo affirmat neminem ante acutissimum Gassendum, cui iste Mercurii per Solem transitus primum Anno 1631 visus sit, // horum Planetarum coniunctionem vidisse quam ille secundus Anno 1661 d[ie] 3. Maii observarit, adeoque Venerem in Sole Anno 1639. 24. Nov[embris] S[tylo] V[etere] ab Jeremia Horroxio visam describens, nec tunc Anno 807. nec postea ab Averroe, qui qui 12[mo] seculo a C[hristo] N[ato] vixerit, nec ab Excellentissimo Keplero Anno 1607. Mercurium in Sole visum esse. Ut ita invisi adhuc phaenomeni istius gloria nostro seculo relicta fuerit.

An excerpt of the history of the 161st comet.

Here we can also see clearly that good things may also follow the apparitions of comets, as well as the eclipses and other common omens which should be rather attributed to natural causes. Because in 807, Jupiter was passed over by the Moon on the last day of January, and both the Sun and the Moon in February were seen to suffer an eclipse. In March, Mercury was seen in the Sun, if we believe Adelmus[11] and

7 [et caetera] : et Sabellico **TC2**.

8 notante **TC2** : notate **BO**.

9 Angelocratore : Angelocratore Chronol[ogia] l[ibro] 6 **TC2**.

10 [sectione] ult[ima] : in fine **TC2**.

11 Lubieniecki probably refers here to Adémar de Chabannes (*c.* 989–1034), French monk, composer, and historian. He described the transfer of Mercury in his chronicle, see Adémar de Chabannes, *Chronique*, ed. by Jules Chavanon (Paris: Alphonse Picard et Fils, 1897), p. 96.

Sabellicus,[12] [a] and as the former[13] and Ado,[14] as well as Regino[15], Bünting[16] and even Eckstorm[17] himself testify, there were three lunar eclipses in that year.

Additions.

[a] 'and even the former and Ado' and so on. In *Ennead* 8, Book 8. The same star was seen in the middle of the Sun in 778, as Angelocrator says.[18]

In the same place, in the last paragraph, 'there were three lunar eclipses that year.' See in the work by an anonymous author on the life of Charlemagne in Pithou, on page 269,[19] and especially in Bünting's *Chronologia*, on page 385,[20] where he proves that Mercury could not be seen in the Sun then (because it was too distant), but there was some new phenomenon. Therefore, the unparalleled Hevelius (together with the most skilled Schickard)[21] attests that no one has seen the conjunction of these planets before the most brilliant Gassendi[22] who saw the transit of Mercury in the

12 Marcus Antonius Coccius Sabellicus (1436–1506), Italian historian. Lubieniecki refers to his *Rhapsodiae historiarum enneadum* [...] (in Parrhisiorum Academia: ab Joanne Parvo, 1509), sig. hhh ii r.

13 Adémar de Chabannes, *Chronique*, p. 96.

14 Ado of Vienne (d. 874), archbishop of Vienne in Lotharingia. Lubieniecki refers here to his *Chronicon*, see S. Adonis Viennensis *Chronicon in aetates sex divisum*, in *Patrologiae cursus completus*, series Latina, vol. 123, ed. by J.-P. Migne (Parisiis: apud Garnier fratres, 1879), col. 131.

15 Regino of Prüm (d. 915), Benedictine monk and chronicler. Lubieniecki refers to his *Chronicon*, see Reginonis *Chronicon*, in *Monumenta Germaniae historica. Scriptorum tomus I*, ed. by Georg Heinrich Pertz (Hannoverae: impensis Bibliopolii Aulici Hahniani, 1826), p. 564.

16 See note to Letter 27. Lubieniecki refers to his *Chronologia* (Servestae: impressum typis Bonaventurae Fabri, 1590), sig. 382v–383r, 385r–386v.

17 Heinrich Eckstorm (see note to Letter 27), *Historiae eclipsium, cometarum et pareliorum* (Helmaestadi: typis heredum J. Lucii, 1621), pp. 140–41. Having referred to a description of lunar eclipses in Büntig's *Chronologia*, Eckstorm concludes that there were three eclipses between September 806 and August 807.

18 Daniel Angelocrator (see note to Letter 66), *Rationis temporum et rerum totius orbis memorabilium ab initio mundi usque ad Rodolphum II Romanorum imperatorem libri novem* (Francofurti: typis Johannis Lancelloti, impensis Jonae Rhodii, 1611), p. 318.

19 Pierre Pithou (1539–1596), French lawyer, scholar, and historian. Lubieniecki referred to a wrong page in his edition the works of early French historians. Information about Mercury seen in the Sun in 807 is in *Annalium et historiae Francorum ab anno Christi DCCVIII ad ann. DCCCCXC scriptores coaetanei XII*, vol. 2 (Parisiis: apud Claudium Chappelet, 1588), p. 92.

20 Heinrich Bünting (see note to Letter 27), *Chronologia* (Servestae: impressum typis Bonaventurae Fabri, 1590), sig. 385v.

21 Wilhelm Schickard (1592–1635), German Hebraist and astronomer, constructor of one of the first mechanical calculators. He was the addressee of Gassendi's letters about the transit of Mercury (published in Pierre Gassendi, *Mercurius in Sole visus et Venus invisa* (Parisiis: sumptibus Sebastiani Cramoisy, 1632)) and answered them in *Pars responsi ad epistolas P. Gassendi insignis philosophi Gallici de Mercurio in Sole viso, et aliis novitatibus uranicis* (Tubingae: typis Theodorici Werlini, 1632). Hevelius quoted a passage from the preface to this book (pp. 4–5) in his *Mercurius in Sole visus* (Gedani: autoris typis et sumptibus, 1662), p. 52.

22 See note to Letter 4A.

Sun for the first time in 1631;[23] and Hevelius was the second to see it on 3 May 1661[24] and, when he describes Venus in the Sun, seen on 24 November 1639[25] old style by Jeremiah Horrocks,[26] he says that Mercury was not seen in the Sun either in 807, or later, by Averroes (who lived in the twelfth century after the birth of Christ), or by the most excellent Kepler in 1607.[27] And so the glory of that phenomenon, unseen before, was left for our century.

23 Hevelius, *Mercurius in Sole visus*, pp. 51–53. See also Alexandre Guy Pingré, *Annales célestes du dix-septième siècle* (Paris: Gauthier-Villars, 1901), pp. 84–86, CJH3 p. 319, Letter 94, des Noyers to Hevelius, 23 January 1661, and p. 329, Letter 97, des Noyers to Hevelius, 14 March 1661.

24 Pingré, *Annales célestes*, pp. 248–49.

25 Pingré, *Annales célestes*, pp. 133–36.

26 Jeremiah Horrocks (1618–41), English astronomer. In 1639, he conducted the first observation of the transit of Venus. His observations, titled *Venus in Sole visa*, were published and annotated by Hevelius in his *Mercurius in Sole visus*.

27 The opinion cited here does not come from Hevelius's annotations to *Venus in Sole visa*, but from Horrocks's text, see Jeremiah Horrocks, *Venus in Sole visa*, in Hevelius, *Mercurius in Sole visus*, p. 119.

67.

2 August 1667, Lubieniecki to Hevelius

BO: C1, v. 8, 1218
TC: 954–55

Salutem Plurimam Nobilissime, Amplissime, Clarissime Vir.

Vide quam conspiremus animis et benevolentia simul etiam certemus! Octavus ab hinc dies fuit,[1] quum uterque litteras nostras daremus. En Tibi meum responsum ad Tuas, quas hodie accepi. Si et Tu ad meas nunc respondeas, erit quo nomine uterque nobis gratulamur. Ego praesertim, qui magno cum desiderio responsum ad binas meas d[ie] 31. Maii et 26. Julii exaratas abs Te exspecto, dum operi meo colophonem impono. Miror quidem nullam Te prioris illius epistolae facere mentionem. Inde suspicor eas[2] vel non fuisse recte Tibi redditas,[3] vel inter tot gravissimos labores oblivioni traditas.[4] Praecipuum, de quo ibi Tecum egi, erat de parallaxibus Cometatum an et quomodo hae Lunari minores ex Tychone contra Aristotelem, et Claremontium aliosque demonstrari possint. Ubi dixi tantae controversiae decisionem abs Te nobis exspectandam esse. In nuperis ad Te datis litteris de Mercurio in Sole Anno 778. viso Te consului. Ad utrumque tribus verbis quamprimum responde, et descriptionem Cometae Anno 1580. visi, brevem licet, adiunge ex Tychone, de cuius observationibus a Doctissimo Curtio editis et ad Te missis, non tantum Tibi, sed et Reip[ublicae] litterariae pluri-//mum gratulor. Quas et quales nobis munificentia Augustissimi Regis Danorum Maximi Musarum Euergetae datura sit, nondum scio. Perfectissimas tamen fore, nullus dubito. Vellam hodie de hac re aurem Cl[arissimo] et Humanissimo Erasmio Bartholino, eique Tuarum litterarum apographum mittam. Salutem pl[urimam] dic a me Cl[arissimo] Büthnero, cui me brevi, si Deus volet, responsurum spero. Salve et Tu et Vale. Dab[am] Hamburgi postr[idie] Kal[endas] Aug[ustas] Greg[orianas] Anno 1667.

Tuus
omni studio et officio

S[tanislaus] L[ubieniecius] Rolitsius
manu propria[5]

1 fuit : agitur **TC**.
2 eas : eam **TC**.
3 redditas : redditam **TC**.
4 traditas : traditam **TC**.
5 Dab[am]...propria *om*. **TC**.

Post Scriptum. Gratulare et Tu mihi de reddita tandem mihi a biennio amplius exspectata *Astronomia Philolaica* Cl[arissimi] Bullialdi, ab eo mihi donata, quae prius Gedani quam Hamburgi fuit.[6] Iterum Vale.

Greetings, most noble, worthy, and most Sir!

See how we agree with each other and how we compete together in benevolence! It was eight days ago when both of us sent letters. Now, I am sending my answer to your letter which I received today. If you are answering my letter now, there would be a reason for both of us to congratulate one another. As I am waiting eagerly for your reply to my two letters, written on 31 May and 26 July, I am about to finish my book. I wonder why you do not mention that earlier letter. I suspect it was not delivered to you at all, or it was given to you while you were preoccupied with your many tedious works. The most important thing about which I wrote there were the parallaxes of comets: if and how they can be proven to be smaller than the lunar ones according to Tycho,[7] and contrary to Aristotle, Claremontius,[8] and others. I said there that we should wait for your decision about such a controversy. In the last letter I sent to you, I asked you about Mercury seen in the Sun in the year 778. Answer briefly to both of these as soon as you can, and enclose a description of the comet of 1580,[9] even a brief one, which was made by Tycho;[10] I congratulate not only you, but the whole Republic of Letters on his observations edited by the most learned Curtz[11] and sent to you. I do not know yet which observations (and of what kind) the generosity of the most august King of Denmark,[12] the greatest benefactor of the Muses, will give us. But I do not doubt they will be the most exquisite ones. I will ask the most renowned and kindest Erasmus Bartholin[13] about it, and I will send him copies of your letters. Greet the most renowned Büthner[14] from me; I will reply to him soon, God willing.

6 a biennio...fuit : *Astronomia Philolaica*, quam ante biennium Cl[arissimus] et Humanissimus Bullialdus munus ad me miserat, et quae prius Gedanum quam Hamburgum perlata erat **TC**.

7 Relying on the observations of the parallax of the comet of 1577, Brahe proved that the comets cannot be situated below the Moon and, therefore, they are astronomical objects. See Tycho Brahe, *De mundi aetherei recentioribus phaenomenis*, in Tychonis Brahe *Opera omnia*, vol. 4, ed. by John Louis Emil Dreyer (Hauniae: in Libraria Gyldendaliana, 1922), pp. 82–134.

8 Probably Scipione Chiaramonti, see note to Letter 63A.

9 C/1580 T1, see Alexandre Guy Pingré, *Cométographie ou Traité historique et théorique des comètes*, vol. 1 (Paris: Imprimerie royale, 1783), pp. 521–40; Gary W. Kronk, *Cometography. A Catalogue of Comets. Volume 1: Ancient – 1799* (Cambridge: Cambridge University Press, 1999), pp. 321–22.

10 Tycho Brahe did not publish his observations of the comet of 1580. For their modern edition see Tychonis Brahe *Opera omnia*, vol. 13, ed. by John Louis Emil Dreyer (Hauniae: in Libraria Gyldendaliana, 1926), pp. 305–33. Hevelius could have learned about his observations from Erasmus Bartholin who worked on their publication, see note to Letter 65.

11 See section 2.4.5 of the Introduction.

12 See note to Letter 38.

13 See section 2.4.1 of the Introduction.

14 See section 2.4.3 of the Introduction.

You too stay in good health, and farewell! Given at Hamburg, on the day after the Kalends of August 1667, Gregorian style.

Yours obligingly,

Stanisław Lubieniecki Rolitsius
by my own hand

Postscript. Congratulate me on the *Astronomia Philolaica* by the most renowned Boulliau[15] which was finally delivered to me; I have been waiting for it for two years; it was given to me by the author himself, and it was earlier in Danzig than in Hamburg. Farewell again!

15 Ismaël Boulliau (see section 2.4.2 of the Introduction), *Astronomia Philolaica* (Parisiis: sumptibus Simonis Piget, 1645).

68.

undated,[1] Lubieniecki to Hevelius

BO: C1, v. 9, 1276
BnF: Lat 10348, v. 9, 41–42

Enclosure:
68A. Letter from Lubieniecki to Lange, 24 July / [3 August] 1667

Salutem Plurimam Amplissime et Clarissime Vir.

Non dubito te toto pectore et mentis corporibusque viribus ad perficiendum praeclarum opus incumbentem, jugiter occupatum esse: ideoque Amicis vacare non posse. Utinam quamprimum prodeat tantopere desideratum opus! Meum quando sit proditurum, ipse nescio, utpote absens, et Hollandis pro arbitrio cuncta agentibus. Paulo ante accepi a Cl[arissimo] Langio, Hafniae nunc agente, literas de argumento Cometico ad me scriptas: ad quas, ut vides, pauca respondi. Cum autem Tua plurimum intersit talia legere et mentibus virorum praeclarorum oborta dubia tollere, en Tibi haec cuncta. Poteris merito harum literarum mentionem facere in Tuo opere, et Veritatem, quam de hoc argumento profers, asserere. Id enim a Te pars orbis litterati melior meritissimo expectat. Vale et fave constanter

Tuo

Lub[ieniecio]
manu propria

Post Scriptum. Saluta pl[urime] Amicos. Manus Langii, ut et Rudbecii ad me tempore commodo redeat. Iterum Vale.

Greetings, most worthy and renowned Sir!

I do not doubt that you work (with all your heart and prowess of body and mind) on finishing your famous book and that you are always busy; hence, you cannot have any time for friends. May such an anticipated book be published soon! I do not know myself when my book will be released since I am not there and the Dutch do whatever

1 The earliest possible *terminus post quem* is 3 August 1667 — the date of the enclosed letter. I presume this letter was not sent much later, as there would be no reason for a delay.

they like.[2] I have recently received a letter about the cometary discussion; it was written to me by the most renowned Lange[3] who lives in Copenhagen now; I have answered it briefly, as you can see. I am sending you all of it because you care so much for reading such things and for freeing the minds of renowned men from the doubts that arose. You can mention this letter in your book and confirm the truth which you present in the discussion. Because this is what is expected from you by a better part of the learned world. Farewell, and always be benevolent towards

Yours,

Lubieniecki
by my own hand

Postscript. Greet all our friends. I would like you to send back to me, when you can, what Lange and Rudbeck[4] wrote. Farewell again!

68A.

Letter from Lubieniecki to Lange, 24 July / [3 August] 1667

BO: C1, v. 9, 1276
TC: p. 945
BnF: Lat 10348, v. 9, 41

Hamburgo Hafniam d[ie] XXIV. Sept[embris] Jul[iani] MDCLXVII.

Multum omnino debeo Tuae humanitati, quod inter tot graves occupationes, quae Te ex Musaeo, imo Athenaeo celeberrimo, in forum translatum, ac amplissimae provinciae, et sic necessitudinibus publicis impositum jugiter tenent, me tam docto et amabili responso bearis. Ad quod tria verba summum manum operi vasto imponens, reponere Tibi necesse habeo. Equidem speraveram Te nobis decisionem illius de nuperis Cometis difficilis quaestionis daturum fuisse. Sed Tibi alibi occupato et necessariis ad tantam rem excutiendis instrumentis et adiumentis defecto Academicorum vestigia legere, vel certe nobis monstrare non[5] placuit. Dixi et ego alicubi, magnis quoque Viris nil certi in Rep[ublica] ut sic dicam, Cometarum videri. Tibi prae caeteris Democritus arrisit. Non pauci sunt (quos inter et ego nomen meum profiteor) // qui sperant

2 Lubieniecki refers here to the problems he had with Francis Kuyper, his publisher, see section 1.3.3 of the Introduction.

3 See section 2.4.10 of the Introduction.

4 See section 2.4.16 of the Introduction.

5 non **TC** *om.* **BO BnF**.

Ingeniosissimi et tam omnibus necessariis scientiae ac instrumentorum adminiculis Instructissimi, quam Solertissimi Hevelii maximis et felicissimis laboribus veritatem, qua Cometarum ortum et motum respicit, ex puteo Democriti extractum iri. Mea quoque vel potius illa, quae mihi multi Viri in Mathematicis disciplinis exercitatissimi, suppeditarunt a me *Theatro Cometico* illata, non parum, ut spero, ad eandem veritatem eruendam facient. Sed nobis orbis litterati judicium de non exiguis profecto conatibus exspectandum est. Atque hoc mihi favens promitto in illo arduo de Cometarum significatione et astrorum operatione argumento. Attamen cuivis integram libertatem relinquo. Ideo et Tibi: cuius ego judicium et eruditionem merito magni facio. Vale et fave.

From Hamburg to Copenhagen, on 24 September 1667 Julian.

I owe much to your kindness because you favoured me with a very learned and pleasing answer among your many serious duties; they have taken you away from your Museum (or even the famous Atheneum) to a public square and keep you constantly overburdened with high offices and public duties. I find it justified to answer you briefly while I am adding a final touch to the great work. Indeed, I hoped that you would give us the solution to that difficult question about the late comets. But you — however otherwise busy and lacking necessary devices and help in conducting such a thing — decided not[6] to follow in the footsteps of the Academicians,[7] or at least to make a demonstration for us. I said somewhere, too,[8] that even great men have no specific opinion about the republic of comets, if I may say so. Democritus smiled[9] at you more than at others. There are many (I confess there is my name among them) who hope that the truth about the origin and the movement of comets will be drawn from Democritus's well[10] by the greatest and most auspicious works of the most ingenious and proficient Hevelius who is both equipped with all things necessary for science and with helpful instruments. My work, too (or rather all that was given to me by many men experienced in mathematics and what I put in the *Theatrum cometicum*) will help to bring that truth to light, I hope. But we should wait for the opinion of the learned world about not so petty efforts. And I have promised myself that it will be favourable in that problematic discussion about the meaning of comets and the influences of stars. But I leave free judgement to everyone — also to you, whose opinion I hold in high esteem. Farewell, and stay benevolent!

6 Negative particle 'non' is present in the printed text in the TC, but not in the manuscripts. The context suggests that it should not be omitted. Lange, however, sent a long letter in response to Lubieniecki's inquiries, TC pp. 942–45.

7 'Following the footsteps of Academicians' is probably a synonym of mathematical and astronomical activities, see note to Letter 2B.

8 Lubieniecki probably refers to his introduction to TC2, pp. 1–3, where he briefly discussed problems related to cometary studies.

9 Perhaps an allusion to a proverbial 'Democritus's laughter', see Seneca, *De ira*, 2, 10.

10 See note to Letter 2B.

69.

10 August 1667, Hevelius to Lubieniecki

BO: C1, v. 8, 1219
TC: 955

Domino
Rolisio
Hamburgi

Illustrissime Vir.

Interrogata Tua adeo brevibus expedire nequeo. Materia namque illa de Parallaxibus adeo ampla ac ardua est, ut coactus fuerim vastissimum volumen ea de re componere. Si tribus verbis tantummodo significandum sit quid sentiam, scias me omnes Cometas aethereos statuere, et nullum unquam Cometam ortum occasumque suum in aere habuisse: at quomodo hoc ipsum indubitate ex parallaxibus demonstretur, fuse ex opere nostro mox mox percipies; brevi enim Divino annuente Numine Ei finem imponam. De Mercurio nihil amplius dicendum habeo, quam quod firmiter existimem, neminem ante Gassendum Mercurium observasse in Sole; ut ut Historicorum nonnemo aliter sentiat. Observationes Tychonicas Cometae 1580, libenter Tibi transmitterem, sed eas nusquam in *Historia Caelesti*, observationes nempe Tychonis, offendi: cum ab anno 1582 primum suam ducant originem. Atvero Cometae 1585 observatio invenitur in Epistola Tych[onis]. Vale et ignosce[1] quod sim in respondendo brevior, quam merito debebam. Non solum enim sum occupatissimus, sed fere negotiis totus obrutus. Dabam Gedani, Anno 1667, die 10 August[i] st[ylo] n[ovo].

Tuae Illustrissimae Dominationis
optime cupiens

J[ohannes] Hevelius

1 brevitati scriptionis *del. post* ignosce **BO**.

To Master
Rolitsius
in Hamburg

Most illustrious Sir!

I cannot answer your questions briefly. Because the knowledge of parallaxes[2] is so vast and complex that I would have to write a large volume about it. If I were to tell you briefly what I think, you should know that I claim that all comets are in Aether and none of them has ever had its birth and its end in the air;[3] and you will very soon learn in detail from my book how it is undoubtedly deduced from parallaxes; because my book will soon be finished, with the help of God's might. I have nothing more to say about Mercury except that I am convinced that no one has seen Mercury in the Sun before Gassendi,[4] however some historians hold a different opinion. I would willingly send you Tycho's observations of the comet of 1580,[5] but I did not find them anywhere in the *Historia coelestis*[6] (that is in Tycho's observations) because it starts in 1582. Nonetheless, the observation of the comet of 1585[7] can be found in Tycho's letter.[8] Farewell, and forgive that I am more concise in writing than I should be. Not only am I very busy but almost entirely buried under my duties. Given at Danzig, on 10 August 1667, new style.

Illustrious Sir,

Your affectionate friend,

Johannes Hevelius

2 See note to Letter 67.

3 See section 2.2.1 of the Introduction.

4 See note to Letter 4A.

5 See note to Letter 67.

6 Albert Curtz (see section 2.4.5 of the Introduction), *Historia coelestis* (Augustae Vindelicorum: apud Simonem Utzschneiderum 1666).

7 C/1585 T1, see Alexandre Guy Pingré, *Cométographie ou Traité historique et théorique des comètes*, vol. 1 (Paris: Imprimerie royale, 1783), pp. 550–54; Gary W. Kronk, *Cometography. A Catalogue of Comets. Volume 1: Ancient – 1799* (Cambridge: Cambridge University Press, 1999), pp. 324–25.

8 Hevelius is probably referring to Brahe's letter to Wilhelm, the Landgrave of Hessen, published in Brahe's *Epistolarum astronomicarum liber primus* (Uraniburgi: ex officina typographica authoris, 1596), pp. 13–20.

70.

16 August 1667, Lubieniecki to Hevelius

BO: C1, v. 9, 1275
TC: 955
BnF: Lat 10348, v. 9, 39–40

Enclosure:
70A. Letter from Bartholin to Lubieniecki, 29 July 1667

Salutem Plurimam Vir Nobilissime Clarissime.

Eadem plane ratio meorum laborum,[1] quae tuorum est. Nam in relegendo et sic perficiendo opere versor: cui ultima manus imponenda magnam requireret curam, si per tot alia negotia necessario ad vitam tuendam expedienda et inevitabilia impedimenta liceret. Satisfactum mihi est Tuo, brevi licet, responso. In novas enim easque solidas spes nos erexisti, dum promittis certa fide, brevi nos praeclarissimum Tuum opus inque eo explicatam de parallaxibus et sede Cometarum doctrinam visuros. En quae nuper accepi a Cl[arissimo] Bartholino. Cui exemplum Tuae epistolae hodie mihi redditae jam jam misi. Brevitati et Tuae do, et meae postulo veniam aequam.[2] Vale et mihi constanter fave.[3]

Tuus
omni cultu, studio et officio

S[tanislaus] L[ubieniecius] Rolitsius
manu propria

Hamb[urgi] 16. Aug[usti] 1667.

Greetings, most noble and renowned Sir!

The situation with my works is the same as yours because I am busy with revising and completing my book; the final stroke has to be very careful if it is at all possible among so many other duties (which have to be done for a living) and inevitable diffi-

1 meorum laborum **BO TC** : laborum meorum **BnF**.

2 aequam **BO BnF** : aequam. Moneo, nos aliquot ab hinc diebus singulis vesperis Andromedaeam illam nebulosam videre. Cuius observationes accuratissimas Tu nobis dubio procul dabis. **TC**.

3 et...fave *om.* **TC**.

culties. Your answer, however brief, satisfied me. You had aroused new and sure hopes in us when you made a particular promise that we would soon have your famous work and the knowledge about parallaxes and places of comets which you discuss in it. I enclose what I have recently received from the most renowned Bartholin.[4] I have sent him a copy of your letter which was delivered to me today. I forgive your brevity and ask you to forgive mine. Farewell, and always be benevolent towards me.

Your obliged admirer,

Stanisław Lubieniecki Rolitsius
by my own hand

In Hamburg, on 16 August 1667.

70A.

Letter from Bartholin to Lubieniecki, 29 July 1667

BO: C1, v. 9, 1276
TC: 960
BnF: Lat 10348, v. 9, 40

Haunia Hamburgum d[ie] 29. Julii Anno 1667.

Ne existimes in animis Nostris exolevisse amicitiam, cogor publicas curas tantisper abrumpere, ut tuis tandem satisfiat precibus. Exspectatae diu Meieri observationes adhuc suspensos nos relinquunt, et forsitan non desinent in posterum relinquere. Demonstrationes Prostaphaereticorum[5] compendiorum, calculi eventus, ut puto satis comprobant; nam reliqua persequi constitui peculiari opere, quod synopsin *Astronomiae Danicae* suggeret. Clariss[imi] Hevelii quoque votis respondere conabor prima occasione: gaudeo enim in publicam jam lucem prodiisse observationes Brahaei, sed doleo vicem Astronomorum, quod taedium tot erronearum observationum ex editionis incuria multum fatigabit, nisi mature cognoverint infinitorum mendorum copiam inutiles publicis usibus reddidisse // desideratos *Historiae caelestis* conatus; ut nihil jam dicam de observationibus annorum deficientibus. Historiam Cometae Ann[i] 1580, quam ex M[anu]s[cript]o Tychonico petis, plenius cognosces, ubi ipsae observationes copiosius et correctius una cum reliquis Brahaei scriptis prodierint,

4 See section 2.4.1 of the Introduction.
5 Prostaphaereticorum **TC** : Protaphaereticorum **BO BnF**.

quippe quae prolixior, quam, ut hae pagellae eam complectantur. Hisce vale et me amare perge. Festinanti calamo scripsit.

Tuus

Erasmius Bartholinus

From Copenhagen to Hamburg, on 29 July 1667.

I am forced to cease my public duties for a while in order to fulfil your wishes, so that you would not think that our friendship has disappeared from my heart. We are still waiting for the observations from Mejer,[6] and probably we will have to wait some more. Tables of prostaphaeretic summaries[7] — the results of calculations — are sufficient proof, in my opinion; I have decided to discuss the rest in a separate book which will present a synopsis of the *Astronomia Danica*.[8] I will try to address the most renowned Hevelius's request at the first opportunity. I am glad that Brahe's observations are already published, but I grieve at the fate of the astronomers who will struggle with many troubles resulting from numerous mistakes in these observations, caused by the negligence in this edition, unless they realize soon that the infinite number of errors made the expected efforts of the *Historia coelestis*[9] useless for the public use; I will say nothing about the lack of twenty years of observations. You will see the description of the comet of 1580[10] (for which you asked), derived from Tycho's manuscript, in detail when the observations are published more voluminously and correctly together with the rest of Brahe's writings; it is longer than these pages can contain.[11] Farewell, and always stay benevolent. Written in haste by

Yours,

Erasmus Bartholin

6 Perhaps Johannes Mejer (1606–74), Danish cartographer and mathematician, since 1647 the Royal Mathematician of Denmark.

7 Prostaphaeresis is a method of simplifying calculations by replacing products and quotiens of trigonometric functions with their sums and differences.

8 It seems that Bartholin did not publish a separate volume devoted to Christian Longomontanus's *Astronomia Danica* (Amsterodami: ex officina typographica Guilielmi I. Caesii, 1622).

9 Albert Curtz (see section 2.4.5 of the Introduction), *Historia coelestis* (Augustae Vindelicorum: apud Simonem Utzschneiderum 1666).

10 See note to Letter 67.

11 See note to Letter 65.

71.

19 August 1667, Lubieniecki to Hevelius

BO: C1, v. 9, 1277
BnF: Lat 10348, v. 9, 42

Salutem Plurimam Vir Nobilissime Humanissime.

Quod proxime inter tot negotiorum moles memoria excidit, hoc nunc addo, Tibi licet, ut credere fas est, notum, ut tamen ingratum futurum, utpote candoris et affectus documentum, me scil[icet] cum M[agistro] Ysenio aliquot ab hinc diebus nebulosam illam in Andromeda, sed nudis oculis vidisse. Cuius observationes accuratissimas abs Te merito exspectamus.

Vale et mihi fave constanter.

Tuus
omni cultu, studio et officio

S[tanislaus] L[ubieniecius] Rolitsius
manu propria

Hamburgi 19. Aug[usti] s[tylo] n[ovo] 1667.

Greetings, most noble and kindest Sir!

Because the last time, among so many duties, I forgot to do it, now I am adding something which you know (as one can presume) but which will not be unpleasant for you, being a sign of my sincerity and goodwill. I mean that Master Ysenius[1] and I, we saw that nebulous star in Andromeda a few days ago, but only with the naked eye. We await your most accurate observations of it.

Farewell, and always be benevolent towards me.

Your obliged admirer,

Stanisław Lubieniecki Rolitsius
by my own hand

In Hamburg, on 19 August 1667, new style.

1 See note to Letter 55.

72.

6 June 1668, Hevelius to Lubieniecki

BO: C1, v. 8, 1233
BO: C1, v. 9, 1278 (BO 2)
BnF: Lat 10348, v. 9, 42–43

Illustr[issimo] Viro
Domino Rolisio
Hamburgi

Illustrissime Vir.[1]

Nec scribendi aliquod argumentum hactenus habui, nec profecto citius ad Te literas exarare volui, quam cum simul Tibi significare possem, me *Cometographiae* meae feliciter colophonem imposuisse. Quod cum nunc tandem Divina annuente gratia feliciter mihi obtigerit nolui diutius officio meo deesse; sed quantocyus Tibi exemplar aliquod transmittere, et quidem per Lubecam, quod ut suo tempore benevole accipias, auctoremque porro commendatum habeas rogo. Adiunxi aliud insuper exemplar, quod Admodum Reverendo P[atri] Alberto Curtio haud gravatim meo nomine perferri curabis. Interea Tuum desideratissimum opus exspecto; non dubito quin istud pariter ad umbilicum jam perduxeris. Salve et vale. Dabam Dantisci Anno 1668 die 6 Junii.

Tuus
omni studio et officio

J[ohannes] Hevelius
manu propria

To the most illustrious Sir,
Master Rolitsius
in Hamburg

Most illustrious Sir!

So far, I have neither had anything to write to you about nor wanted to send you a letter before I could tell you that I have successfully finished my *Cometographia*.

1 Vir Illustrissime Vir (*primum* Vir *del.*) **BO** : Hamburgi...vir *om.* **BO 2**.

With God's help, I finally succeeded. I did not want to neglect my duty any longer but rather to send you a copy as soon as possible; I am sending it through Lübeck, so that you can receive it favourably in due time, and I ask you to keep good opinion about the author. I have enclosed another copy which I would like you to deliver to the most Reverend Father Albert Curtz[2] on my behalf. Meanwhile, I await the very expected book of yours; I do not doubt you have already finished your work, too. Be in good health and farewell! Given at Danzig, on 6 June 1668.

Yours obligingly,

Johannes Hevelius
by my own hand

2 See section 2.4.5 of the Introduction.

73.

19 June 1668, Lubieniecki to Hevelius

BnF: NAL 1641, 78r
BnF: Lat 10348, v. 9, 170 (BnF 2)

Salutem Plurimam Nobilissime, Amplissime, Humanissime Vir.

Ante triduum accepi recte litteras Tuas, quae me mirum in modum exhilararunt. Nam non tantum maxime sinistrum rumorem (quem utinam auctor vitae tibi largiter bona valetudine et vita longaeva bono publico penset!) de Te sparsum refutarunt, sed et nuncium laetum de *Cometographia* Tua perfecta attulerunt. Gestio quamprimum videre hoc eximium opus Tuum. Cuius exemplum Curtio destinatum, mittam diligenter ubi domum Deo dante reversus fuero. Hic enim a trimestri ago in Aula expediens tarde mea negotia, cum ob religionem Urbe Hamburgo excedere jussus sim. Spero tamen me intra paucos dies Deo volente hinc abiturum. Prima operis mei exemplaria paulo ante hic accepi et S[acrae] R[egiae] M[aiesta]ti, tum Seren[issimis] Principibus Regiis et Primoribus nonnullis obtuli. Nec amicos praeterii. Cl[arissimus] Bartholinus edidit nuper *Specimen recognitionis nuper editarum observationum Tychonicarum*, quem librum munus ad Te misit. Ingratum hoc opusculum erit Curtio. Vidi ego autographa Tychonis et Longomontani, quorum impressionem Elzevirio proposui, et porro editionem tam eximii operis pro virili juvabo. Meum opus tandem absolutum. Mittam Tibi gratus exemplar, ut et Cl[arissimo] Büthnero, quem pl[urima] salute[1] impertio, quamprimum domum rediero et exemplar aliquod ex Hollandia accepero, Deo adiuvante. Cuius protectioni Te Tuosque Tuaque Vir Eximie, commendo. Dab[am] Hafniae d[ie] 9.19. Junii 1668

Tuus
omni studio et officio

S[tanislaus] L[ubieniecius] Rolitsius

Greetings, most noble, worthy, and kindest Sir!

Three days ago, I received your letter, which made me very happy. Because not only did it refute a very sinister rumour[2] about you which is circulating (may the Creator of life richly repay you for it with good health and long life for the public good!), but

1 pl[urima] salute : pl[urimam] salutem **BnF 2**.
2 Probably a rumour about Hevelius's sickness, or even death, unattested elsewhere.

it also brought joyful news about your completed *Cometographia*. I cannot wait to see this exceptional work of yours. I will diligently send a copy to Curtz[3] when I am back home, if God allows it. I have been here at the court for three months, doing my business, because I was ordered to leave the city of Hamburg due to my religion.[4] But I hope that in the next few days, God willing, I will leave this place. I have recently received the first copies of my book, and I have sent them as gifts to His Sacred Royal Majesty,[5] to the princes, and to the most distinguished men in the kingdom. I have not ignored my friends. The most renowned Bartholin recently published the *Specimen recognitionis nuper editarum observationum* by Tycho[6] which he sent to you as a gift. This book will be unpleasant for Curtz. I have also seen Tycho's and Longomontanus's manuscripts; I suggested their publication to Elsevier,[7] and I will support the edition of such an exceptional book in everything I can. My book is finally finished. I will willingly send a copy for you and for the most renowned Büthner[8] (for whom I pass many greetings) as soon as I get back home and receive a copy from the Netherlands, with God's help. I entrust you and your family, exceptional Sir, to His protection. Given at Copenhagen, on 9/19 June 1668.

Yours obligingly,

Stanisław Lubieniecki Rolitsius

3 See section 2.4.5 of the Introduction.

4 See section 1.1.3 of the Introduction.

5 Most probably Frederick III, see note to Letter 38.

6 Erasmus Bartholin (see section 2.4.1 of the Introduction), *Specimen recognitionis nuper editarum observationum astronomicarum nobilissimi viri Tychonis Brahe* (Hafniae: prostat apud Danielem Pauli, literis Henrici Gödiani, 1668).

7 Since 1654, Elsevier publishing house in Amsterdam was run by Lodewijk Elsevier the younger (1604–70) and Daniel Elsevier (1628–80).

8 See section 2.4.3 of the Introduction.

74.

12 March 1669, Lubieniecki to Hevelius

BO: C1, v. 9, 1360
BnF: Lat 10348, v. 9, 168–69

Seal with Lubieniecki's coat of arms (Rola coat of arms), address: 'A Monsieur / Monsieur Jean Hevelius / Conseiller de la ville / à Dantzic' ('To *Monsieur* Johannes Hevelius, city Councillor, in Danzig')

Nobilissime Amplissime Clarissime Vir.

Non puto me in aere Tuo esse, quod ad litterarum commercium attinet. Nam quas ad me aestate praeterita mittendo mihi exemplar Tuae *Cometographiae* dederas, ad eas respondi, et de munere tam eximio et praestanti gratias Tibi egi, prout etiamnum ago. Alterum exemplar Alb[erto] Curtio destinatum penes me adhuc haeret. Nulla enim occasio se offert illud Neoburgum promovendi. Adhaec non ita pridem domum redii ex Hollandia, quo ex Dania eram profectus, atque ita per integrum fere annum a domo abfui. Pecuniae nonnihil pro exemplaribus aliquot *Prodromi* Tui posterioris a me partim donatis, partim venditis numerari Tibi tunc feceram. Tuum erit de reliquis statuere quid de illis fieri velis. Doleo utriusque nostrum vicem, nos incidisse in seculum a studiis, praesertim Mathematicis, aversum. Nam me praesente Amsterdami nec Tuum nec meum opus Cometicum distrahi feliciter potuisse, ex Bibliopolis cognoveram. Spero tamen me Tibi hic aliquos emtores conciliaturum. Unum jam habeo. Velim ut et Tu eandem officii vicem mihi in urbe Vestra praestes. Exemplaria quo // felicius distrahantur, minui pretium eorum et reduxi illud ad 6. thal[eros] Imper[iales]. Hoc pretio liber meus Amsterodami comparari poterit apud meos Amicos, si forte Bibliopolae augeant nimis pretium. Elzevirius puto vix aliquid amplius exiget. Accepisse Te et Cl[arissimum] Büthnerum exemplaria Vobis a me destinata, Elzevirio commendata, ex illo ipso libens intellexi. Boni et aequi consulite munus amicum, et veniam date aequi et faciles tot erroribus. Assiduae meae et graves occupationes, absentia a loco ubi opus imprimebatur, aliaque impedimenta apud aequos lectores me excusabunt. Precor ut hoc libro incolumes quam diutissime utamini. Iterum fama novum Cometem in Aquitania prope Burdegalam apparuisse vulgavit. Sed a nullo Mathematico vel Viro docto de eo quidquam habeo. Cl[arissimo] Buthnero nisi hodie respondere possim, respondebo alias Deo volente. Salutem ei pl[urimam] interim adscribo. Caeterum velim juves Tuo studio et auctoritate, qua in urbe patria polles, Jacobum Grotium Guil[ielmi] filium, Hugonis fratris filium, Virum egregie doctum, in linguis variis, etiam Germanica Vestra, exercitatum, insigni rerum // civilium cognitione instructum, magnorum Virorum et Legatorum Hagae Comitis (ubi et

ille, et frater eius Joannes, insignis Jurisconsultus, degit) amicitiis et familiaritate pollentem, ut nuper defuncto D'Aitzema[1] in officio, quod in Aula Hagana nomine urbium Hanseaticarum obtinuit, subrogetur. Facietis rem pergratam Batavis, qui ei bene cupiunt, praecipue vero Consobrino eius, qui jam Legati munere in Svecica Aula fungitur. Cl[arissimus] Kirstenius Te pl[urime] salutat. Vale, Vir Excellentissime, et mihi constanter fave qui sum

Clarissimi Nominis Tui
studiosissimus
observantissimus

S[tanislaus] L[ubieniecius] Rolitsius
manu propria

Altenaviae ad Hamburgum
12. Martii s[tylo] n[ovo] 1669.

Post Scriptum. Litteras Tuas, ubi ad meas respondere voles (quod ut quamprimum facias, rogo et obsecro) dirige Hamburgum, ut ante.

Grotius iste jam munere Consiliarii et Residentis Elect[oris] Palatini fungitur.

Most noble, worthy, and renowned Sir!

I do not think that I am indebted to you as far as epistolary communication is concerned. Because I did answer your letter of last summer, written when you sent me a copy of your *Cometographia*; I thanked you then for such an exceptional and great gift as I thank you now. Another copy, meant for Albert Curtz,[2] is still stuck with me, as there was no opportunity to send it to Neuburg. Moreover, I have recently returned home from the Netherlands, where I had travelled from Denmark; thus, I was not at home for almost entire year. Then I sent you some money for a few copies of your second *Prodromus*; some of them were given away by me, some sold. You will have to decide about the rest, that is what you want me to do with them. I grieve at our fate that we happen to live in the century which turned its back on the sciences, especially those mathematical ones. Because when I was in Amsterdam, I learned from the booksellers that neither your cometary book nor mine could be successfully distributed. I hope, however, that I will find here a few buyers for you. I have already got one. I would like you to return the favour for me in your city. In order to distribute the copies more successfully, I lowered their price and reduced it

1 D'Aitzema *corr. ed.* : D. Aitzema **BO BnF**.
2 See section 2.4.5 of the Introduction.

to 6 Reichstalers. In Amsterdam, if it happens that the booksellers raise the price too much, my book can be bought for this price from my friends. Elsevier, I think, does not demand more. I learned from him that you and the most renowned Büthner[3] had received the books I meant for you and entrusted to him.[4] Use that friendly gift honestly and sufficiently, and forgive sufficiently and calmly so many mistakes. My constant and serious duties, absence from the printing place of the book, and other obstacles will excuse me in the eyes of just readers. I pray that you will use this book for a very long time, staying in good health. A rumour has it that a new comet was seen in Aquitania, near Bordeaux.[5] But I have nothing about it from any mathematician or learned man. If I cannot answer the most renowned Büthner today, I will answer him another day, God willing. Meanwhile, I am sending him many greetings. Moreover, I would like you to help (with your efforts and respect which you hold in your native city) Jacob de Groot[6] (who is the son of Willem,[7] who was Hugo's[8] brother, and who is a man of great learning, experienced in various languages — including your German — educated in public affairs, enjoying the friendship and confidence of great men and delegates in The Hague, where both he and his brother Jan,[9] an excellent lawyer, live) to be elected in the place of recently deceased van Aitzema[10] for the office he occupied at the court in The Hague on behalf of Hanseatic cities. You will do a great favour for the Dutch who wish him well, especially for his cousin[11] who is a delegate at the Swedish court. The most renowned Kirsten[12] sends you many greetings. Farewell, most excellent Sir, and always be benevolent towards me, who is

renowned Sir,

Yours devoted and obliged,

Stanisław Lubieniecki Rolitsius
by my own hand

3 See section 2.4.3 of the Introduction.

4 I.e. to Elsevier. There is no surviving correspondence between him and Lubieniecki.

5 There are no records of a comet seen in 1669.

6 Jacob de Groot (1628–94).

7 Willem de Groot (1597–1662).

8 Hugo Grotius (de Groot, 1583–1645), Dutch lawyer, considered to be the founder of the modern international law.

9 Jan de Groot (1625–78).

10 Lieuwe van Aitzema (1600–1669), Dutch diplomat and historian. Since 1629, he was a resident of the Hanseatic towns in The Hague. His *Saken van Staet en Oorlogh* ('Affairs of state and war'), published in 1657–68 in twelve *quarto* volumes (reprinted with changes in 1669–72 in seven *folio* volumes), discuss the history of the Netherlands in 1621–69 and present many official documents.

11 Pieter de Groot (1615–78), Dutch politician and diplomat, ambassador in Sweden in 1668–70.

12 See section 2.4.9 of the Introduction.

In Altona near Hamburg
on 12 March 1669 new style.

Postscript. Send your letters to Hamburg as before when you decide to answer me (I hope and ask you to do it as soon as possible).

That de Groot is a counsellor and resident of the Elector Palatine.[13]

13 Charles I Louis (1617–80), Elector Palatine since 1648.

75.

20 April 1669, Hevelius to Lubieniecki

BO: C1, v. 9, 1362
BnF: Lat 10348, v. 9, 170–71

Domino
Rolisio
Hamburgi

Illustrissime Vir.

Tandem opus Tuum praestantissimum, longeque desideratissimum obtinui, pro quo ingentes Tibi ago et habeo gratias; gratulorque Tibi de tantis molestissimis exantlatis laboribus, qui profecto Viris praesertim penitiorum rerum deditis et Eruditis displicere vix possunt. Quod mei vero toties ac toties adeo honorificam in dicto Tuo opere mentionem habueris, grata mente agnosco, eoque magis, quo minus hunc Tuum erga me Benevolum affectum et amorem meritus unquam fuerim. Literas Tuas, quibus significasti, Te opus meum Cometographicum accepisse necdum accepi; gratissimum tamen accidit, quod modo intelligam Tibi pagellas meas recte esse oblatas; optarem ut occasio detur, alterum illud exemplar Rev[erendo] P[atri] Curtio transmittendi, ne plane ingratus esse videar. Exemplaria illa residua *Mantissae* meae, quae apud Te reservantur, aut mihi Dantiscum remittere, aut pro lubitu unum aut alterum exemplar Tui *Theatri Cometici*, loco eorum transmittere poteris. De caetero lubentissime emptores hic Tibi, si potero, conciliaturus sum: quippe hic, ut ubique vix paucissimi inveniuntur rerum Coelestium cultores; quicquid tamen sit, sat feliciter haud pauca exemplaria *Cometographiae* meae hucusque distracta sunt. Novum denuo Cometam in Aquitania nuper apparuisse valde dubito: // cum nullus rerum Mathematicarum peritus illum (quantum sciam) ullibi observaverit. Cl[arissimus] Dominus Buthnerus Te plurimum salutat. Vale, et amore pristino me porro prosequere, qui sum

Tuae Illustrissimae Dominationis
Omni officiorum studio
observantissimus

J[ohannes] Hevelius
manu propria

Dabam Dantisci
Anno 1669, die 20 Aprilis.

To Master
Rolitsius
in Hamburg

Most illustrious Sir!

I have finally received your excellent and long-expected book for which I am very grateful; I congratulate you on such unpleasant and tedious works which certainly cannot be unpleasant, especially for the men devoted to more detailed studies and for the learned. I thankfully acknowledge that you mentioned me so many times in such an honorific way in the aforementioned book of yours, especially because I have never deserved your benevolence and goodwill. I have not received your letter in which you mentioned that you had received my *Cometographia*; it is very pleasant for me finally to learn that my little pages were successfully delivered; I wish there were an opportunity to send another copy to the Reverend Father Curtz[1] so that I would not seem unthankful. Those remaining copies of my *Mantissa* you can either send back to me to Danzig or, if you wish, send me one or two copies of your *Theatrum cometicum* instead. As far as the rest is concerned, I am eager to find you some buyers here, if I can; because here, like everywhere else, there are hardly any worshippers of heavenly things; nevertheless, quite a number of copies of my *Cometographia* has been distributed so far. I doubt that a new comet appeared in Aquitania because no one who is experienced in mathematics has seen it anywhere (as far as I know).[2] The most renowned Master Büthner[3] sends you many greetings. Farewell, and still, as before, be benevolent towards me, who is

illustrious Sir,

Yours most obliged,

Johannes Hevelius
by my own hand

Given at Danzig
on 20 April 1669.

1 See section 2.4.5 of the Introduction.
2 See note to Letter 74.
3 See section 2.4.3 of the Introduction.

76.

25 June 1669, Lubieniecki to Hevelius

BO: C1, v. 9, 1358
BnF: Lat 10348, v. 9, 221–22

Nobilissime Amplissime Clarissime Vir.

Pergratum mihi fuit ex Tuis ultimis de valetudinis et rerum Tuarum statu aliquid cognoscere. Horum laetorum nunciorum participes Amicos Parisienses, Hafnienses, Holmienses feci. Nunc Te de meis rebus certiorem facio: me scil[icet] commoda per gratiam Dei valetudine frui, et pro mea virili parte etiamnum aliquid per commercium litterarum quod cum Viris Doctis colo, ad promovenda commoda publica conferre. Fateor tamen sumtu me operis mei ultra vires gravatum esse, praesertim, cum quae conditio huius seculi est, exemplarium distractio lente procedat. Equidem et ad divendenda *Mantissae Prodromi* Tui exemplaria hic et illic omnem adhibui operam. Sed ubique frustra fui. Aliquot dono obtuleram, et eo nomine Tibi nonnihil nummorum numerari curaveram. Idem et in posterum facere sum paratus, ut et cuncta Tibi grata officia praestare. Experieris id // reipsa: etiamsi *Cometographiae* Tuae exemplar Senatui Hamburgensi offerri velis. Velim autem ut nunc id officii mihi apud Amplissimos Collegas Tuas praestes. Mitto exemplar decenter compactum pro Amplissimo Senatu. Mitto alterum Magnifico Domino Capitaneo Mirachoviensi. Tertium offeres alicui ex Ampliss[imis] Dominis Consulibus. Dabit Tibi et quartum (si sit opus) meus Patruelis, per quem haec exemplaria mitto, utque eum consilio et opera Tua juves, obnixe rogo. Sal[utem] pl[urimam] Cl[arissimo] Buthnero adscribo. Vale Vir Cl[arissime] et Humanissime, mihique porro fave. Dabam Altenaviae d[ie] 15./25. Junii 1669.

Tuus
studiosissimus
observantissimus

S[tanislaus] L[ubieniecius] Rolitsius
manu propria

Most noble, worthy, and renowned Sir!

It was most pleasant for me to learn from your last letter something about your health and the state of your affairs. I wrote to friends in Paris, Copenhagen, and Stockholm about this happy news.[1] Now, I am informing you about my affairs, viz. that due to God's grace I am in good health and, as much as I can, I continue epistolary communication with learned men, in order to help the public good a little. I admit, however, that I am overburdened beyond my power with the publication of my book, especially because it sells slowly, which is caused by the condition of our century. I also did everything to distribute the copies of your *Mantissa Prodromi cometici* in many places. But it was all in vain. I gave a few of them as gifts, and I took care to send you money for those. I am ready to do so in the future and to fulfil all the duties which are kind to you. You will experience it indeed, even if you wanted to give a copy of your *Cometographia* to the city council of Hamburg. But I would like you to fulfil such a duty on my behalf for your excellent colleagues. I am sending a well-bound copy for the greatest city council. I am sending another one for the noble governor of Mirachowo.[2] I would like you to give the third one to one of the greatest mayors. You can also get the fourth one (if there is such a need) from my cousin,[3] through whom I am sending you these copies; I ask you kindly to help him with your advice and other efforts. I convey many greetings for the most renowned Büthner.[4] Farewell, most renowned and kindest Sir, and keep being benevolent towards me! Given at Altona, on 15/25 June 1669.

Yours obliged and devoted,

Stanisław Lubieniecki Rolitsius
by my own hand

1 See note to Letter 73.

2 Probably Ferdynand Garbiel Szczepański (*c.* 1620–90), governor (*starosta*) of Mirachowo (in northern Poland) from after 1667 until his death.

3 I could not determine to which of his cousins Lubieniecki refers here.

4 See section 2.4.3 of the Introduction.

77.

12 July 1669, Lubieniecki to Hevelius

BO: C1, v. 9, 1359
BnF: Lat 10348, v. 9, 222–23

Seal with Lubieniecki's coat of arms (Rola coat of arms)

Salutem Plurimam Vir Nobilissime Amplissime Humanissime.

Cum jam patruelis meus feliciter per Dei gratiam in urbem Vestram pervenerit, iterum iterumque rogo, velis eum consilio et opera Tua juvare, ut cuncta apud Vos ex voto et bona existimatione perficere possit. Quod si mutua mea opera in his oris opus habeas, libere impera, experiere paratum. Difficilia sunt profecto haec tempora, quibus cuncti bonarum artium amantes Virorum bonorum et prudentum consilio et opera opus habent. Accedit quod operis mei editionem Ephorus eius nimis preciosam et tardam reddiderit, ita ut optimum tempus exemplarium distrahendorum neglexerit. Sed haec inter nos dicta sunto. Tu qua polles prudentia, dexteritate et auctoritate, et qua es erga me benevolentia ita facies cuncta, ut conveniant exspectationi et existimationi meae. Habes ibi aliquot exemplaria distribuenda. Misissem plura: sed necdum quae exspecto ex Hollandia accepi. Exspecto praeterea quid consilii ea in re mihi sis daturus. Faciam tamen cuncta ex mente Tua. Novi Regis indigenae // electio nobis hic sed et cunctis aliis perplacet. Accepi ego litteras aliquot ab Illustribus Viris de hac re ad me datas. Doleo tamen multum multa huic spei Ducem Neoburgicum prole numerosa gravatum impendisse, nullo laeto eventu. Sed 'superanda omnis fortuna ferendo est.' De statu amicorum, quod silent, nil ad Te scribere possum. Cl[arissimus] Oldenburgius ante aliquot hebdomades ad me literas dederat benevolentiae plenas: sed nil memoratu dignum perscripserat.[1] Tu si quid tale habes, imperti more tuo. Salutem adscribo officiosam Magnifico Domino Capitaneo Mirachoviensi, Te vero, Vir Cl[arissime] cum Tuis bene et diu valere cupio ex animo. Dab[am] Altenaviae, 12. Julii Gregor[iani] 1669.

Tuus
omni officio et studio

S[tanislaus] L[ubieniecius] Rolitsius
manu propria

1 sed...perscripserat *om.* **BnF**.

Greetings, most noble, worthy, and kindest Sir!

Because, thanks to God's grace, my cousin has successfully come to your city, I ask you to help him with your advice and other duties so that he can achieve everything he wants there. If you have any business in which I can help you in this country, feel free to tell me — you can consider it done. These times are indeed difficult when everyone who loves good arts needs advice and help from good and prudent men. Moreover, the custodian of the printing of my book[2] made its publication too expensive and delayed, so he missed the best time for selling it. But may it be said between us. Due to your prudence, experience, and respect, and because you are benevolent towards me, you will do everything that meets my expectations and my wishes. Here you have a few copies to distribute. I would have sent more, but I have not received yet what I am expecting from the Netherlands. Moreover, I expect your advice on this matter. I will do anything following your opinion. Here, we (as all others) — are delighted with the election of the new King of the native origin.[3] I have received a few letters from renowned men concerning this issue. I grieve, however, that the Duke of Neuburg, overburdened with such a numerous offspring, made so much effort due to that hope without any success.[4] But 'endurance must master every fortune.'[5] I cannot write to you about what is happening with friends who are silent. A few weeks ago, the most renowned Oldenburg[6] sent me a letter full of benevolence, yet he wrote nothing noteworthy. If you have something of this kind, pray, tell me, as you usually do. I send greetings for the noble governor of Mirachowo;[7] I wish for you, most renowned Sir, and for your family, good health. Given at Altona, on 12 July 1669 Gregorian.

Yours obligingly,

Stanisław Lubieniecki Rolitsius
by my own hand

2 Frans Kupyer (1629–92), Dutch printer. He was known for publishing controversial books on religious topics (including atheism). He was Lubieniecki's correspondent; in the TC3 pp. 1–20, there are 8 letters they exchanged (4 from Kuyper, 4 from Lubieniecki).

3 Michał I of the Wiśniowiecki family (1640–73, King of Poland and Grand Duke of Lithuania since 1669). He was the first elective king who was not a member of a foreign royal family. This fact was strongly emphasized by his supporters.

4 Philip William (1615–90), Count Palatine of Neuburg, stood for the election for the Polish and Lithuanian throne in 1669 but he lost to Michał Wiśniowiecki.

5 Virgil, *Aeneis*, 5, 710, transl. by H. Rushton Fairclough, in Virgil, *Eclogues. Georgics. Aeneid I–VI* (Cambridge, Mass., London: Harvard University Press, 2004).

6 See section 2.4.12 of the Introduction.

7 Ferdynand Gabriel Szczepański, see note to Letter 76.

78.

22 April / [2 May] 1671, Lubieniecki to Hevelius

BO: C1, v. 10, 1447/81
BnF: Lat 10348, v. 10, 145–47

Seal with Lubieniecki's coat of arms (Rola coat of arms), address: 'Viro Nobilissimo / Amplissimo Clarissimo / Domino Joanni Hevelio / Consuli veteris urbis / Dantiscanae / Mathema- / ticorum Principi / Amico / pl[urime] colendo.' ('To the most noble and greatest Master Johannes Hevelius, city Councillor of the Old Town in Danzig, the first among mathematicians, much-esteemed friend')

Vir Amplissime, Clarissime, Humanissime.

Silentium meum diuturnum, postquam scil[icet] ad ultimas Tuas responderam, et novis Te litteris per Patruelem meum, qui exemplaria operis mei ad Vos pertulit, compellaram, aequus, ut spero excusabis. Tanto vero magis, postquam acceperis me diversa, eaque longinqua itinera obiisse, tum meis meorumque morbis implicitum Amicis vacare non potuisse. Nam ne Te quid rerum mearum lateat, bina itinera in Daniam, unum in Hollandiam, tum in Marchiam et terram Lunaeburgicam perfeci. Sed anno praeterito Hafniae gravi calculo, demum morbo articulari correptus aegre adulto autumno domum redii. Dehinc tota hyeme una cum uxore et filiola aegrotavi. Multa praeterea votis adversa qua publice, qua privatim evenere: praesertim post obitum Opt[imi] Max[imi] Principum Divi Friderici III. Patroni mei Clemen[tissimi]. Sed necdum cessant arthricorum dolorum reliquiae. Exercent me etiamnum et alia adversa. Nec tamen me ideo de statu bonae mentis deiici patior. Vide quid ante annum sub abitum in Marchiam et Daniam perfecerim. *Monumentum* est *Radzivilianum*, quod tamen ad *Theatrum* meum *Cometicum* refertur. Exemplaria quae superant[1] a Tuo, quod ut boni et aequi consulas, oro, offeres Magnificis et Amplissimis Dominis Praeconsulibus, sed et Amplissimo Senatui unum offeras velim, ut sit additamentum, licet exiguum et // nudum, exemplaris illius a Patrueli meo oblati. Jam sub manu habeo *Theatrum Politicum* ex *Theatro Cometico* fabrefactum per Quaestiones, Decisiones, Consultationes Mathematicas, Physicas, Historicas, Chronologicas, Geographicas, Politicas, Ethicas, Juridicas, Theologicas, Oeconomicas, Philologicas et Criticas, usum eius multiplicem cuivis hominum generi exhibens, addito eius Indice Locupletissimo, ut ita Vir Politicus omnium prope scientiarum Locos Communes habeat. Sed me vis morbi et fortunae iniuria retardat, et vix mentem ad altiora attollere permittit. Adhaec nosti quale sit hoc seculum, bonis scil[icet] artibus adversum, saltem eas fastidiens, paucis exceptis earum amantibus. Jam video non esse amplius

1 et *del. post* superant **BO**.

pudori Principibus beneficentia vinci a privatis, ac ingrati animi labe adspergi. Si tamen aliqua ratione honestos conatus nostros Reip[ublicae] litterariae et Christianae consecratos juvari illis posse putes, facies quod aequitas et humanitas suadebunt. Si habetis aliquem Forsterum redivivum, posset is aliquid in hanc rem facere. Exemplaria Tua posteriora maiori ex parte dono Amicis obtuli. Nemo enim ullum emere volebat, nec vult. Ita remanent adhuc penes me aliquot. Accipies et tu Amsterodamo V. exemplaria operis mei et experieris quid pro Amici commodo facere possis. Id quidem quod poteris, libenter Te facturum, nota Tua promittit mihi humanitas. Quam ego omni meliori ratione demereri conabor. Vale Vir undecunque Nobilissime ac mihi porro fave. Dab[am] Altenaviae, d[ie] XXII. Aprilis Juliani Anno MDCLXXI.

Tuus
omni studio et officio

S[tanislaus] L[ubieniecius] Rolits[ius]
manu propria

Most worthy, renowned, and kindest Sir!

You will justly excuse my long silence, I hope, viz. since I answered your last letter and sent you a new letter through my cousin (who brought you copies of my book), especially when you learn that I had various long journeys and I could not have any time for friends, as both my family and I were ill. Not to hide anything of my businesses from you, I made two journeys to Denmark, one to the Netherlands, to the March,[2] and to the province of Lüneburg.[3] But last year (I was suffering from severe calculi then and, later, from the sickness of joints in Copenhagen) I returned home in late autumn. Then I was ill all winter, together with my wife and our little daughter.[4] Moreover, many disturbing things happened (both public and private), especially after the death of the best and greatest of lords, divine Frederick III, my most clement patron. The remnants of gout have not stopped yet. Some other unpleasant issues bother me as well. But I do not suffer from any worse condition of my mind. See what I did before my journey to the March and Denmark. It is the *Monumentum Radzivilianum*,[5] though it is still related to my *Theatrum cometicum*. So that you could render me a good and just service, I ask you to give the additional copies to the noble

2 The New March (see section 1.1.3 of the Introduction).

3 Little is known about Lubieniecki's travels outside Hamburg and Altona, and it cannot be determined what the exact purpose of these journeys was. It may be speculated, however, that he went to the New March to visit Stefan Niemirycz (see section 1.1.3 of the Introduction) and his travel to Lüneburg was somehow related to his contacts with late Johann Ernst von Rautenstein.

4 See note to Letter 60.

5 Stanisław Lubieniecki, *Monumentum Radzivilianum* (Hamburgi: ex officina Georgii Rebenlini, 1670).

and greatest mayors, and also one to the greatest city council since it is a supplement (however little and miserable) to that copy given by my cousin. I have the *Theatrum politicum*, based on the *Theatrum cometicum* — with questions, opinions, and discussions in mathematics, physics, history, chronology, geography, politics, ethics, law, theology, economics, philology, and criticism; it shows universal use of the *Theatrum cometicum* to all kinds of people; there is also a comprehensive index added, so that a political man has the commonplaces of almost all sciences.[6] But the strength of my illness and the injustice of my fate cause the delay and barely let me lift my mind to higher things. Moreover, you know what our century is like, viz. opposing all good arts, or despising them at least, except for a few people who love them. Now I see there is no more shame for a private person to get beneficence from the people of importance than to be disgraced by an offence of an ill-disposed mind. If you think that you can help our honest efforts, devoted to the Christian Republic of Letters, in any way, you will do what justice and humanity suggest. If you had a revived Forster[7] he could do something for this cause. I gave most of the copies of your latest book to friends as gifts. Because no one either wanted or wants to buy any. Therefore, some of them are still with me. You will receive five copies of my book from Amsterdam, and you will see what you can do as a friendly favour. Your well-known kindness promises me that you will readily do everything you can. I will try to earn your affection in all possible ways. Farewell, most noble Sir, and always be benevolent towards me! Given at Altona, on 22 April 1671 Julian.

Yours obligingly,

Stanisław Lubieniecki Rolitsius
by my own hand

6 Lubieniecki did not publish this *Theatrum politicum* or any other index to the TC (except for the *Monumentum Radzivilianum*, see section 1.3.4 of the Introduction). I could not find this manuscript and I cannot say whether it was completed.

7 Lubieniecki may refer here to Georg Förster (*c.* 1615–60), publisher from Danzig. About him see CJH3 p. 152, note to Letter 5, Hevelius to des Noyers, 8 April 1647.

79.

4 August 1671, Hevelius to Lubieniecki

BO: C1, v. 10, 1448/84
BnF: Lat 10348, v. 10, 149–50

Hamburgi

Illustri Viro
Domino Rolisio
J[ohannes] Hevelius Salutem

Non Tibi, sed mihi incumbit diuturnum silentium excusare; sed cum sciam non ex frequentibus literis, verum ex affectu Te amicum cognoscere, non est, quod ea de re pluribus verba faciam. Doleo autem vehementer adversam Tuam fortunam, quam hactenus es expertus; Deus O[ptimus] M[aximus] Tibi sit propitius, ut eam superare feliciter possis; laetor vero maximopere, quod Te de statu bonae mentis deiici non patiaris, sed in laudabili proposito, Reipublicae Literariae prodesse, constanti animo pergas, non dubito, quin Omnipotens Tibi optime prospiciet, quo omnia in Nominis Sui Gloriam, etiam hoc seculo nobis artibus licet adverso, ex voto Literatorum omnium in lucem proferre non nequeas; id quod et mihi meisque studiis bono publico susceptis ut obtingat, animitus exopto. Ego in eo totus jam sum deditus, quo possim *Machinam* meam *Coelestem*, observationesque universas, cum novo Catalogo Fixarum, quantocyus in publicum proferre; Pars istius operis jam sub praelo sudat. Possent quidem studia mea alacrius progredi, si Juvenem aliquem ab auxiliis, et a calculo haberem, sed in decem et amplius jam annis plane mihi solo omnia peragere incumbit. Idcirco si studiosum quendam probum ac Matheseos cultorem nosses, qui in calculo sphaerico et astronomico optime versatus esset, vellem illum evocare Dantiscum; forte Cl[arissimus] Müllerus, quem officiose saluto, mihi quen-//dam conciliare poterit, aut Hamburgi, aut alibi locorum, facietis mihi rem gratissimam. Nam vix plurimi,[1] omnes ponere calculos, rem omnem conscribere, tot ac tot observationes peragere, describere, nec non Typographicos labores corrigere sufficiunt, ne dicam unus solus, praesertim qui aliis insuper negotiis nunquam non est occupatissimus. Exemplaria *Theatri Cometici* Tui ad me transmissa non adeo pridem bene accepi; quantum in me erit, dabo operam ut distrahi possint, quanquam vix aliquem hic novi, si illos excipias, qui libro illo jam gaudent, qui materiam illa sibi comparabit; vel ea delectatur; interea ne graveris pretium istius operis mihi indicare, quo eo melius Tuo satisfacere possim desiderio. Vale, et amore Tuo me porro honorato. Dabam Dantisci Anno 1671 die 4 Augusti.

1 pluribus *del.* BO.

In Hamburg

To the most illustrious Sir,
Master Rolitsius,
Johannes Hevelius sends greetings

It is not yours but my duty to excuse the prolonged silence; yet, since I do not know from the letters but from your affection that I can recognize you as a friend, there is no reason why I should talk more about it. I grieve much at your miserable fate which you have experienced so far; may the best and the greatest God be favourable to you, so that you can happily overcome all of it; I am glad, however, that you did not suffer any loss in the condition of your mind, but that you continue your laudable intention to help the Republic of Letters; I do not doubt that the Almighty will protect you, so that you can publish everything for the glory of His Name — even in this century which opposes the good arts — according to the wishes of all learned men; it is my sincere wish that it could also happen to me and to my studies which I had begun for the public good. Now I make every effort to publish my *Machina coelestis*[2] together with a new star catalogue[3] as soon as possible; a part of this book is already in press. My studies could proceed faster if I had a young man for assistance and calculations, but for over ten years I have had to do everything on my own. If you know anyone who is a proven, learned man, knowledgeable in mathematics, and well experienced in spherical and astronomical calculations, I would like to call him to Danzig; perhaps the most renowned Müller,[4] for whom I send greetings, can find such a person for me either in Hamburg or elsewhere; thus you would do a most pleasant thing for me. Because if a group of people is barely enough to make all the calculations, to write the whole thing, to conduct and describe so many observations, and also to help with proofreading, so what can be said about one man, especially if he is busy with other duties? I have received the copies of the *Theatrum cometicum* you sent me only recently. I will do whatever I can, to the best of my ability, to distribute them, but I do not know anyone here, except for those who already have this book, who finds this subject suitable for him or who likes it; meanwhile, do not hesitate to tell me the price of this book so that I can fulfil your wish better. Farewell, and keep honouring me with your affection! Given at Danzig, on 4 August 1671.

2 Johannes Hevelius, *Machinae coelestis pars prior* (Gedani: Auctoris typis et sumptibus, 1673); Johannes Hevelius, *Machinae coelestis pars posterior* (Gedani: in Aedibus Auctoris, 1679).

3 Hevelius's star catalogue was published posthumously: Johannes Hevelius, *Prodromus astronomiae* (Gedani: typis Johannis Zachariae Stolli, 1690).

4 See section 2.4.11 of the Introduction.

80.

12 October 1671, Lubieniecki to Hevelius

BO: C1, v. 11, 1481/6
BnF: Lat 10348, v. 11, 13–14

Salutem Plurimam Vir Nobilissime, Amplissime, Clarissime.

Facile Tibi veniam do silentii diuturni. Nam et ipse jam ea opus habeo, postquam nondum respondi ad litteras tuas d[ie] 4. Aug[usti] datas. Morbo per hoc tempus conflictatus sum solito, nempe articulari, qui me aliquandiu lecto affixum tenuit, nec usque adeo tamen, per Dei gratiam. Sed praeter alia scriptionis impedimenta plura erat et hoc, quod cum ad id toto pectore incumberem, ut tibi de tali studioso, qualem habere desideras, prospicerem, responsum super hac re ab Amico tam cito habere non potuerim. Id jam habeo, nec ex voto tamen, quod doleo. Ille namque juvenis, quem mihi commendabat, aliam conditionem vitae nactus est. Video paucos his Mathematum esse amantes. Omnes ad Theologiam, Jurisprudentiam et Medicinam provolant,[1] quod istae scientiae sint pingues. Reliqua studiorum ferme negliguntur. Verum tamen liceat mihi bona cum venia Tua, et quidem Tibi soli, proponere, annon velles opera Cl[arissimi] Placentini uti. Ille namque jam pristinae valetudini restitutus est, et in Mathematicis, credo, Tibi satisfaceret. Sed et si ille, quod adhuc sciam, conditionem nullam // obtineat, nescio tamen an istam accepturus esset. Spero tamen id facturum. Dubio enim procul a Te honeste haberetur. Sed an Tu vicissim in aedes Tuas eum sis recepturus, nisi mihi indicaveris, scire non possum. Si in id consenseris, indicabis mihi quibus legibus id fieri velis. Illo namque ego amico utor, imo haud ita pridem per litteras me rogitavit, ut ei de aliqua honesta et conventiente vitae conditione prospicerem. Amicus olim utriusque nostrum Mullerus fatis ante aliquot menses cessit. Ita ut cuncta in orbe, et nos tendimus ad centrum quietis. *Theatri* mei *Cometici* exemplaria recte Tibi reddita gaudeo. Si aliquod pecuniae ex illorum venditione ad me redire queat, fac quaeso ut voti reddar compos, nunc praesertim, cum me adversa fortuna premit. Quae tamen me de statu bonae mentis nunquam, Deo juvante, deiiciet, ut nec ipse morbus. Illud solum doleo, omne ferme tempus studiis impendendum extrahi laboribus molestis vitae sustentandae destinatis. Inclusas Tuae humanitati commendo, ut et memet meaque omnia. Vale eximium seculi, licet bonis artis adversi, decus et salve

a Tuo

S[tanislao] L[ubieniecio] Rolits[io]
manu propria

Altenaviae 12. Oct[obris]
1671.

1 provolant : pervolant **BnF**.

Greetings, most noble, worthy, and renowned Sir!

I have no problem with forgiving your prolonged silence. I need your forgiveness myself as I have not answered your letter sent on 4 August yet. During that time, I was struck by constant illness, viz. the sickness of joints which held me in bed for some time but, thanks to God's grace, not anymore. Apart from other obstacles in writing, I was also trying with all my heart to look for such a learned man as you would like to find; I could not receive an answer from a friend[2] about that so soon. I have it now, but I grieve that it does not meet your expectations. Because that young man, who was recommended to me, chose different employment. I see there are not many people who love mathematics. All of them flee for theology, law, and medicine because these fields are profitable. Other studies are usually neglected. But excuse me, please, and let me ask if you would like to use the help of the most renowned Placentinus.[3] Since he recovered to his former health, I believe he could provide enough assistance in mathematics for you. But even though he does not have any permanent post, as far as I know, I do not know if he would accept this one. Still, I hope he will. Undoubtedly, he would be treated fairly by you. But I cannot know if you would welcome him in your home unless you tell me so. If you agree to it, tell me under which conditions you would like it to happen. Because he is my friend, and not so long ago he asked me to look for just and suitable employment for him. Müller,[4] an old friend of ours, died a few months ago. Like all other things in this world, we head for the utmost peace, too. I am glad that the copies of my *Theatrum cometicum* were successfully delivered to you. If you could send me some money from selling them, you would make me very obliged, especially now when I am oppressed by hostile fate. Thanks to God's help, neither this (nor the illness) caused any detriment in the condition of my mind. I only grieve that almost all the time I need for studies is taken away by this unpleasant struggle for making a living. I commit the enclosed to your kindness,[5] as well as me myself and everything mine. Farewell, exceptional glory of this age (even though it is hostile towards the good arts), and take wishes

from your

Stanisław Lubieniecki Rolitsius
by my own hand

In Altona, on 12 October 1671.

2 I could not determine to whom Lubieniecki refers here.

3 See section 2.4.14 of the Introduction.

4 See section 2.4.11 of the Introduction.

5 There is no information about the contents of these enclosures.

81.

5 December 1671, Hevelius to Lubieniecki

BO: C1, v. 11, 1482/7
BnF: Lat 10348, v. 11, 14–15

Domino
Rolisio
Hamburgi

Illustrissime Vir.

Gratias Tibi habeo debitas, quod sollicitus fueris de studioso quodam, qui mihi possit esse a calculis. Et cum alium hac vice non inveneris, quaeris ex me, quid mihi videatur de Cl[arissimo] Placentino, jam pridem optime mihi noto. Sed haec conditio, ipsi non est conveniens, ut facile intelligis, nec adeo ampla, ut possit talem egregium Virum per tot jam annos Matheseos Professorem sustentare. Adhaec ob calculum molestissimum ac toediosissimum nunquam non peragendum haud[1] ipsi[2] suaderem ut officium hocce susciperet. Quare, ut libere dicam, rebus nostris melius prospicietur, si ille habeat lautiorem conditionem, a tantis laboribus liberiorem, et ego Juvenem quendam, qui mihi possit in quibusvis esse ab auxiliis. Idcirco exspectandum adhuc erit paullulum, donec Deus etiam hac in parte nobis prospiciat, quo possimus[3] ea omnia in Nominis Divini Gloriam suscepta, etiam felicissimo successu ad finem perducere; inprimis *Machinam* nostram *Coelestem*, quae jam a Mensibus aliquot sub praelo sudat. Vale Vir Illustrissime, precatus Tibi et novum, sic multo feliciorem annum. Dabam Gedani Anno 1671 die 5 Decembr[is] St[ylo] n[ovo].

Tuae Illustrissimae Generositatis
Officiosissimus

Johannes Hevelius
manu propria

1 puto *del. post* haud **BO**.
2 fore hocce offi *del. post* ipsi **BO**.
3 hac in parte *del. post* possimus **BO**.

To Master
Rolitsius
in Hamburg

Most illustrious Sir!

I am very grateful for your efforts to find a learned man who could help me in calculations. And because you could not find anyone this time, you asked what I thought about the most renowned Placentinus[4] whom I had got to know well before. But this employment is not suitable for him, as you surely understand, nor paid well enough to provide a decent living for such a great man who used to be a professor of mathematics for many years. Moreover, because calculations, complicated and toilsome, seem to have no end, I could not ask him to undertake this task. Therefore, let me say it frankly, it would be better for our businesses if he had more proper employment without such duties; and I have a young man who could help me in various things. Hence, we will have to wait a little longer until God becomes favourable to us also in this matter, so that we could successfully bring to an end everything we have begun for the glory of His Name: first of all, our *Machina coelestis*[5] which has been in the press for a few months. Farewell, most renowned Sir; I pray the new year will be much happier for you! Given at Danzig, on 5 December 1671, new style.

Illustrious and generous Sir,

Yours obligingly,

Johannes Hevelius
by my own hand

4 See section 2.4.14 of the Introduction.
5 Johannes Hevelius, *Machinae coelestis pars prior* (Gedani: Auctoris typis et sumptibus, 1673).

82.

9/[19] January 1672, Lubieniecki to Hevelius

BO: C1, v. 11, 1443/19
BnF: Lat 10348, v. 11, 29–31

Vir Nobilissime, Amplissime, Clarissime.

Debeo Tibi responsum, debeo et novi anni votum. Utrumque differre necesse habui ob gravia impedimenta, quae adversam sortem comitari solent. Quanquam et infirma valetudo saepius manum iniicit menti in officia amicis debita et alia honesta agenda pronae. Melius tamen per Dei gratiam valeo et habeo, quam ante annum. Eandem cum novo anno auctum iri spero, postquam non exigua patientiae et modestiae adhuc dedi experimenta. Iste ut et Tibi Tuisque laetus decurrat ac pluribus caelestium donorum copia stipatis praecurrat, precor ex animo. Tu enim ex privatis vel solus, vel profecto unus de paucis es, qui maximo non tantum labore, sed et sumtu studia Mathematica, proh dolor! hoc depravato et ferreo seculo, contemta promoves. Mirarer aversum a bonis artibus seculi genium, nisi ad malas bellorum et iniquitatis artes conversum viderem. Nunquam ab orbe condito visa ambitio et arrogantia animos Principum, ut non tantum medioxumi, sed et minimi, maximis se comparent, occupavit. Accessit ei tuendae et fovendae ac augendae destinata opum, hinc et alieni, cupido. Atque inde tot fallaciae, fluxa fides, vis rapax, et innumerae iniustitiae ac iniquitatis artes provenere. // Tot monstra debellare, hoc virtutis opus vere Herculeum ac Heroicum prorsus. Equidem robur virtutis non tantum invictum, sed et victrix semper fuit. Hoc induti quae recta, quae laudabilia sunt, agamus assiduo, certi mercedis a Deo paratae, et quam grata posteritas cuique rependet. Progredere itaque, Vir Magne, in Tuis praeclaris laboribus exantlandis. Tu nostri seculi Hipparchus, tu alter Atlas caeli Mathematici tanto justius audies, postquam Tuam *Machinam caelestem* ad optatam metam promoveris. Et tanto quidem maior manet Te gloria, quanto plures labores molestiasque devoraveris, quanto minus subsidii aliunde nactus fueris. Ego quidem perquam libenter Tibi prospicerem de aliquo studioso, rei Mathematicae perito. Qua de re cum aliquot Viris Doctissimis hic contuli: cumprimis cum Cl[arissimis] Kirstenio et Sivero. Sed communis est querela, Mathematica studia sperni, saltem non esse ea lucrosa, ideo eorum cultores animum ad alia studia, vitae tuendae magis idonea, applicare. Atque id mihi de quodam Hadlero, cui patria Hamburgum, speciatim dictum est. Novi eum de facie. Sed compellare eum a multo tempore non licuit. Et scio in eo haec et illa desiderari. // De Cl[arissimo] Placentino Tibi consentio. Nil ab illo tempore litterarum ab eo accepi, quove in statu sit, juxta cum ignarissimis scio. Me adhuc molesti quidam labores, vitae tuendae caussa necessario subeundi, detinent, ut nequeam meis studiis vacare. Sed Deus meliora! Mullerus olim noster ante aliquot menses fatis cessit. Ita et Martinus von der Eyghen, qui antehac litteras

meas curabat. Sed nomen meum praefectis tabellariorum satis notum est. Cl[arissimo] Titio plur[imas] sal[utes] adscribo, et has Cl[arissimo] Buthnero tradi, officiose peto. Vale Vir Cl[arissime] diu felix ac mihi constanter fave. Dabam Altenaviae ad Hamburgum, d[ie] 9. Jan[uario] Juliani 1672.

Clarissimi Nominis Tui
officiosus cultor

S[tanislaus] L[ubieniecius] Rolitsius
manu propria

Post Scriptum. His diebus prodiit hoc scriptum curiosum de nova urbes muniendi ratione, quod Tibi mitto, utque id boni et aequi consulas ac cum Cl[arissimo] Buthnero communices, rogo.

Most noble, worthy, and renowned Sir!

I owe you an answer; I owe you also best wishes for the new year. I had to delay both of them due to severe problems which usually accompany the opposing fate. And my poor health hindered my mind generally tending towards the duties owed to friends, and to other honest things. But I am in a better state than a year before, thanks to God's grace. It will be even better in the new year, I hope, since I have already given proof of my patience and modesty. I sincerely pray that the new year will pass happily for you and your family and that it will shower you with heavenly gifts. You are the only one, or certainly one of the few who (not only with so much struggle but also with own expenses) propel mathematical studies, despised — alas! — in this degenerated and iron age. I would wonder at the spirit of the century which turns away from good arts, if I did not see it turning towards bad arts of wars and inequity. Since the beginning of the world, never have such a bold ambition and arrogance occupied the minds of rulers, when not only the mediocre ones but also the smallest compare themselves to the greatest. There is also an eagerness for keeping, treasuring, and increasing possessions, even those belonging to others. And this is the origin of many frauds, ambiguous fidelity, greedy force, and countless examples of injustice and inequity. To defeat so many monstrosities is a Herculean and heroic work. Because the strength of virtue has always been not only invincible but also victorious. Equipped with what is right and what is laudable we shall always act, being sure of the reward prepared by God which will be granted to everyone by the grateful posterity. Continue, great Sir, your famous and arduous works! You will even more justly hear that you are the Hipparchus of our age and another Atlas of mathematical heavens after you finish this anticipated *Machina coelestis*[1] of yours. And the greater glory

1 Johannes Hevelius, *Machinae coelestis pars prior* (Gedani: Auctoris typis et sumptibus, 1673).

awaits you, the more works and struggles you will have to swallow, and the less help you will get from elsewhere. I would willingly take care of finding for you a student experienced in mathematics. I discussed this thing here with some learned men, especially with the most renowned Kirsten[2] and Sivers.[3] But there is a common dispute that mathematical studies should be despised, or at least that they are not profitable, so those who follow them turn their minds towards some other studies, more suitable for earning a living. And I was told especially about a certain Hadlerus who comes from Hamburg. I know him by sight. But for a long time I could not meet him. And I know some things are expected of him. I agree with you about the most renowned Placentinus.[4] Since then, I have received no letter from him, and I know about his situation as much as those who know nothing. I am preoccupied with some tedious works, necessary for earning my living; therefore, I cannot have any time for my studies. But God will make everything better! Our Müller[5] died a few months ago. And so did Martin von der Eyghen,[6] who took care of my correspondence. But my name is well-known to the superiors of messengers. I send many greetings for the most renowned Titius,[7] and I ask you to pass them to the most renowned Büthner.[8] Farewell, most renowned Sir, be happy and always benevolent towards me! Given at Altona near Hamburg, on 9 January 1672, Julian style.

Renowned Sir,

Your devoted admirer,

Stanisław Lubieniecki Rolitsius
by my own hand

Postscript. An interesting text about a new method of fortifying cities[9] has been recently published; I am sending it, asking you to use it well and to show it to the most renowned Büthner.[10]

2 See section 2.4.9 of the Introduction.

3 See note to Letter 20.

4 See section 2.4.14 of the Introduction.

5 See section 2.4.11 of the Introduction.

6 See note to Letter 1.

7 See note to Letter 46.

8 See section 2.4.3 of the Introduction.

9 Probably: Allain Manesson-Mallet, *Les travaux de Mars ou la fortification nouvelle* (Paris: chez l'autheur, chez Jean Henault et chez Claude Barbin, 1671).

10 See section 2.4.3 of the Introduction.

83.

11 March 1672, Hevelius to Lubieniecki

BO: C1, v. 11, 1494/20
BnF: Lat 10348, v. 11, 31–32

Enclosures:
83A. Letter from Hevelius to Boulliau, 9 March 1672
83B. Description of the comet of 1672

Domino
Rolisio
Hamburgi

Illustrissime Vir.

Cum nullum argumentum scriptionis hactenus habuerim, responsum ad Tuas mihi gratissimas die 9 Januarii datas hucusque distuli. Jam vero cum novum Sidus Crinitum in aethere appareat, volui nonnulla de eo quantocyus Tibi, Rerum Coelestium officioso admiratori, ac praeconi, communicare; sed generalia tantum paucula, quae ex una alterave observatione hucusque obtinui; specialiora[1] exspectabis, si plura impetravero. Quae autem fuerint[2] observata brevibus ad Cl[arissimum] Dominum Bullialdum perscripsi; quarum literarum copiam hisce quoque transmitto. Si ita videtur poteris et amicis communibus ea communicare, inprimis Cl[arissimis] Domino Kirstenio, Domino Fogelio, Domino Sivero, Domino Mich[aeli] Francis[co] Hacky nec non Cl[arissimo] Domino Bartholino et R[everendo] P[atri] Curtio. Quid alii de eodem phaenomeno annotaverint pariter a Te exspecto. Vale Vir Illustrissime[3] et saluta omnes nobis bene cupientes: inprimis sollicita responsum a Domino Fogelio ad meas 23 Januarii datas.

Illustrissimae Tuae Dominationis
officiosa voluntate

Johannes Hevelius
manu propria

Gedani Anno 1672
die 11 Martii

1 plura *del.* **BO**.
2 qui hucusque *del. post* fuerint **BO**.
3 et me permane? *del. post* Illustrissime **BO**.

To Master
Rolitsius
in Hamburg

Most illustrious Sir!

Since there was nothing to write about so far, I have delayed my answer to your most pleasant letter, sent on 9 January, until now. But because a new hairy star appeared in the sky,[4] I have decided to write, as soon as possible, something about it to you — an ardent admirer and herald of celestial things; but I have only a few basics which I have gathered so far from one or two observations; you can expect more details when I get more. I wrote briefly about what I had seen to the most renowned Master Boulliau;[5] I am sending here a copy of that letter. If you wish, you can share it with our mutual friends, chiefly with the most renowned Kirsten,[6] Fogel,[7] Sivers,[8] Michał Franciszek Hacki,[9] and also with the most renowned Bartholin,[10] and the Reverend Father Curtz.[11] I am waiting for you to tell me what these men observed about that phenomenon. Farewell, most renowned Sir, and greet all who wish us well; first of all, encourage Master Fogel to answer my letter of 23 January.

4 C/1672 E1, see Alexandre Guy Pingré, *Cométographie ou Traité historique et théorique des comètes*, vol. 2 (Paris: Imprimerie royale, 1784), pp. 23–24; Gary W. Kronk, *Cometography. A Catalogue of Comets. Volume 1: Ancient – 1799* (Cambridge: Cambridge University Press, 1999), pp. 362–64. See also CJH3 p. 429, Letter 156, Hevelius to des Noyers, 11 March 1672, and subsequent letters.

5 See section 2.4.2 of the Introduction.

6 See section 2.4.9 of the Introduction.

7 Martin Fogel (1634–75), German physician and linguist. He studied oriental languages and Finnish. In 1675, he was the professor of metaphysics and logic at the *gymnasium academicum* in Hamburg. Fogel was a correspondent of Hevelius and they exchanged 18 letters (12 from Fogel, 6 from Hevelius) between 1663 and 1674, cf. Klaus-Dieter Herbst, 'Alphabetical list of the correspondents', CJH1 p. 260. See also CJH3 p. 421, note to Letter 151, Hevelius to des Noyers, 16 September 1671.

8 See note to Letter 20.

9 Hevelius mixed up the given names of the Hacki brothers: Jan Franciszek (1637–96), Jesuit theologian and polemist, professor of Jesuit college in Stare Szkoty near Danzig, and Michał Antoni (1620–1703), Cistercian monk, in 1666–73, the chaplain of Queen Christina, since 1683, the abbot in the Oliwa monastery near Danzig. It is almost certain that the addressee was Michał Antoni Hacki, because his brother, Jan Franciszek, lived in Danzig at the time of the correspondence. Michał Antoni Hacki was Hevelius's correspondent and they exchanged 32 letters (26 from Hacki, 6 from Hevelius) between 1654 and 1685, cf. Klaus-Dieter Herbst, 'Alphabetical list of the correspondents', p. 262. See also CHJ3, note to Letter 151, Hevelius to des Noyers, 16 September 1671. There are no traces of Lubieniecki's contacts with either of the Hacki brothers.

10 See section 2.4.1 of the Introduction.

11 See section 2.4.5 of the Introduction.

Illustrious Sir,

Your affectionate friend,

Johannes Hevelius
by my own hand

In Danzig, on 11 March 1672

83A.

Letter from Hevelius to Boulliau, 9 March 1672

BO: C1, v. 11, 1490/17
MNW: ms 518/3
BnF: Lat 10348, v. 11, 27–28

Domino
Bullialdo
Parisiis

Vir Illustris

Interea dum responsum tuum ad meas 5 Decemb[ris] datas avidissime exspecto, hasce literulas ad te dari volui quibus Tibi quantocyus significarem, Cometam recens exortum[12] hic Gedani a 2 Martii conspectum esse, quem autem, cum per aliquot dies peregre domo abfuerim, ante 6 Martii vespere observare Instrumentis haud licuit.[13] Conspicitur tam tempore vespertino, quam matutino Helespontium versus; exiguae magnitudinis est,[14] caudam[15] non nisi unius[16] et dimidii grad[us][17] prae se fert; maiorem sine dubio exhiberet si non circa crepusculum existeret, tum Luna adesset. Versatur nunc[18] circa stellas in brachio et cingulo[19] Andromedae:[20] quan-

12 recens exortum novum (novum *del.*) **BO** : recens exortum **BnF** novum **MNW**.
13 potui licuit (potui *del.*) **BO** : licuit **BnF** potui **MNW**.
14 magnitudinis est **BO BnF** : est magnitudinis **MNW**.
15 modo *del. post* causam **BO**.
16 unius vel unius (unius vel *del.*) **BO**.
17 grad[us] **BO BnF** : grad[us] modo **MNW**.
18 modo nunc sub (modo *et* sub *del.*) **BO**.
19 brachio et cingulo **BO BnF** : dext[ero] brachio **MNW**.
20 supra scapulas *del. post* Andromedae **BO** : supra scapulas **MNW**.

tum ex una alterave observatione diiudicare adhuc possum tendit supra[21] lucidam cinguli circ[iter] versus Andromedae, et quidem motu directo diurno duorum fere[22] graduum in suo circulo. Die 6 Martii vespere Hor[a] 7 40' versabatur in 7 grad[u][23] Arietis sub latitudine 35° Boreali[24] quantum rudiori minerva ex globo coniicere licet. Die 7 mane H[ora] 3. 30 Longi[tudo] eius erat fere 8° Arietis[25] cum[26] paulo minori latitudine Bor[eali]; vespere eiusdem diei Longi[tudo] extitit 10° Arietis et la[titudo] 34° fere.[27] Die 8 Martii mane H[ora] 4 Longi[tudo] Cometae erat 12 grad[us] Arietis et lati[tudo] 33°. Die 10 vesp[ere] hor[a] fere 8. Longi[tudo] eius extitit 19° circa et lati[tudo] Bor[ealis] 32 grad[us]. Die 12 24° Long[itudo] Arietis et Latitud[o] 34° fere. Die 14 Maii Long[itudo] 28° et Latitud[o] 28 1/2° fere. Die 15 Maii in 1° Tauri et 27° Lat[itudinis] Bor[realis][28] sed ruditer tantum ut recte capias; observationes namque[29] meas calculo adhuc subiicere[30] nequivi.[31] Caudam direxit initio stellas versus in manu Andromedae, deinde ad stellas Cassiopeiae.[32] Specialiores observationes meas quas summa diligentia maioribus meis Instrumentis 8 et 9 ped[um][33] quoties licet perago, proxima occasione Vobis transmittam.[34] Si placet poteris haec qualiacunque Ill[ustrissimis] et Clariss[imis] Viris utpote Dominis[35] Capellano, Cassino et Picardo praevia nostra salute communicare.[36] Quid Vos[37] vel alii de hoc phaenomeno[38] notastis magno desiderio exspectamus. De caetero scias me novam stellam sub capite Cygni die 7[39] Martii mane denuo observasse sed vix ac

21 supra *om.* **MNW**.
22 circiter duorum fere (circiter *del.*) **BO** : duorum fere **BnF** circit[er] duorum **MNW**.
23 grad[u] **BO BnF** : grad[u] fere **MNW**.
24 latitudine...Boreali **BO BnF** : latitud[ine] Boreal[i] 35 grad[us] **MNW**.
25 Arietis *om.* **BnF**.
26 a *del. post* cum **BO**.
27 fere **BO BnF** : fere Arietis **MNW**.
28 Die 10...Lat[itudine] Bor[rali] **BnF** *in marg.* **BO** *om.* **MNW**.
29 namque **BO BnF** : enim **MNW**.
30 adhuc subiicere **BO BnF** : subiicere adhuc **MNW**.
31 Vespere fidimus ita Deo favente me iterum spero me illum observaturum. *del. post* nequivi **BO**.
32 Caudam...Cassiopeiae **BO BnF** : Vespere Deo favente spero me iterum illum observaturum **MNW**.
33 maioribus...ped[um] *om.* **MNW**.
34 transmittam **BO MNW** : lubens communicabo *supr.* **BO** lubens transmittam **BnF**.
35 Dominis : Domino **BnF**.
36 Si placet...communicare **BnF** *in marg.* **BO** *om.* **MNW**.
37 circum illum *del. post* Vos **BO**.
38 de...phaenomeno **BO BnF** : circum illum **MNW**.
39 7 **BO BnF** : 6 **MNW**.

ne vix nudis oculis adhuc conspicitur.[40] Vale et me ama. Dabam raptim[41] Gedani Anno 1672 die 9 Martii.

Tuus
toto animo

J[ohannes] Hevelius
manu propria[42]

To Master
Boulliau
in Paris

Illustrious Sir!

While I am avidly awaiting your answer to my letter sent on 5 December, I have decided to send you this small letter in order to tell you as soon as possible that a comet, recently risen,[43] has been seen here, in Danzig, since 2 March; because for a few days I was not at home, I could not observe it with my instruments before the evening on 6 March. It is being seen both in the evening and in the morning, towards the north-east; it is small; it has a tail, only one and a half degree long, up the front; it would certainly be longer if it were not at dusk and if there were no Moon. It is right now near the stars of the Arm and the Belt of Andromeda; as far as I can say, relying on one or two observations, it is moving over the bright star of the Belt towards Andromeda, and with its daily direct motion it makes almost 2 degrees in its circle. On 6 March, at 7.40 in the evening, it was in the seventh degree of Aries at latitude 35° north, as far I could roughly guess from the globe. On the seventh, at 3.30 in the morning, its longitude was 8 degrees of Aries, and the northern latitude was slightly smaller; on the same day, in the evening, it was in the tenth degree of Aries, and its latitude was almost 34°. On 8 March, at 4 o'clock in the morning, the longitude of the comet was 12 degrees of Aries and the latitude 33°. On the tenth, at almost 8 o'clock in the evening, its longitude was about 19° and the northern latitude was 32 degrees. On the twelfth, the longitude was 24° of Aries and the latitude almost 30°. On 14 May, the longitude was 28° and the latitude almost 28 1/2°. On 15 May, it was in 1° of Taurus and at latitude 27° north; but these are only raw numbers so that you can understand it correctly; so far, I could

40 conspicitus **BO BnF** : conspicitur. Ill[ustrissimos] Dominos Capellanum, Cassinum, et Picardum salutes meo nomine iisque si placet haec nova significabis, a quibus responsum ad meas avidissime exspecto. **MNW**.

41 Dabam raptim **BO BnF** : Raptim ut vides **MNW**.

42 manu propria : Die 6 Martii vesp[ere] **MNW** *om.* **BnF**.

43 See note to Letter 83.

not make the calculations based on my observations. Its tail was initially directed towards the stars in the Hand of Andromeda, then towards the stars of Cassiopeia. Whenever I can, I diligently conduct more detailed observations with my bigger instruments — which are 8 and 9 feet long; I will pass my results to you at the first opportunity. If you wish, you can give my petty letter to the most illustrious and renowned men, such as Masters Chapelain,[44] Cassini,[45] and Picard,[46] together with our greetings. We are avidly expecting what you and others observed about this phenomenon. As far as the rest is concerned, know that I saw again the new star under the Head of Cygnus,[47] on 7 March in the morning; so far, it can barely be seen with the naked eye. Farewell, and keep loving me! Given at Danzig, on 9 March 1672.

Yours endearingly,

Johannes Hevelius
by my own hand

44 Jean Chapelain (1595–1674), French poet and a founding member of the Académie française. Chapelain was Hevelius's correspondent and they exchanged 41 letters (20 from Chapelain, 41 from Hevelius) between 1663 and 1673, cf. Klaus-Dieter Herbst, 'Alphabetical list of the correspondents', CJH1 p. 257. On Hevelius's contacts with Chapelain see also Chantal Grell 'Introduction', CJH2 pp. 21–33 *et passim*.

45 See note to Letter 8.

46 Jean Picard (1620–82), French astronomer and geodesist. He was the first to make exact calculations of the Earth's size. In 1671–72, he travelled to Tycho Brahe's former observatory in Uraniborg in order to establish its exact geographic coordinates. For more information about him see *Jean Picard et les débuts de l'astronomie de précision au XVIIe siècle. Actes du Colloque du tricentenaire*, ed. by Guy Picolet (Paris: Editions du CNRS, 1987). Picard was Hevelius's correspondent and they exchanged 6 letters (3 from Picard, 3 from Hevelius) between 1671 and 1679, cf. Klaus-Dieter Herbst, 'Alphabetical list of the correspondents', p. 269. See also CJH2 p. 335, note to Letter 89, Hevelius to Perrault, 25 August 1673; CJH3 p. 420, note to Letter 151, Hevelius to des Noyers, 16 September 1671.

47 The nova CK Vulpeculae. For the information about it see Michael M. Shara, Anthony F. J. Moffat, 'The Recovery of CK Vulpeculae (1670) — the Oldest "Old Nova"', *The Astrophysical Journal*, 258 (1982), L41–L44; John C. Barentine, *The Lost Constellations. A History of Obsolete, Extinct, or Forgotten Star Lore* (Heidelberg: Springer, 2016), pp. 35–37.

83B.

Description of the comet of 1672

BO: C1, v. 11, 1491
BnF: Lat 10348, v. 11, 28–29

Generalis Cometae descriptio

Validi quidem rumores[48] hoc et praeterito anno orti sunt de variis Cometis hinc inde visis, nunc Romae, Viennae, nunc Lubecae et alibi, sed omnino falsi fuerunt[49] modo hunc, modo[50] illum planetam modo Paraselenas ut Lubecae factum imperitiores[51] loco Cometae arripuerunt[52] cum[53] nullus genuinus cometa ab anno 1665 hucusque quantum ipse aliique rerum Caelestium Cultores mecum observarunt si novas illas stellas excipias prodiit. Nunc vero sidus crinitum a[54] die 2 Martii hic Gedani observatur, quid autem

J[ohannis] Hevelii
Generalis[55] Cometae descriptio
Anno 1672, Mense Martio exorti[56]

General description of the comet

In the last two years, there have been certain rumours about various comets, seen in many places — as in Rome and Vienna, so in Lübeck and elsewhere; but all of them were false; unexperienced people considered some of the planets or paraselenes (as happened in Lübeck) to be comets; because no new comet (apart from some new stars) appeared since 1665, as far as other worshippers of celestial things and I observed. But now a star with hair is being observed here in Danzig since 2 March; this is

Johannes Hevelius's
General description of the comet
which appeared in March 1672.[57]

48 haecte *del. post* rumores **BO**.
49 fuerunt : extiterunt *supr.* **BO** extiterunt **BnF**.
50 aut *del.* **BO**.
51 pro *del. post* imperitiores **BO**.
52 Siquidem *del. post* arripuerunt **BO**.
53 ullus omnino *del. post* cum **BO**.
54 ab *del.* **BO**.
55 Cometae *del. post*. Generalis **BO**.
56 J[ohannis]...exorti : et c[etera], et c[etera] **BnF**.
57 See note to Letter 83.

84.

5 April 1672, Lubieniecki to Hevelius

BO: C1, v. 11, 1518/46
BnF: Lat 10348, v. 11, 59–61
Olhoff: 131–32

Enclosure:
84A. Lubieniecki's poems for Louis XIV

Nobilissime, Amplissime, Clarissime Vir.

Litteras Tuas, novi iri caelo hospitis nuncias, non tantum habui gratissimas, sed et cum Amicis in hac urbe degentibus, et talium rerum curiosis, communicavi. Atque hi etiam Tibi ingentes eo nomine agunt gratias, quod eos per me de isto Cometa reddideris certiores, quem alias ignorassemus, partim quod instrumentis ad observanda caelestia sydera et phaenomena destituamur, partim quod dum Tu totus in caelo versaris, nos immersi terrae, necessitati parendo, incuriosam caeli agamus vitam, imo et a vulgatae saepius famae, novorum Cometarum nunciae, erroribus vindicaris. Plurimum profecto Tibi, velut assiduo caeli custodi et astrorum metatori, Resp[ublica] debet litteraria, quod nihil te eorum fugiat, quae in caelo geruntur: quae alioquin et nobis circa terrena occupatis, silentio transirent, et posteritati ignota forent, nisi ea Tua singularis diligentia et nobis indicaret, et posterorum memoriae commendaret. Ac proinde insignem illam Tuam *Machinam caelestem*, qua ut novus Atlas caelum, sed et ingenia caeli studiosa fulcies, ac terram caelo feliciter coniunges, priscis syderibus in ordinem redactis, quin et novis ac hucusque ignotis in censum relatis, cum maximo exspectamus desiderio. Nullum enim dubium est nos insignem ex tam arduo opere tuo, ut et ex aliis praeclaris[1] operibus tuis Mathematicis, fructum capturos. Sed redeo ad Cometam. De isthoc nil prorsus ab aliis habeo. Iccirco merito dixi, nullam nos eius notitiam sine te habituros fuisse. Caeterum nobis et sero de apparitione eius monitis, et necessario instrumentorum paratu defectis, tam beatis esse non licuit, ut eum observaremus, et vel minime armatis oculis adspiceremus. Cl[arissimus] Siverus, industrius caeli inspector, Te monitore semel atque iterum caelo sereno editissimam turrim aedi Catherinae impositam, ut tanto melius illustrissimos totius horizontis hospites, et lumen vibrantes ignes cum Luna minores, et si quis forte novus, inprimis caudatus vel crinitus, inter eos compareret, intueri posset, ascenderat. Sed nil prorsus vidit: praeter exiguam quandam stellam situ Cassiopeam et Perseum inter medio. Quid illud phaenomeni fuerit, nec ipse dicere habet. Cl[arissimus] Fogelius d[ie] 2. Martii versus mediam noctem insolens quoddam sydus in coelo conspexit, quod Cometen fuisse putat.

1 praeclaris **BO Olhoff** : claris **BnF**.

Cuius responsum tandem obtinui. Miserat quidem // illud proxime elapsa die Veneris. Sed tunc mihi non licuit epistolam hanc finire, et ei illud includere. Descriptionem generalem Cometae novi abs Te factam communicavi Amicis hic et alibi degentibus: a quibus si quid observationum accepero vicissim, Tibi mittam. Interim quae tu observata de isto phaenomeno ad calculum revocasti, a Te exspectamus. Portenta bellorum et aliorum malorum, quae jam in terra et mari conspiciuntur, ut postremum Cometam anni 1665. serius sunt secuta, ita hunc magna ex parte antecesserunt, ut ita ex apparitione eiusmodi syderum crinitorum vel caudatorum nil necessitatis malis, quae in orbe, sive illa in caelo appareant, sive minus, geruntur, adscribi possit. Qua aequitate erga illustres eiusmodi caeli hospites utitur et Serenissimus Elector Palatinus in nuperis litteris clementissime ad me datis: quo, utpote Principum Europae, quod sciam, Doctissimo, suffragatore merito gloriamur. Quod reliquum est, Vir Humanissime, facio te certiorem, me a sesquianno in arcto constitutum et tantum non ictibus adversantis fortunae prostratum, eximia Regis Christianissimi, Incomparabili hac aetate Musarum Patroni, munificentia erectum, et designata annua CCCC. thalerorum Imperialium pensione meliori fortunae ac genio, divina benignitate cuncta mea disponente, redditum. Quod quidem beneficii post Deum eximiae ac singulari Illustrissimi Arnaldi Pomponii Dandillaei, Viri in tantum laudandi, in quantum tanti Nominis probitas et laus assurgere potest, humanitati et benevolentiae, gratus debeo. Mitto Tibi carmina quae Opt[imo] Maximo Regum per Illustriss[imum] Pomponium misi: ut Te, qui pariter insigni Regis illius beneficentia gaudes, faciam gaudii et cultus participem. Amicorum nonnemo pretium *Selenographiae* Tuae, quam sibi comparare vult, indicari sibi per me postulat.[2] Bene Vale et feliciter age, illustrissimum caeli litterarii decus. Hamburgi d[ie] 5. Aprilis Gregoriani Anno 1672.

Tuus
omni officio et studio

S[tanislaus] L[ubieniecius] Rolitsius
manu propria

Post Scriptum. Sal[utem] pl[urimam] velim ex me dicas Cl[arissimis] Viris Titio et Büthnero: quibus poteris copiam facere horum carminum, quae fecit Poeta nec natus nec factus, sed Poetam faciebat amor.[3]

Most noble, worthy, and renowned Sir!

Not only have I found your letter — a message about a new visitor going through the sky[4] — very pleasant, but I have also shared it with our friends who live in this

2 Sed redeo ad Cometam...postulat **BO BnF** : et c[etera] **Olhoff**.
3 Post Scriptum...amor *om.* **Olhoff**.
4 See note to Letter 83.

city and who are curious about such things. And they are very grateful, too, for the information you sent through me about that comet; otherwise, we would have overlooked it, partly because we lack the instruments to watch celestial stars and phenomena, and partly because whereas you are entirely preoccupied with the sky, we are leading our lives uninterested in the heavens, being tied to the ground, and earning our living. And you even protested against widespread misconceptions about new comets. The Republic of Letters is undoubtedly indebted to you, as you are always watching the sky and measuring the stars; thus, nothing in the heavens could escape your attention; otherwise, some phenomena would pass us by in silence (as we are too much down to earth) and they would be unknown for the posterity, had it not been for your singular diligence which showed them to us and recommended for the memory of future generations. And, therefore, we are impatiently awaiting the great *Machina coelestis*[5] of yours, in which (as the new Atlas) you will support both the sky and the minds interested in the sky, and you will successfully combine the earth with the sky, putting the old stars in their proper places and bringing forward the new and so far unknown ones. There is no doubt that we will have a great profit both from this toilsome work of yours and from your other mathematical books. But I return to the comet. I have nothing more about it from anyone. Therefore, I was right to say that we would have no information about it without you. Moreover, we were informed late about its appearance, and we lacked the necessary equipment, so we were not lucky enough to watch it, or even to notice it with the naked eye. The most renowned Sivers,[6] who watches the sky diligently, climbs up the highest tower of the Saint Catherine's church[7] (according to your recommendation) every time when the sky is clear, in order to look for the brightest visitors over the horizon and some smaller light-emitting fires which emanate less light — as well as the Moon — if there occurs something new among them, especially with a tail or hair. But he saw nothing except for a little star between Cassiopeia and Perseus. He cannot say what that phenomenon was. The most renowned Fogel[8] saw an extraordinary star in the sky around midnight on 2 March; he thinks it was a comet. I have finally received an answer from him. He sent it last Friday. But then I could not finish this letter and enclose it.[9] I shared a general description of the new comet, written by you, with our friends in various places; if I receive any observations in return, I will send them to you. Meanwhile, we are waiting for your calculations based on the observations of that phenomenon. Wars and other misfortunes on land and sea followed the last comet of 1665 and, mostly, they happened before this one; therefore, there is no evident influence caused by the appearance of such stars with hair or tail on the misfortunes which occur in the world, regardless of the fact whether the comets

5 Johannes Hevelius, *Machinae coelestis pars prior* (Gedani: Auctoris typis et sumptibus, 1673).

6 See note to Letter 20.

7 The tower of the Saint Catherine's church is about 115 m high.

8 See note to Letter 83.

9 Fogel's letter to Hevelius, dated 22 March / 1 April 1672 (BO C1, v. 11, 1521/1471/48) was only passed by Lubieniecki to Hevelius. Therefore, it is not considered to be an enclosure.

appear in the sky or not.[10] This is also the approach to such shining visitors in the sky which is preferred by the most serene Elector Palatine[11] in his recent letter to me; we are justly proud to have him as a protector because he is — as far as I know — the most learned of the European rulers. As far as the rest is concerned, kindest Sir, let me inform you that since a year and a half — after having been in an unfortunate situation and defeated not only by the blows of the opposing fate — I am now uplifted by the great munificence of the most Christian king,[12] the incomparable patron of arts in our age, and I am restored to better luck and condition having been granted the annual salary of 400 Reichstalers; the divine grace watches over all of my things. After God, I owe this aid to the great and singular humanity and benevolence of the most illustrious Arnauld de Pomponne d'Andilly[13] — a man who should be as much praised, as only the integrity and praise of such a name can be extolled. I am sending you the poems which I sent for the best and the greatest of kings through the most illustrious de Pomponne, in order to make you a partner in my joy and worship, because you enjoy the great benevolence of the king as well. Some of my friends ask me about the price of your *Selenographia* which they would like to buy. Farewell, and be successful, the most famous ornament of the learned sky! In Hamburg, on 5 April 1672, Gregorian

Yours obligingly,

Stanisław Lubieniecki Rolitsius
by my own hand

Postscript. I would like you to pass many greetings from me to the most renowned Titius[14] and Büthner,[15] and to make for them a copy of those poems which were neither composed by somebody who was born or made a poet, but whom love made a poet.[16]

10 See section 3.3 of the Introduction.

11 See note to Letter 74.

12 See section 1.1.3 of the Introduction.

13 Simon Arnauld de Pomponne d'Andilly (1618–99), French diplomat, ambassador in Sweden and in the Netherlands; from 1672 until 1679, minister of foreign affairs (*secrétaire d'état des affaires étrangères*). See also CHJ2 p. 335, note to Letter 89, Hevelius to Perrault, 25 August 1673.

14 See note to Letter 46.

15 See section 2.4.3 of the Introduction.

16 An allusion to the saying: 'nascuntur poetae, fiunt oratores' ('poets are born, speakers are made'), attributed to Cicero, Horace, or Quintilian.

84A.

Lubieniecki's poems for Louis XIV

BO: C1, v. 11, 1518/46
BO: C1, v. 11, 1519/47
BnF: Lat 10348, v. 11, 62–63

In Solem
Regis Christianissimi
Augustissimum emblema.

Viderat insignem Parisinam lumine Phoebus
Stare urbem, et terra ac ponere jura mari.
Nunc, Neptune, mihi Venetam, nunc, Juppiter, urbem
Obiice Romuleam, quas celebraris, ait.
Sequana si pelago et Tybri praeponitur; urbes
Cernite: Sol hic est, astra videntur ibi.

Soli Galliarum
Ludovico XIV. Opt[imo] Max[imo]
Christianissimo Regi.
Solis Augustissimum emblema gerenti
Sole novo annum MDCLXXII. auspicante.

Tendit ad occiduas semper Sol mobilis oras;
Occidit at nunquam Sol, Ludovice, tuus.
Sol beat adversas et nostras lumine terras;
Sed replet radiis gloria cuncta tuis.
Percurrit duodena anno Sol aureus astra;
Omnia percurris tu tua regna simul.
Sol meat huc illuc vagus et vaga lumina fundens;
Omnia tu fixo lumine solus agis.
Maximus ante alios luces hoc orbe Planetas,
Atque minor solo tu potes esse Deo.
Cedite lucifugae volucres, et cornua Lunae,
Phoebus adest rutilus, lilia Phoebus amans.
Cedite tristitiae noctes, quas protulit Arctos,
Sol novus hic noctes non sinit esse graves.
Omine felici Sol laetos explicat annos
Sol magis his laetos, tu Ludovice, dabis.
Nescius occasus jubar immortale refundes,

SOL SOLus talis Rex Ludovicus erit.

gratulabundus, supplex
lusit
Stanislaus Lubienietzki

The most noble emblem
On the Sun
of the most Christian king.

Phoebus saw that the great city of Paris stands in the light and gives its laws on the land and at the sea. He said 'Now, Neptune and Jupiter, leave the cities of Venice and Rome which you hold dear! If the Seine is above the sea and the Tiber, behold, cities! This is the Sun, the stars will be seen there!'

For the Sun of France,
Louis XIV, the best and the greatest,
the most Christian king,
who bears the Sun as the most noble emblem,
with the beginning of the year 1672 with the new Sun

The mobile Sun is always tending westwards, but your Sun, Louis, never sets. The Sun gladdens with its light both foreign lands and ours, but it is your glory which fills everything with your rays. The golden Sun runs through the twelve constellations within a year; you and your rule run through everything. The vague Sun walks everywhere, spilling its vague light; you always act with your steady light. Your shine is greater than that of all the other planets in the world, and you are perhaps only smaller than God. Go away, light-avoiding birds and horns of the Moon! There is a red Phoebus who loves lilies. Go away, the nights of sadness brought by Ursa! The new Sun does not allow sad nights. The Sun concludes happy years with a good omen; you, Louis, as the Sun, will give us even happier ones. You will spill the immortal light which never sets; only Louis will be the Sun-King.

composed
with the best wishes and humility
by Stanisław Lubieniecki

85.

11 June 1672, Hevelius to Lubieniecki

BO: C1, v. 11, 1520/48
BnF: Lat 10348, v. 11, 63

Illustri Viro
Domino Rolisio

Nunc quoque, hac occasione data, pleniorem aliquam recentioris Cometae descriptionem, cum accurata eiusdem delineatione Tibi transmitto; quam nuper typis exscribere curavi, quo eo promptius amicis nostris communibus offerri ac communicari possit. Quem in finem aliquot exemplaria perferri curavi, ut ea[1] amicis meo nomine distribuantur[2] imprimis vero Excell[entissimo] Domino Kirstenio, R[everendo] Hackio, R[everendo] P[atri] Curtio; residua[3] vel Tibi reserves vel quibus videbitur ea offeras. Gratulor ex animo Te in numerum eorum susceptum esse quos Rex Christianissimus Sua Gratia ac Protectione dignos habet, foveoque ut[4] Deus O[ptimus] M[aximus] Te quam diutissime conservet, quo possis adhuc plura praeclarissimi Tui Ingenii, Literarum monumenta in publicum proferre. Dabam Gedani, Anno 1672, die 11 Junii st[ylo] n[ovo].

Tuae Illustrissimae Dominationis
observantissimus et amantissimus

J[ohannes] Hevelius
manu propria

To the illustrious Sir
Master Rolitsius

Finally, having the opportunity, I am sending you a more detailed description of the latest comet,[5] along with the illustration; I had it printed recently, so it can be easily given away and shared with our mutual friends. To that end, I took care to send a few copies to you to be given to friends on my behalf — first of all, to the most excellent

1 pro lubitu *del. post* ea **BO**.
2 possis *del. post* distribuantur **BO**.
3 vero *del. post* residua **BO**.
4 Te *del. post* ut **BO**.
5 Johannes Hevelius, *Epistola de cometa 1672 anno MDCLXXII Gedani observato* (Gedani: authoris typis et sumptibus imprimebat Simon Reiniger, 1672); see note to Letter 83.

Master Kirsten,[6] reverend Hacki,[7] and the Reverend Father Curtz;[8] keep the rest for yourself or give them to anyone you want. I sincerely congratulate you on becoming one of those whom the most Christian king found worthy of his grace and protection;[9] and I pray that almighty God will keep you long in good health, so that you can publish more learned monuments of your famous mind. Given at Danzig, on 11 June 1672, new style.

Illustrious Sir,

Yours lovingly devoted,

Johannes Hevelius
by my own hand

6 See section 2.4.9 of the Introduction.
7 Probably Michał Antoni Hacki, see note to Letter 83.
8 See section 2.4.5 of the Introduction.
9 See section 1.1.3 of the Introduction.

86.

12/22 July 1672, Lubieniecki to Hevelius

BO: C1, v. 11, 1537/67
BnF: Lat 10348, v. 11, 93–94

Nobilissime, Amplissime, Clarissime Vir.

Recte accepi ante quatriduum litteras Tuas die 11. Junii datas una cum 7. exemplaribus epistolae Tuae, quam de nuperrimo Cometa ad Cl[arissimum] Oldenburgium scriptam publici juris fecisti. Ego quidem de meo exemplari, quod pro solita tua, quae profecto eximia est, humanitate ac benevolentia, mihi munus misisti, gratias Tibi ingentes ago. Idem faciunt Amici, quibus ea Tuo nomine tradere licuit. Reverendi Viri Alb[ertus] Curtius (de quo a multo tempore nil certi accepi) et Hatzkius noster, qui Romam a vobis redux factus concessit brevi ad nos rediturus accipient illa quae ipsis destinasti, quamprimum id fieri poterit. Plurimum Tibi gratulor, Vir Eximie, de hoc quoque labore feliciter per gratiam divinam perfecto: sed et Inclutae Urbi Verstrae ac universae Reip[ublicae] litterariae, quarum utramque praeclarissimis ingenii Tui monumentis illustras, cumprimis autem rei Mathematicae studiosos juvas. Et ita quidem egregium hic est Tuum studium, ut vix unus atque alter Cometam istum observarit. Quos inter est Reverendus et Doctissimus Picardus, qui huius Cometae observationes in excitato pro tempore super Uraniburgi Tychoniani ruderibus observatorio factas, // nec non alia ibidem, ubi per aliquot menses studiis caelestibus deditus Vir ille Praeclarus ac studiosus caeli metator degit, adnotata rei Astronomicae inservitura, Tychonianis operibus una Lutetiae Parisiorum cum adiuncta Uraniburgi eius percelebris Typographica historia et tabula, edendis, quae hic inter manus eius vidi cuncta, addet. Ut ita eximium quidpiam illinc nobis sit exspectandum. Honorificam ille Tui mentionem saepius apud me fecit, Te plur[ima] salute impertire, et se Tibi scripturus brevi Te certiorem reddere jussit. Vir certe est perhumanus, qui cum egregiae eruditioni morum probitatem coniunxerit, nil vulgare ab eo exspectare debemus. Sistet ille aliorum opera adiutus orbi litterario redivivum ex cineribus Tychonem, ut sit Regis nostri Christianissimi munificentia velut ales illa

Quae super Eoos extincta renascitur ignes.

Meam qualemcunque industriam in propaganda re litteraria sors adversa adhuc deprimit, imo aliorsum ad victum cum familia quaerendum, ad labores, inquam praegraves convertit. Spero tamen eam brevi, juvante Deo, magis propitiam. Cl[arissimis]

Viris Titio et Büthnero plur[imam] sal[utem] adscribo. Salve et Tu ac Vale diu feliciterque Vir Maxime. Hamburgi 12/22 Julii 1672.

Tui Clarissimi Nominis
observantissimus

Stanislaus Lubienietzki Rolitsius
manu propria

Most noble, worthy, and renowned Sir!

Four days ago, I successfully received your letter sent on 11 April, together with seven copies of your epistle written to the most famous Oldenburg which you decided to make public.[1] I am very grateful for the copy you sent as a gift for me out of your usual great kindness and benevolence. And so are the friends to whom I could give copies on your behalf. The Reverend Fathers Albert Curtz[2] (about whom I have not received anything certain) and our Hacki[3] (who after his return to Rome from your land agreed to return to us soon) will receive what you destined for them as soon as possible. I congratulate you very much, exceptional Sir, on this work which you finished successfully thanks to God's grace; I commend also the famous city of yours and the whole Republic of Letters because you honour them with the renowned monuments of your talent; and you especially help those who study mathematics. Thanks to your great effort, some people could see that comet. Among them, there is the reverend and learned Picard who added his observations of the comet (conducted in a temporarily constructed observatory upon the ruins of Tycho's Uraniburg) to Tycho's works; he also added some others which he took in that place (where that renowned and learned man, devoted to celestial matters, a surveyor of the heavens, lived for a few months) for the sake of astronomers. Tycho's works are going to be published in Paris together with the typographical history and a map of the famous Uraniburg; I have seen all of it in his hands.[4] We may expect something extraordinary from it. He always mentioned you to me with respect; he told me to pass many greetings to you and to assure you that he will soon write to you himself. He is certainly a very kind person who combines his great learning with uprighteous conduct, so we can expect nothing crude from him. With the help from others, for the sake of

1 Johannes Hevelius, *Epistola de cometa 1672 anno MDCLXXII Gedani observato* (Gedani: authoris typis et sumptibus imprimebat Simon Reiniger, 1672). The book has a form of a letter to Oldenburg (see section 2.4.12 of the Introduction). On the comet see note to Letter 83.

2 See section 2.4.5 of the Introduction.

3 Probably Michał Antoni Hacki, see note to Letter 83.

4 Finally published as Jean Picard (see note to Letter 83A), *Voyage d'Uranibourg* (Paris: de L'Imprimerie Royale, 1680).

the learned world, he raises the revived Tycho from the ashes so that he may become — thanks to the generosity of the most Christian King — like that bird

> which, when dead, revives in the flames of the east.[5]

So far, the opposing fate hindered my moderate effort to promote sciences, and it even made me turn elsewhere to, let me say so, difficult works, in order to earn a living for my family. But I hope that fate will soon become more benevolent with God's help. I pass many greetings for the most renowned Titius[6] and Büthner.[7] Farewell, greatest Sir, and stay in good health for long! In Hamburg, on 12/22 July 1672.

Renowned Sir,

Yours obliged,

Stanisław Lubieniecki Rolitsius
by my own hand

5 John Barclay, *Satyricon* (Augustae Trebocorum: apud Joannem Hessium, 1623), p. 262.
6 See note to Letter 46.
7 See section 2.4.3 of the Introduction.

87.

24 January 1673, Lubieniecki to Hevelius

BO: C1, v. 11, 1572/86
BnF: Lat 10348, v. 11, 125–26

Nobilissime, Amplissime, Clarissime Vir.

Inter strepitum armorum dum silent cuncta jura et studia, quorum cultores quietis amantes furor belli abripuit multos, plures metu perculit, vel ad incitas redegit, silent etiam litterarum commercia. Nam de re litteraria nil prorsus ad me perscribunt amici, etiam illi qui me talibus delectari norunt. Vestram quoque Prussiam, ut cum dolore accepi, discordantis Reip[ublicae] mala lacessunt: quibus spero vel ipsam duram necessitatem, Turcici novi belli procella patriae ingruente, finem impondituram. Sed enim pietas in patriam id potius efficere deberet, etsi nullo traballino necessitatis clavo permota, vel ullo minantis periculi metu sollicitata. Te toto pectore ad perficiendam Tuam Historiam caelestem, plusquam Augustam, incumbere ex Cl[arissimo] Fogelio nuper accepi. Quo nomine ut publico gratulor, ita et Tibi, Vir Maxime, cui novum hunc annum et plures alios subsequentes ac vitae longaevae indiscretum comitem prosperam et constantem vale-//tudinem, Herculeis laboribus consecrandam, ex animo precor voveoque. Cl[arissimis] Titio et Buthnero cum officiosa salute idem votum perpetuae amicitiae tesseram per Te mitto: et peto *Theatri* mei *Cometici* exemplaria duo tradi praesentium exhibitori D[omino] Casparo Richtero, amico meo. Vale Vir Incomparabilis ac me constati favore prosequere, qui sum aeternum

Clarissimi Nominis Tui
cultor studiosissimus

S[tanislaus] L[ubieniecius] Rolitsius
manu propria

Hamburgi 24. Jan[uarii] 1673.

Most noble, worthy, and renowned Sir!

Among the noise of arms, when all the laws and studies are silent[1] — as many of their worshippers, who loved peace, were taken away by the madness of war, and much

1 An allusion to a famous quotation from Cicero, *Pro Milone*, 4: 'silent enim leges inter arma' ('the laws are silent among the arms').

more were stricken with fear or driven to poverty — epistolary contacts are silent too. My friends do not write to me about scholarly things anymore, even those who know I like such matters very much. Your Prussia, as I learned with grief, is also harassed by the misfortunes of our quarrelling country; I hope that the necessity — when the tempest of the new war with the Turks approaches the homeland — will end them.[2] But the love of the homeland itself should have caused it, moved by no urgent necessity or forced by no fear of threatening danger. I have recently learned from the most renowned Fogel[3] that you work with all your heart on finishing your celestial history,[4] a very noble book. I congratulate everyone and you, greatest Sir, on publishing it; I pray for you, and sincerely wish you — for this new year and for many more to come — much health, the sure companion of long life which you can devote to your Herculean works. For the most renowned Titius[5] and Büthner,[6] I send through you the same wishes with the kindest greetings as a sign of our enduring friendship; and I ask you to give two copies of my *Theatrum cometicum* to the one who delivered you this letter — to Master Caspar Richter,[7] my friend. Farewell, incomparable Sir, and always be benevolent towards me, because I am forever

renowned Sir,

Your obliged admirer,

Stanisław Lubieniecki Rolitsius
by my own hand

In Hamburg, on 24 January 1673.

2 Lubieniecki refers here to the conflict between supporters and opponents of King Michał I which led to a series of political unrests. See also CJH3 p. 426, note to Letter 154, des Noyers to Hevelius, 19 February 1672. The Polish-Turkish war took place in 1672–76.

3 See note to Letter 83.

4 Either *Machina coelestis* (*Machinae coelestis pars prior* (Gedani: Auctoris typis et sumptibus, 1673); *Machinae coelestis pars posterior* (Gedani: in Aedibus Autoris, 1679)) or Hevelius's star catalogue, finally published in *Prodromus astronomiae* (Gedani: typis Johannis Zachariae Stolli, 1690).

5 See note to Letter 46.

6 See section 2.4.3 of the Introduction.

7 Perhaps Caspar Richer (1641–94), royal postmaster in Danzig.

88.

24 April 1673, Lubieniecki to Hevelius

BO: C1, v. 11, 1582/91
BnF: Lat 10348, v. 11, 128–29

Nobilissime, Amplissime, Clarissime Vir.

Non equidem peccarem in Tua tempora caelestibus dicata curis inutili scriptione, nec ut anser olori obstreperem, praesertim cum nullum Te dignum scribendi argumentum habeam. Sed salva officiosae benevolentiae ratione fieri nequit quin has ad Te dem perferendas per fidissimam coniugem meam, quae cum bono Deo mare ad vos volare et ibi sororem meam belli Turcici procella ex Russia pulsam invisere, qua etiam levidense munusculum, Tibi, ut spero, gratum futurum, utpote manu eius non sine jucundo artificio elaboratum, ac typographiae stramineae nondum visum specimen exhibens, Tibi me jubente offeret. Quod ut boni et aequi consulas, et in memoriam nostri serves, etiam atque etiam rogo et obsecro. Conatus mei promovendo bono publico dicati iniquitate huius temporis, bonis artibus perquam adversi, sufflaminati. Spondeo tamen mihi amiciora Musis tempora. Qua si divina benignitate affulserint, habebo non pauca // de argumento nostro Cometico, etiam quod ad figuras attinet, Mathematices studiosis non ingrata nec inutilia futura. Caelestissimi ingenii tui divinum opus perfectum esse jam credo. Quo nomine Tibi et publico postquam id accepero, jure meritoque gratulabor. Cl[arissimis] Viris Titio et Buthnero officiosam salutem adscribo, scribendi argumento defectus nullas novas litteras ad eos do, vicem responsi ab eis exspectans. Poteris pro Tua insigni humanitate et devoto erga Regem nostrum Christianissimum studio, hosce versiculos illis, utpote insignibus Poetis, sed et aliis Amicis, compositos a Vestrate Sandio, communicare. Si qua porro in re mea opera Tibi usui esse possit, utere ea mei candoris et promtissimae voluntatis certus; utere et jure tuo, quod tibi apud me et omnes bonarum litterarum cultores praestantissima Tua merita compararunt. Vale Vir Magne diu ac feliciter age bono publico: ac me constanti favore prosequere. Dabam Hamburgi, XXIV. Aprilis MDCLXXIII.

Clarissimi Nominis Tui
Cultor perpetuus

S[tanilaus] Lubienietzius Rolitzius

Most noble, worthy, and most renowned Sir!

I would not sin against your time (devoted to celestial matters) with my useless writing, and I would not interrupt you as a goose does to a swan,[1] especially as I had nothing to write which was worth your attention. However, obeying the principles of kind benevolence, I could not miss an opportunity to send this letter to you through my faithful wife who travels to you with God, by sea, to visit my sister,[2] driven away from Ruthenia by the storm of the war with the Turks.[3] I asked my wife to deliver to you a little gift which, I hope, will be pleasant for you, because it is an example of meticulous craftsmanship and elegant typography, never seen before.[4] I heartily ask you to use it well and to keep us in your memory. My efforts to promote the public good have been ceased by the injustice of this age which opposes good arts. But I promise to myself that there will be times which are more friendly towards the Muses. When these times beam with God's grace, I will have something which concerns our discussion about comets, even something related to their images, which will be pleasant and useful for those who study mathematics. I believe the work of your celestial mind[5] has finally been completed. I will congratulate you and the public on it when I receive it. I send kind greetings for the most renowned Titius[6] and Büthner;[7] as I have nothing to write about, I do not send them new letters and I do not expect any answers. Out of your great kindness and devotion towards the most Christian king, you can send those poems, written by your Sandius,[8] to prominent poets and other friends.[9] If I can provide you with any help, use it, assured about my honesty and goodwill; use it as you wish, because your outstanding merits deserve it from me and all the worshippers of good arts. Farewell, great Sir, and be successful for the public good, and always be benevolent towards me! Given at Hamburg, on 24 April 1673.

Renowned Sir,

Your admirer,

Stanisław Lubieniecki Rolitsius

1 A reference to a passage from Giovanni del Virgilio's (*fl.* 1st half of the 14th century) poem for Dante Alighieri (*c.* 1265–1321): 'in temerity the goose is cackling at the clear-toned swan', translation in Philip H. Wicksteed, Edmund G. Gardner, *Dante and Giovanni del Virgilio* (Westminster: Archibald Constable & Company, Ltd., 1902), p. 151.

2 Probably Zofia Przypkowska.

3 See note to Letter 87.

4 There are no more details about this gift.

5 Probably an allusion to Hevelius's *Machina coelestis* (*Machinae coelestis pars prior* (Gedani: Auctoris typis et sumptibus, 1673); *Machinae coelestis pars posterior* (Gedani: in Aedibus Autoris, 1679)).

6 See note to Letter 46.

7 See section 2.4.3 of the Introduction.

8 Christopher Sandius (1644–80), German Arian writer from Königsberg. He worked in the Netherlands as a publisher of Arian and Socinian texts. His best-known book is *Bibliotheca antitrinitariorum* (Freistadii: apud Johannem Aconium, 1684), a catalogue of Arian and Socinian writers, published posthumously.

9 I could not find the enclosure to this letter.

89.

2 May 1673, Lubieniecki to Hevelius

BO: C1, v. 11, 92
BnF: Lat 10348, v. 11, 129

Vir Nobilissime, Amplissime, Clarissime.

Spero dilectissimam meam uxorem favente per triduum vento portum vestrum jam attigisse. Per hanc scripsi ad Te ut officio defungerer. Nunc paucis Te compello in negotio Cl[arissimi] Domini Fogelii communis Amici. Is petit officiose mitti sibi per meam Herbarium Sireni. Tu nosti Bibliopolam, qui pro eo XX. thal[eros] postulat. Speramus aliquid pretio detractum iri. Quod jubebis uxor mea persolvet. Haec tibi nuncianda Amici communis gratia volui, debui. Vale diu et feliciter bono publico. Hamburgi d[ie] 2. Maii 1673.

Tui Nominis Clarissimi
cultor perpetuus

S[tanislaus] L[ubieniecius] Rolitsius
manu propria

Most noble, worthy, and renowned Sir!

I hope that my beloved wife has already reached your port, thanks to three days of favourable wind. I wrote to you using her as a messenger in order to fulfil my duty. Now, I am briefly inquiring on behalf of the most renowned Fogel,[1] our mutual friend. He asks you kindly to send him a copy of Syreński's herbarium[2] through my wife. You know a bookseller who wants 20 thalers for it. We hope that the price will be reduced a little. My wife will pay you whatever you want. I wanted to communicate these to you on behalf of the mutual friend, as I should. Farewell, and stay in good health for the public good! In Hamburg, on 2 May 1673.

Renowned Sir,

Your admirer,

Stanisław Lubieniecki Rolitsius
by my own hand

1 See note to Letter 83.

2 Szymon Syreński (Sirenius, Syreniusz, *c.* 1540–1611), Polish physician and botanist, professor of the Jagiellonian University. Lubieniecki refers to his *Zielnik* (Cracoviae: w drukarni Bazylego Skalskiego, 1613).

90.

29 August 1673, Hevelius to Lubieniecki

BO: C1, v. 11, 1596/111
BnF: Lat 10348, v. 11, 158–59

Domino
Rolisio
Hamburgi

Illustrissime Vir

Benevolum Tuum erga me affectum, quem saepius abunde es contestatus, demereri quidem omnimodo hac vice non possum; nihilominus tamen, ut quandantenus reciprocum meum amorem, gratissimumque animum Tibi rursus declarem, en ecce Tibi *Machinam* meam *Coelestem* nunc primum editam: quam in optimam Auctoris recordationem benevole accipere, meque porro amare non dedigneris, rogo. In primis rem facies mihi gratissimam si Tuum aliorumque clarissimorum Virorum judicium de hocce quali quali opusculo, nostrisque inventiunculis suo tempore libere ac sincere exponas. Rev[erendo] P[atri] Alberto Curtio lubenter etiam exemplar quoddam transmitterem, dummodo scirem, an adhuc in vivis esset, ubi degat, et qua via istud perferri, ut[1] hocce turbulentissimo rerum statu, Ei exhiberi recte[2] possit. Quod superest DEUM O[ptimum] M[aximum] oro, ut Te Tuosque salvos quam diutissime tueatur. Dabam Gedani Anno 1673, die 29 Augusti.

Tuae Illustrissimae Dominationis[3]
Omni officioso
cultu
paratissimus

J[ohannes] Hevelius
manu propria

1 ut : ut recte **BnF**.
2 recte *om.* **BnF**.
3 Tuae Illustrissimae Dominationis *om.* **BnF**.

To Master
Rolitsius
in Hamburg

Most illustrious Sir!

This time I cannot return your benevolent attitude towards me which you have often testified; nevertheless, to show my mutual affection and thankfulness for you, I am sending you my *Machina coelestis*[4] which has just been published. Please, accept it in order to keep the author in good memory and to be always benevolent towards me. First of all, you will do a very pleasant thing for me, if — in due time — you tell me freely and sincerely the opinion of yours and of other renowned men about that little book and my small inventions. I would willingly send a copy to the Reverend Father Albert Curtz,[5] if I only knew if he is still alive, where he lives and how I can send it so that it can be correctly delivered to him in such turbulent times. As far as the rest is concerned, I pray to the almighty God to keep you and your family in good health for very long. Given at Danzig, on 29 August 1673.

Illustrious Sir,

Your admirer to serve,

Johannes Hevelius
by my own hand

4 Johannes Hevelius, *Machinae coelestis pars prior* (Gedani: Auctoris typis et sumptibus, 1673).
5 See section 2.4.5 of the Introduction.

91.

8 September 1673, Lubieniecki to Hevelius

BO: C1, v. 11, 1610/128
BnF: Lat 10348, v. 11, 182

Magnifice, Amplissime, Clarissime
Domine Amice Colendissime

Postquam accepi hasce inclusas a Cl[arissimo] Vogelio, et certior ab eo sum redditus, Te Proconsulem Vestrae Neapoleos esse creatum, non potui quin Te paucis hisce compellarem, et tam novam hancce eamque amplissimam in Urbe patria dignitatem Tibi gratularer, quam et auctionem omnium dignationum, et prosperam in plures annos, insigni non tantum Inclutae Vestrae, sed et universae litterariae Reip[ublicae] ornamento et emolumento,[1] valetudinem tibi precarer. Officii enim ratio id a me exigit, ut et gratias Tuae eximiae humanitati, quam Uxori et filiolae testatus es, debitas agam.[2] Filiola, de qua mihi gratularis // strenue per Dei gratiam in linguis exoticis ediscendis pergit. Puto enim id ei solatio et praesidio fore tam calamitosis temporibus, dum nullus est bellorum finis. Quae utinam cessent, et eis bonae artes succedant! Sal[uta] pl[urime] amicos et Vale diu feliciterque, ac fave mihi constanter. Hamb[urgi] 8. Sept[embris] 1673.

Tuae Magnificentiae[3]
Omni cultu, studio, officio

S[tanislaus] L[ubieniecius] R[olitsius]
manu propria

Most magnificent, worthy, and renowned Sir, dear friend!

Having received the enclosed letter from the most renowned Fogel,[4] and after I had learned from him that you were made the mayor of that Naples of yours,[5] I had to

1 et emolumento *om.* **BnF**.

2 agam **BnF** *om.* **BO**.

3 Tuae Magnificentiae *om.* **BnF**.

4 Fogel's (see note to Letter 83) letter to Hevelius, dated 28 August / 7 September 1673 (BO C1, v. 11, 1613/131) was only passed by Lubieniecki to Hevelius. Therefore, it is not considered to be an enclosure.

5 Information that in 1673 Hevelius was elected the mayor of Danzig or entrusted with another high office in his city is not mentioned in his biographies. On Hevelius's public offices see Chantal Grell, 'Hevelius en son temps', CJH1 p. 61.

write to you briefly to congratulate you on that new and great dignity in your home city, as well as to wish you even greater dignities and good health, as you are the ornament and asset not only of your famous city but of the whole Republic of Letters as well. The sense of duty requires deserved thanks from me for your exceptional kindness which you showed towards my wife and our little daughter. Our daughter, on whom you congratulate me, keeps learning exotic languages, thanks to God's grace.[6] I think it will be comforting and helpful for her, in times full of misfortunes, when there is no end to wars. May they end, and may they be followed by good arts! Greet all friends, be in good health and prosperous, and always benevolent towards me! In Hamburg, on 8 September 1673.

Magnificent Sir,

Your most devoted admirer,

Stanisław Lubieniecki Rolitsius
by my own hand

6 Katarzyna Salomea Lubieniecka (*c.* 1666–75). Nothing is known about her education.

92.

22 December 1673, Lubieniecki to Hevelius

BO: C1, v. 11, 1623/141
BnF: Lat 10348, v. 11, 198–200
Olhoff: 137–38

Nobilissime, Amplissime, Clarissime Vir.

Accepi recte hisce diebus *Machinam* Tuam *caelestem*, quam mihi misisti munus acceptissimum, non tantum quod tanto magis grata et cara sint munera, auctor quae pretiosa facit, sed quod opus hoc in seipso sit percarum et pretiosum. In quo mirerne prius organorum Tuorum splendorem et accuratissimum examen nec pili transversi discrimen patiens: an vero potius opus nitidissime impressum, nativa suavitate oculos et mentem Lectoris capiens, non facile dixero. Nam verba quidem a divino ingenio, quod caeli Metatori Solertissimo conveniat, profecta sunt: sed acumen et certum eorum a demonstratione, quae organis perficitur, dependet. Ita cum alioqui ars sit naturae aemula, hic magistra et dux magistrae evasit. Non sum falsus, Magne Heveli, quod in meo *Theatro Cometico* pronuntiavi, Te alterum Hipparchum hac aetate velut a caelo missum esse. Haec caelestium machinarum et siderum consummata descriptio, correctus stellarum, etiam fixarum, situs, aucta insuper illa caelestis lucida familia, Virum Te caelestem ac verum caeli incolam esse, evincunt. Sed desino elogiorum, quantumvis meritis tuis debitorum, et parco encomiis, ut Tuae verecundiae parcam. Dixisse satis est: habere et nostram aetatem, quod per Te priscorum opponat solertiae. Pauca me, fateor, intra triduum hocce, inter spissas occupationes, quae mentem ad sublimiora se erigere conantem deprimunt // ad terram, ex Tuo isto praeclaro opere legisse. Visne ut omnia uno verbo dicam? Cuncta mire placuerunt, et si forte de nonnullis a Te docendus sim, utar in Te libertate, quam jubendo dedisti, et quam cunctis Reip[ublicae] litterariae incolis Alma tribuit Philosophia. Vita profecto non in ore et cordibus hominum, quae te certissime manet, sed in oculis eorum perpetua dignissimus fores, qui cunctis propemodum ad alias minus nobiliores (quid enim caeli scientia nobilius, imo quid aeque nobile?) saepius vanas, imo et noxias occupationes dilabentibus, et res in terra bene vel male gestas bene vel male describentibus, solus aut certe cum paucis res in caelo gestas in commentarios referas et seculorum aeternitati commendes. At ego apud Te, Tua tamen bona cum venia, S[enatum] P[opulum]q[ue] Reip[ublicae] litterariae meritissimo compello ad exemplum nescio cuius Romanorum: Si quid Magno Hevelio humanitus evenerit, quem Ei substituetis? Dicetisne ut tunc, 'Te Quinte Catule'? Ubi vero iste est? Decet autem et Te, Cl[arissime] Heveli, ut cum legibus mortalitatis aeque cum ignarissimis caeli subiectus vivas, quam optime caelestem familiam ordines, et supremarum tabularum // fidos custodes et executores, ac idoneum aliquem successorem posteris relinquas. Cum Rever[endo] Curtio nullum mihi jam

intercedit litterarum commercium quod ille et antehac aegre et curto stilo, nomini respondens potius quam amicorum exspectationi, excolebat. Certior tamen de eo per Amicum Coloniae degentem reddar: teque vicissim certiorem de eo reddam. Porro Caesareanis hyberna in Westphalia sumere volentibus, tanto magis omnes aditus obsidebuntur, et omnia civilis societatis commercia impedientur. Cl[arissimus] Fogelius gratias Tibi dubio procul aget hodie, id enim se facturum certo heri promisit.[1] Vale, Vir Excellentissime, caelifero par Atlanti, impertitus plurima salute cum Tuis a me meisque,[2] et age feliciter in novum mox inchoandum annum, et plures alios, quantum quidem mortalium fert conditio. Hamburgi XXII. Dec[embris] MDCLXXIII.

Clarissimi Nominis Tui
et Amplissimae Dignitatis
observantissimus

Lubieniecius Rolitsius
manu propria

Most noble, worthy, and renowned Sir!

I have recently received your *Machina coelestis*[3] which you sent me as a most welcomed gift, not only because gifts are nicer and dearer when the person of the author makes them more precious, but because that book is very dear and precious in itself. I am unable to decide if I admire more the beauty of your instruments and their precision which does not allow a difference of a crooked hair, or the excellently printed book which allures the readers' eyes and minds with its charm. Because the words came from the divine talent, typical of the diligent surveyor of the sky, but their sharpness and proof depend on the demonstration which originates from the instruments. Even though art is an imitator of nature, here it became the teacher and the teacher's leader. I was right, great Hevelius, when I declared in my *Theatrum cometicum* that you are the second Hipparchus of our age,[4] sent from heaven, as it were. That comprehensive description of the celestial machines and stars, the corrected placement of stars (even those fixed ones) and that enlarged bright family of the sky prove that you are a heavenly man and a real inhabitant of the sky. But now I am ending my eulogies, however much your merits deserve them, and I am saving praises in order to protect your modesty. It is enough to say that, thanks to you, our age has something to compare to the talents of the ancients. I admit that I have read only a little from that famous book

1 Cum Rever[endo] Curtio...promisit **BO BnF** *om.* **Olhoff**.
2 impertitus...meisque **BO BnF** *om.* **Olhoff**.
3 Johannes Hevelius, *Machinae coelestis pars prior* (Gedani: Auctoris typis et sumptibus, 1673).
4 I could not find such phrase in the *Theatrum cometicum* (but see Letter 82). Even though references to Hevelius and his works are abundant throughout the book, Lubieniecki usually calls him 'magnus' (great), 'amplissimus' (the greatest) or 'doctissimus' (the most learned).

of yours in the last three days because of many duties which keep the mind, tending upwards, down to Earth. Do you want me to say everything in one word? I liked all of it, and if there were some things which I would like to ask you about, I would use that prerogative which you gave as an order and which calm philosophy granted to all the citizens of the Republic of Letters. You would surely be worthy of eternal life, not only in the mouths and hearts of people (where you certainly are) but in their eyes, because you are the only one, or one of the few, who writes comments about what happens in the sky and who commends it for the eternity; meanwhile, many others often get lost in other less noble businesses (because what can be nobler than the knowledge of the sky, or at least equally noble?) — vain and even pernicious ones, and they describe well or poorly what happened well or poorly on Earth. And now in your presence, if you allow, I am asking the senate and the people of the Republic of Letters as once one of the Romans did: if something typically human happens to the great Hevelius, with whom will you replace him? Will you say like them: 'with you, Quintus Catulus'?[5] But where is he? It would be good, if you, most renowned Hevelius (living by the same laws of mortal men like those who know nothing about the sky) just as you manage the celestial family, left faithful guardians and makers of the best tables and a proper successor for the posterity. I do not have epistolary contacts with the reverend Curtz[6] anymore; even earlier he answered unwillingly and briefly, as would suit his name[7] but not the expectations of his friends. I will learn about him, however, from a friend from Cologne, and then I will inform you. Since the imperial army wants to winter in Westphalia,[8] all the ways will be closed, and every civilian activity will be hindered. The most renowned Fogel[9] will undoubtedly thank you today, as he yesterday promised me to do. Farewell, most excellent Sir, equal to Atlas who holds the heavens! Accept many greetings for you and your family from my family and myself, and be successful in the new year, which will soon begin, and in many more to come, as long as the human condition allows it! In Hamburg, on 22 December 1673.

Most renowned and distinguished Sir,

Yours obliged,

Lubieniecki Rolitsius
by my own hand

5 Quintus Catulus Lepidus (d. *c.* 60 BC), Roman politician and military commander, consul in 78 BC. Lubieniecki alludes to a story from Valerius Maximus's *Factorum et dictorum memorabilium libri novem* (8, 15, 9) where the people of Rome proclaimed Quintus Fabius Catulus a Pompey's equal.

6 See section 2.4.5 of the Introduction.

7 Pun based on the similarity of the Latin version of Curtz's name, Curtius, and the adjective *curtus* (short, cut). Lubieniecki used it in his letter to Curtz, 23 September 1665, TC p. 223.

8 Lubieniecki refers to an episode of the Franco-Dutch war (1672–78); Spain and the Empire took the side of the Dutch.

9 See note to Letter 83.

APPENDICES

Appendix 1.

Lubieniecki's Extract from Hevelius's *Prodromus cometicus*

TC: 388–93

Observationes Dantisco Hevelianae ex *Prodromo* eiusdem *Cometico* in compendium redactae

Observatus est eximius hic Cometa primum a me (Johanne Hevelio) hic Gedani Anno 1664. Die Solis 14. Decembris St[ylo] Gregor[iano] sub horam quintam, in quadrante Orientali ad Euronotum prope rostrum Corvi, vix ab eo semi gradu distans, Favonium versus, altus 9. grad[us] ab Horizonte in 8. gr[adu] Librae, sub Latitudine Australi 22. grad[uum]. Caput erat satis conspicuum, subflavi coloris, luminisque debilioris, et aliquanto obtusioris, ut reliquarum stellarum, planetarumque. In meditullio eius lucidius quoddam lumen trium vel quatuor minutorum in diametro,[1] stipato alio obscuriori ac tenuiori sive capillitio, apparebat. Meditullium corporis animadversum ex uno solo clariori nucleo, sed ex pluribus particulis sive corpusculis, tum densioribus et lucidioribus, tum tenuioribus et obscurioribus, maioribus et minoribus constans, interspersa alia diversae densitatis subtiliori materia, ad instar effigiei sub Numero 1. in Iconismo A. delineatae. Cauda 14. gr[aduum] longa, a capite tenuior, sub extremitate vero magis magisque divaricabatur, decurrens inter stellam, basin Crateris dictam, et eam borealiorem sub Cratere, propius tamen basi. Die 15. Dec[embris] Hora 4. matutina, deprehensus est versari in 7. gr[adu] 20'. Librae sub. 22. gr[adu] 20'. latit[udine] Australis: eratque Meridiana altitudo ad 20. min[uta] imminuta, progressus cursu retrogrado, (hoc est, contra Signorum seriem) Tropicum versus Capricorni spatio 24. horarum per integrum pene gradum. Caput vix quicquam immutatum videbatur, nisi quod nucleus et materia universa corporis aliquanto fere constipatior lucidiorque per tubum apparuit: ut facies sub Numero 2. commonstrat. Cauda vero paulo brevior 11. graduum circiter, specie tamen satis notabili et luminosa, ascendere basin versus crateris visa est. Die 16 et 17. Decemb[ris] totum caelum erat nubilum. Sed

1 diametro *corr. ed.* : diametto [*sic*] **TC**.

Die 18. Decemb[ris] Ex distantiis eius (Cometae) denuo captis, didicimus Cometam hora 4. mane in tertio gradu, 45'. Librae sub Latit[udine] Australi 25. g[radus] 30'. extitisse, atque spatio trium dierum 4. gr[adus] 10'. motu retrogrado progressum esse in suo tramite, nec non altitudinem meridianam 1. gr[adus] 25. min[utorum] imminutam esse, Tropicum Capricorni jam supergressus erat. Caput aliquantillum maius clariusque apparebat, cuius et nuclei materia pariter plus plusque condensari videbatur; prout Numero 3. delineavimus. Cauda 12. gr[aduum] aeque lucida ac die 15. protendebatur versus Cor Hydrae et Procyonem. Diameter totius Capitis aequabat 15. min[uta] quae prima die apparitionis vix 12. minuta expleverat.

Die 19 et 20. Decemb[ris] Nubibus caelum erat tectum. At

Die 21. Decembris hora 4. 15. min[utis] visus est Cometa sub Cratere in Hydra in 28. gr[adu] Virginis et latit[udine] Austr[ali] 30. gr[aduum]. Altitudo eius Meridiana a die 18. decreverat 2. gr[adus] 4'. siquidem tantum 9. graduum 16'. 30". adhaerente refractione reperiebatur. Unde manifestum erat, Cometam motu proprio 6 gr[adus] 31'. confecisse. Capitis diameter tota ad 18. minuta excreverat, videbatur quoque fere clarius. Cauda hac die plane indolens erat: non solum quod ultra 22. gr[adus] in Longum, et in latum 1. gr[adum] 30'. imo ad 2. gr[adus] fere excurrebat, sed quod, si accurate oculos ad eam intenderes, ex parte circa cuspidem incurvesceret, et quidem concavitate deorsum, Horizontem scilicet versus: instar illius Anni 1618. aliorumque. Dirigebatur inter secundam et quartam a corde Hydrae; ita tamen, ut curvitate in secundam a corde magis propenderet.

Die 22. Dec[embris] Nubila obstiterunt.

Die 23. Dec[embris] Haerebat sub Hydra in 22. gr[adu] Virginis et latit[udine] Australi 34 gr[aduum] 30'. eiusque altitudo Merid[ionalis] ad 2. gradus decreverat: quandoquidem non 7. gradus eo sic excedebat. Motus vero proprius a die 21. 7. gr[adus] fere confecerat. Cauda vero insignis 20 gr[aduum] circiter exporrigebatur in directum rursus ad primam a corde Hydrae occasum versus.

Die 24, 25, 26, 27. Ne semel quidem Cometa emicuit. Verum

Die 28. Dec[embris] hora 1. illum in Austro prope Meridianum vix 6. gr[adus] supra Finitorem elevatum detexi, situm in puppi sub clypeo navis in 4. gr[adu] Leonis sub Latit[udine] 49. gr[aduum] 30'. Australi, postquam confecerat in suo tramite immensum itineris spatium 38. graduum. Diameter eius 22'. fere deprehensa est. Caput crispo ac denso cinctum erat capillitio, eiusque nuclei adhuc densiores. Cauda longitudine 7. gr[aduum] fere instar caudae pavonis, velut Iconismus sub Numero 5. figurae A. clare ostendit, ad Septentrionem omnino protendebatur.

Die 29. Decemb[ris] sub vespera hora 9. 30'. Cometa rursus 10. gr[adus] altus eluxit, infra sinistrum pedem anteriorem Canis Maioris, in Longitudine 28. gr[aduum]

40'. Geminorum, et Latitudine 47. gr[aduum] altit[udinis] Merid[ionalis] ex globo (caelum enim nubibus obumbratum erat) colligebatur jam ultra 12. gr[adus] excrevisse, motus in sua orbita spatio 43. horarum 23. gr[adus] et 40'. fere.

Cauda vel potius barba recta fere sursum porrigebatur 1. gr[adum] propemodum. Caput praeterea hirsutum, cuius diameter 24'. in sui meditullio lucidissimum et densissimum nucleum colore fere igneo exhibebat, qui alia diversae densitatis materia, nucleisque minoribus intermixtis circumdabatur, ut habet delineatio sub Numero 6.

Die 30. Dec[embris] circa 10. vespertinam Cometa in collo Leporis deprehensus est perquam lucidus et conspicuus, barbam 1. gr[adum] a circulo verticali paululum recedentem spargens Euronotum versus ad pedem dextrum Orionis. Caput idem, nisi quod nucleus ille rotundus intermedius ex centro quasi dimoveri ad limbum ortivum videretur. Exstitit in 12. gr[adu] 40'. Geminorum in Latit[udine] Australi 40. gr[aduum] 10'. Ascensio Rect[a] 76. gr[aduum] 30'. atque Declin[atio] 17. gr[aduum] 10'. altus in meridie 18. gr[adus] 27'.

Die 31. Dec[embris] Vesperi hora 6. Cometa jam ad 14. gr[adum] supra Horizontem ascenderat, corpore aliquanto imminuto quod ad diametrum, colore vero albicanti, ac satis splendido. Vix Regel, vel sinistro pedi Orionis vel quicquam cedere videbatur. Faciem nonnisi largior nuclei secessus a centro, limbum versus ortivum, mutaverat. Cauda longitudine 1. gr[adus] fere versum medium cinguli Orionis proiiciebatur. Qualem sub Numero 8. adumbratum vides. Haerebat hora 9. vespertina in 2. gr[adu] 40'. Geminorum, et Latit[udinis] Austr[alis] fere 34. gr[aduum]. In motu suo aliquanto remissior factus, siquidem spatio 34. horarum, tantum 11. gr[adus] 10'. transierat. Ascensio itaque recta prodiit 66. gr[aduum] Declin[atio] 12. gr[aduum] 30'. Austr[alis] existente eius altitudine meridiana hoc tempore 23. gr[aduum] 13'. 50".

Anno 1665. Die 1. Januarii hora 8. vespertina Cometa infra stellulam in Eridano, dictam alias tertiam, a primo flexu ad dimidium gradum fere versus Libonotum incedebat. Caput fere idem, ut et materia interioris corporis: nucleus tamen ille maior, solidior, lucidiorque, jam aliquantulum a suo pristina splendore remiserat. Locum terminabat Longit[udo] 24. grad[uum] 20'. Tauri, et Latit[udo] Austr[ali] 27. gr[aduum] 45'. Ascens[io] Rect[a] 58. et Declin[atio] 8. gr[aduum] 20'. Motus itaque diurnus Cometae extitit 9 gr[adus].

Die 2. Januarii totus erat turbidus.

Die 3. Jan[uarii] Sidus illud crinitum emicuit cauda 3. gr[aduum] circiter, sparsa ad mediam fere stellam trium informium ad mandibulam Ceti existentium. Versabatur autem in 14. gr[adu] 26. Tauri et Latit[udine] Austr[ali] 19. gr[aduum]. Peregerat in suo itineris ductu 12. gr[adus] 30'.

Die 4. Januarii caliginosa nox nullas observationes dedit.

Die 5. Januarii occidente Sole hora 4. 30. min[utis] affulgente mandibula Ceti, simul Cometam, quod ad popularem aspectum id temporis ab inferiore parte dictae stellae 30'. remotum occidentem versus conspeximus. Hora 6 et 7. antequam Luna oriretur, Caput, licet ex parte imminutum, perquam tamen erat conspicuum, nec non cauda ad 14. gr[adus] excurrens, praesertim circa eductionem notabili lumine ac densitate, quasi vix hactenus unquam gaudebat. Extendebatur autem ipsa coma quod ad directionem ad 4. in clypeo Orionis. Nuclei vero illius clarioris vix vestigium amplius supererat, ut sub Numero 11. fig[urae] A. delineavimus. Circa horam 9. mandibula Ceti, Cometa et illa in ore Ceti sub linea fere recta conspecta est: eratque Longitudo phaenomeni 8. gr[adus] Tauri Latit[udine] Austr[ali] 13'. Asc[ensio] Rect[a] 40. gr[aduum] 20'. Decl[inatio] 2. gr[aduum] 20'. Boreal[is]. Motus vero proprius a die 3. ad 5. 8. gr[aduum] 20'. et altitudo Meridiana 38. gr[aduum] 8'. 45". observata est.

Die 6. Jan[uarii] ab hora 6. ad 11. vespertinam, Caput Cometae probe adhuc lucidum colore subargenteo 8'. duntaxat apparuit. Nuclei interiores materiaque dissolvi deprehendebantur. Quod ad situm constituit Cometa eo tempore cum mandibula Ceti et illa in ore triangulum propemodum aequilaterum. Cauda 15. gr[aduum] egregie clara decurrens inter stellas in pede et paleari Tauri, ad dextrum scilicet genu, nonnunquam quasi scintillabat fluctuabatque, ac si per intervallum comprimeretur, rursus dilataretur, novosque veluti radios vibraret ac evomeret. Cometa ipse versabatur in long[itudine] 7. gr[aduum] Tauri, et latit[udine] 10. gr[aduum] 36'. Austr[ali] existente. Ascens[io] rect[a] 38. gr[aduum] et Declin[atio] 4. gr[aduum] 4'. Boreal[i]. Peregit itaque uno die 3. gr[adus] in suo tramite, sub latitudine Meridiana 39. gr[aduum] 41'.

Die 7. Jan[uarii] paulo post horam 4. vespertinam Cometam vidimus capite decrescente, nihilo tamen minos abunde claro, coloris argentei et albicantis: quin etiam nucleis in languidiorem quasi massam coeuntibus: sicuti sub Numero 13. delineatus est. Coma insigniter longa 13. gr[aduum] et Lata minimum 2. gr[aduum] et amplius, proiecta est ad stellam, quae sequitur tertiam in armo Tauri. Longitudo autem Cometae detecta est in 5. gr[adu] 30'. Tauri. Latitudo Australis 8. gr[aduum] 30'. Ascens[io] rect[a] 35. gr[aduum] 30'. Declin[atio] Boreal[is] 5. gr[aduum] 23. atque altit[udo] Merid[ionalis] 40. gr[aduum] 59'. 4".

Die 8. Jan[uarii] Nubila caelum totum subtexerant.

Die vero 9. Jan[uarii] hora 6. vespertina caput phoenomeni fere in pristino deprehendimus statu, nec non caudam, nisi quod aliquanto longior 15. gr[aduum] excurreret ad secundam in armo Tauri, occupabat Longit[udinem] 2° 50'. Tauri, Latit[udinem] 5. gr[aduum] 40'. Austr[alem] Asc[ensio] Rect[a] 32 gr[aduum] et Declin[atio] 7 gr[aduum] 23'. intervallo vero 2. dierum 3 gr[adus] 45'. emensus.

Die 10. Jan[uarii] post 5. vespertinam observatus materiam interiorem capitis plus plusque dissolutam, prout sub Numero 15. adumbravimus. Caudam vero et rariorem et breviorem, 12 grad[uum] tantum inter supremam et secundam in armo Tauri decurrentem exhibebat. Longit[udo] eius tum extitit 2. gr[adus] Tauri. Latit[udo] 4. gr[aduum] 40'. Austr[alis]. Ascen[sio] rect[a] 31 gr[aduum] et Decl[inatio] 8 gr[aduum] motu vero proprio, spatio 24. horarum 1 gr[adum] 31'. absolvit.

Die 11. Jan[uarii] Cometa caput pallidum et subobscurum, minusque ratione Diametri: caudam vero 18 gr[aduum] circiter ad supremam et secundam fere in armo Tauri vibratam protulit. Versabatur eo tempore in 1 gr[adu] 29'. Tauri, Latit[udine] Austr[ali] 3 gr[aduum] 30'. confecto itineris intervallo a die hesterna hucusque 1 gr[adum] 20'.

Die 12 Jan[uarii] Semel quidem inter nubes Cometa emicuit cum cauda ad septem octove gradus rariori, sed protinus a nubibus abreptus est, ut nihil amplius certi instrumentis deprehendi potuerit.

Die 17 Jan[uarii] vesperi Corpus Cometicum aliquanto diminutum, tum rariori et tenuiori materia praeditum: sicuti effigies ad Num[erum] 16. exhibet, commorabatur in Longit[udine] 28 gr[aduum] 20'. Arietis, et Latit[udine] 1 gr[adus] sed boreali: postquam a die 10. ad 17. usque cursu suo 6 gr[adus] 14'. confecisset. Cauda constans 8 gr[aduum] circiter, subobscuris et rarissimis radiis tendebat ad oculum boreum Tauri.

Die 18 Jan[uarii] Caelum nobis minime arrisit. At vero

Die 19 Jan[uarii] cadente Sole, observationes rursus continuatae sunt. Caput multum decreverat, lucidius tamen ac scintillans apparuit, situm autem nucleorum materiaeque interioris facies sub Num[ero] 17. commonstrat. Cauda debilis 8 gr[aduum] et frequenter pariter radians. Long[itudo] Cometae animadversa est 27 gr[aduum] 40'. Arietis Latit[udo] 1 gr[adus] 45'. boreal[is]. Ascen[sio] Rect[a] 25 gr[aduum] Declin[atio] 12 gr[aduum] 36'. boreal[is]. Ratione motus proprii jam ad tantam redactus erat lassitudinem, ut a die 17. ad 19. usque tantum 10. gradus 4'. traiecerit.

Die 20 Jan[uarii] hora 7. vespertina Caput valde erat diminutum. Cauda pertenuis et albescens 7. gradus circiter ad stellam praecedentem in collo Tauri sub Pleadibus exporrecta. Commorabatur Cometa id temporis in Long[itudine] 27 gr[aduum] 25'. Arietis, et in Latit[udine] boreal[i] 2 gr[aduum] 12'. in motu vero suo proprio vix hac die 27'. peregit.

Die 21 Jan[uarii] vesperi Cometa observatus est capite minore et obscuriore, tum cauda breviori, duobus tantummodo gradibus longa exititit, Luna tamen praesens, maxima parte, eo tempore istam obscuritatem palliditatemque Cometae induxerat. Nam

protinus post Lunae occasum, et caput lucidius et cauda longior 4. gr[adibus] fere, ad primam in eductione caudae Arietis directa, observabantur; et quidem in 27 gr[adu] 20'. Arietis in Latit[udine] 2 gr[aduum] 36'. Bor[eali] confectis in motu proprio, 24. horarum intervallo 24. min[utis].

Die 22 Jan[uarii] totus exstitit nebulosus. At vero

Die 23 Jan[uarii] Caelo arridente Cometa noster denuo animadversus est, corpore admodum tenuis debilisque: caudae vero vix vestigium apparuit: prout ex Fig[ura] B. sub Numero 20. patet. Quod ad popularem adspectum autem, cauda, ut dixi, pene caruit: attentiori tamen observatori comam 30'. vel 40'. Hellespontum versus proiicere visus est. At Telescopio consideratus, caput insigni lumine tum spectabili magnitudine adhuc gaudebat: in sinu suo diversos nucleos, materiamque diversae densitatis fovens, lumine debiliori cinctam.

Discus autem, quod notabile, minime absolutam prae se ferebat rotunditatem, sed asperam, laceratamque, cuspidibus hinc inde leviter eminentibus. Commorabatur in 27 gr[adu] Arietis in Latit[udine] Bor[eali] 3. gr[aduum] motu proprio duobus diebus 41' consumserat.

Die 24, 25, et 26. Januarii continuo subobscurae noctes extiterunt.

Die 27 Januarii hora octava vespertina ex unica distantia, quantum acquirere licuit, cognovi Cometam spatio quatuor praecedentium dierum 1 gr[adum] 1' progressum esse, tum lineam istam rectam, ex secunda per primam Arietis ductam, Eclipticam versus jam transcendisse, atque nunc in 26 gr[adu] 40'. Arietis, et Latit[udine] Bor[eali] 4 gr[aduum] haerere, admodum obscurum et omni cauda orbatum, propter Lunam vicinam.

Die 28 Januarii circa horam sextam, utut caelo admodum annuente, adeo tamen imminutus pallidusque, inermibus videbatur oculis, ut vix aegre per rimulas pinnacidiorum deprehendere Cometam potuerimus. Verumtamen nonnullae distantiae extortae sunt, ex quibus didicimus a 23 hucusque 1 gr[adum] 20' tranasse: existente videlicet in Long[itudo] 26 gr[aduum] 30' Arietis, et Latit[udine] 4 gr[aduum] 12' fere. Quali fere facie interior pars capitis tum resplenduerit, delineatio ad Num[erum] 21 Fig[urae] B. aperte ostendit.

Die 29, 30, 31. et 1. Februarii nihil certi penitus annotari concessum fuit.

Die 2 Febr[uarii] vero hora quinta vespertina Cometa iterum prospere detectus est egregia specie. Animadvertimus itaque a die 23 Jan[uarii] ad 2 Februarii phoenomenon traiectione sua 2 gr[adus] 8' 25" et hac die ultima vix 8' promotum: in Eclipticae vero Longitudine 26 gr[aduum] 20'. Arietis, et latit[udine] Bor[eali] 5 gr[aduum]

8' versari. Videbatur quod ad longit[udinem] Stationarius, maxime in latum jam protendens motum. Caput 3' imo 4' capillitio pene exutum, claritate nihil cedebat primae stellae Arietis. Materia interior erat rarior, et nuclei fere deleti. Cauda duorum graduum cum dimidio, interdum trium graduum sed pertenuis, tendens ad stellulam in collo Arietis sub cornu.

Die 3 Febr[uarii] hora septima vespertina iterum conspectus eadem fere specie qua hesterno die, spatio 24. horarum, tantum 5' 3" progressus erat, et haerens in eadem long[itudine] 26 gr[aduum] 20' vel saltem 26 gr[aduum] 19' Latitud[inem] 5 gr[aduum] 13' obtinebat: constitutus in una eademque recta cum capite Andromedae et mandibula Ceti.

Die 4 Febr[uarii] hora septima vespertina Cometa vix ultra 4' progressus, caput 3' vel 4' referebat, lumine rariori Latit[udo] Bor[ealis] auctior evaserat 3' scilicet 5 gr[aduum] et 16'.

Die 5, 6, 7, 8, 9. Februarii ob continuas nives, aerisque turbas nulla observatio facta est. At

Die 10 Febr[uarii] hora septima vespertina Cometa Tubo Optico conspectus videbatur 26 gr[aduum] 40' Longit[udinem] et 6 gr[aduum] 20' fere Lat[itudinem] Bor[ealem] occupare, et viam reflectere, caput Arietis versus.

Die 11 Febr[arii] Caelum nubilum nullam observationem concessit.

Die 12 Febr[uarii] vesperi hora 6 1/2 Cometa in conspectum prodiit admodum diminutus, pallidus et subobscurus, ut vix stellae quintae magnitud[inis] aequiparari potuerit. Loco caudae radiolum quendam pertenuem et obscurum primam Arietis et collum versus spargere videbatur. Situs erat ut cum eadem Arietis et collo sub cornu triangulum fere aequicrurum constitueret: cuius basin Cometa et stella in collo, reliquae vero binae distantiae crura referebant, ita ut ducta recta per secundam et primam Arietis in basis medium incideret. Longit[udo] deprehensa est 27 gr[aduum] Arietis, et Latit[udo] bor[ealis] 6 gr[aduum] 30' motus autem proprius a 2 Febr[uarii] ad hunc 12 Febr[uarii] 1 gr[radus] 45'. Ex quibus elucet, quod jam motu velociori incesserit, et sub ductu suo incurvato uno die 13' confecerit.

Die 13 Febr[uarii] Tubo deprehensus est Cometa 1 gr[adum] vel paulo plus adhuc a prima Arietis distare, haerens in 27 gr[adu] 13'. Arietis: in Latit[udine] 6 gr[aduum] 35' confectis motu proprio 25. horarum spatio 14'.

Die 14 Febr[uarii] hora sexta 30 min[utis] Cometa quidem conspectus est, sed ob nubila locari nequivit.

Die 15, 16 et 17, nihil animadvertimus.

Die vero 18 Febr[uarii] hora septima vesperi Cometam offendi prope ipsam primam stellam Arietis, atque ad eius sinistram, mediam scilicet Capitis Arietis et Pleiadas versus, vix 5' vel 6' ab ea stella prima remotum. Hinc Longitudo eius constabat 28 gr[adus] Arietis atque Latit[udo] 7 gr[adus] 10'. Boreal[is] perrexerat motu suo directo a 12 ad hanc 18 Febr[uarii] per 1 gr[adum] 40' sive 1 gr[adum] 42' sex scilicet diebus fere tantum caeli spatium emensus, quantum praecedentibus 10. Atque ita motus diurnus ultimo observatus exstitit 21' cum die 4 Febr[uarii] tantum 5' procederet. Respectu capitis erat pallidissimus: nihilominus 2' circiter in diametro, sed cauda prorsus exutus: prout sub schemate B. Numero 28 eum delineavimus.

A die 19 Febr[uarii] usque ad 7 Martii per nubes observare non licuit: hoc vero die vesperi hora octava Cometam quidem omni diligentia inter astra quaesivimus, sed nullus amplius conspectus fuit.

Haec sunt quaecunque hic Dantisci, divina affulgente ope, de hoc Cometa observare nobis licuit: non quidem ipsae observationes sextante et quadrante Orichalcicis amplissimis debita sollicitudine et diligentia captae, quae ultra 450 excurrunt, atque in ipsam *Cometographiam* reservantur, sed ea solummodo ratione, qua Cometae loca beneficio maioris globi, motumque proprium ex distantiis deduximus. Quae tamen nihilo serius, ut puto, satis sunt accurata; prout suo tempore calculus noster docebit.

Haec ut facilius et certius comprehendere quivis possit ac contemplari, tramitem istum cursumque Cometae delineatum dedimus. Ubi notandum, quod nudi isti circelli in orbita delineati loca Cometae ad eos dies, quibus hic Gedani a nobis observari haud potuerunt, ad proportionis ductum motus diurni, horam scilicet octavam cuiusvis diei vespertinam referant: ut ita velocitas diversa et motus decrementum, tum ubi stationarius, simul directus fuerit, percipiatur.

Hevelius's observations from Danzig from his *Prodromus cometicus* made into a summary

I (Johannes Hevelius) saw that notable comet for the first time here, in Danzig, on 14 December 1664 Gregorian around 5 o'clock, in the eastern quarter of the sky, towards the south-east, near the Beak of Corvus — about a half degree from it towards the west — 9 degrees over the horizon, in the eighth degree of Libra, at latitude 22 degrees south. Its head was visible enough, yellowish in colour; its light was weaker and a little more obscure than that of other stars and planets. In its centre, there was some kind of light, 3' or 4' in diameter, surrounded by darker and thinner hair. The centre consisted of only one clear nucleus, but there were also many little parts or bodies, bigger and smaller (some of them were denser and brighter, some thinner and darker), with some

more subtle matter of various density mingled with it; it is depicted at number 1 in Illustration A. The tail was 14 degrees long, thinner than the head; it was expanding at its end; it was directed towards the star which is called the Base of Crater, and the one which is more towards the north under Crater, closer to the base. On 15 December, at 4 o'clock in the morning, it was seen to be in 7° 20' of Libra, at latitude 22° 20' south; its elevation in the meridian was about 20 minutes smaller; with its retrograde movement (viz. against the order of signs), within 24 hours it moved almost an entire degree towards the Tropic of Capricorn. Its head seemed almost unchanged, except that its nucleus and the whole matter of the body looked a little thicker and brighter through a telescope; it is shown in Illustration number 2. The tail was a little shorter, about 11 degrees; it looked quite notable and shining, and it was seen to reach towards the Base of Crater. On 16 and 17 December, the whole sky was cloudy. But

On 18 December, based on its (the Comet's) distance measured again, we learned that at 4 o'clock in the morning the comet was in 3° 45' of Libra at latitude 25° 30' south; within three days it moved 4° 10' with retrograde movement on its route, and the altitude in the meridian diminished by 1° 25', and it had already crossed the Tropic of Capricorn. Its head looked a little bigger and brighter; the matter of the head and the nucleus seemed equally denser — we showed it in Illustration 3. The tail, 12 degrees long, equally bright as on the fifteenth, stretched towards the Heart of Hydra and Procyon. The diameter of the entire head was 15 minutes, while on the first day when it appeared it was barely 12 minutes.

On 19 and 20 December, the sky was covered by clouds. But

On 21 December, at 4.15 in the morning, the comet was seen below Crater, in Hydra, in the twenty-eighth degree of Virgo and at latitude 30 degrees south. Its altitude in the meridian since the eighteenth decreased by 2° 4' and was only 9° 16' 30", and there was refraction to consider. Hence, it was clear that the comet moved 6° 31' with its own motion. The diameter of the entire head increased to 18 minutes, and it seemed brighter. The tail on that day was extraordinary; not only was it longer than 22 degrees and wider than 1° 30' (almost 2 degrees), but also, when observed correctly, its part around its end was curved, and the concavity was down, viz. towards the horizon; it was like the comet of 1618[2] and other similar ones. It was directed towards the second and the third star from the Heart of Hydra, and the curve bent more towards the second one from the heart.

On 22 December, the clouds were an obstacle.

On 23 December it was below Hydra in the twenty-second degree of Virgo, at latitude 34° 30' south; its altitude in the meridian decreased by 2 degrees and, therefore,

2 See note to Letter 1F.

on that day it did not exceed 7 degrees. Since the twenty-first, its own movement was almost 7 degrees. A significant tail, about 20 degrees long, stretched again towards the first star from the Heart of Hydra, towards the west.

On 24, 25, 26, 27, the comet did not show itself even once. But

On 28 December, at 1 o'clock, I found it in the south, near the meridian, elevated barely 6 degrees over the horizon, placed in the stern below the Shield of Navis in the fourth degree of Leo, at latitude 49° 30' south, after it had passed on its route a great distance of 38 degrees. Its diameter was seen to be almost 22'. The head was encircled by curly and dense hair, and its nuclei were denser. The tail stretched towards the north; it was almost 7 degrees long and similar to a tail of a peacock, as shown at number 10 in Illustration A.

On 29 December, at 9.30 in the evening, the comet shone again at altitude 10 degrees, below the Left Forefoot of the Canis Maior, at longitude 28° 40' of Gemini and latitude 47 degrees; it seemed to exceed 12 degrees of the latitude in the meridian; its movement in orbit within 43 hours was almost 23° 40'.

The tail (or rather the beard) was straight and stretched about 1 degree towards the top. The head was hairy, and its diameter was 24'; in its centre, it showed a very bright and very dense nucleus of fiery red, which was surrounded by the matter of various density and some smaller nuclei; it is shown in the Illustration at number 6.

On 30 December, around 10 o'clock in the evening, the comet — very bright and visible — was seen in the Neck of Lepus, stretching its beard (which was slightly bent from the vertical circle) about 1 degree towards the south-east, to the Right Foot of Orion. The head was the same, except that the round nucleus in the middle seemed to have moved from the centre towards the edge. It was in 12° 40' of Gemini, at latitude 40° 10' south. Its right ascension was 76° 30' and declination was 17° 10'; the altitude in the meridian was 18° 27'.

On 31 December, at 6 o'clock in the evening, the comet was elevated about 14 degrees over the horizon; its body was a little smaller in diameter, the colour was white and bright enough. It seemed to be slightly smaller than Rigel, or the Left Foot of Orion. It did not change its face, except that the large nucleus from the centre moved towards the eastern edge. The tail, almost 1 degree long, was stretching towards the middle star in the Belt of Orion. It is depicted at number 8. At 9 o'clock in the evening, it was in 2° 40' of Gemini, at latitude almost 35 degrees south. Its movement became a little slower, as within 34 hours it moved only 11° 10'. Its right ascension was 66 degrees, and declination was 12° 30' south; at that time, its altitude in the meridian was 23° 13' 50".

Year 1665. On 1 January, at 8 o'clock in the evening, the comet was moving below the little star in Eridanus, also known as the third one, about one degree from the first curve, towards the south-west. The head was the same, and so was the matter of the inner body. The nucleus (even though it was bigger, denser, and brighter) lost something of its former splendour. Its place was at longitude 24° 20' of Taurus and latitude 27° 45' south. Its right ascension was 58° and declination 8° 20'. The daily movement was 9 degrees.

On 2 January, the sky was turbulent.

On 3 January, that hairy star shone with a tail 3 degrees long, directed towards the middle one of the three unformed stars[3] in the Jaw of Cetus. It was in 14° 20' of Taurus, at latitude 19 degrees. It moved 12° 30' on its way.

On 4 January, the misty night did not bring any observations.

On 5 January, when the Sun was setting at 4.30, and the Jaw of Cetus was shining, we saw the comet because it was then commonly seen to be 30' towards the west from the lower part of that star. At 6 and 7 o'clock (before the Moon rose), its head (however smaller) was clearly visible and it had a tail 14 degrees long, especially bright and dense at its base; so far, it had no such tail. The tail stretched towards the fourth star in the Shield of Orion. There was only a trace of that bright nucleus, as depicted at number 11 in Illustration A. About 9 o'clock, the Jaw of Cetus, the comet, and the star in the Mouth of Cetus were seen to be almost in a straight line; the longitude of the phenomenon was 8° 40' of Taurus, latitude 13' south, right ascension 40° 20', declination 2° 20' north. Its own movement from the third until the fifth was 8° 20' and the altitude in the meridian was 38° 8' 45".

On 6 January, from 6 o'clock until 11 o'clock in the evening, the head of the comet looked still silverish and had 8' in diameter. It could be seen that the internal nuclei and the matter were dissolved. As far as its place is concerned, the comet at that time made an almost equilateral triangle with the Jaw of Cetus and the one in the Mouth. The tail — 15 degrees long and very bright — stretching towards the stars in the Leg and the Dewlap of Taurus (viz. towards the right knee), was wider again as if it were fluctuating and emitting new rays. The comet itself was at the longitude of the seventh degree of Taurus and latitude 10° 36' south. The right ascension was 38 degrees and declination was 4° 4' north. During one day, it moved 3 degrees on its route; the altitude in the meridian was 39° 41'.

On 7 January, soon after 4 o'clock in the evening, we saw the comet with a smaller head (however not less bright), in silver and white; the nuclei were merging into a

3 See note to Letter 8A.

kind of an inert mass; it is depicted at number 13. The tail was exceptional — 13 degrees long and over 2 degrees wide; it was directed towards the star which follows the third one in the Back of Taurus. The longitude of the comet was measured to be 5° 30' of Taurus, latitude 8° 36' south, right ascension 35° 30', declination 5° 23' north and the altitude in the meridian 40° 59' 4".

On 8 January, the clouds covered the whole sky.

But on 9 January, at 6 o'clock in the evening, we saw the head of the phenomenon the same as before, and the tail too, except that it was a little longer (15 degrees) and stretched towards the second star in the Back of Taurus; the comet was at longitude 2° 50' of Taurus, latitude 5° 40' south, right ascension 32 degrees and declination 7° 23'; within three days, it moved 3° 45'.

On 10 January, after 5 o'clock in the evening, the matter of the inner head was seen to become more and more dissolved, as presented at number 15. The tail was thinner and shorter (only 12 degrees) stretching towards the place between the first and the second star in the Back of Taurus. The longitude was then 2 degrees of Taurus, latitude 4° 40' south, right ascension 31 degrees and declination 8° 11'; with its own motion, within 24 hours, it passed 1° 31'.

On 11 January, the comet was pallid and darkish, and its diameter was smaller; it stretched its tail, which was about 18 degrees long, towards the higher and the second star in the Back of Taurus. At that time, it was in 1° 29' of Taurus, at latitude 3° 30' south; since yesterday, so far it has passed 1° 20' on its path.

On 12 January, the comet shone once with a thinner tail, 7 or 8 degrees long, but it was soon covered by the clouds, so it was impossible to see anything certain with instruments.

On 17 January, in the evening, the comet's body was a little smaller, and its matter was thinner and weaker; it is shown at number 16; it was at longitude 28° 20' of Aries and latitude 1 degree south; since 10 and 17 January it has passed 6° 14'. The tail was steady, about 8 degrees long, with its darkish and thinner rays it stretched towards the Northern Eye of Taurus.

On 18 January, the sky was unfavourable to us. But

On 19 January, at sunset, the observations were resumed. The head became much smaller, but seemed brighter and twinkling; the place of nuclei and the internal matter is presented at number 17. The tail was weak, 8 degrees long, and equally radiating. The longitude of the comet was seen to be 27° 40' of Aries, latitude 1° 45' north, right ascension 25 degrees, declination 12° 36' north. As far as its proper motion is

concerned, it became so slow that from the seventeenth until the nineteenth it passed only 10° 4'.

On 20 January, at 7 o'clock in the evening, the head was very small. The tail was very thin and white, about 7 degrees long, stretched towards the foregoing star in the Neck of Taurus, below the Pleiades. The comet was then at longitude 27° 25' of Aries, at latitude 2° 12' north; on that day, it moved barely 27' with its own motion.

On 21 January, the comet was seen in the evening with a smaller and darker head, and the tail was shorter, only 2 degrees long; but the Moon was seen in its largest part, and at that time it caused the darkness and paleness of the comet. Because right after the moonset, the head was seen brighter and the tail longer, directed towards the first star at the Base of the Tail of Aries; it was in 27° 20' of Aries, at latitude 2° 36' north; with its own movement, it passed 24' within 24 hours.

On 22 January, the whole sky was cloudy. But

On 23 January, the sky was favourable, and the comet was seen again, with its body thin and weak; there was only a trace of the tail, as it is shown in Illustration B, at number 20. According to what was commonly noticed, as I said, there was no tail, but a more accurate observer could see a braid, 30' or 40' long, stretching towards the north-east. And when observed with a telescope, the head still had some considerable light and a substantial size; in its centre, surrounded with weaker light, there were various nuclei and a matter of various density.

What is notable is that the disc was not perfectly round, but it was uneven, cut, and full of small thorns. It was in the twenty-seventh degree of Aries, at latitude 3° north; within 2 days, it passed 41' with its own motion.

On 24, 25 and 26 January, there were only dark nights.

On 27 January, at 8 o'clock in the evening, on the basis of only one measurement I could take, I saw that the comet moved 1° 1' within last four days and it travelled along the straight line from the second star to the first star of Aries towards the Ecliptic; it was in 26° 40' of Aries, at latitude 4 degrees north, very dark and without a tail, because of the proximity of the Moon.

On 28 January, about 6 o'clock, although the sky was helpful, the comet seemed so small and pallid to a naked eye that we could barely see it through the holes in vanes. But a few measurements were taken, from which we learned that since the twenty-third it passed 1° 20'; it was at longitude 26° 30' of Aries, and latitude almost 4° 12'. Image number 12 in Illustration B clearly shows how the interior of the head looks like.

On 29, 30, 31 January and on 1 February, it was impossible to observe anything specific.

But on 2 February, at 5 o'clock in the evening, the comet was found great again. We saw that from 23 January until 2 February, the phenomenon moved on its way 2° 8' 25", and 8' on the last day; it was at the ecliptic longitude 26° 20' of Aries, latitude 5° 8' north. In its longitude, it seemed stationary, whereas it was moving fast in latitude. The head, 3' or 4' in diameter, was almost without hair; its brightness was no less than that of the first star of Aries. The internal matter was thinner, and there were almost no nuclei. The tail was 2 1/2 degrees long — sometimes 3 degrees, but very thin — stretching towards the star in the Neck of Aries below the horn.

On 3 February, at 7 o'clock in the evening, it was seen again, and its appearance was almost the same as the day before; within 24 hours, it passed only 5' 3"; it was at the same longitude of 26° 20', or at least 26° 19', and at latitude 5° 13'; it was in the straight line with the Head of Andromeda and the Jaw of Cetus.

On 4 February, at 7 o'clock in the evening, the comet moved about 4'; the head was 3' or 4' in diameter, with some thinner light; the latitude was a 3' bigger, viz. 5° 16' north.

On 5, 6, 7, 8 and 9 February, no observation was made due to constant snowfall and the confusion in the air. But

On 10 February, at 7 o'clock in the evening, the comet, seen with a telescope, seemed to be at longitude 26° 40' and latitude almost 6° 20' north; it also seemed to curve its route towards the Head of Aries.

On 11 February, the cloudy sky did not allow any observations.

On 12 February, at 6.30 in the evening, the comet looked smaller, pale, and darkish, so it could barely equal the stars of the fifth magnitude. Instead of the tail, it seemed to have had a little ray — very thin and dark — stretching towards the first star and the Neck of Aries. Its position was such that with that star in Aries and the neck under the horn it made almost an isosceles triangle: its base was the comet and the star in the neck, and two other lines were its legs; the straight line through the second and the first star of Aries divided the base in half. The longitude was twenty-seventh degree of Aries and latitude 6° 30' north; its own movement from 2 February until 12 February was 1° 45'. Therefore, it is clear that its movement became faster and that on its curved path it passed 13' in one day.

On 13 February, with a telescope, the comet was seen to be removed 1 degree or slightly more from the first star of Aries; it was in 27° 13' of Aries at latitude 6° 36'; with its own motion, it passed 14' in 25 hours.

On 14 February, at 6.30, the comet was seen, but it was impossible to determine its position due to clouds.

On 15, 16, and 17 February, we saw nothing.

But on 18 February, at 7 o'clock in the evening, I found the comet near the first star of Aries, on its left, viz. towards the middle one in the Head of Aries and the Pleiades; it was 5' or 6' from the first star. Therefore, its longitude was the twenty-eighth degree of Aries and latitude 7° 10' north; with its direct motion, from 12 until 18 February, it passed 1° 40' or 1° 42' in six days — the distance which took it ten precedent days to pass. And so the daily movement, when observed last time, was 21' while on 4 February it was only 5'. The head was barely visible; it was not smaller than 2' in diameter, but it had no tail; we presented it at number 28 in Illustration B.

From 19 February until 7 March, it was impossible to conduct any observations due to clouds; but on that day, at 8 o'clock in the evening, we were diligently searching for the comet among the stars, but we found no trace of it.

This is what we observed about that comet here, in Danzig, with God's help; not only did we cautiously and diligently conduct the observations with a brass quadrant and sextant (there are more than 450 observations which are kept for the *Cometographia*), but we also determined its positions with the help of a bigger globe and the comet's own motion taken from the measurements. In my opinion, they are equally exact, as, in due time, will be clear from our calculations.

In order to make it easier and more certain to understand for everyone, we provided the illustration of the trajectory and the route of the comet. Therefore, it should be remarked that those empty circles drawn in orbit are the places of the comet in those days when we could not observe it here in Danzig; they refer to the ratio of its daily movement, at 8 o'clock in the evening; thus, the changes in speed and the decrease of the motion (when it was stationary, and when direct) will be understood.

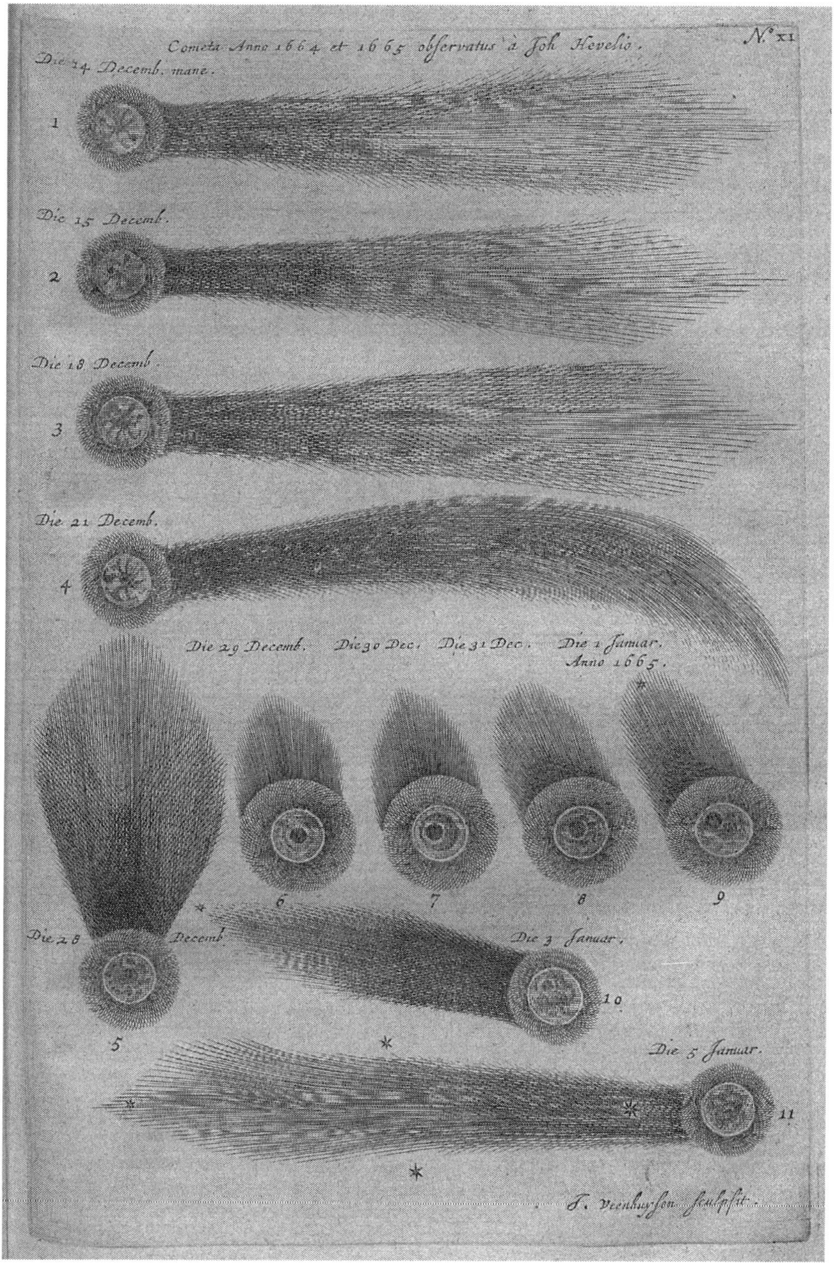

Figure 16. Hevelius's observations of the shape of the winter comet of 1664–65 in Lubieniecki's *Theatri cometici pars prior*. Biblioteka Narodowa, Warsaw, SD XVII.4.2660.

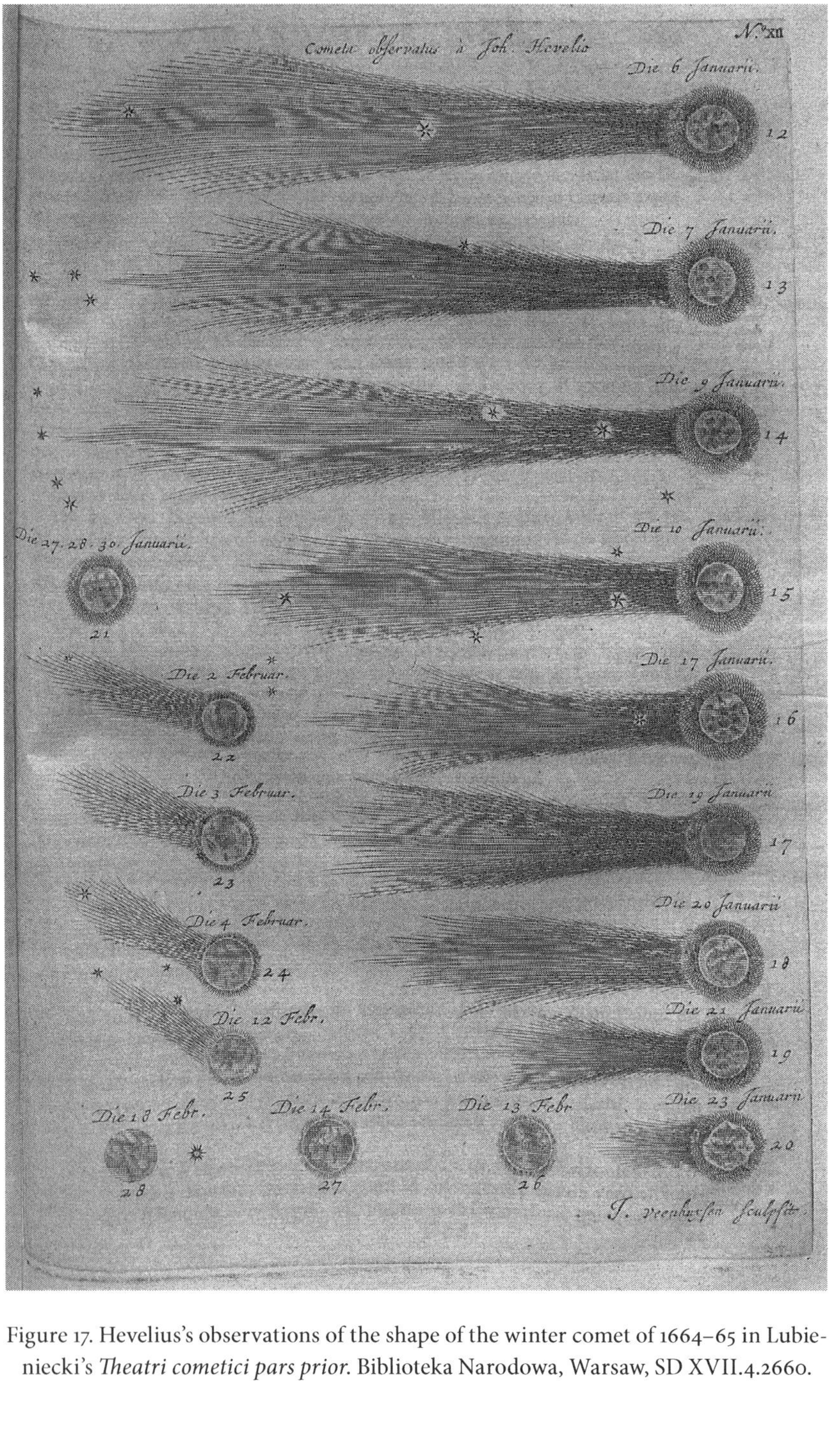

Figure 17. Hevelius's observations of the shape of the winter comet of 1664–65 in Lubieniecki's *Theatri cometici pars prior*. Biblioteka Narodowa, Warsaw, SD XVII.4.2660.

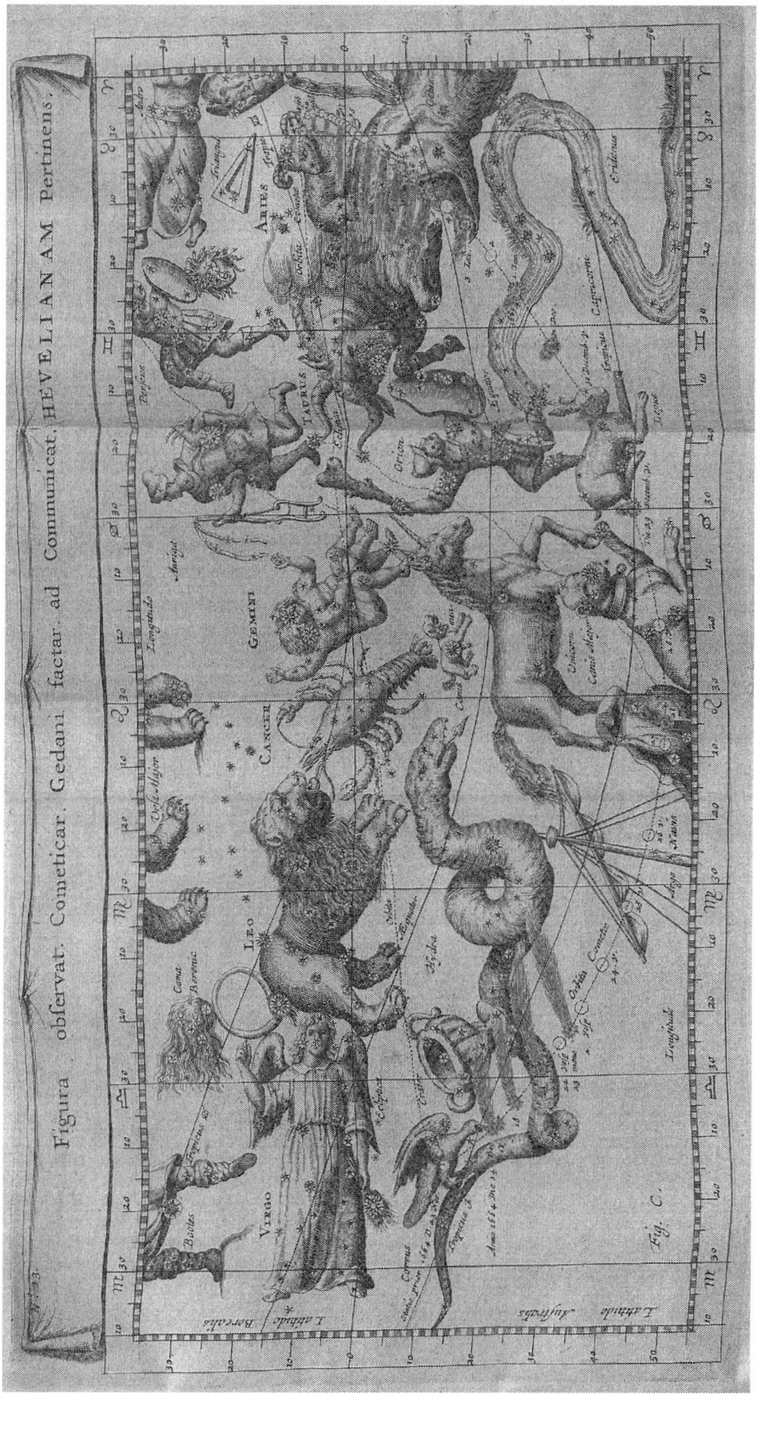

Figure 18. Hevelius's depiction of the route of the winter comet of 1664–65 in Lubieniecki's *Theatri cometici pars prior*. Biblioteka Narodowa, Warsaw, SD XVII.4.2660.

Appendix 2.

Lubieniecki's Extract from Hevelius's *Descriptio cometae*

TC: 949–50

Generalis Cometae qui apparuit Anno 1665. mense Apr[ili] descriptio Danisco Heveliana ab Auctore in compendium consentiente Ampliss[imo] et Cl[arissimo] Domino Hevelio redacta.

Cometa hic quem longe maturius illuxisse puto, atque d[ie] 5. Aprilis, hic Gedani primum ab Excubiis Civitatis animadversus est, innotuit demum mihi (Joanni Hevelio) 6. Aprilis, ipso die Lunae, mane hora 1. 30'. fere circa Hellespontium supra nostrum Finitorem assurgens. Corpus huius albicans et ex parte subflavum lucidius multo, quanquam minus apparuit ac prioris Cometae. Tubus in meditullio unicum rotundum nucleum, ad instar auri splendescentem offerebat, quem alia materia tenuior ae aequabilis excipiebat ac cingebat, prout ex prima delineatione Fig[urae] D. clare liquet. Existebat tum temporis in ipso pectore Pegasi in 14°. 22'. Piscium, et quidem lat[itudine] bor[eali] 26°. 30'. sub duabus videl[icet] illis stellis in pectore Pegasi.

Die 7. denuo observatus est, distantiae ab iisdem fixis fere eadem proportione,[1] ut initio immutabantur, progressus est intervallo 14 horarum per 4°. 6'. Et in longitud[ine] 4°. 30'. via quae nobis innotuit ex detectis utriusque diei Cometae locis per distantias in globo et quidem ampliori delineatis, usque ad horizontem ortivum deductis, tetendit per Pegasum sub capite Andromedae Piscem Boreum versus. Verum cum nullus Cometa accurate tramitem circuli maximi observet, suasor sum, ne huic explorationi viae itineris circa eius nempe extremitates nimium fidamus, atque ita longa satis a vero quod ad intersectionis punctum attinet, aberremus. Lumen et splendor nihil: magnitudo vero pallulum videbatur mutata, existente tamen illo unico nucleo in meditullio. Cauda 12°. vel 13°. pergebat versus stellam in extremitate alae austrinae cygni.

1 proportione *corr. ed.* : porportione [*sic*] **TC**.

Die 8. April[is] hora secunda mane Cometa cum Scheat Pegasi et stella praecedenti in pectore pegasi triangulum propemodum aequilaterum constituit, progressus 4°. circiter, caput eximie luminosum, nucleus tamen vix maior et lucidior quam hactenus apparuit. Cauda per praecedentem in pectore pegasi incidens ad sinistrum usque genu adaequavit fere 15°. cui superius adhaesit sequens in pectore Pegasi. Caput integrum maculae lunari quae Sicilia vocatur 6': Nucleus vero insulae Corsicae 12" et 13", aequalia deprehensa sunt. Vel nucleus collatus cum Saturno nondum ad magnitudinem corporis huius intermedii rotundi omnino accedebat.

Die 9. Aprilis pluvia.

Die 10. hora 1°. 54'. exoriens adhuc lucidior et maior apparebat oculo nudo; at nucleus ope telescopii nec maior nec luminosior erat. Cauda similis, ad 20°. excurrebat versus ungulam sinistri pedis boreal[is] Pegasi, Scheat superiori sua parte ferme attingens. Longitudo 2° Arietis, Latitudo et motus decreverant.

Die 11. April[is] summo mane Cometa denui insigniter clarus colore haud dissimilis Saturno capite nihil immutato, caudam in ipso exortu ad 22°. diluculo vero ingravescente ad 17°. tantum extendebat.

Die 17. Adversa tempestas.

Die 13. Cometa hora 1. 50'. fere primum emergebat cum cauda praecedente 25°. quae tamen paulo post 20°. tantum aestimabatur, excurrens ad sinistrum usque genu Pegasi.

Die 14. Aprilis nubes observationem non admiserunt. Die 15. Cometa magis adhuc splendidus ac vegetus exoriebatur, tum ipse tubi beneficio aliquanto maior videbatur. Cauda ante ortum Cometam 25°. a capite superiori sua parte stringebat scapulam sinistram Andromedae, ut habet typus ad Num[erum] 7. Hoc die tria singularia notata sunt I. caput ex solo nucleo constans et reliqua materia rariori circundata pene omni exutum. II. Caput aut cauda scintillantia et radios quasi vibrantia. III. Cauda ad eductionem spatio aliquo obscuriori fissa perinde ac si nucleus ille umbram proiiceret. Situs erat Cometa in cap[ite] Piscis Boreal[is] in 19°. 27'. scilicet Arietis sub latit[udine] 19°. et 38'. Boreal[i].

Die 16. April[is] Cometa ipso Saturno lucidior fere magis quam hactenus in oculos incurrit, ut et ipse nucleus telescopio maior animadversus est, sine ullo crine, nisi qui versus Solem in apicem fastigiatus fuit, uti ex delineatione ad Num[erum] 8. aperte liquet. Adhaec diameter eius Jovis diametro minor, Saturni vero (quod ad corpus intermedium) aequalis detecta est, ut et umbra per caudam mediam (quae fuit 20°.) ut pridie. Cauda aeque vibrans radios ac caput, dirigebatur per dextram scapulam versus

tres illas in brachio Andromedae, et sub auctiori ac intensiori diluculo plane decurtabatur, ut habet Iconismus ad Num[erum] 9.

Die 17. intemperies aeris observare non permisit.

Die 18. Cometa idem deprehensus est et facie et nucleo; Cauda tam debiliori ob nubeculas horizontem obsidentes. Haec versus superioran stellam in manu Andromedae ad sinistram videl[icet] 1 lucida cinguli 1°. vel 1°. et 30'. tetendit longitudine circiter 15°. et quotidie magis magisque sursum elevata est ratione situs solaris.

Die 19. ob vapores horizonti adhaerentes nihil propemodum observatum est.

Die 20. hora tertia 43'. in altitud[ine] 3°. 45'. observatus est Cometa a quibusdam fixis, et hora 4 adhuc visus est nudis oculis satis distincte in altitudine 6°. 30': locus autem eius sub cuspide trianguli paulo supra cornua Arietis viam quasi versus cornu dextrum seu lucidam Arietis dirigens in 1°. 18'. Taur[i] et latit[udine] 13°. 14'. annotatus est. quod ad motum proprium vero attinet, spatio 48. hor[arum] 4° 51'. traiecit sic ut uno die posteriori 2°. 23'. tantum absolverit. Cauda primum ob nubes circa horizontem conspecta non est, deinde tamen 3°. vel 4°. visa, eius cuspis verticalem jam transcenderat pauxillum Septentrionem versus deflectens; ac si coniunctio Solis et Cometae jam celebrata esset. Materia illa dilutior, in frontispicio capitis nulla; ad utrumque latus vero satis dilatata fuit, ut habet Num[erus] 10.

Die 21. Evaporationes circa Horizontem.

Die 22. Vero saeva et turbida tempestas omnem observationem turbavit.

At die 23. April[is] licet coelum circa horizontem esset nitidissimum, nec nudis nec armatis oculis Cometam usquam deprehendere amplius potuimus, atque ita tantum 14. diebus a 6. ad 20. Aprilis nobis conspicuus fuit, quo tempore 46°. in suo orbe absolvit. Et haec est Generalis descriptio Cometae posterioris: specialis videatur in ipsa descriptione a Clar[issimo] Hevelio edita.

NOTA

Tabula haec G. eum in finem confecta ab Ampliss[imo] Hevelio, ut discrimina appareant inter observationes Cl[arissimorum] Virorum Auzutii et Gottinii cum suis collatas pro dirimenda lite, quae est inter ipsum (Hevelium) et D[ominum] Auzutium. Statuit enim Auzutius Cometam iter suum continuasse ad secundam Arietis semperque tardiorem super eam incessisse, ultimo visum a se die 17. Martii quemadmodum exhibet delineatio Cl[arissimi] Petiti seorsim ad sinistrum marginem superiorem posita: provocatque ad Gottinianas observationes, quas infra similiter seorsum datas cum suis omnino congruere dicit. Contra Cl[arissimus] Hevelius Co-

metae postremum die 18. Februarii a se visi iter ad primam Arietis deducit, eumque (Cometam) post 4. Februarii eadem proportione in signorum consequentia cursus celeritate crevisse probat, ac antea contra horum seriem decreverit. Deinde postquam Auzutii observationes tum a suis maxime discrepare ostendit (dum eius 4. Februarii cadat inter sextum et septimum suarum observationum, item illius sextus Febr[uarii] respondet suo nono, et c[etera]) Tum a Gottinianis observationibus pariter recedere, ut et a duabus istis Clar[issimi] Rudbekii supra ad dextram convenienti loco positis, pluribus rationibus tandem commonstrat istos omnes post 15. Febr[uarii] necessario longe aliud phaenomenon vidisse diversum ab eo quod per Jan[uarium] et primam Feb[ruarii] partem conspexerint. Vide eius *Mantissam prodromi Cometici*, in qua Mathematicarum rerum studiosus cuncta ista, quae huc pertinent accuratissima et incomparabili industria, operosis computationibus et firmissimis rationibus deducta, probata et demonstrata singulari cum oblectatione et fructu videbit.

A general description of the comet that appeared in April 1665, prepared by Hevelius in Danzig, made into a summary by the author with the permission of the greatest and most renowned Master Hevelius.

I think that this comet was shining much earlier and it was seen for the first time here in Danzig on 5 April by the city guards; I (Johannes Hevelius) was informed on Monday, 6 April, that at 1.30 in the morning it was rising near our horizon towards the north-east. Its body was white and in part brightly yellowish, but it seemed smaller than the previous comet. The telescope showed one round nucleus in its centre, shining like gold; it was kept and surrounded by some matter — thinner and smooth — as it is plainly seen in the first image in Illustration D. At that time, it was in the very Chest of Pegasus, in 14° 22' of Pisces, at latitude 26° 30' north, viz. under those two stars in the Chest of Pegasus.

On the seventh, it was seen again; the distance from those stars changed at almost the same ratio as initially; within 24 hours it moved by 4° 6'. And in longitude it moved 4° 30' on the way which we deduced — judging by the position of the comet on both dates, and the distances drawn on a larger globe and extended to the eastern horizon — to lead through Pegasus under the Head of Andromeda towards the North Fish. But because no comet kept its route exactly on a great circle, I am sure that we should not rely too much on the analysis near the ends of its route; in this way, we may make an error concerning the point of the intersection. The comet did not change its light or brightness; its size seemed to change a little, and there was one nucleus in the centre. The tail was 12° or 13° long and pointed towards the star at the end of the Southern Wing of Cygnus.

On 8 April, at 2 o'clock in the morning, the comet made an almost equilateral triangle with Scheat in Pegasus and the preceding star in the Chest of Pegasus; it moved about 4°; the head was very bright, and the nucleus seemed slightly smaller and brighter than before. The tail was directed through the precedent star in the Chest of Pegasus towards its knee and was 15° long; above it, there was the subsequent star in the Chest of Pegasus. The head was seen to be equal to the lunar spot which is called Sicilia[2] (6'), and the nucleus to the lunar Corsica[3] (12" or 13"); or, the nucleus compared to Saturn did not reach the size of the round body in its centre.

On 9 April, it was raining.

On 10 April, at 1.54, the comet was rising and seemed bigger and brighter to the naked eye; but the nucleus seen with a telescope was not bigger or brighter. The tail was similar, 20° long, and stretched towards the Hoof of the Left, Northern Leg of Pegasus, almost touching Scheat with its upper part. The longitude was 2° of Aries, the latitude and motion were smaller.

On 11 April, very early in the morning, the comet was seen again; the comet was especially bright, and its colour was no different from that of Saturn, and the head was not changed; the tail stretched up to 22° when it was rising, and at dawn it diminished to 17°.

On the twelfth, the weather was unfavourable.

On the thirteenth, at 1.50 in the morning, the comet rose with the preceding tail, 25° long, which soon after became 20° long, and was directed towards the Left Knee of Pegasus.

On 14 April, the clouds did not allow any observations. On the fifteenth, the comet rose very bright and fresh, and with the telescope, it was seen to be slightly bigger. Before the comet appeared, the tail was 25° long from the head, and it touched the Left Shoulder of Andromeda with its upper part, as is printed at number 7. On that day, three notable things were observed: 1. the head consisted of the nucleus only, and it had almost no surrounding thinner matter; 2. the head or the tail was twinkling and emitting rays; 3. because a part of the tail at its base was darker, there was a gap, as if the nucleus were casting a shadow. The place of the comet was on the head of the North Fish, in 19° 27' of Aries at latitude 19° 38' north.

2 Crater Copernicus and its nimbus, see Ewen A. Whitaker, *Mapping and Naming the Moon. A History of Lunar Cartography and Nomenclature* (Cambridge, New York: Cambridge University Press, 1999), p. 207.

3 Crater Timocharis, see Whitaker, p. 203.

On 16 April, the comet seemed brighter than Saturn and almost as big as before, and the nucleus, seen with the telescope, was bigger, without any hair, except for a thorn towards the Sun, as it is clear from the image at number 8. Moreover, its diameter was smaller than the diameter of Jupiter, but was seen to equal that of Saturn (viz. that of its internal body); the shadow on the middle part of the tail (which was 20° long) was the same as yesterday. The tail was emitting rays equally, just like the head, and it was directed through the Right Shoulder to the three stars in the Arm of Andromeda; at dawn, which was bright and intense, the tail was entirely cut off, as can be seen in the illustration at number 9.

On the seventeenth, the turbulent air did not allow any observations.

On the eighteenth, the sight of the comet and the nucleus were the same; the tail was weaker due to some little clouds near the horizon. It was directed towards the upper star in the Hand of Andromeda, viz. towards the left one, and it was 1° or 1° 30' from the bright star in the Belt; it was about 15° long, and it was getting longer each day because of the position of the Sun.

On the nineteenth, due to vapours near the horizon, almost nothing was observed.

On the twentieth, at 3.43, the comet was seen at altitude 3° 45' near some stars, and at 4 o'clock it was seen clearly with the naked eye at altitude 6° 30'; it was placed under the vertex of Triangulum, a little over the Horns of Aries; its route seemed to be towards the right horn or the bright star of Aries; it was seen in 1° 18' of Taurus, at latitude 13° 14'. As far as its own movement is concerned, within 48 hours it passed 4° 51' and during the last day it moved 2° 23' only. Due to clouds near the horizon, initially, the tail was visible; later it was 3° or 4° long, and its thorn was vertical, slightly curved towards the north because the conjunction of the Sun and the comet had already passed. In the face of the head, there was no thinner matter; the comet was dilated towards both sides, as at number 10.

On the twenty-first, there were vapours near the horizon.

On the twenty-second, the fiery and turbulent weather hindered all observations.

But on 23 April, although the whole sky was clear, we could not find the comet anywhere — either with the naked or the armed eye — so it was visible for 14 days only, from 6 until 20 April; in that time, it passed 46° on its orbit. And this is the general description of the second comet; details are in the description published by the most renowned Hevelius.

Note

Ilustration G was made by the greatest Hevelius in order to present the discrepancies between the observations of the most renowned Masters Auzout and de Gottignies,[4] when compared to his very own.[5] It was meant to end the quarrel which arose between him (Hevelius) and Master Auzout who claimed that the comet continued its route to the second star of Aries, steadily slowing down, and he saw it for the last time on 17 March. It is presented in the illustration by the most renowned Petit, placed separately in the left margin; he also appealed to the observations by de Gottignies (which are given separately below) and said that they agreed with his own. On the contrary, the most renowned Hevelius outlined the route of the comet (which was seen by him for the last time on 18 February) to the first star of Aries and proved that after 4 February it (the comet) moved faster on its way according to the order of signs at the same ratio, and earlier — against the order of signs — it became slower. Afterwards, he showed that the observations by Auzout disagreed much with his own (Auzout's 4 February is between Hevelius's sixth and seventh, Auzout's sixth is equal to Hevelius's ninth, and so on) and that they are equally different from the observations by de Gottignies, as well as from those two by the most renowned Rudbeck[6] (placed above, in the convenient place, on the right). Therefore, with many arguments, he proved that after 15 February they all saw a different phenomenon, other than the one which was seen in January and in the first part of February. See his *Mantissa Prodromi cometici*, where anyone learned in mathematics will see everything that concerns this matter, analysed by the most precise and unmatched diligence, with painstaking calculations and the most certain arguments, proven and presented with singular pleasure and profit.

4 See note to Letter 16A.

5 See section 2.2.4 of the Introduction.

6 See section 2.4.16 of the Introduction.

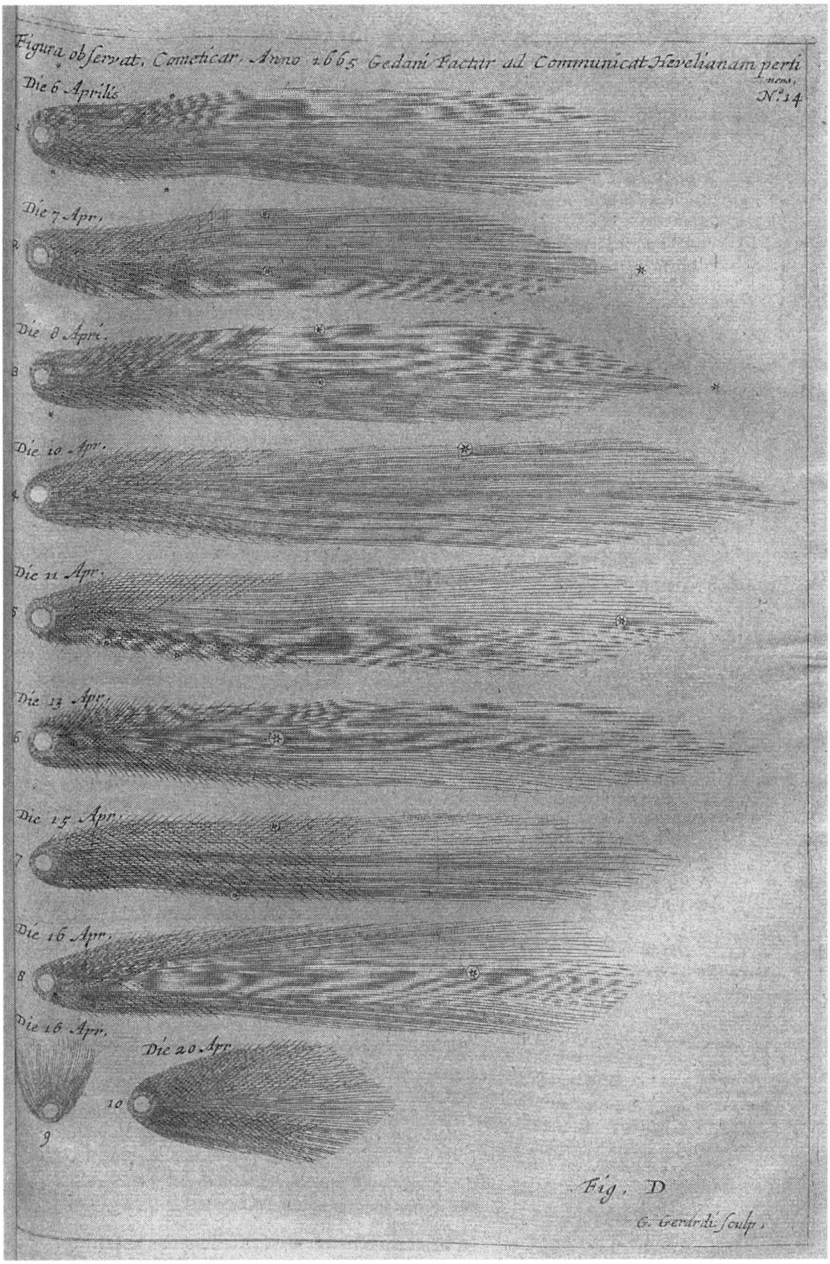

Figure 19. Hevelius's observations of the shape of the spring comet of 1665 in Lubieniecki's *Theatri cometici pars prior*. Biblioteka Narodowa, Warsaw, SD XVII.4.2660.

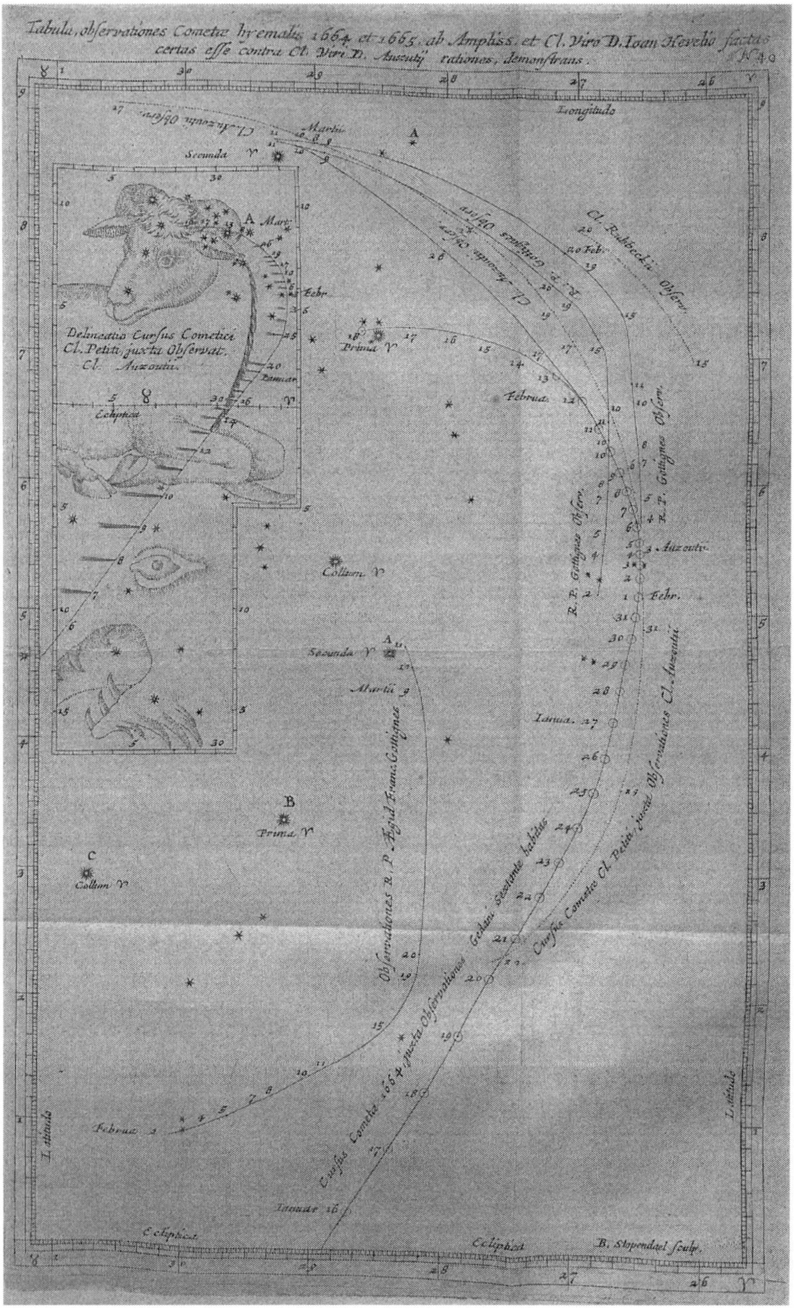

Figure 20. The comparison of the observations of the comet in February 1665 by Hevelius, Auzout, de Gottignies, Petit, and Rudbeck in Lubieniecki's *Theatri cometici pars prior*. Biblioteka Narodowa, Warsaw, SD XVII.4.2660.

Appendix 3.

Lubieniecki's Opinion on the Meaning of Comets[1]

TC3: 41–43

FUNDAMENTA
Sententiae contra determinatam Astrorum efficentiam et pro indifferenti ac aequa bonis malisque Cometarum significatione morali disputantis.

I. Astrologicas quidem praedictiones determinatas assidua refutat experientia, juvat tamen ex occasione Cometarum, quibus illi plurimum tribuunt plures rationes in hanc rem adferre.

II. Omnes divinationes in genere, et praedictiones ex astris in specie, damnantur absolute in Script[ura] S[acra] Levit[ico] XIX. 27. Deut[eronomio] XVIII. 10, 11. Esa[ia] XIX. 12, XLIV. 25. XLVII. 9, 12, 13. Itaque sine periculo saltim talia curiosa tractari non possunt.

III. Si haec damnantur in Chaldaeis et Aegyptiis, tantominus ferenda sunt in Christianis, quos ad omnem sanctitatis perfectionem adspirare decet, nisi admodum sobrie illa tractentur, ignis instar.

IV. Providentiae divinae, omnia libere regenti, sidera quoque et humana cuncta pro arbitrio juste temperanti, per fatum Astrologicum detrahitur.

V. Per siderum efficientiam fatalem, vel tollitur vel laeditur libertas hominis.

VI. Ex multorum sententia sidera lucent tantum, non agunt, saltim non eo modo, quo communiter agere putantur in terrena, utpote immane quantum remota. Sol solus agit, tum demum Luna, sed mutuatitio lumine: quanquam nec aliud omnibus stellis quidam tribuant, et hoc solo Planetas (quos solos Astrologi crepare solent, nisi quod et situs eorum ad fixas rationem etiam aliquam habeant) gaudere certum vi-

1 See section 3.3.3 of the Introduction.

deatur. Vid[e] Kepler[um] l[ibro] 6. *Epit*[*omes*] *Astron*[*omiae*] *Coper*[*nicanae*] Weigel Comm[entariis] de Cometis 1664 et 1665. c[apitibus] VI. VII. et seq[uentibus] et Clarissimus Ballialdus l[ibro] 1. *Astronom*[*iae*] *Philol*[*aicae*] c[apitibus] X. XI. et seq[uentibus]. Ubi quidem Vir Doctissimus probare nititur Planetas proprium quoque et ingenitum lumen, praeter illud adventitium habere, imbecillius tamen esse ad oculos non facile pervenire et efficaciter pupillam non ferire fatetur. Et, ut nil dicam de sententia Kepleri, Magni profecto Mathematici, imo inter maximos merito ponendi, et aliorum qui Planetas ullum lumen congenitum obtinere negant, et illos diversos planetarum a Sole, secundum certos proportionum gradus illuminatorum, colores diversae constitutioni illorum corporum, tum et diverso situi tribuunt, ac res luce propria carentes si a Sole illuminentur diversos colores exhibentes in exemplum adferunt, ex ipsa Viri Doctissimi confessione, Planetas dominatum in terras, multo minus in mentem hominis, non exercere, et Astrologicas praedictiones a Planetarum efficientia deductas corruere satis patet.

VII. Etsi operatio siderum in inferiora concedatur, tamen specialis eorum virtus determinari et effectus certi ex ea praedici, ob varia impedimenta huius rei explorationem turbantia nequeunt.

VIII. Servire non saevire astris datum Gen[esi] 1. nec quicquam in historia creationis de maligna vi eorum in inferiora influente legimus.

IX. Cometa proprie, et stricte loquendo non est signum, multominus naturale, (licet sit naturale phoenomenon) et necessarium, sed signum supernaturale, arbitrarium, memoriale, doctrinale, quod pro diversitate hominum illud contemplantium et usu vel abusu donorum et operum Dei huiusque ipsius Phoenomeni, bona vel mala Ethice significat et praenunciat.

X. Cometa si signum naturale foret, bona potius quam mala significaret, cum natura bonum intendat et illud efficiat, nisi impediatur. Impeditur vero hic vitio hominum multorum (non tamen omnium) officii negligentium, bonis Dei abutentium: licet haec bona in se semper maneant.

XI. Cometa, etsi sit aereus, non necessario agit in aerem (in quo naturam sine ulla Cometarum apparitione agere assiduo videmus) multominus in remotiora, quae in terra sunt, omnium mimine in liberam mentem hominis, quae a solo DEO conditore pendet.

XII. Cometae sunt phoenomena illustria, lucida, jucunda visu, rariora et praestantiora caeteris phoenomenis temporariis, adeoque specialia, ut ideo plura ad praeviam eorum apparitionem non tantum tristia sed et laeta Ethice saltim referri possint, tam generali quam etiam speciali DEI voluntate, quam omnia eius opera, edicta justitiam inculcantia docent, considerata.

XIII. 'Futurorum malorum ignoratio utilior quam scientia.' Cic[ero] I. *De div*[*inatione*], saltim si utilis scientia malorum utilis et bonorum.

XIV. Cometas laeta quoque et funesta significare quidam veteres Philosophi et multi recentiores statuunt, ut alibi probavimus.

XV. Pro vario caeli positu Cometas faustos Astrologi ipsi esse fatentur: Lunares enim in occasu, et Joviales, omnia bona, eos vero qui caudam dirigant in occasum, optimam anni constitutionem praenunciare dicunt. Ita et Orientales et Occidentales respectu caudae Cometas laetos admittunt. Imo vero teste Bertio initio Geograph[iae] et meridionales Cometae, iique matutinis oppositi, fausti judicantur.

XVI. In variis caeli plagis conspiciuntur, et modo in aere, modo in aethere, ut multorum fert sententia, generantur Cometae. Ideo nil certi in specie ex eorum apparitione colligi potest.

XVII. Nec illis, quibus verticales sunt, nocent (quod illustri illius Cometae anno 1618. visi aliorumque exemplo probatur) nec illis in quos caudam dirigunt, nec etiam istis, qui subiecti sunt illis signis quae Cometae percurrunt, quod itidem multa exempla testantur.

XVIII. Experientia saepius fallax est, non quidem in se, sed exploratorum vitio vel observationum difficultate.

XIX. Annales Cometarum non raro discrepant, ita ut exinde nil certi depromi possit, cum tamen ex primo apparitionis tempore et duratione, Comerarum quoque effectus formentur.

XX. Saepe Cometomantes et Cometarii qui vocantur[2] sibimet contradicunt, non tantum de origine (de qua nil certi constat) sed et de Situ, motu et effectis Cometarum.

XXI. Vix ad[3] triennium perpetuorum siderum operationem extendunt Astrologi, ut ait Clarissimus Bullialdus: Cometomantes vero saepius ad plures annos Cometarum effectus extendunt.

XXII. Ante Cometas tristia, post hos laeta, sine illorum apparitione utraque, eveniunt multa, ita ut nil certi in alterutram partem determinari queat.

2 vocantur *corr. ed.* : vocantut [*sic*] **TC3**.
3 ad *corr. ed.* : ad ad **TC3**.

XXIII. Cometae cum vel omnes vel plerique in aethere generentur, tollitur efficientia, quae eis, ut phaenomenis aereis, in vicina tribuitur.

XXIV. Cometas semper existere, licet non semper appareant, multi veteres et recentiores philosophi docent. Quod si verum est, arguit eos inter portenta proprie dicta non esse reputandos.

XXV. Tutissimum est in tanto sententiarum incerto et discrimine, mediam viam tenere, ut Cometas, aliave signa caeli, nec temnamus cum impiis, nec timeamus cum superstitiosis, nec despiciamus, nec suspiciamus, sed ut incitamenta virtutis contemplemur.

XXVI. Mihi sufficit Cometam ad mentem Ethices et institutionis quae secundum pietatem est, quae morum Sanctimoniam ex quavis occasione commendat et inculcat, tanquam signum arbitrarium, et memoriale ob oculos bonis malisque, cum utrisque necessaria sint talia monita et documenta undecunque conquisita, posuisse, sive laeta, sive tristia sive (quod potius verum est ut assidua docet experientia) laeta hic, tristia illic Cometae apparitionem sequantur: Ethici enim officium occasione Cometae defungi volo.

XXVII. Cum bonis omnia, etiam mala, in bonum cooperentur ut S[acra] Scriptura docet, et malis, qui bonis DEI abutuntur, bona quoque cedant in malum, merito dico Cometam bona bonis, etam si tristia et mala eis obtingant, mala vero malis, licet bona eis obveniant, significare et denunciare. Sapiens arte industria, pius virtute constanti mala corrigit, et in bonum, DEO auxiliante, vertit. Insipiens inertia ignava, Impius malitia temeraria et pervicaci bona corrumpit, et in malum, DEO permittente vertit. Tam illi quam huic officia in memoriam occasione Cometarum, quibus plerique cuncta mala adscribunt, et quos naturali rerum serie laeta et tristia sequuntur, revoco.

XXVIII. Accurate proprie et stricte, loquendo, et respectu unius subiecti eodem modo se habentis, una res duo contraria significare non potest. Sed citra ἀκρίβειαν Grammaticam et Logicam, Politici, Historici, ac Ethici, saltim Oratoris potius quam Physici more, improprie et latius accepta significandi et portendendi voce, Cometa tanquam mutus Concionator vel Consiliarius tam bonis bona, quam malis mala significat et portendit. Nam et una res pro diversitate subiecti in quod agit (ut Sol, Pluvia et c[etera]) contraria non tantum significat, sed et operatur ac efficit: si vel diverso modo, ut quidam contendunt, id facere dicatur.

XXIX. Integra DEI aequitas et bonis bona et malis mala semper et ubique tam explicite quam implicite (ut omne contrarium est index sui et obliqui) denunciat. Huius itaque voluntati pietatis amans accommodare se in vita et doctrina debet.

XXX. Cuncta signa ex instituto Dei laeta piis futura, ex natura contrariorum tristia sunt et erunt impiis, ut iris et signum Filii hominis in ultimo illo die, adeoque hoc singulari privilegio gaudent muti isti, ut sic dicam, concionatores, ut simul contraria denuncient, quod vocales praestare non possunt, qui contraria contrariis verbis efferunt: Sic et Scriptura Sacra in universum considerata bona bonis, mala malis: sed hic vel ille Propheta et Apostolus his vel illis verbis seorsum bona bonis, mala malis in specie denunciat. Clarius dabo exemplum Esa[ias] c[apite] 3. Paulus Apost[olus] Rom[anos] II. bonis bona, malis mala universim denunciat: Sed in specie bonis bona Esa[ias] 3. 10. Paulus Rom[anos] 2. v[ersu] 7. malis mala, Esa[ias] v[ersu] 11. Paulus v[ersibus] 8, 9. Idem toties fit alibi. Sic etiam signum et adventus Filii hominis Matt[hiae] XXIV. XXV. et Luc[ae] XXI. bonis bona, malis mala universim denunciat: Sed speciatim bona bonis Matt[hiae] XXIV. 31. XXV. 34, 35, 36. Lucae XXI. v[ersibus] 28. malis mala Matt[hiae] XXIV. v[ersu] 30. XXV. v[ersis] 41, 42, 43, quin et coniunctim utrisque convenientia Matt[hiae] XXIV. v[ersibus] 40, 41. XXV. v[ersu] 46. Ita igitur et Cometa signum improprie dictum, arbitrarium, doctrinale, Concionatori et Consiliario comparatum, simul bonis bona, malis mala, unica apparitione denunciat, nisi forte quis eum, fulgore grato bona bonis, cauda vero minaci, virgae simili, mala malis denunciare, dicere malit. Qua libertate eum frui ego facile feram, eandemque cuivis morum probitatem hinc et illinc allatis rationibus et exemplis commendare volenti integram relinquo, eodemque aequitatis jure meam mihi relinqui postulo. Quod reliquum est, cum omnia in se habeat, quem penes est virtus, Plauto judice, haecque in omnium animos lumen suum immittat, adeo ut etiam qui non sequuntur eam, videant, Seneca teste, ex animo opto, ut bruta quoque animantia non tantum intelligere (quod Aristo Chius in votis habuit apud Plutarchum) sed et eloqui possint verba ad virtutem incitantia. Caeteroquin caeli enarrant gloriam DEI Psal[mo] XIX. eique benedicunt omnia opera eius Psal[mo] CXLVIII. Dan[iel] 3. et alibi.

The foundations of the opinion which argues against the predetermined influence of the stars and for the moral meaning of comets which is indifferent and alike both for the good people and for the bad.

I. As frequent experience shows, predetermined astrological predictions are refuted; on the occasion of comets, to which astrologers ascribe much meaning, it is helpful to present more arguments in this matter.

II. All manners of foretelling the future in general (and especially astral predictions) are entirely condemned in the Holy Scripture: Leviticus 19. 27,[4] Deuteronomy 18. 10,

4 Lubieniecki probably meant here Leviticus 19. 26: 'Ye shall not eat any thing with the blood: neither shall ye use enchantment, nor observe times.'

11,[5] and Isaiah 19. 12,[6] 44. 25,[7] 47. 9, 12, 13.[8] Therefore, such a curious matter cannot be considered without an element of risk.

III. If such doings are condemned among the Chaldeans or the Egyptians, there should be less of them among the Christians, who should aspire to the perfection of their sanctity, unless such predictions are dealt with very considerately, as one should deal with fire.

IV. Due to astrological fate, divine providence — which freely rules over everything and arranges the stars and all human things justly according to its wish — is offended.

V. Because of the fatal influence of the stars, human freedom is abolished or diminished.

VI. The opinion of many is that the stars only shine and do not act — at least not in the way in which they are commonly considered to act — as they are extremely far away. Only the Sun acts, and then the Moon — but it acts only with borrowed light; and even though some people ascribe it to other stars as well, only the planets (the astrologers are used to talking only about them, unless their position is somehow related to the fixed stars) seem to certainly have it. See Kepler in Book 6 of the *Epitome astronomiae Copernicanae*,[9] Weigel in Chapters 6, 7 and following of his commentary about the comets of 1664 and 1665,[10] and the most renowned Boulliau in Book 1, Chapters 10, 11 and following of his *Astronomia Philolaica*.[11] This most learned man tries there to prove that also the planets have their own, natural light, apart from

5 'There shall not be found among you any one that maketh his son or his daughter to pass through the fire, or that useth divination, or an observer of times, or an enchanter, or a witch, or a charmer, or a consulter with familiar spirits, or a wizard, or a necromancer.'

6 'Where are they? where are thy wise men? and let them tell thee now, and let them know what the LORD of hosts hath purposed upon Egypt.'

7 'That frustrateth the tokens of the liars, and maketh diviners mad; that turneth wise men backwards, and maketh their knowledge foolish.'

8 'But these two things shall come to thee in a moment in one day, the loss of children, and widowhood: they shall come upon thee in their perfection for the multitude of thy sorceries, and for the great abundance of thince enchantments. [...] Stand now with thine enchantments, and with the multitude of thy sorceries, wherein thou hast laboured from thy youth; if so be thou shalt be able to profit, if so be thou mayest prevail. Thou are wearied in the multitude of thy counsels. Let now the astrologers, the stargazers, the monthly prognosticators, stand up, and save thee from these things that shall come upon thee.'

9 Johannes Kepler, *Epitomes astronomiae Copernicanae libri V, VI, VII* (Francofurti: sumptibus Godefridi Tampachii, 1621), pp. 824–25.

10 Erhard Weigel (1625–99), German mathematician and astronomer. Lubieniecki refers to his *Fortsetzung des Himmels-Spiegels* (Jena: in Verlagung Thomas Matthias Götzen, 1665), pp. 52–69.

11 Ismaël Boulliau (see section 2.4.2 of the Introduction), *Astronomia Philolaica* (Parisiis: sumptibus Simonis Piget, 1645), pp. 17–20. The illumination of the planets by the Sun is discussed in Chapters 10 and 11 only.

the one which they receive, but he also admits that it is weaker and it does not reach our eyes easily, and it does not hit the pupils efficiently. Leaving aside the opinion of Kepler, a great mathematician, deservedly placed among the greatest, and of those who deny that the planets have any natural light and who ascribe various colours of the planets (which are illuminated by the Sun to a certain extent) to their various constitution and various place, and who, giving it as an example, say that things deprived of their own light, when illuminated by the Sun, present various colours — it is clear from the declaration of the most learned man that the planets have no rule over Earth, and even less over human minds, and that astrological predictions (drawn from the planetary influence) collapse.

VII. Even if it were assumed that stars affect the things below them, it would be impossible to determine this exact faculty of theirs and to predict any specific effects due to various obstacles that hinder the investigation of this matter.

VIII. Stars were given the faculty of serving, not of frightening (Genesis, Chapter 1),[12] and in the whole history of creation, we read nothing about their evil influence on the things below them.

IX. Precisely and strictly speaking, a comet is not a sign, and especially not a natural (despite the fact it is a natural phenomenon) and necessary one, but a supernatural, arbitrary, mnemonic and doctrinal sign; because of the diversity of people who contemplate it, and because they use or abuse the gifts and works of God — and this phenomenon, too — it signifies and announces good or bad things, in the ethical sense.

X. If a comet were a natural sign, it would signify good things rather than bad, because nature tends to good and causes it, if it is not disturbed. And there it is hindered by the mistakes of many people (but not all) who neglect their duties and abuse the good received from God; this good, however, is in itself intact.

XI. Even if a comet were of airy nature, it would not have to act in the air (we see that the nature acts there all the time without any presence of comets), and even less in the further places which are on Earth, and the least of all it would act in the free mind of a man, which is dependent on God the Creator only.

XII. Comets are shining and bright phenomena, pleasant to see, less frequent and more excelling than other temporary phenomena; they are so exceptional that not only many sad things but also those fortunate ones — at least in the ethical sense —

12 Genesis 1. 14–15: 'And God said, Let there be lights in the firmament of the heaven to divide the day from the night; and let them be for signs, and for seasons, and for days, and years: and let them be for lights in the firmament of the heaven to give light upon the earth: and it was so.'

can be related to their previous appearances, if we consider both the general and the special will of God and how His works are said to teach justice.

XIII. 'Ignorance of future ills is more profitable than the knowledge of them,' Cicero, Book 1 *De Divinatione*[13]; if the knowledge of ills is profitable, so is the knowledge of fortunes.

XIV. Ancient philosophers and many contemporary ones state that comets signify both auspicious and pernicious things; we proved it elsewhere.

XV. Astrologers admit that comets are fortunate because of their place in the sky; they say that the lunar ones in the west, much like the jovial ones, foretell all the good things; and those which direct their tails towards the west foretell the best condition of the year. And so they admit that both eastern and western comets (in relation to their tails) are fortunate. And, as Bertius testifies at the beginning of his geography,[14] both the southern comets and those opposite the morning ones are considered auspicious.

XVI. Comets are seen in various parts of the sky and come into being in the air and in Aether, as many people say. Therefore, no certain details about their appearance can be stated.

XVII. They neither harm those over whom they are vertical (which is proven by the example of the comet seen in 1618[15] and others), nor those to whom they direct their tails, not even those who are subjected to the signs which the comets convey, which is again proven by many examples.

XVIII. Experience is often fallacious, not in itself but due to the mistakes of the observers and the difficulty in observing.

XIX. The cometary annals often differ, so nothing certain can be found in them, while the effects of a comet are based on the time of its first appearance and on its duration.

XX. The cometary diviners and the cometarians (as they are called) often contradict one another, not only about the origin of comets (about which nothing certain is known), but also about their place, movement, and effects.

13 Cicero, *De divinatione*, 2, 9, 23 ('Book 1' in Lubieniecki's text is a mistake), in Cicero, *De senectute, De amicitia, De divinatione*, transl. by William Armistead Falconer (Cambridge, Mass., London: Harvard University Press, 1923). Quoted with minor changes.

14 Petrus Bertius (Pieter de Bert, 1565–1629), Flemish philosopher, historian, and geographer. Lubieniecki may refer here to his edition of Ptolemy's *Geography* — *Theatri geographiae veteris tomus prior* (Amstelodami: ex officina Iudoci Hondii, 1618) — but I could not find the exact passage.

15 See note to Letter 1F.

XXI. Astrologers extend the influence of permanent stars almost to three years, as the most renowned Boulliau says; but the cometary diviners often extend the effects of comets to even more years.

XXII. Many sad things happen before comets appear, many fortunate things happen afterwards, and many of both happen even if no comets appear; therefore, nothing certain can be concluded about any of these views.

XXIII. Because all comets, or most of them, come into being in Aether, their influence on neighbouring objects (which is ascribed to them as to airy phenomena) does not exist.

XXIV. Many ancient and modern philosophers teach us that comets always exist, even if they do not always appear. If this is true, it is an argument that they should not be considered to be omens *per se*.

XXV. In the face of uncertainty and so many contradicting opinions, it is safest to be in the middle — not to scorn the comets or other celestial signs (as the impious do), not to fear them (as the superstitious do), neither to despise nor to admire them, but to consider them as incentives to virtue.

XXVI. For me, it is enough to say that a comet is placed as an arbitrary and mnemonic sign with ethical and pious meaning; given a chance, this piety advises and recommends to us the sanctity of our conduct; the comet is placed both in front of good and bad people alike, because they need such admonitions and examples from everywhere, regardless of whether the comet is followed by fortunate events, or by sad ones, or (which — as experience often teaches us — is probably true) fortunate events in some places, sad ones in other places; on the occasion of a comet, I want to fulfil my ethical duty.

XXVII. Everything — even bad things — work with the good ones for the sake of good, as the Holy Scripture teaches us, and for bad people who abuse good things from God even good things turn to bad; therefore, I am right to say that a comet signifies and announces good things to good people (even if sad and bad things happen to them), and bad things to bad people (even if good things happen to them). With God's help, a wise man with his talent and diligence, and a pious man with his constant virtue correct the bad and turn it into good. With God's admittance, an unwise man with his slothful idleness, and an impious man with his reckless and persistent malice destroy what is good and turn it into bad. On the occasion of comets to which many people ascribe all the evil and which, by the natural order of things, are followed by fortunate and sad things, I remind those men of their duties.

XXVIII. Precisely, correctly, and strictly speaking, in respect to one subject, one thing cannot signify two contradicting things. But leaving aside the accuracy of grammar and logic, when we speak in the guise of a politician, historian, and ethicist — or at least of an orator, rather than in the guise of a physicist — and when we are imprecisely and too broadly defining the meanings of 'signifying' and 'portending', a comet, as a mute preacher or counsellor, signifies and portends both good things to good people and bad things to the bad. Because due to the diversity of the objects on which it acts, one thing (such as the Sun, rain and so on) not only signifies something, but also influences and works on these objects, even if, as some people claim, it does so in a different way.

XXIX. The absolute justice of God announces both good things to good people and bad things to the bad — always and everywhere, both explicitly and implicitly (because every contradiction indicates at the same time one thing and its opposite). Therefore, the one who loves piety should adapt his life and beliefs to His will.

XXX. Because of God's decision, all signs which are fortunate for the pious — due to the nature of the opposition — are and will be sad for the impious, like the rainbow or the signs of the Son of Man on the last of days; therefore (let me say so) those mute preachers enjoy this prerogative to announce contradictory things simultaneously, which cannot be achieved by a human voice which utters opposite things with opposite words. And the Holy Scripture, generally considered, announces good things to good people and bad things to the bad; but in detail, in many places, a prophet or an apostle says so — good things to good people and bad things to the bad — separately, with particular words. I will give a clear example: Isaiah, Chapter 3, and Paul the Apostle, Romans, Chapter 2, announce good things to good people and bad things to bad people in general. But in detail, Isaiah, 3. 10,[16] and Paul, Romans, 2. 7,[17] announce good things to good people; and Isaiah in verse 11,[18] Paul in verses 8, 9[19] — bad things to bad people. And it happens many times there. And Matthew 24, 25 and Luke 21 say that the sign and the coming of the Son of Man announce good things to good people and bad things to bad people in general. But in detail, Matthew 24. 31,[20]

16 'Say ye to the righteous, that it shall be well with him: for they shall eat the fruit of their doings.'

17 'To them who by patient continuance in well doing seek for glory and honour and immortality, eternal life.'

18 'Woe unto the wicked! it shall be ill with him: for the reward of his hands shall be given him.'

19 'But unto them that are contentious, and do not obey the truth, but obey unrighteousness, indignation and wrath, tribulation and anguish, upon every soul of man that doeth evil, of the Jew first, and also of the Gentile.'

20 'And he shall send his angels with a great sound of a trumpet, and they shall gather together his elect from the four winds, from one end of heaven to the other.'

25. 34, 35, 36,[21] and Luke 21. 28[22] say about good things to good people, and Matthew 24. 30,[23] 25. 41, 42, 43[24] — about bad things to bad people, and Matthew 24. 40, 41,[25] 25. 46[26] — about the joined equivalence of both. And, therefore, a comet, improperly called an arbitrary and doctrinal sign, compared to a preacher and a counsellor, with its single appearance announces both good things to good people and bad things to bad people, unless someone prefers to say that with its pleasant brightness it announces good things to good people, and with its threatening tail, similar to a rod, it announces bad things to bad people. I will easily let him do as he wishes and I let the intact liberty to anyone who wants to recommend the probity of conduct with various reasonings and examples, and I ask for such a prerogative for myself. As far as the rest is concerned, because a comet contains everything in itself which is close to virtue, as Plautus says,[27] and this virtue sends its light to the souls of all, so that even those who do not follow it see it, as Seneca testifies,[28] I hope that even mindless creatures will not only understand (which was hoped by Ariston of Chios in Plutarch),[29] but they will also speak the words which incite people to virtue. Moreover, the sky speaks the glory of God — Psalms 19,[30] and all His works bless Him — Psalms 148,[31] Daniel 3,[32] and elsewhere.

21 'Then shall the King say unto them on on his right hand, Come, ye blessed of my Father, inherit the kingdom prepared for you from the foundation of the world; for I was an hungred, and ye gave me meat: I was thirsty, and ye gave me drink: I was a stranger, and ye took me in: naked, and ye clothed me: I was sick, and ye visited me: I was in prison, and ye came unto me.'

22 'And then these things begin to come to pass, then look up, and lift up your heads; for your redemption draweth nigh.'

23 'And then shall appear the sign of the Son of man in heaven: and then shall all the tribes of the earth mourn, and they shall see the Son of man coming in the clouds of heaven with power and great glory.'

24 'Then shall he say also unto them on the left hand, Depart from me, ye cursed, into everlasting fire, prepared for the devil and his angels: for I was an hungred, and ye gave me no meat: I was thirsty, and ye gave me no drink: I was a stranger, and ye took me not in: naked, and ye clothed me not: sick, and in prison, and ye visited me not.'

25 'Then shall two be in the field; the one shall be taken, and the other left. Two women shall be grinding at the mill; the one shall be taken, and the other left.'

26 'And these shall go away into everlasting punishment: but the righteous into life eternal.'

27 An allusion to Plautus, *Amphitruo*, Act 2, Scene 2, verse 21, 'Virtus omnia in se habet: omnia assunt bona, quem penes est virtus'.

28 Seneca, *De beneficiis*, 4, 17.

29 Plutarch, *Maxime cum principibus philosopho esse disserendum*, 776c. Ariston of Chios (*fl.* 1st half of the 3rd century BC) was a Stoic philosopher. His development of Zeno's teaching, focusing chiefly on ethics, was discarded by later Stoics.

30 Psalms 19. 1, 'The heavens declare the glory of God; and the firmament sheweth his handywork.'

31 Psalms 148 *passim*.

32 Daniel (deuterocanonical/apocryphal) 3. 57 (*The Song of the Three Holy Children*, 35), 'O all ye works of the Lord, bless ye the Lord: praise and exalt him above all for ever.'

Bibliography

Manuscripts

Bibliothèque de l'Observatoire, Paris, ms C1, v. 6–11.
Bibliothèque nationale de France, ms Lat 10348 v. 9–11, FF. 13026, FF. 13031.
Universiteitsbibliotheek, Amsterdam, ms Ar 50.
Universiteitsbibliotheek, Leiden, ms HUG 45, PAP 15.
Muzeum Narodowe, Warsaw, ms 518.

Early Printed Books

Alsted Johann Heinrich, *Thesaurus chronologiae* (Herbornae Nassoviorum: [s.e.], 1650).

Angelocrator Daniel, *Rationis temporum et rerum totius orbis memorabilium ab initio mundi usque ad Rodolphum II Romanorum imperatorem libri novem* (Francofurti: typis Johannis Lancelloti, impensis Jonae Rhodii, 1611).

Annalium et historiae Francorum ab anno Christi DCCVIII ad ann. DCCCCXC scriptores coaetanei XII, vol. 2, ed. by Pierre Pithou (Parisiis: apud Claudium Chappelet, 1588).

Auzout Adrien, *Lettre de monsieur Auzout du 17 Juin à monsieur Petit* (Paris: [s.e.], 1665).

Aventinus Johannes, *Annalium Boiorum sive veteris Germaniae libri VII*, ed. by Nicolaus Cisner, (Francofurti: impensis Ludovici Regis, 1627).

Barclay John, *Satyricon* (Augustae Trebocorum: apud Joannem Hessium, 1623).

Bartholin Erasmus, *De cometis anni MDCLXIV et MDLXV opusculum* (Hafniae: apud Danielem Paulli, 1665).

Bartholin Erasmus, *Specimen recognitionis nuper editarum observationum astronomicarum nobilissimi viri Tychonis Brahe* (Hafniae: prostat apud Danielem Pauli, literis Henrici Gödiani, 1668).

Bartholin Thomas, *De cometa consilium medicum cum monstrorum nuper in Dania natorum historia* (Hafniae: apud Matthiam Godicchenium, 1665).

Bayer Johann, *Uranometria* (Augustae Vindelicorum: excudis Christophorus Magnus, 1603).

Bayle Pierre, *Pensées diverses écrites à un docteur de Sorbonne à l'occasion de la Comète qui parut au mois de décembre 1680* (Rotterdam: chez Reinier Leers, 1683).

Beauvais Vincent of, *Speculi maioris* [...] *tomus quartus, qui Speculum historiale iscribitur,* (Duaci: ex officina typographica Baltazaris Belleri, 1624).

Beckmann Berhnad, *Methodus conciliandi* (Francofurti ad Oderam: typis Eichornianis, 1663).

Bibliotheca Heinsiana, sive catalogus librorum quos magno studio et sumtu, dum viveret, collegit Nicolaus Heinsius (Lugduni Batavorum: apud Johannem de Vivie, 1682).

Bock Friedrich Samuel, *Historia Socinianismi Prussici* (Regiomonti: typis et impensis Io. Henr. Hartungii, 1754).

Bodin Jean, *Universae naturae theatrum* (Francofurti: apud heredes Andreae Wecheli, 1597).

Boulliau Ismaël, *Ad astronomos monita duo. Primum de stella nova, quae in collo ceti ante annos aliquot visa est. Alterum de nebulosa in Andromeda cinguli parte borea ante biennium iterum orta* (Parisiis: apud Sebastianum Mabre-Cramoisy, 1667).

Boulliau Ismaël, *Astronomia Philolaica* (Parisiis: sumptibus Simonis Piget, 1645).

Brahe Tycho, *Epistolarum astronomicarum liber primus* (Uraniburgi: ex officina typographica authoris, 1596).

Brahe Tycho, Curtz Albert, *Historia coelestis* (Augustae Vindelicorum: apud Simonem Utzschneiderum 1666).

Buchanan George, *Opera omnia*, vol. 2, part 1 (Lugduni Batavorum: apud Johannem Arnoldum Langerak, 1725).

Buchholzer Abraham, *Index chronologicus* (Francofurti: excudebat Wolfgangus Hoffmannus, sumptibus Viduae Jonae Rosae, 1634).

Bünting Heinrich, *Chronologia* (Servestae: impressum typis Bonaventurae Fabri, 1590).

Calvisius Sethus, *Opus chronologicum* (Francofurti: impensis Johannis Thymii, 1620).

Cardano Girolamo, *In Cl. Ptolemaei de astrorum iudiciis, aut (ut vulgo appellant) Quadripartitae Constructionis Lib. IIII,* (Basileae: ex officina Henricpetrina, 1578).

Cassini Giovanni Domenico, *De aliis Romanis observationibus macularum Martis* (Bononiae: ex typographia HH. de Ducciis, 1666).

Cassini Giovanni Domenico, *De periodo quotidiana revolutionis Martis* (Bononiae: typis HH. de Ducciis, 1666).

Cassini Giovanni Domenico, *Lettera astronomica sopra l'ombre de pianetini Medicei in Giove* (Roma: appresso Fabio de Falco, 1665).

Cassini Giovanni Domenico, *Lettere astronomiche sopra la varietà delle macchie osservate in Giove, e loro diurne rivoluzioni* (Roma: [s.e.], 1665).

Cassini Giovanni Domenico, *Martis circa axem proprium revolubilis observationes* (Bononiae: ex typographia HH. de Ducciis, 1666).

Castelli Bartolomeo, *Lexicon medicum Graeco-Latinum* (Basileae: impensis Joh. Jacobi Genathi, 1628).

Creutzer Peter, *Außlegung uber den erschröcklichen Cometen, so im Westrich und umbligenden grentzen erschinen, am XI. tag Weinmonats, des M.D.XXVII. jars* (Nürnberg: durch Georg Wachter, [1528]).

Chyträus David, *In Genesin enarratio* (Vitebergae: excudebat Johannes Crato, 1561).

Clarorum virorum epistolae centum ineditae (Amstaeledami: prostant apud Sebastianum Petsoldum, 1702).

Cysat Johann Baptist, *Mathemata astronomica de loco, motu, magnitudine et causis cometae qui sub finem anni 1618 et initium anni 1619 in coelo fulsit* (Ingolstadii: ex typographeo Ederiano, 1619).

Descartes René, *Discours de la méthode pour bien conduire sa raison et chercher la vérité dans les sciences, plus la dioptrique, les météores et la géométrie qui sont des essais de cette méthode* (Leyde: à l'imprimerie de Ian Maire, 1637).

Eckstorm Heinrich, *Historiae eclipsium, cometarum et pareliorum* (Helmaestadi: typis heredum J. Lucii, 1621).

Erastus Thomas et al., *De cometis disserationes novae* ([Basel]: ex officina Leonardi Ostenii, 1580).

Funck Johann, *Chronologia* (Basileae: excudebat Jacobus Parcus sumptibus Joannis Oporini, 1554).

Galilei Galileo, *Il saggiatore* (Roma: appresso Giacomo Mascardi, 1623).

Gassendi Pierre, *Mercurius in Sole visus et Venus invisa* (Parisiis: sumptibus Sebastiani Cramoisy, 1632).

Gassendi Pierre, *Nicolai Claudii Fabricii de Peiresc, senatoris Aquisextiensis, vita* (Parisiis: sumptibus Sebastiani Cramoisy, 1651).

Gemma Cornelius, *De prodigiosa specie, naturaque cometae, qui nobis effulsit altior lunae sedibus, insolita prorsus figura, ac magnitudine, anno 1577 plus septimanis 10. apodeixis tum physica tum mathematica* (Antverpiae: ex officina Christophori Plantini, 1578).

Georgii Cedreni annales (Basileae: per Ioannem Oporinum, 1566).

Graevius Johann Georg, *Oratio de cometis contra vulgi opinionem cometas esse malos nuncios* (Traiecti ad Rhenum: typis et officina Gisberti a Zyll, 1665).

Grandami Jacques, *Le Parallèle des deux comètes qui ont paru les années 1664 et 1665* (a Paris: chez Sebastien Cramoisy, et Sebastien Mabre-Cramoisy, 1665).

Guericke Otto von, *Experimenta nova (ut vocantur) Magdeburgica de vacuo spatio* (Amstelodami: apud Joannem Janssonium a Waesberge, 1672).

Hájek Tadeáš, *Dialexis de novae et prius incognitae stellae inusitatae magnitudinis et splendidissimi luminis apparitione, et de eiusdem stellae vero loco constituendo*, (Francofurti ad Moenum: [s.e.], 1574).

Ἡσιόδου Ἀσκραίου τὰ εὑρισκόμενα. Hesiodi Ascraei quae exstant, ed. by Johann Georg Graevius (Amstelodami: apud Danielem Elzevirium, 1667).
Helvicus Christophorus, *Theatrum historicum, sive chronologiae systema novum* (Gissae Hessorum: excudebat Nicolaus Hampelius, 1609).
Hevelius Johannes, *Descriptio cometae anno aerae Christianae MDCLXV exorti* (Gedani: autoris typis et sumptibus, imprimebat Simon Reiniger, 1666).
Hevelius Johannes, *Epistola de cometa 1672 anno MDCLXXII Gedani observato* (Gedani: authoris typis et sumptibus imprimebat Simon Reiniger, 1672).
Hevelius Johannes, *Epistulae II. Prior de motu lunae libratorio. Posterior de utriusque luminaris defectu anni 1644* (Gedani: sumptibus autoris, typis Andreae Julii Molleri, 1654).
Hevelius Johannes, *Epistulae IV. I. De observatione deliquii Solis anno 1649 habita. II. De eclipsi Solis anno 1652 observata. III. De motu lunae libratorio. IV. De utriusque luminaris defectu anni 1644* (Gedani: sumptibus autoris, typis Andreae Julii Molleri, 1654).
Hevelius Johannes, *Firmamentum Sobiescianum* (Gedani: typis Johannis Zachariae Stolli, 1690).
Hevelius Johannes, *Machinae coelestis pars prior* (Gedani: Auctoris typis et sumptibus, 1673).
Hevelius Johannes, *Machinae coelestis pars posterior* (Gedani: in Aedibus Autoris, 1679).
Hevelius Johannes, *Mercurius in Sole visus* (Gedani: autoris typis et sumptibus, 1662).
Hevelius Johannes, *Prodromus astronomiae* (Gedani: typis Johannis Zachariae Stolli, 1690).
Hevelius Johannes, *Prodromus cometicus* (Gedani: autoris typis et sumptibus, imprimebat Simon Reiniger, 1665).
Hodierna Giovanni Battista, *De systemate orbis cometici deque admirandis coeli characteribus* (Panormi: typis Nicolai Bua, 1654).
Hooke Robert, *Lectiones Cutlerianae* (London: printed for John Martyn, printer to the Royal Society, at the Bell in S. Pauls Church-yard, 1679).
Jöcher Christian Gottlieb, *Allgemeines Gelehrten-Lexicon*, 4 vols (Leipzig: in Johann Friedrich Gleditschens Buchhandlung, 1750–51).
Jungius Joachim, *Logica Hamburgensis* (Hamburgi: sumptibus Bartholdi Offermans, 1638).
Keckermann Bartholomaeus, *Systema physicum* (Hanoviae: apud Guilielmum Antonium, 1610).
Kepler Johannes, *De cometis libelli tres* (Augustae Vindelicorum: typis Andreae Apergeri, sumptibus Sebastiani Mylii, 1619).
Kepler Johannes, *Epitomes astronomiae Copernicanae libri V, VI, VII* (Francofurti: sumptibus Godefridi Tampachii, 1621).
Kircher Athanasius, *Itinerarium exstaticum* (Romae: typis Vitalis Mascardi, 1656).

Kirsten Michael, *Cometa, poema ad Hevelium* (Hamburgi: apud Zachariam Hertelium, 1665).
Lambeck Peter, *Liber primus prodromi historiae literariae* (Hamburgi: sumptibus autoris, typis Michelis Piperi, 1659).
Lavater Ludwig, *Cometarum omnium fere catalogus* (Tiguri: per Andream Genesum, [1556]).
Leinberer Wolfgang, *Theoria cometae mense Decembri anni MDCLXIV Januario item ac Februario anni MDCLXV Ingolstadii observati* ([Ingolstadt]: typis Joannis Ostermaieri, 1665).
Libavius Andreas, *Singularium* [...] *pars secunda*, (Francofurti: impressa typis Ioannis Saurii, impensis Petri Kopfii, 1599).
Liber secundus C. Plinii de mundi historia, ed. by Jakob Milich (Halae Suevorum, ex officina Petri Brubachii, 1537).
Liberati François, *Discours de la comete commencee a apparoir sur Paris le XI, iour de Nouembre, mil cinq cens septante-sept, à six heures du soir* (Paris: Pour Iean de Lastre, libraire, demouran en la rue S. Iean de Latran, pres le College de Cambray, 1577).
Lipsius Justus, *Politicorum sive civilis doctrinae libri sex* (Amsterdami: apud Guilielmum Bleau, 1632).
Longomontanus Christian, *Astronomia Danica* (Amsterodami: ex officina typographica Guilielmi I. Caesii, 1622).
Lubieniecki Stanisław, *Historia reformationis Polonicae* (Freistadii [i.e. Amsterdam]): apud Johannem Aconium, 1685).
Lubieniecki Stanisław, *Monumentum Radzivilianum* (Hamburgi: ex officina Georgii Rebenlini, 1670).
Lubieniecki Stanisław, *Theatri cometici exitus* (Amstelodami: apud Franciscum Kuyperum, 1668).
Lubieniecki Stanisław, *Theatri cometici pars posterior* (Amstelodami: apud Franciscum Kuyperum, 1666).
Lubieniecki Stanisław, *Theatri cometici pars prior* (Amstelodami: apud Franciscum Kuyperum, 1668).
Lycosthenes Conrad, *Prodigiorum ac ostentorum chronicon* (Basileae: per Henricum Petri, 1557).
Manesson-Mallet Allain, *Les travaux de Mars ou la fortification nouvelle* (Paris: chez l'autheur, chez Jean Henault et chez Claude Barbin, 1671).
Marius Simon, *Mundus Jovialis anno MDCIX derectus ope perspicilli Belgici* ([Nürnberg]: sumptibus et typis Johannis Lauri, 1614).
Mathesius Johannes, *Sarepta* (Nürnberg: durch Johann von Berg und Ulrich Newher, 1562 and Nürnberg, [s.e.], 1578).
Montucla Jean-Étienne, *Histoire des mathématiques*, vol. 2 (Paris: chez Ch. Ant. Jombert, 1758).
Müller Friedrich, *Cometologiae libri tres* (Giessae Hassorum: imprimebat Fridericus Kargerus, 1665).

Olearius Adam, *Tertius interveniens, gesetzt zwischen denen, so den Cometsternen allzu große Kraft und Wirckung zuschreiben und daher die Leute allzu furchtsam machen. Und denen, die den Cometen gar keine Krafft zueigen, ihre Bedeutung ganz verwerffen und die Leute gar zu sicher machen wollen* ([s.l.]: [s.e.], 1665).

Olhoff Johann Eric, *Excerpta ex litteris illustrium et clarissimorum virorum ad nobilissimum, amplissimum et consultissimum dominum Johannem Hevelium* (Gedani: ex officina Janssonio Waesbergiana, 1683).

Paprocki Bartosz, *Gniazdo cnoty* (Kraków: z drukarniey Andrzeia Piotrkowczyka, 1578).

Petrucci Gioseffo, *Fisiologia nuova della natura delle comete* (Roma: nella stamperia del Varese, 1665).

Picard Jean, *Voyage d'Uranibourg* (Paris: de L'Imprimerie Royale, 1680).

Pingré Alexandre Guy, *Cométographie ou Traité historique et théorique des comètes*, 2 vols (Paris: Imprimerie royale, 1783–84).

Placentinus-Kołaczek Jan, *Astronomica et astrologica observatio cometae terribilis, mense Februario huius currentis anni adhuc spectati* ([s.l.]: auff Kosten des Autoris, Drucks Eramus Rösner, 1665).

Platonis *Omnium quae extant operum tomus tertius Graece et Latine* ([s.l.]: [s.e.], 1578).

Poetae Minores Graeci (Cantabrigiae: apud Joann. Field, 1667).

Praetorius Johannes, *De cometis, qui antea visi sunt, et de eo, qui novissimus mense Novembri apparuit* (Noribergae: in Officina Typographica Catharinae Gerlachin et Haeredum Johannis Montani, 1578).

Riccioli Giovanni Battista, *Almagestum novum*, pars posterior tomi primi (Bononiae: ex Typographia Haeredis Victorii Benatii, 1651).

Riccioli Giovanni Battista, *Astronomiae reformatae tomi duo* (Bononiae: ex typographia Haeredis Victorii Benatii, 1665).

Riccioli Giovanni Battista, *Chronologia reformata* (Bononiae: ex typographia Haeredis Dominici Benatii, 1669).

Riccioli Giovanni Battista, *Geographiae et hydrographiae reformatae libri duodecim* (Bononiae: ex typographia Haeredis Victorii Benatii, 1661).

Rockenbach Abraham, *De Cometis Tractatus Novus Methodicus* (Witebergae: ex Officina Cratoniana, sumptibus Henningii Grosii, 1602).

Rolevinck Werner, *Fasciculus temporum* (Coloniae Aggripinae: [s.e.], 1474).

Sabellicus Marcus Antonius, *Rhapsodiae historiarum enneadum* [...] (in Parrhisiorum Academia: ab Joanne Parvo, 1509).

Scaliger Julius Cesar, *Exotericarum exercitationum liber quintus decimus* (Lutetiae: ex officina typographica Michaelis Vascosani, 1557).

Schickard Wilhelm, *Pars responsi ad epistolas P. Gassendi insignis philosophi Gallici de Mercurio in Sole viso, et aliis novitatibus uranicis* (Tubingae: typis Theodorici Werlini, 1632).

Schimpffer Bartholomäus, *Kurtze Beschreibung deß dunckelen Cometen so anno 1652 den 8. Decembr. erschienen* (Hall in Sachsen: Johann Rappoldt, 1652).

Schmid Ludolph Hinrich, *Versuch einer historischen Beschreibung der an der Elbe belegenen Stadt Altona* (Altona, Flensburg: in Verlag der Gebrüder Korte, 1747).
Schuler Johannes, *Cometologia, sive de Cometis Disquisitio Philosophica* (Hagae-Comitis: ex typographia Adriani Ulaco, 1665).
Schwenckfeld Caspar, *Stirpium et fossilium Silesiae catalogus* (Lipsiae: impensis Davidis Alberti, 1600).
Sigonius Carolus, *Historiarum de occidentali imperio libri XX* (Francofurti: apud Heredes Andreae Wechelii, 1593).
Sponde Henri de, *Annales ecclesiastici ex XII tomis Caesaris Baronii in epitomen redacti* (Lutetiae Parisiorum: sumptibus Dionysii de la Noue, 1613).
Syreński Szymon, *Zielnik* (Cracoviae: w drukarni Bazylego Skalskiego, 1613).
Trew Abdias, *Gründtlicher Bericht von deme im Ende verwichenen Januarii Alten und Anfang deß Februarii Neuen Calenders erschienenen Cometen* (Nürnberg: verlegt von Johann Taubern, 1661).
Weigel Erhard, *Fortsetzung des Himmels-Spiegels* (Jena: in Verlagung Thomas Matthias Götzen, 1665).

Modern Editions

Adémar de Chabannes, *Chronique*, ed. by Jules Chavanon (Paris: Alphonse Picard et Fils, 1897).
Aristotelis qui ferebantur librorum fragmenta, ed. by Valentinus Rose (Lipsiae: in aedibus B. G. Teubneri, 1886).
Brahe Tycho, *Tychonis Brahe opera omnia*, 15 vols, ed. by John Louis Emil Dreyer (Hauniae: in Libraria Gyldendaliana, 1913–29).
Cicero, *De officiis*, transl. by Walter Miller (Cambridge, Mass., London: Harvard University Press, 1913).
Cicero, *De senectute, De amicitia, De divinatione*, transl. by William Armistead Falconer (Cambridge, Mass., London: Harvard University Press, 1923).
Cicero, *The Letters to his Friends*, vol. 3, transl. by W. Glynn Williams (London, Cambridge, Mass.: Harvard University Press, 1958).
Cicero, *The Orations of Marcus Tullius Cicero*, transl. by C. D. Yonge (London: George Bell and Sons, 1903).
Correspondance de Johannes Hevelius. Tome I, prolégomènes critiques, ed. by Chantal Grell (Turnhout: Brepols, 2014).
Correspondance de Johannes Hevelius. Tome II: Correspondance avec la cour de France et ses agents avec un dossier sur la querelle de la comète de 1664–1665, ed. by Chantal Grell (Turnhout: Brepols, 2017).
Correspondance de Johannes Hevelius. Tome III: Correspondance avec Pierre des Noyers, secrétaire de la Reine de Pologne, 1646–1686, ed. by Chantal Grell in collaboration with Damien Mallet (Turnhout: Brepols, 2020).
Diels Hermann, Kranz Walther, *Die Fragmente der Vorsokratiker*, vol. 2 (Berlin: Weidmannsche Verlagsbuchhandlung, 1959).

'Dwa listy Wiszowatego do S. Lubienieckiego', introduction by Zbigniew Ogonowski, transl. by Irmina Lichońska, *Studia i Materiały z Dziejów Nauki Polskiej*, 4 (1956), 347–66.

Evelyn John, *Sylva, or a Discourse of Forest Trees. A Reprint of the Fourth Edition in Two Volumes*, vol. 1 (London: published by Arthur Doubleday & Company Limited, 1908).

Guericke Otto von, *Neue (sogennante) Madgeburger Versuche über den leeren Raum*, ed. and transl. by Hans Schimank et al. (Düsseldorf: VDI-Verlag, 1968).

Horace, *The Odes and Carmen Saeculare*, transl. by John Conington (London: George Bell and Sons, 1882).

Horace, *The Works of Horace*, transl. by C. Smart, Theodore Alois Buckley (New York: Harper & Brothers, 1863).

Horace, *The Works of Horace Translated Literally into English Prose*, vol. 2, transl. by C. Smart (Philadelphia: Joseph Whetham, 1836).

Lubieniecki Stanislaus, *Compendium veritatis primaevae*, ed. by K. E. Jordt Jørgensen, 2 vols (København: Akademisk Forlag, 1982).

Monumenta Germaniae historica. Scriptorum tomus I, ed. by Georg Heinrich Pertz (Hannoverae: impensis Bibliopolii Aulici Hahniani, 1826).

Monumenta Germaniae historica. Scriptorum tomus II, ed. by Georg Heinrich Pertz (Hannoverae: impensis Bibliopolii Aulici Hahniani, 1829).

Œuvres complètes de Christiaan Huygens. Tome cinquième, correspondance 1664–1665 (La Haye: Martinus Nijhof, 1893).

Œuvres complètes de Christiaan Huygens. Tome sixième, correspondance 1666–1669 (La Haye: Martinus Nijhof, 1895).

Oeuvres de Descartes, vol. 6, ed. by Charles Adam and Paul Tannery (Paris: Léopold Cerf, 1902).

Patrologiae cursus completus, series Latina, vol. 123, ed. by J.-P. Migne (Parisiis: apud Garnier fratres, 1879).

Pliny, *Natural History in Ten Volumes*, vol. 1, transl. by H. Rackham (Cambridge, London: Cambridge University Press, 1967).

Poetae lyrici Graeci, vol. 2, ed. by Theodorus Bergk (Lipsiae: in aedibus B. G. Teubneri, 1882).

Seneca, *Ad Lucilium epistulae morales*, vol. 1, transl. by Richard M. Gummere (Cambridge, Mass., London: Harvard University Press, 1979).

Seneca, *Ad Lucilium epistulae morales*, vol. 2, transl. by Richard M. Gummere (London, New York: William Heinemann, G.P. Putnam's Sons, 1920).

Seneca, *Naturales quaestiones. Books IV–VII*, transl. by Thomas H. Corcoran (Cambridge, Mass., London: Harvard University Press, 2004).

Tacitus, *Complete Works of Tacitus*, transl. by Alfred John Church and William Jackson Brodribb (New York: Random House Inc., 1873).

The Bible. Authorized King James Version, ed. by Robert Carroll and Stephen Prickett (Oxford: Oxford University Press, 2008).

The Correspondence of Henry Oldenburg. Vol. III, 1666–1667, ed. and transl. by A. Rupert Hall and Marie Boas Hall (Madison: University of Wisconsin Press, 1966).

The Polish Brethren. Documentation of the History and Thought of Unitarianism in the Polish-Lithuanian Commonwealth and in the Diaspora, 1601–1685, 2 vols, ed. by George Huntston Williams (Missoula: Scholars Press, 1980).

Virgil, *Eclogues. Georgics. Aeneid I–VI*, transl. by H. Rushton Fairclough (Cambridge, Mass., London: Harvard University Press, 2004).

Virgil, *Aeneid VII–XII. Appendix Vergiliana*, transl. by H. Rushton Fairclough (Cambridge, Mass., London: Harvard University Press 2002).

Secondary Literature

Åkerman Susanna, *Queen Christina of Sweden and her Circle. The Transformation of a Seventeenth-Century Philosophical Libertine* (Leiden, New York: Brill, 1991).

Allen Richard Hinckley, *Star-Names and their Meanings* (New York: G. E. Stechert, 1899).

Allgemeine Deutsche Biographie, 56 vols (Leipzig: Verlag von Duncker & Humblot, 1875–1912).

Andriesse C. D., *Huygens. The Man Behind the Principle* (Cambridge: Cambridge University Press, 2005).

Archenhold Friedrich Simon, 'Johannes Hevelius. Ein Gedankblatt zum 300. Geburtstag', *Das Weltall*, 11 (1911), no. 10, 137–60.

Athanasius Kircher. The Last Man Who Knew Everything, ed. by Paula Findlen (New York, London: Routledge, 2004).

Barentine John C., *The Lost Constellations. A History of Obsolete, Extinct, or Forgotten Star Lore* (Heidelberg: Springer, 2016).

Barnes Robin B., *Astrology and Reformation* (New York: Oxford University Press, 2016).

Barrett A. A., 'Observations of Comets in Greek and Roman Sources before A.D. 410', *Journal of the Royal Astronomical Society of Canada*, 72 (1978), no. 2, 81–106.

Barta Gábor, 'Die Anfänge des Fürstentums und erste Krisen (1526–1606)', in *Kurze Geschichte Siebenbürgens*, ed. by Béla Köpeczi (Budapest: Akadémiai Kiadó, 1990), pp. 243–301.

Bersohn Mathias, *Kilka słów o Janie Heweliuszu, astronomie gdańskim z wieku XVII-go oraz o jego korespondencyi* (Warszawa: druk P. Laskauera & W. Babickiego, 1898).

Bieńkowska Barbara, *Kopernik i heliocentryzm w polskiej kulturze umysłowej do końca XVIII wieku* (Wrocław: Zakład Narodowy im. Ossolińskich, Wydawnictwo Polskiej Akademii Nauk, 1971) (*Studia Copernicana*, vol. 3).

Boas Hall Marie, *Henry Oldenburg. Shaping the Royal Society* (Oxford: Oxford University Press, 2002).

Boner Patrick J., *Kepler's Cosmological Synthesis. Astrology, Mechanism and the Soul* (Leiden, Boston: Brill, 2013).

Chmaj Ludwik, 'Jan Placentinus-Kołaczek. Nieznany kartezjanin XVII wieku', *Archiwum Historii Filozofii i Myśli Społecznej*, 1 (1957), 71–81.

Dansk biografisk leksikon, vol. 13, ed. by C.F. Bricka et al. (København: J. H. Schlutz Forlag, 1938).

Davies Norman, *God's Playground. A History of Poland. Volume 1: The Origins to 1795* (Oxford: Oxford University Press, 1981).

de Groot Mart, Sterken Chris, 'P Cygni 2000: 400 Years of Progress', *Publications of the Astronomical Society of the Pacific*, 113 (2008), 258–59.

Delambre Jean-Baptiste Joseph, *Histoire de l'astronomie moderne*, vol. 2 (Paris: Mme Ve Courcier, 1821).

Dictionary of Scientific Biography, 16 vols, ed. by Charles Coulston Gillispie (New York: Charles Scribner's Sons, 1970–80).

Dobrzycki Jerzy, Włodarczyk Jarosław, *Historia naturalna gwiazdozbiorów* (Warszawa: Prószyński i S-ka, 2002).

Elmqvist Söderlund Inga, *Taking possession of astronomy. Frontispieces and illustrated title pages in 17th-century books on astronomy*, (Stockholm: The Center for History of Science at the Royal Swedish Academy of Sciences, 2000).

Encyklopedia Gdańska, ed. by B. Śliwiński i et al. (Gdańsk: Fundacja Gdańska, 2012).

Gadomski Jan, *Zarys historii astronomii polskiej* (Kraków: Polska Akademia Umiejętności, 1948).

Graney Christopher M., *Setting Aside All Authority. Giovanni Battista Riccioli and the Science against Copernicus in the Age of Galileo* (Notre Dame, Indiana: University of Notre Dame Press, 2015).

Gualandi Andrea, *Teorie delle comete. Da Galileo a Newton* (Milano: FrancoAngeli, 2009).

Hans Nicholas, 'Polish Protestants and their Connections with England and Holland in the 17th and 18th Centuries', *The Slavonic and East European Review*, 37 (1958), no. 88, 196–220.

Hatch Robert A., *The Collection Boulliau (BN, FF. 13019–13059). An Inventory* (Philadelphia: American Philosophical Society, 1982).

Heidarzadeh Tofigh, *A History of Physical Theories of Comets, from Aristotle to Whipple* (New York: Springer, 2009).

Hellman C. Doris, *The Comet of 1577: its Place in the History of Astronomy* (New York: AMS Press, 1971).

Historia astronomii w Polsce, vol. 1, ed. by Eugeniusz Rybka (Wrocław: Zakład Narodowy im. Ossolińskich, Wydawnictwo Polskiej Akademii Nauk, 1975).

Isaac Vossius (1618–1689) betwen Science and Scholarship, ed. by Eric Jorink and Dirk van Miert (Leiden, Boston: Brill 2012) (*Brill's Studies in Intellectual History*, vol. 214).

Jasiński Maciej, 'Stanisław Lubieniecki and Johannes Hevelius: (Extra)ordinary "Men of Letters"', *Quarterly Journal of the History of Science and Technology*, 64 (2019), no. 1, 125–37.

Jasiński Maciej, *Stanisław Lubieniecki i astronomia kometarna XVII stulecia* (Warszawa, Gdańsk: Instytut Historii Nauki im. L. i A. Birkenmajerów PAN, PAN Biblioteka Gdańska, Oficyna Wydawnicza ASPRA-JR, 2017).

Jasiński Maciej, 'Otto von Guericke's cometary theory in Stanisław Lubieniecki's correspondence,' *Journal for the History of Astronomy*, 51 (2020), no. 2, 131–51.

Jean Picard et les débuts de l'astronomie de précision au XVIIe *siècle. Actes du Colloque du tricentenaire*, ed. by Guy Picolet (Paris: Editions du CNRS, 1987).

Jørgensen K. E. Jordt, *Stanislaw Lubieniecki. Zum Weg des Unitarismus von Ost nach West im 17. Jahrhundert* (Göttingen: Vandenhoek & Ruprecht, 1968).

Kamieński Michał, *Rola astronomów polskich w nauce światowej. The Role of Polish Astronomers in World Science* (Warszawa: [s.e.], 1933).

Kot Stanisław, '"Descriptio gentium" di poeti polacchi del secolo XVII', *Ricerche slavistiche*, 6 (1958), 151–84.

Kowalik Janina, 'Stanisław Lubieniecki: företrädare för socinianerna och svensk korrespondent', *Karolinska förbundets årsbok* (1981/2), 148–73.

Kowalik Janina, 'Szwedzka korespondencja Stanisława Lubienieckiego', *Odrodzenie i Reformacja w Polsce*, 47 (2003), 177–90.

Kronk Gary W., *Cometography. A Catalogue of Comets. Volume 1: Ancient – 1799* (Cambridge: Cambidge University Press, 1999).

Meinel Christoph, *In physicis futurum saeculum respicio. Joachim Jungius und die Naturwissenschaftliche Revolution des 17. Jahrhunderts* (Göttingen: Vandenhoeck & Ruprecht, 1984).

Meinel Christoph, *Die Bibliothek des Joachim Jungius. Ein Beitrag zur Historia litteraria der frühen Neuzeit* (Göttingen: Vandenhoeck & Ruprecht, 1992).

Milani Nausicaa Elena, 'The Prodromus Cometicus in the Académie des Sciences and the Royal Society: The Hevelius-Auzout Controversy', in *Johannes Hevelius and his Gdańsk*, ed. by Marian Turek (Gdańsk: Gdańskie Towarzystwo Naukowe, 2014), pp. 195–208.

Montiglio Silvia, *Silence in the Land of Logos* (Princeton: Princeton University Press, 2000).

Mosley Adam, 'Past Portents Predict: Cometary *Historiae* and Catalogues in the Sixteenth and Seventeenth Centuries', in *Celestial novelties on the eve of the scientific revolution, 1540–1630*, ed. by Dario Tessicini and Patrick J. Boner (Firenze: Leo S. Olschi Editore, 2013), pp. 1–32.

Müller Hans-Joachim, *Irenik als Kommunikationsreform. Das Colloquium Charitativum von Thorn 1645* (Göttingen: Vandenhoeck & Ruprecht, 2004).

Nellen Henk J.M., *Ismaël Boulliau (1605–1694). Astronome, épistolier, nouvelliste et intermédiaire scientifique*, (Amsterdam: APA-Holland University Press, 1994).

Nouhuys Tabitta van, *The Age of Two-Faced Janus. The Comets of 1577 and 1618 and the Decline of the Aristotelian World View in the Netherlands* (Leiden: Brill, 1998).

Ogonowski Zbigniew, *Socynianizm. Dzieje, poglądy, oddziaływanie* (Warszawa: Instytut Historii Nauki im. L. i A. Birkenmajerów PAN, Oficyna Wydawnicza ASPRA-JR, 2015).

Omodeo Pietro D., 'Central European Polemics over Descartes: Johannes Placentinus and His Academic Opponents at Frankfurt on Oder (1653–1656)', in *History of Universities*, vol. XXIX/1, ed. by Mordechai Feingold (Oxford: Oxfrord University Press, 2016), pp. 29–64.

Otto Maria, *Prace Jana Heweliusza*, in: *Jan Heweliusz*, ed. by Maria Pelczar and Jarosław Włodarczyk (Radom: Instytut Technologii Eksploatacji – Państwowy Instytut Badawczy w Radomiu, 2011), pp. 243–47.

Pingré Alexandre Guy, *Annales célestes du dix-septième siècle* (Paris: Gauthier-Villars, 1901).

Polski Słownik Biograficzny, vols 1–, ed. by. Władysław Konopczyński et al. (Kraków et al.: Polska Akademia Umiejętności et al., 1935–).

Pomian Krzysztof, 'De la lettre au périodique: la circulation des informations dans les milieux des historiens au XVIIIe siècle', *Organon*, 10 (1974), 25–43.

Przypkowski Tadeusz, 'Pierwsze poprawne widoki Jowisza i Saturna z r. 1664 ogłoszone przez Stanisława Lubienieckiego', *Sprawozdania z Czynności i Posiedzeń Polskiej Akademii Umiejętności*, 52 (1951), no. 6, 536–37.

Remmert Volker R., *Picturing the Scientific Revolution*, transl. by Ben Kern (Philadelphia: Saint Joseph's University Press, 2011).

Robinson James Howard, *The Great Comet of 1680. A Study in the History of Rationalism* (Northfield: press of the Northfield News, 1916).

Roşu Felicia, *Elective Monarchy in Transylvania and Poland-Lithuania, 1569–1587* (Oxford: Oxford University Press, 2017).

Ruffner J. A., 'The Curved and the Straight: Cometary Theory from Kepler to Hevelius,' *Journal for the History of Astronomy*, 2 (1971), 178–94.

Rybka Przemysław, *Instrumentarium astronomiczne Heweliusza (geneza i rozwój konstrukcji)* (Wrocław: Zakład Narodowy im. Ossolińskich, Wydawnictwo Polskiej Akademii Nauk, 1987) (*Monografie z dziejów nauki i techniki*, vol. 139).

Saffrey Henri-Dominique, *Ἀγεωμέτρητος μηδεὶς εἰσίτω. Une inscription légendaire*, 'Revue des Études Grecques', 81 (1968), no. 384–85, 67–87.

Saridakis Voula, 'The Hevelius-Hooke Controversy in Context: Transforming Astronomical Practice in the Late Seventeenth Century' in *Johannes Hevelius and His World. Astronomer, Cartographer, Philosopher and Correspondent*, ed. by Richard L. Kremer and Jarosław Włodarczyk (Warsaw: Institute for the History of Science, Polish Academy of Sciences, Copernicus Center for Interdisciplinary Studies, 2013) (*Studia Copernicana*, vol. 44), pp. 103–35.

Schechner Genuth Sara, *Comets, Popular Culture, and the Birth of Modern Cosmology* (Princeton: Princeton University Press, 1997).

Schöder Hans, 'Genealogische und biographische Nachrichten von der Familie Qualen', *Nordalbingische Studien. Neues Archiv der Schleswig-Holstein-Lauenburgischen Gesellschaft für Vaterländische Geschichte*, 3 (1846), 103–45.

Schultz Daniel, *The Andromeda Galaxy and the Rise of Modern Astronomy* (New York: Springer, 2012).

Shapin Steven, *A Social History of Truth. Civility and Science in Seventeenth-Century England* (Chicago, London: The University of Chicago Press, 1994).

Shara Michael M., Moffat Anthony F. J., 'The Recovery of CK Vulpeculae (1670) — the Oldest "Old Nova"', *The Astrophysical Journal*, 258 (1982), L41–L44.

Sibbald John A., 'The *Heinsiana* — Almost a Seventeenth-Century Universal Short Title Catalogue', in *Documenting the Early Modern Book World. Inventories and Catalogues in Manuscript and Print*, ed. by Malcolm Walsby and Natasha Constantinidou (Leiden, Boston: Brill, 2013), pp. 141–59.

Simpson W. J., *A treatise on Plague, dealing with the Historical, Epidemiological, Clinical, Therapeutic and Preventive aspects of the Disease* (Cambridge: at the University Press, 1905).

Stone Daniel, *The Polish-Lithuanian State, 1386–1795* (Seattle, London: University of Washington Press, 2001).

Studia nad arianizmem, ed. by Ludwik Chmaj (Warszawa: Państwowe Wydawnictwo Naukowe, 1959).

Tazbir Janusz, 'Diariusz Stanisława Lubienieckiego (młodszego)', *Odrodzenie i Reformacja w Polsce*, 5 (1960), 201–21.

Tazbir Janusz, *Stando lubentius moriar. Biografia Stanisława Lubienieckiego* (Warszawa: Iskry, 2003).

Tazbir Janusz, *Stanisław Lubieniecki. Przywódca ariańskiej emigracji* (Warszawa: Państwowe Wydawnictwo Naukowe, 1961).

Tazbir Janusz, 'The Fate of Polish Protestantism in the Seventeenth Century', in *A Republic of Nobles. Studies in Polish History to 1864*, ed. by J. K. Fedorowicz et al. (Cambridge: Cambridge University Press, 1982), pp. 198–217.

Tazbir Janusz, 'Wiersze Stanisława Lubienieckiego', *Przegląd Humanistyczny*, 2 (1965), 113–22.

The Cambridge History of Poland from the Origins to Sobieski (to 1696), ed. by W. F. Reddaway et al. (Cambridge: Cambridge University Press, 1950).

Thoren Victor E., *The Lord of Uraniborg. A Biography of Tycho Brahe* (Cambridge: Cambridge University Press, 1990).

Visser Piet, 'Blasphemous and pernicious: the role of printers and booksellers in the spread of dissident religious and philosophical ideas in the Netherlands in the second half of the seventeenth century', *Quaerendo*, 26 (1996), no. 4, 303–26.

Wedel Theodore Otto, *The Mediaeval Attitude toward Astrology* (New Haven, London, Oxford: Yale University Press, Humphrey Milford, Oxford University Press, 1920).
Weidemann Volker, *Theatrum Cometicum. Hamburg und Kiel im Zeichen der Kometen von 1664 und 1665*, (Göttingen: Vandenhoeck & Ruprecht, 1987).
Wilbur Earl Morse, *A History of Unitarianism. Socinianism and its Antecedents* (Cambridge, Massachusetts: Harvard University Press, 1946).
Whitaker Ewen A., *Mapping and Naming the Moon. A History of Lunar Cartography and Nomenclature* (Cambridge, New York: Cambridge University Press, 1999).
Wicksteed Philip H., Gardner Edmund G., *Dante and Giovanni del Virgilio* (Westminster: Archibald Constable & Company, Ltd., 1902).
Wojciechowski Piotr Henryk, *Il "colloquium charitativum" di Torun del 1645* (Pelplin: Bernardinum, 2000).
Woleński Jan, 'Polish-English (British) Philosophical Contacts and Comparisons from the Fifteenth through the Eighteenth Century, in *Britain and Poland-Lithuania. Contacts and Comparison from the Middle Ages to 1795*, ed. by Richard Unger (Leiden, Boston: Brill, 2008), pp. 391–400.
Yeomans Donald K., *Comets. A Chronological History of Observation, Science, Myth, and Folklore* (New York: Wiley Science Editions, 1991).

Links

Alidade (Accès en Ligne aux Instruments, Documents et Archives De l'astronomiE), alidade.obspm.fr [accessed 10 August 2019].
Gallica, gallica.bnf.fr [accessed 10 August 2019].
Schröder Hans, *Lexikon der hamburgischen Schriftsteller bis zur Gegenwart*, 8 vols (Hamburg: Perthes-Besser u. Mauke 1851–83), access online: schroeder.sub.uni-hamburg.de [accessed 19 June 2018].

INDICES

Numbers in italics denote the sections of the Introduction. Numbers in regular script denote the numbers of letters or enclosures. Appendices are noted as 'Ap' with the number of appendix. The critical apparatus is not indexed. Footnotes are indexed, except for the authors and the titles of works which are not the subject matter. Index rerum indexes only anonymous works. Correspondents mentioned in the footnotes are not indexed. The names 'Johannes Hevelius' and 'Stanisław Lubieniecki' are not indexed.

Index nominum

Index rerum